GEOTECHNICAL MEASUREMENTS AND MODELLING

PROCEEDINGS OF THE INTERNATIONAL SYMPOSIUM ON GEOTECHNICAL MEASUREMENTS AND MODELLING, 23–25 SEPTEMBER 2003, KARLSRUHE, GERMANY

Geotechnical Measurements and Modelling

Edited by

O. Natau
Institute of Soil Mechanics and Rock Mechanics, University of Karlsruhe, Germany

E. Fecker
Geotechnisches Ingenieurbüro Prof. Fecker & Partner GmbH, Ettlingen, Germany

E. Pimentel
Institute of Soil Mechanics and Rock Mechanics, University of Karlsruhe, Germany

A.A. BALKEMA PUBLISHERS   LISSE / ABINGDON / EXTON (PA) / TOKYO

Published by: A.A. Balkema, a member of Swets & Zeitlinger Publishers
www.balkema.nl and www.szp.swets.nl

ISBN 90 5809 603 3

Printed in The Netherlands

Preface

Field measurements and laboratory experiments play an important role in the understanding of geotechnical processes and the complex three dimensional behaviour of soil and rock masses.

To date research has been focused on the development and improvement of measuring techniques and equipment in order to characterise the relevant dominant processes and phenomena. Empirical associations and constitutive laws were derived and verified, and in conjunction with the rapid development of information technology and numerical modelling, more complex conditions could be analysed.

Currently the application of the experimentally determined results in context of the verification of the numerical methods and the problem of scale is proving to be extremely important. With this two-fold approach to geomechanical problems (modelling and experimental investigation) coupled with a scale appreciation, more reliable tools for the geotechnical engineer can be developed.

This fundamental relationship is equally valid for the three main phases of any engineering consideration, i.e.:

– site investigation and planning,
– construction and
– operation.

The aim of the symposium is to provide a plenum for the exchange of ideas and experience for both of the key approaches, modelling and measurement for geotechnical engineering.

The scope of the symposium is reflected by the following topic areas covered in the proceedings and also in the conference: underground openings, underground waste disposals, slope engineering and retaining structures, risk assessment and management, new technologies for laboratory and field tests and recent developments in modelling. For each of the conference sessions, the Organising Committee asked two prime specialists to be chairmen.

Last but not least, the organisers are grateful to the authors for their contributions and having shared their knowledge and experience with the participants of the symposium.

Editors
Otfried Natau, Edwin Fecker & Erich Pimentel

Organisation

The symposium was organised by the Chair of Rock Mechanics of the University of Karlsruhe, Germany in co-operation with Geotechnisches Ingenieurbüro Prof. Fecker & Partner GmbH, Ettlingen, Germany.

Organising Committee

Prof. Otfried Natau (Chairman)
Chair of Rock Mechanics, Institute of Soil Mechanics and Rock Mechanics, University of Karlsruhe, Germany

Prof. Edwin Fecker
Geotechnisches Ingenieurbüro Prof. Fecker & Partner GmbH, Ettlingen, Germany

Dr. Erich Pimentel
Chair of Rock Mechanics, Institute of Soil Mechanics and Rock Mechanics, University of Karlsruhe, Germany

Advisory Board

Dr. Peter Blümling
National Cooperative for the Disposal of Radioactive Waste (NAGRA), Wettingen, Switzerland

Prof. Günter Borm
GeoForschungszentrum (GFZ), Potsdam, Germany

Prof. Peter Egger
Rock Mechanics Laboratory, Ecole Polytechnique Fédérale de Lausanne (EPFL), Switzerland

Dr. Bernhard Fröhlich
Gesellschaft für Baugeologie und -messtechnik mbH (GBM), Limburg, Germany

Prof. Gerd Gudehus
Institute of Soil Mechanics and Rock Mechanics, University of Karlsruhe, Germany

Prof. Richard A. Herrmann
Institut für Geotechnik-Grundbau, Siegen University, Germany

Prof. Stefan Heusermann
Bundesanstalt für Geowissenschaften und Rohstoffe (BGR), Hannover, Germany

Prof. Dimitrios Kolymbas
Institute of Geotechnical and Tunnel Engineering, University of Innsbruck, Austria

Prof. Klaus Kühn
Institut für Bergbau, Technical University of Clausthal, Germany

Dr. Wolfgang Minkley
Institute of Rock Mechanics GmbH (IfG), Leipzig, Germany

Prof. Michael Moser
Institut für Geologie und Mineralogie, University Erlangen-Nürnberg, Germany

Prof. Wulf Schubert
Institute for Rock Mechanics and Tunnelling, Graz University of Technology, Austria

Dr. Georg-Michael Vavrovsky
Austrian Society for Geomechanics, Salzburg, Austria

Acknowledgements

Special thanks for close co-operation should be paid to the chairmen of the conference sessions: Prof. Peter Egger, Prof. Wulf Schubert, Dr. Peter Blümling, Prof. Klaus Kühn, Prof. Gerd Gudehus, Prof. Michael Moser, Prof. Günter Borm, Dr. Bernhard Fröhlich, Prof. Stefan Heusermann, Prof. Richard Herrmann, Prof. Dimitrios Kolymbas and Dr. Wolfgang Minkley.

Special thanks should also be paid to the ideal supporters of the symposium:

– German Society for Geotechnics e.V., represented by its president Prof. Manfred Nußbaumer,
– Austrian Society for Geomechanics, represented by its president Dr. Georg-Michael Vavrovsky, and
– Swiss Society for Soil and Rock Mechanics, represented by its president Prof. François Descoeudres.

For the generous financial support special thanks should be paid to the sponsors of the symposium:

– Glötzl, Gesellschaft für Baumeßtechnik mbH, Rheinstetten
– Südwestdeutsche Salzwerke AG, Heilbronn
– K + S Aktiengesellschaft, Kassel
– Ercosplan Ingenieurgesellschaft Geotechnik und Bergbau mbH, Erfurt
– IfG Institut für Gebirgsmechanik GmbH, Leipzig
– K-UTEC, Sondershausen GmbH, Sondershausen

Table of Contents

Geotechnical measurements and modelling in slope engineering and retaining structures

Risk assessment and management

New technologies for laboratory and field tests

Recent developments in modelling (Workshop)

Geotechnical measurements and modelling in underground openings

Geotechnical Measurements and Modelling, Natau, Fecker & Pimentel (eds)
© 2003 Swets & Zeitlinger, Lisse, ISBN 90 5809 603 3

Tunnelling in Germany: State of the art – Tunnel driving methods

M. Nußbaumer & U. Hartwig
Ed. Züblin AG, Stuttgart, Germany

ABSTRACT: In the following paper, various examples are presented to demonstrate the high technical state of tunnelling technology in Germany. Tunnel projects which have been carried out in Germany under widely differing technical and economical parameters are described, where shotcrete linings have been successfully used in both rock and soft ground. The construction of the world's largest tunnel passing under a river in soft ground – the 4th Elbe Tunnel Tube – serves as an example of the use of a slurry shield under the most severe geological and hydrogeological conditions.

Following this, a look is taken at the possible direction of further developments in Germany. On one hand, this is the endeavour to use the shotcrete method for the excavation of large cross-sections in soft ground full-face. On the other hand, due to the advances in equipment technology and monitoring methods for shield tunnelling, the scope of where it can be applied has been widened, especially in urban areas.

All the tunnel projects presented here have been constructed using underground driving methods. Tunnels constructed in open cut have not been included. Even with this limitation, only a small selection of projects can be covered. Besides the tunnel projects described here, many other equally important and challenging projects are currently being realised. The large number of tunnels being constructed abroad with German participation, which are not covered her, also deserve a mention.

1 INTRODUCTION

The undiminished increase in road traffic in Germany since the 1960s is increasingly accompanied by aspects resulting from the general rise in the importance of the standard of living quality for our society. This affects mainly the requirements regarding speed, comfort and cost of the respective means of transport. Environmental considerations also play an important role.

In order to meet these requirements for road and rail transport, an important point is the transfer of part of the traffic underground, that means within tunnels.

Not only does construction in tunnel reduce the land take and noise emission, but also the almost free choice of route if the tunnel will be driven by underground methods plays an important role. Due to the progress made in tunnel construction methods the effect of the geological parameters on the choice of route has decreased in influence.

The following paper presents a summary of the current state of the tunnelling art in Germany, illustrated by a choice of current projects. The parameters of each project will be examined, and the construction method chosen will be described.

Due to the large number of projects, only those which concern mined tunnels for road and rail purposes will be considered. Only projects within Germany have been considered, which should not detract from the successful projects carried out abroad with German participation.

2 EXAMPLES OF CURRENT TUNNEL PROJECTS

2.1 *General*

Tunnel construction methods by mining for road and rail tunnels can be divided into the following categories:

— conventional driving methods/shotcrete methods
  • simple excavation
  • drill and blast
  • excavation with hydraulic hammer
  • excavation with road header
  • ... also in combination with methods of systematically securing the face in advance

- mechanical driving methods
  - driving with tunnel boring machines
  - shield tunnelling

The choice of method is dependent upon the following criteria:

- geological and hydrogeological conditions (e.g. rock or soft ground, head of ground water)
- size and geometry of the cross-section (e.g. structure gauge)
- location of the route (depth underground, existing buildings)
- construction costs, construction time

Increasingly, shield tunnelling and mining methods compete on routes which, in the recent past, would have been uneconomic for shield use. This is for the construction time and operational reasons. However, environmental aspects or the minimisation of geotechnical risks e.g. the reduction of subsidence in urban areas often give shield tunnelling the advantage.

2.2 Conventional driving methods/shotcrete methods

Rennsteig Tunnel (1998 to 2003)

The Rennsteig Tunnel is on the route of the motorway A71 between Erfurt and Schweinfurt, forming part of the "Traffic Project German Unity No. 16". In the course of crossing the ridge of the Thüringen forest,

this involves 4 tunnels, with a total length of 12.6 km, as well as 3 bridges over valleys. The Rennsteig Tunnel passes under the main ridge of the forest, and consists of 2 bores of approx. 85 m^2 cross-section, each of 2 lanes, with a length of approx. 7.9 km, making it Germany's longest motorway tunnel.

Besides the 2 bores, the construction works include 12 emergency lay-bys (cross-section 120 m^2), 25 cross passages, 2 caverns for ventilation equipment (LAZ, cross-section 200 m^2, 55 m long), 2 air extraction shafts (6.2 m diameter, 25 m long) and 2 ventilation galleries (cross-section 55 m^2, 300 m long).

The tunnel passes in parts through sedimentary rocks (mudstones, sandstones and conglomerates), but is mainly situated in igneous rock (quartzporphyry). The tunnel crosses 2 valleys, where the depth of overburden sinks to below 10 m. The soil conditions here consist of strongly weathered rock or slope debris. As a rule the groundwater table lies above the tunnel.

After each round, the face and walls were sealed using shotcrete (mesh reinforcement, lattice arches, wet spray concrete, thickness 18 cm to 22 cm). Provision of anchors was made according to the geological conditions encountered.

An especial challenge was presented by the crossing of the 2 line Brandleite Railway Tunnel, built in 1884, which is used by about 50 trains per day. The minimum space between the bores of the Rennsteig Tunnel and the Brandleite Tunnel was 5 m and 6.5 m respectively. Here the most important measures taken were the

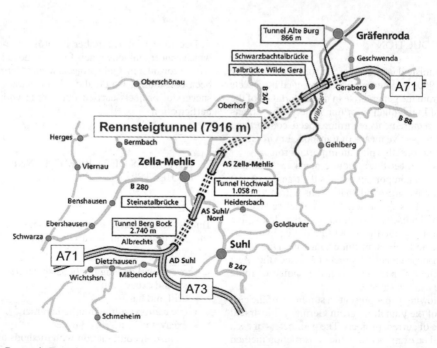

Figure 1. Rennsteig Tunnel on the motorway A71 between Erfurt and Schweinfurt (Prax et al. 2000).

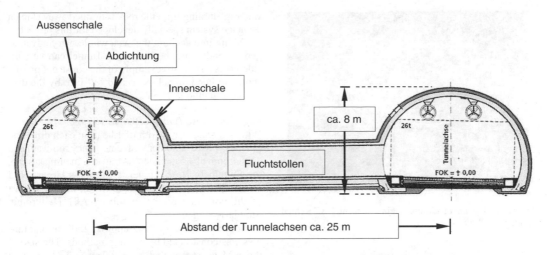

Figure 2. Diagram of tunnel construction between 2 emergency lay-bys (Prax et al. 2000).

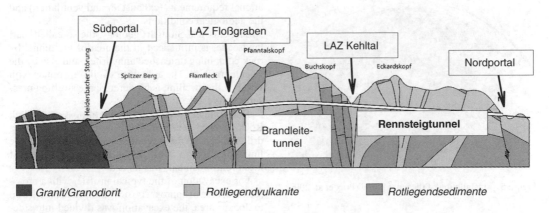

Figure 3. Geological longitudinal section in the area of the Rennsteig Tunnel (Prax et al. 2000).

change to partial face driving, with the provision of a pilot tunnel and the placing of rock bolts (32 mm diameter) in the remaining rock before the bench and invert were excavated. The other special area was the Kehltal valley crossing, where the overburden consisted of less than 10 m of slope debris, making additional support measures, such as the provision of a pipe umbrella necessary. The challenges presented by the construction of the 2 ventilation equipment chambers LAZ Kehltal and LAZ Floßgraben also deserve a mention. These each consist of 2 enlargements with cross-section of 200 m^2, a cross-tunnel, an exhaust air shaft and a ventilation gallery. The enlargements are approx. 14 m wide, 16 m high and 55 m long.

The average monthly rate of advance was approx. 150 m per bore, i.e. 300 m of tunnel. During the 2 years it took to complete the tunnel drives, around 1.4 million m^3 of rock were removed, 140,000 rock bolts were installed, 250 km of lattice arches and 200,000 m^3 of shotcrete were placed, and 1000 tons of explosive were detonated.

The 2 bores of the tunnel were driven simultaneously – i.e. parallel – and also as a rule full-face. The advance per round for the 3 drives varied between 1.2 m and 2.2 m depending on the ground conditions encountered. The drill and blast method was used, with each round breaking out between 90 m^3 and 120 m^3 of solid rock.

The inner lining of the horseshoe-shaped tunnel was generally supported on benches, with an open base. In geological fault zones, a flat base slab was placed. The mass concrete inner lining generally has a thickness of 30 cm. The blocks have a length of 12 m. Sealing of the lining is done using a 2 mm thick polyolefin sheet. The

5

Figure 4. Placing of additional reinforcement, LAZ Kehltal (Prax et al. 2000).

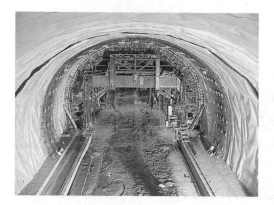

Figure 5. Installation of the inner lining (Prax et al. 2000).

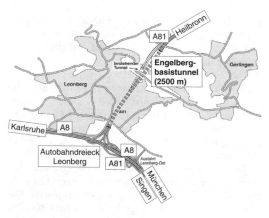

Figure 6. Engelberg Base Tunnel on the motorway A81 (Hering & Walliser 1998).

water infiltrating from the rock is drained away using a drainage system specially developed for this project.

Taking into account the, as a whole, very advantageous conditions, the Rennsteig Tunnel can be considered as a standard example of the construction of a shotcrete lined tunnel excavated full face by the drill and blast method.

Engelberg Base Tunnel (1995 to 1999)

With a carriageway width of 13.5 m in each direction (3 lanes, 1 hard shoulder with emergency footway) and a headroom of 4.5 m under the traffic signalling system, the Engelberg Base Tunnel is Europe's largest road tunnel. The twin bore tunnel forms an important part of the widening of the motorway A81 Heilbronn–Stuttgart–Singen.

Approx. 1780 m of the 2 approx. 2500 m long tunnels was constructed by mining methods. The size of the excavation face, between 200 m^2 and 310 m^2 was established by the carriageway geometry, the operational requirements (exhaust air and ventilation) and the geological conditions.

The classical Stuttgart geology, the so-called marl landscape, is influenced in the area of the tunnel by rock containing unleached anhydrites, and also by the Engelberg fault. The anhydrite swells on contact with water. If the swelling is hindered, large swelling pressures can result.

Both tunnel bores were driven almost in parallel. Rock removal was carried out by excavator or using drill and blast. After steel arches had been placed, the lining of 30 cm to 50 cm of shotcrete, with 2 layers of mesh reinforcement, was placed.

For excavation of the typical profile, with a cross-sectional area of approx. 200 m^2, and special profiles up to 265 m^2 area, the excavation was divided into sidewall, crown, bench and base headings. The load-bearing capacity of the rock was increased using a system of 4 m to 8 m long rock bolts (SN-anchor) and also in places using grouted anchors.

The swelling of the anhydrite, caused by infiltrating water, resulted in partial damage to the shotcrete lining already during construction time. Due to this a buffer zone between 1.7 m and 2.7 m thick, together with an additional anchoring of the base, was provided in this area.

Thus the floor heave which occurred before the placing of the inner lining, which is capable of resisting the swelling pressure, was reduced to a minimum. In order to accommodate the buffer zone beneath the designed 3 m thick base of the inner lining, it was necessary to enlarge the tunnel cross-section to 310 m^2.

The excavation of both tunnel drives took a total of 27 months. Placing of the inner lining could commence before completion of the tunnel drives, as the cross-passages and connecting tunnel, with a spacing of 350 m, could be used for the spoil disposal.

6

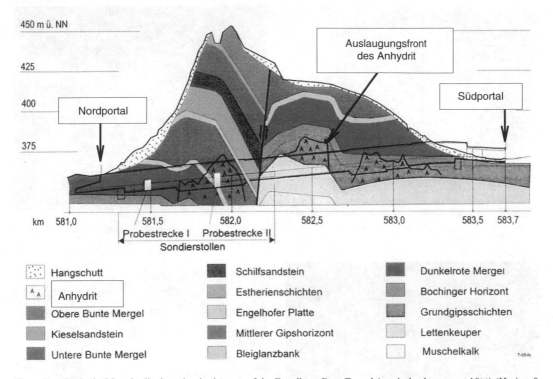

450 m ü. NN

425

400

375

Nordportal

Auslaugungsfront des Anhydrit

Südportal

km 581,0   581,5   582,0   582,5   583,0   583,5 583,7

Probestrecke I   Probestrecke II
Sondierstollen

Hangschutt
Anhydrit
Obere Bunte Mergel
Kieselsandstein
Untere Bunte Mergel

Schilfsandstein
Estherienschichten
Engelhofer Platte
Mittlerer Gipshorizont
Bleiglanzbank

Dunkelrote Mergel
Bochinger Horizont
Grundgipsschichten
Lettenkeuper
Muschelkalk

Figure 7. Geological longitudinal section in the area of the Engelberg Base Tunnel (vertical enlargement 10×) (Hering & Walliser 1998).

Ulmenstollen und Kalotte

Strosse

Sohle

Figure 8. Excavation cross-section (Hering & Walliser 1998).

The Engelberg Base Tunnel plays an exceptional role compared to other tunnel structures even only in respect of the cross-sectional area. In addition there are the special challenges presented by the swelling rock, which were met by a combination of resistance and yielding principles.

Siegauen Tunnel (1998 to 2002)

The Siegauen Tunnel forms the heart of the 5.2 km long Section 2.4/St. Augustin on the new ICE highspeed railway line NBS Köln-Rhein/Main between Cologne and Frankfurt. Coming from the north, it passes beneath the flood plain of the Sieg river, the Sieg river, the motorway A560, the community centre St. Martinus, with the parish church and graveyard, and the flood plain of the Pleis brook.

Due to the existing buildings mentioned above, the 370 m long section, the most difficult from a geological point of view, had to be constructed using mining methods.

In a simplified form, the soil strata consists of 3 layers. Underneath a non-cohesive fill layer of sands and gravels there exists quaternary layers of lightly silty gravels and sands containing stones and boulders. The underlying tertiary layer consists mainly of clays

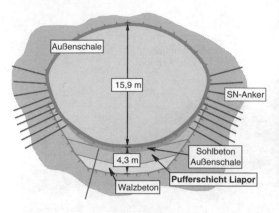

Figure 9. Typical cross-section in the area containing anhydrite.

Figure 10. Enlargement of the cross-section in the area containing anhydrite (Hering & Walliser 1998).

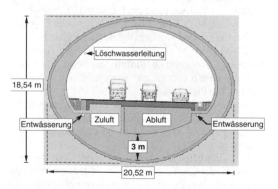

Figure 11. Inner lining section in the area containing anhydrite (Hering & Waliser 1998).

Figure 12. Construction of the arch-form base (Hering & Walliser 1998).

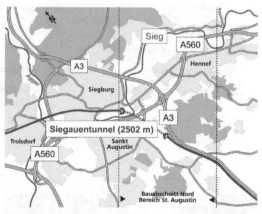

Figure 13. Siegauen Tunnel as section of the new high-speed railway line NBS Köln-Rhein/Main between Cologne and Frankfurt (Sänger et al. 2000).

and silts, which however contain differing amounts of fine sand. The groundwater level in the quarternary layer corresponds to the level of the Sieg river, which can vary by up to 5 m.

The following boundary conditions had to be taken into account in the concept for the tunnel drive:

- cross-sectional area of 152 m² (without enlargements)
- existing buildings susceptible to settlement damage
- groundwater level up to 6 m above crown of tunnel
- areas of unstable ground, prone to flowing

As a result the technical concept chosen included the following main points:

- installation of a pilot tunnel (internal dia. 2.8 m) using pipe jacking

Figure 14. Siegauen Tunnel passing beneath the parish church (view in northerly direction) (Sänger et al. 2000).

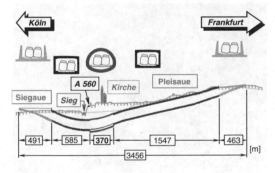

Figure 15. Longitudinal section of Section 2.4 of the NBS Köln-Rhein/Main (Sänger et al. 2000).

- installation and operation of vertical quaternary wells, inclined well points and vacuum deep level wells
- side-wall drift excavation
- placing of spiles in front of the face in quaternary soils and the transition to tertiary soils (up to 100 spiles per round)
- max. round depth 0.8 m
- compensation grouting beneath the community centre and the parish church

- observation program for water levels and displacements

Commencing at the south end, 220 m of tunnel could successfully be completed using this method. Following this, an extremely heterogeneous alternating layer of sands and gravels was encountered, which only had a negligible silt content.

This channel fill in the tertiary layer was up to 3.5 m thick, and extended along the line of the tunnel for about 85 m. Due to the high, and strongly anisotropic permeability, coupled with the very high velocity of the water ingress, it was not possible to stabilise the situation in this channel, despite many extra wells.

After consideration of all appropriate tunnelling methods, it was decided to proceed using compressed air tunnelling. In order to reduce the loss of air through the quaternary layer and to guard against sudden loss of air pressure through blow-outs, jet grouting from the pilot tunnel was used to create a grout roof. The vacuum deep level wells were used in parallel in the tertiary layer to reduce the required air pressure.

The remaining 150 m could then be successfully completed under compressed air. A driving rate of up to 1.0 m per day was achieved, and the compressed air consumption varied between approx. 60 m^3 and 400 m^3 intake air per minute.

The Siegauen Tunnel could only be constructed using shotcrete methods with the help of extensive additional measures. In view of the geotechnical parameters encountered, which were not fully known beforehand, the use of shotcrete should be characterised as a borderline case for the technical feasibility. The tunnel length of 370 m is however too short to allow the economic use of a shield drive.

Frankfurt Interchange Tunnel (1995 to 2001)
This project consisted not only of routing part of new ICE line between Cologne and Frankfurt beneath the busiest motorway interchange in Germany, which is passed by 300,000 vehicles daily, but also of redesigning the interchange to increase its capacity. The works included diverse bridges and interchange structures, as well as the rebuilding of the Zeppelinheim station and the driving of a 281 m long tunnel beneath the motorway A3.

The section PA36 of the new ICE line is located in the northern upper Rhine valley, in the area of the Kelsterbach deep drift. The soil consists of dense to very dense medium to coarse sands with varying gravel content and sandy gravels. There are also quartz and sandstone intrusions, together with thin layers of silt and clay.

During the tunnelling works, the groundwater level was about 6 m above the base of the tunnel. Dewatering was not permitted.

Due to the shallow depth, only 7 m to 15 m below the A3 motorway, the existing ground conditions made

9

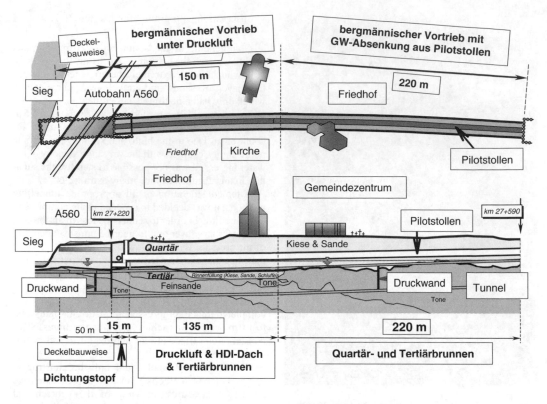

Figure 16.   Mined tunnel (plan view and longitudinal section) (Sänger et al. 2000).

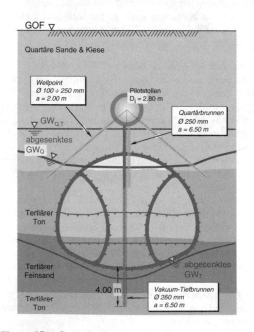

Figure 17.   Groundwater control from the pilot tunnel (Sänger 2000).

Figure 18.   Well construction from the pilot tunnel (Sänger 2000).

extensive crown and base jet grouting necessary for the use of a shotcrete tunnelling method. Top heading excavation was carried out within the protection of a jet grout umbrella, each section having a length of 14.5 m and an overlap of 3.5 m.

The part of the tunnel cross-section in groundwater was sealed using a 1.5 m thick jet grout arch. Lining of the tunnel was carried out using impermeable concrete without additional waterproofing.

The peculiarity of the tunnel beneath the Frankfurt interchange is the shallow overburden of noncohesive soil and the strict conditions concerning limits for subsidence at ground level. In order not to hinder or interrupt the heavy flow of traffic on the A3 motorway, ground displacements were monitored using fully automatic digital levels linked online.

Due to the extensive preparatory works and the construction method chosen, the maximum measured subsidence was about 4 cm, less than the permitted maximum of 5 cm.

Figure 19. Ground breaking wall "south".

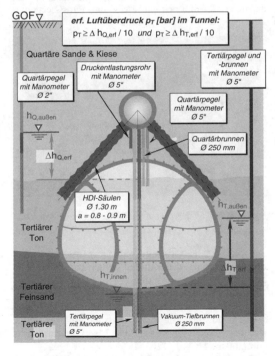

Figure 20. Measurement equipment during compressed air tunnelling (Sänger 2000).

2.3 Shield tunnelling

4th Elbe Tunnel Tube (1995 to 2003)

The capacity of the existing 6-lane Elbe Tunnel, opened in 1975, was no longer sufficient to cope with the more than 100,000 vehicles per day using the A7 motorway. An improvement to the situation could

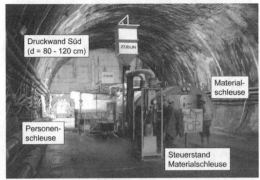

Figure 21. Bulkhead with man lock and material lock.

Figure 22. Frankfurt interchange – the junction of the A3 and A5 motorways at Frankfurt airport.

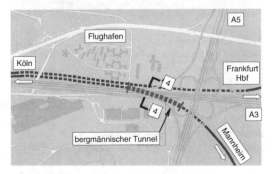

Figure 23. Plan view of mined tunnel.

11

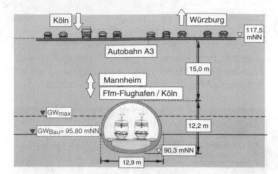

Figure 24.   Cross-section of mined tunnel (4-4).

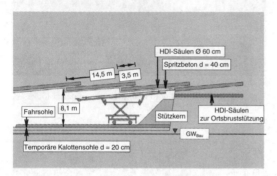

Figure 25.   Construction of the horizontal jet grout umbrella and excavation of the crown (longitudinal section).

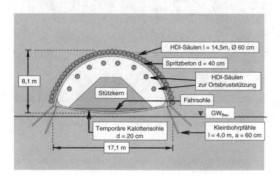

Figure 26. Construction of the horizontal jet grout umbrella and excavation of the crown (cross-section).

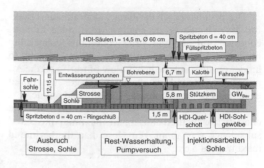

Figure 27.   Construction of the arch-formed jet grout base and excavation of bench and base heading (longitudinal section).

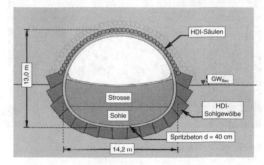

Figure 28.   Construction of the arch-formed jet grout base and excavation of bench and base heading (cross-section).

Figure 29.   Construction of the horizontal grout umbrella.

only be achieved by constructing a 4th tunnel bore. The works were divided into 4 sections, with Section 2 including the main 2561 m long shield drive.

The ground which the tunnel must pass through is distinctly heterogeneous, consisting mainly of depositions from various ice ages (alternating layers of sand/gravel and boulder clay, with sand lenses and boulders of up to 2 m diameter), as well as underlying hard tertiary mica clay strata, surcharged by ice

age deposits. The average water level is 40 m above the deepest part of the tunnel, which can increase to 50 m during high storm tides.

A shield machine with slurry support to the face was used, having a diameter of 14.2 m and weighing about 2600 tons. The total length including machine skin (12.9 m, 2000 tons) and the 2 back-up trains was some 60 m.

The 5 main spokes of the cutter head were hollow, allowing access. They carried 31 roller bits with

Figure 30. Excavation of the crown (Hofmeister & Roth 1999).

Figure 31. Construction of the 4th Elbe Tunnel Tube in the west of the existing tunnel (Lohrmann & Miemietz 2002).

double discs to crush the boulders previously mentioned. 111 cutters were located at the edges of the main spokes, as well as on the 5 secondary spokes. The cutting tools could be accessed over the 5 main spokes, allowing them to be exchanged under atmospheric pressure.

The driving rate achieved at a speed of approx. 1 revolution per minute varied according to the ground conditions between 2 mm (boulders) and 25 mm to 30 mm (sands and gravels). An advance of up to 12 m per day was achieved.

The cutter head could be pushed forward up to 80 cm. In the centre of the main cutter head, an independent mix shield, with a diameter of 3 m, was located. This could also be pushed forward in order to relieve the main head.

For the first time, sensors located in the main spokes were able to provide seismic pre-exploration of the ground ahead, using sound waves. Thus boulders of more than 60 cm diameter could be identified at a distance of up to 15 m in front of the face.

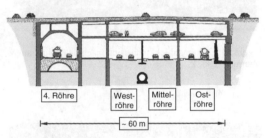

Figure 32. Tunnel cross-section (Lohrmann & Miemietz 2002).

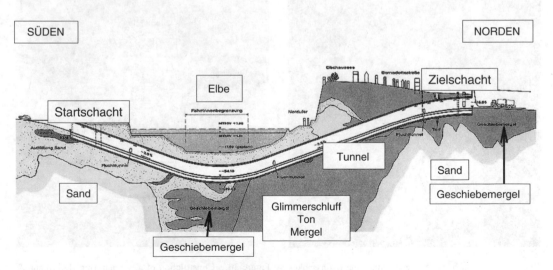

Figure 33. Geological longitudinal section in the area of the Elbe Tunnel (Lohrmann & Miemietz 2002).

The reinforced concrete tunnel segments have an average width of 2 m and an average thickness of 70 cm. A ring consists of 8 standard segments plus the keystone. Each segment has an arc length of 5.2 m and weighs 18 tons. A total of 11,500 segments were installed.

The 4th Elbe Tunnel Tube represents world-wide the largest shield tunnel passing under a river in soft ground. This project demonstrates the state of the art for shield tunnelling.

Figure 34. Slurry-shield tunnelling machine (Lohrmann & Miemietz 2002).

3 PROSPECTS

3.1 *General*

Following the above summary of recently completed tunnel projects, which show the current state of the art, some future developments will be presented. Developments in conventional driving methods/shotcrete methods as well as in shield tunnelling will be shown.

3.2 *Conventional driving methods/shotcrete methods*

Apart from developments in concrete technology and machine technology, it is an endeavour for operational and construction time reasons, in Germany as well as in other European countries, to achieve full-face excavation of large cross-section tunnels in soft ground using the shotcrete method. The main advantage is an improvement of operational constraints and thus as a result in saving of construction time compared to a staged excavation. Because of the early ring closure right behind the face it is possible to control the surface subsidence as well as the bending stresses in the shotcrete lining.

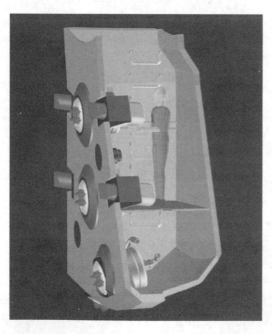

Figure 35. Tools being changed within the main spokes (Herrenknecht 2001).

Figure 36. Construction of a segment ring (Lohrmann & Miemietz 2002).

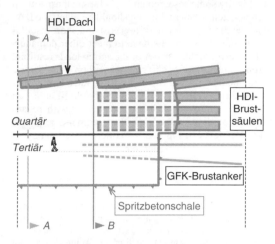

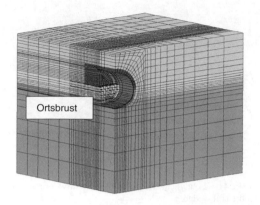

Figure 39. Numerical analysis of face stability: 3 dimensional finite element net.

Figure 37. Possible arrangement of safety measures for full-face excavation (longitudinal section).

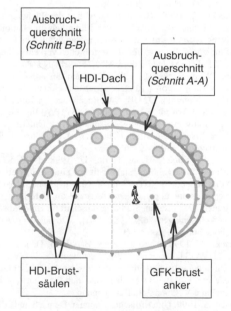

Figure 38. Possible arrangement of safety measures for full-face excavation (cross-section).

In order to increase the face stability the installation of a jet grout umbrella preceding the excavation is useful. This measure reduces the loading coming from the overlying soil and acting on the top of the soil wedge potentially sliding in the tunnel section. Due to the support of the soil wedge the driving forces will be reduced. On the other hand to increase the supporting forces additional safety measures could be installed through the face, e.g. jet grout columns and ground anchors.

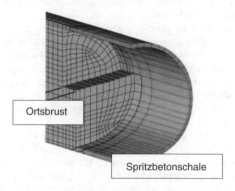

Figure 40. Numerical analysis of face stability: 3 dimensional finite element net – detail face.

Type, diameter and length depend among other things on the geometrical and geotechnical constraints.

In order to achieve an economic design a realistic calculation of the face stability will be carried out with different models based on analytical or numerical methods (e.g. finite element method). Actual research activities are going to improve these calculation methods. In any special case it has to be weighed up if the additional efforts and measures which are necessary to maintain the face stability compare economically to the operational advantages resulting from the full-face excavation.

3.3 Shield tunnelling

A main aspect in the validation process of shotcrete methods against shield tunnelling is that in general a new tunnelling machine has to be manufactured for each tunnel project. The additional time needed for construction, transport and installation of the machine and the costs for construction and commissioning

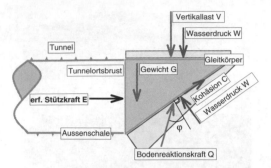

Figure 41. Analytical analysis of face stability: forces acting on the soil wedge.

require a minimum length and a substantial increase of the advance rate. The validation of the construction costs will differ for each project.

In order to choose the right tunnelling method, aspects which have no direct economic value have to be taken into consideration. This could be e.g. an increase in political acceptance or a reduction in geotechnical risks. Also environmental reasons can lead to shield tunnelling, e.g. because dewatering is prohibited or in order to keep the ground subsidence small with an active face support.

In urban areas shield tunnelling is used more and more in cases where in former times tunnelling in an open cut was chosen or even no tunnelling was possible. The problems which have to be solved in urban shield tunnelling are linked in a direct manner with the protection of buildings along the route: the subsidence due to shield tunnelling has to be limited.

Recent concepts combine measurements accompanying the construction (displacements, stresses), simultaneous numerical simulations and a steering of the tunnelling machine based on these results. For example the displacements of buildings respectively subsidence of the ground surface, the soil stresses, supporting pressures at the face and the pressure distribution inside the grout material in the area of the shield tail will be measured. The shield tail grouting prevents the surrounding soil from moving into the annulus resulting from the overbore. For that the pressure in the grout has to be maintained until it hardens.

The results of the numerical calculations will be compared with the results of the measurements and the calculations will be calibrated. On the basis of these results the machine will be controlled in a direct way to minimise the settlements due to the shield tunnelling. To achieve this either the face pressure will be increased or the advance rate will be reduced. Also compensation grouting in parallel to tunnelling can be controlled by using this data base.

Finally some research and development aspects on single-shell segmental lining should be mentioned. As well as the requirement of low cost and short installation time, the segments have to meet other demands. These involve the transfer of the acting loads resulting from transportation, soil stresses and pressure forces, with the guarantee of waterproofing and a high resistance during fire. An optimisation should be achieved with an improved segment design e.g. with regard to the coupling of the individual segments and with developments in the area of concrete technology or the use of steel fibre concrete.

BIBLIOGRAPHY

Bernhardt, K. & Rock, D. 1995. Der Engelbergbasistunnel: Bau und Finanzierungskonzept eines Autobahntunnels mit außergewöhnlichen Querschnittsabmessungen in schwierigen Gebirgsformationen. Weltneuheiten im Tunnelbau: Vorträge World Tunnel Congress/STUVA-Tagung '95 in Stuttgart, Studiengesellschaft für Unterirdische Verkehrsanlagen e.V. (STUVA), Düsseldorf, Alba, (Forschung + Praxis; 36).

Bielecki, R. 1998. Besonderheiten bei der Planung und Ausführung der 4. Röhre des Elbtunnels in Hamburg. Neue Akzente im unterirdischen Bauen: '97 in Berlin, Studiengesellschaft für Unterirdische Verkehrsanlagen e.V. (STUVA), Düsseldorf, Alba, (Forschung + Praxis; 37).

Dietz, W. & Becker, C. 1995. Kriterien zur Auswahl und Bewertung von Tunnelvortriebsmaschinen – Eine Empfehlung des DAUB –. Weltneuheiten im Tunnelbau: Vorträge World Tunnel Congress/STUVA-Tagung '95 in Stuttgart, Studiengesellschaft für Unterirdische Verkehrsanlagen e.V. (STUVA), Düsseldorf, Alba, (Forschung + Praxis; 36).

Dietz, W., Spuler, B. & Härle, D. 1998. NBS Köln-Rhein/Main: Erhöhte Sicherheit bei einem Vortrieb mit vorauseilendem HDI-Schirm durch kontinuierliche Setzungsmessungen. Neue Akzente im unterirdischen Bauen: '97 in Berlin, Studiengesellschaft für Unterirdische Verkehrsanlagen e.V. (STUVA), Düsseldorf, Alba, (Forschung + Praxis; 37).

Haack, A. 1995. Unterirdisches Bauen in Deutschland. STUVA-/ITA-Tagung '95, Stuttgart, STUVA – Studiengesellschaft für Unterirdische Verkehrsanlagen e.V. und DAUB – Deutscher Ausschuß für unterirdisches Bauen e.V., Gütersloh, Bertelsmann.

Haack, A. 1999. Unterirdisches Bauen in Deutschland 2000. STUVA-Tagung '99, Frankfurt, STUVA – Studiengesellschaft für Unterirdische Verkehrsanlagen e.V. und DAUB – Deutscher Ausschuß für unterirdisches Bauen e.V., Gütersloh, Bertelsmann.

Hering, S. & Walliser, T. 1998. Ohne Stau durch Europas größten Straßentunnel. ZÜBLIN-Rundschau 30, Stuttgart

Herrenknecht, M. 1998. Innovationen bei Tunnelvortriebsmaschinen – dargestellt am Beispiel der 4. Röhre Elbtunnel. Neue Akzente im unterirdischen Bauen: '97 in Berlin, Studiengesellschaft für Unterirdische Verkehrsanlagen e.V. (STUVA), Düsseldorf, Alba, (Forschung + Praxis; 37).

Herrenknecht, M. 2001. Praktische Beispiele zur Anpassung von Tunnelvortriebsmaschinen an die Erfordernisse beim

Vortrieb. Unterirdisches Bauen 2001. Wege in die Zukunft: 2001 in München, Studiengesellschaft für Unterirdische Verkehrsanlagen e.V. (STUVA), Gütersloh, Bertelsmann, (Forschung + Praxis; 39).

Hirsch, D., Graf, R. 2000. Besondere Verfahrenstechniken und Stabilisierungsmaßnahmen beim Auffahren des Siegtunnels der NBS Köln-Rhein/Main. Unterirdisches Bauen 2000. Herausforderungen und Entwicklungspotentiale: '99 in Frankfurt am Main, Studiengesellschaft für Unterirdische Verkehrsanlagen e.V. (STUVA), Düsseldorf, Alba, (Forschung + Praxis; 38).

Hofmeister, G. & Roth, H. 1999. 300.000 Baustellenbesucher kommen täglich mit dem Auto. ZÜBLIN-Rundschau 31, Stuttgart.

Kleffner, H.-J. & Denzer, G. 2000. Bauvorbereitung und – durchführung der Kammquerung des Thüringer Waldes im Zuge der A71/73. Unterirdisches Bauen 2000. Herausforderungen und Entwicklungspotentiale: '99 in Frankfurt am Main, Studiengesellschaft für Unterirdische Verkehrsanlagen e.V. (STUVA), Düsseldorf, Alba, (Forschung + Praxis; 38).

Lohrmann, W. & Miemietz, J. 2002. Seismischer Vortrieb steigert Leistung. ZÜBLIN-Rundschau 34, Stuttgart.

Lorscheider, W. & Kuhnhenn, K. 1998. Bautechnische Besonderheiten beim Bau des Engelbergbasistunnels. Neue Akzente im unterirdischen Bauen: '97 in Berlin, Studiengesellschaft für Unterirdische Verkehrsanlagen e.V. (STUVA), Düsseldorf, Alba, (Forschung + Praxis; 37).

Mayer, P.-M., Hartwig, U. & Schwab, C. 2003. Standsicherheitsuntersuchungen der Ortsbrust mittels Bruchkörpermodell und FEM. (to be published).

Prax, K., Groten, A. & Decker, A. 2000. Deutschlands längster Autobahntunnel. ZÜBLIN-Rundschau 32, Stuttgart.

Sänger, C. 2000. Neubaustrecke Köln-Rhein/Main Los 2.4 – Siegauen-Tunnel. Vorträge der Baugrundtagung 2000 in Hannover, Deutsche Gesellschaft für Geotechnik (DGGT), Verlag Glückauf, Essen.

Sänger, C., Roth, H. & Hofmann, S. 2000. Bergmännischer Tunnelvortrieb in schwierigster Geologie. ZÜBLIN-Rundschau 32, Stuttgart.

Geotechnical Measurements and Modelling, Natau, Fecker & Pimentel (eds)
© 2003 Swets & Zeitlinger, Lisse, ISBN 90 5809 603 3

Tunnelbau in Deutschland: Stand der Technik – Vortriebsverfahren

M. Nußbaumer & U. Hartwig
Ed. Züblin AG, Stuttgart, Deutschland

KURZFASSUNG: Im vorliegenden Artikel wird anhand ausgewählter aktueller Beispiele der hohe Stand der Tunnelbautechnik in Deutschland dargestellt. Es werden innerhalb Deutschlands unter verschiedensten technischen und wirtschaftlichen Randbedingungen realisierte Tunnelbauprojekte vorgestellt, bei denen die Spritzbetonbauweise sowohl im Fels als auch im Lockergestein erfolgreich eingesetzt wurde. Die Herstellung des weltweit größten Tunnels im Lockergestein unter einem Fluss – der 4. Röhre Elbtunnel – dient als Beispiel für den erfolgreichen Einsatz eines Hydro-Schildes unter schwierigsten geologischen und hydrogeologischen Randbedingungen.

Im daran anschließenden Ausblick wird beschrieben, in welche Richtung weitere Entwicklungen der Tunnelbautechniken in Deutschland gehen können. Einerseits ist dies bei der Spritzbetonbauweise das Bestreben, auch im Lockergestein Großquerschnitte im Vollausbruch aufzufahren. Bei Schildvortrieben wird andererseits durch Entwicklungen in der Maschinen- und Messtechnik deren Anwendungsbereich vergrößert, insbesondere im innerstädtischen Bereich.

Bei den vorgestellten Projekten wurden die Tunnel sämtlich in geschlossener Bauweise aufgefahren. In offener Bauweise hergestellte Tunnel werden hier nicht behandelt. Selbst bei dieser Einschränkung kann nur von einer kleinen Auswahl an Projekten berichtet werden. Neben den hier vorgestellten Tunnelbauprojekten wurden bzw. werden zur Zeit einige ähnlich wichtige und anspruchsvolle Projekte realisiert. Nicht ohne Erwähnung bleiben sollen zudem die große Zahl der mit deutscher Beteiligung im Ausland erfolgreich aufgefahrenen Tunnel, auf die hier ebenfalls nicht eingegangen werden kann.

1 EINLEITUNG

Das seit den 60er Jahren des letzten Jahrhunderts unvermindert wachsende Verkehrsaufkommen in Deutschland wird zunehmend von Aspekten begleitet, die direkt aus den allgemein wachsenden Ansprüchen der Gesellschaft an ihre Lebensqualität resultieren. Im Wesentlichen betrifft dies Anforderungen an die Schnelligkeit, die Bequemlichkeit und die Preise der jeweiligen Verkehrsmittel. Dazu kommen die Anforderungen des Umweltschutzes.

Im Straßen- und Schienenverkehr gilt als wesentlicher Ansatz zur Erfüllung der o.g. Ansprüche der Gesellschaft die Verlegung eines Teils des Verkehrs in den Untergrund, d.h. innerhalb von Tunneln. Neben der Verringerung des Landschaftsbedarfes und der Schallemissionen spielt dabei die bei geschlossener Tunnelbauweise mögliche, nahezu freie Trassenwahl eine wesentliche Rolle. Insbesondere bei der Trassenwahl treten dabei die vorhandenen geologischen Randbedingungen – aufgrund der Fortschritte in der Tunnelbautechnik – zunehmend in den Hintergrund.

Nachfolgend soll anhand ausgewählter Beispiele aktueller Tunnelbauprojekte ein Überblick über den derzeitigen Stand der Tunnelbautechnik in Deutschland gegeben werden. Es werden jeweils die Randbedingungen der Projekte erläutert und die gewählten Bauweisen vorgestellt.

Dabei erfolgt aufgrund der Vielzahl der Projekte eine Beschränkung auf solche mit geschlossenen Bauweisen bei Tunneln der Verkehrsinfrastruktur (Schiene und Straße). Zudem wird hier nur von Projekten berichtet, die innerhalb Deutschlands ausgeführt wurden.

2 BEISPIELE AKTUELLER TUNNELBAU-PROJEKTE

2.1 *Allgemeines*

Die geschlossenen Tunnelbauverfahren für die Herstellung von Straßen- und Schienen-verkehrstunneln

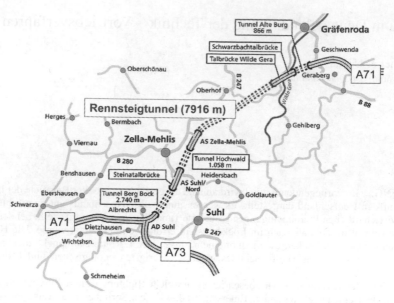

Bild 1. Rennsteigtunnel im Verlauf der Autobahn A71 (Prax et al. 2000).

können i. a. folgendermaßen unterschieden werden:

– konventioneller Vortrieb/Spritzbetonbauweise
  • Sprengvortrieb
  • Bagger-Sprengvortrieb
  • Baggervortrieb
  • Vortrieb mit Teilschnittmaschine
  • … auch in Kombination mit einer systematisch voreilenden Sicherung
– maschineller Vortrieb
  • Vortrieb mit Tunnelbohrmaschine
  • Vortrieb mit Schildmaschine

Die Wahl des Vortriebsverfahrens richtet sich i. W. nach folgenden Kriterien:

– geologische und hydrogeologische Verhältnisse (z.B. Fest- oder Lockergestein, Grundwasserhöhe)
– Größe und Geometrie des Querschnitts (z.B. Lichtraumprofil)
– Lage der Trasse (Tiefe unter Gelände, vorhandene Bebauung)
– Baukosten, Bauzeit

Dabei konkurriert der Schildvortrieb mit dem bergmännischen Vortrieb zunehmend auch auf Strecken, für die ein Schildvortrieb noch vor kurzer Zeit als unwirtschaftlich angesehen worden wäre. Gründe sind vor allem terminlicher und baubetrieblicher Art. Jedoch sprechen häufig auch Aspekte des Umweltschutzes oder der Minimierung geotechnischer Risiken wie z.B. die Reduzierung von Senkungen im innerstädtischen Bereich für einen Schildvortrieb.

2.2 Konventionelle Vortriebe/Spritzbetonbauweise

2.2.1 Rennsteigtunnel (1998 bis 2003)

Der Rennsteigtunnel ist Teil des als Verkehrsprojekt Deutsche Einheit Nr. 16 bezeichneten Teilstücks der Autobahn A71 zwischen Erfurt und Schweinfurt. Im Verlauf der Kammquerung des Thüringer Waldes entstehen vier Tunnel mit einer Gesamtlänge von 12,6 km sowie drei Talbrücken. Der Rennsteigtunnel unterquert mit 2 zweispurigen Röhren von jeweils ca. 7,9 km Länge und einer Querschnittsfläche von jeweils ca. 85 m² den Hauptkamm des Thüringer Waldes. Damit ist er der längste Autobahntunnel Deutschlands.

Die Bauarbeiten umfassten neben den beiden Tunneln 12 Pannenbuchten (Querschnittsfläche 120 m²), 25 Querungen, 2 Kavernen für Luftaustauschzentralen (Querschnittsfläche 200 m², Länge 55 m), 2 Abluftschächte (Durchmesser 6,2 m, Länge 25 m) und 2 Zuluftstollen (Querschnittsfläche 55 m², Länge 300 m).

Der Tunnel liegt z.T. in Sedimenten (Tonsteine, Sandsteine und Konglomerate), jedoch hauptsächlich in vulkanischem Gestein (Quarzporphyr). Der Tunnel kreuzt 2 Täler, bei denen die Tunnelüberdeckung auf unter 10 m sinkt. Der dort anstehende Baugrund besteht aus stark verwittertem Fels bzw. Hangschutt. Der Grundwasserspiegel liegt überwiegend höher als die Tunneltrasse.

Nach dem Ausbruch erfolgten die Versiegelung der Ortsbrust und die Sicherung der Tunnelleibung mit Spritzbeton (Betonstahlmatten und Gitterbögen, Nassspritzbeton, Dicke 18 cm bis 22 cm). Der Einbau der

20

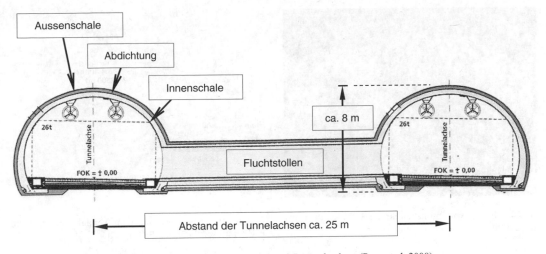

Bild 2.   Systematischer Aufbau des Rennsteigtunnels zwischen 2 Pannenbuchten (Prax et al. 2000).

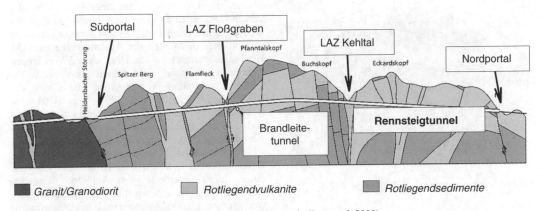

Bild 3.   Geologischer Längsschnitt im Bereich des Rennsteigtunnels (Prax et al. 2000).

Systemankerung erfolgte jeweils angepasst an die angetroffene Geologie.

Besondere Anforderungen stellte zum einen die Überquerung des 1884 fertiggestellten Brandleitetunnels dar, einem zweigleisigen Eisenbahntunnel, der täglich von ca. 50 Zügen passiert wird. Die minimalen Abstände zwischen den Tunnelröhren des Rennsteigtunnels und dem Brandleitetunnel betrugen nur 5 m bzw. 6,5 m. Als wesentliche Maßnahmen erfolgten hier das Auffahren in Teilquerschnitten bzw. die Herstellung eines Pilotstollens und das zwischenzeitliche Setzen von GEWI-Ankern (Ø 32 mm) in der verbliebenen Gebirgsfeste zwischen den Tunneln vor dem nachfolgenden Auffahren von Strosse und Sohle. Zum anderen waren im Bereich der Querung des Kehltals, in dem der Tunnel von Hangschutt mit einer Dicke von weniger als 10 m überdeckt wird, zusätzliche Sicherungsmaßnahmen, wie z.B. der Einbau eines Rohrschirms, erforderlich.

Zu erwähnen sind zudem noch die besonderen Anforderungen bei der Herstellung von 2 Luftaustauschzentralen LAZ Kehltal und LAZ Floßgraben. Diese bestehen jeweils aus 2 Aufweitungen mit einer Querschnittsfläche von jeweils 200 m², einer Überfahrt, einem Abluftschacht und einem Zuluftstollen. Die Aufweitungen sind etwa 14 m breit, 16 m hoch und 55 m lang.

Die beiden Tunnelröhren wurden sowohl gleichzeitig – d.h. parallel – als auch grundsätzlich im Vollausbruch aufgefahren. Die Abschlagslängen der insgesamt 3 Vortriebe variierten abhängig von der angetroffenen Geologie zwischen 1,2 m und 2,2 m. Das Lösen des Gesteins erfolgte im Sprengverfahren. Pro Sprengung wurden dabei zwischen 90 m³ und 200 m³ Fels gelöst.

Die mittlere monatliche Vortriebsleistung lag bei ca. 150 m je Röhre, d.h. 300 m Tunnel. Insgesamt wurden in ca. 2 Jahren Vortriebszeit ungefähr 1,4 Mio. m³

Bild 4. Einbau von Zusatzbewehrung, LAZ Kehltal (Prax et al. 2000).

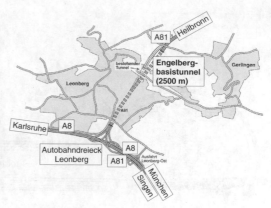

Bild 6. Engelbergbasistunnel im Verlauf der Autobahn A81 (Hering & Walliser 1998).

Bild 5. Einbau der Innenschale (Prax et al. 2000).

Gestein gelöst, 140.000 Anker und 250 km Gitterträger gesetzt, 200.000 m³ Spritzbeton verarbeitet und 1.000 t Sprengstoff zur Explosion gebracht. Die Innenschale mit hufeisenförmigem Profil wurde in der Regel auf Banketten mit offener Sohle gegründet. In geologischen Störzonen wurde eine flache Sohlplatte hergestellt. Die Innenschale ist im Regelprofil 30 cm dick und unbewehrt. Die Blocklängen betragen 12 m. Die Abdichtung gegen Wasser erfolgt mit einer Abdichtungsfolie auf Polyolefin-Basis mit einer Dicke von 2 mm. Anfallendes Bergwasser wird mit einem speziell für dieses Projekt entwickelten Drainagesystem abgeleitet.

Bei den hier vorliegenden, insgesamt sehr günstigen Verhältnissen kann der Rennsteigtunnel als Standardbeispiel für die Herstellung eines Tunnels in Spritzbetonbauweise bei einem Sprengvortrieb mit Vollausbruch betrachtet werden.

2.2.2 Engelbergbasistunnel (1995 bis 1999)

Der Engelbergbasistunnel stellt mit einer Fahrbahnbreite von 13,5 m je Richtung (3 Fahrspuren und 1 Standspur mit Notgehweg) und einer lichten Höhe der verkehrslenkenden Einrichtungen von 4,5 m Europas größtes Straßentunnel dar. Der zweiröhrige Tunnel ist wesentlicher Bestandteil der Ausbaustrecke der A81 Heilbronn-Stuttgart-Singen. Die 2 ca. 2.500 m langen Röhren wurden auf einer Länge von jeweils ca. 1.780 m bergmännisch hergestellt. Die Größe der Ausbruchsquerschnitte von 200 m² bis 310 m² je Röhre wurde sowohl von der Verkehrsraumgeometrie und den betrieblichen Sicherheitsvorgaben (Zu-und Abluft) als auch durch die Geologie bestimmt.

Die klassische Stuttgarter Geologie, die sog. Keuperlandschaft, ist im Bereich des Tunnels zum einen durch unausgelaugtes, Anhydrit führendes Gebirge und zum anderen durch die Engelbergverwerfung geprägt. Der Anhydrit neigt bei Wasserzutritt zum Quellen. Bei Behinderung der Quellerformungen entstehen z.T. große Quelldrücke.

Beide Tunnelröhren wurden nahezu parallel aufgefahren. Das Lösen des Gebirges erfolgte mit dem Bagger bzw. durch Sprengen. Nach dem Stellen von Stahlbögen wurde die zweilagig mattenbewehrte Spritzbetonsicherung mit Dicken von 30 cm bis 50 cm aufgebracht.

Bei einem Regelprofil mit ca. 200 m² und Sonderprofilen bis 265 m² Querschnittsfläche erfolgte der Ausbruch unterteilt in Ulmenstollen, Kalotte, Strosse und Sohle. Die Gebirgstragfähigkeit wurde durch eine Systemankerung mit 4 m bis 8 m langen SN-Ankern sowie durch Injektionsbohranker erhöht.

Aufgrund bereits während der Bauzeit durch den Zutritt von Bergwasser im Bereich des Anhydrits aufgetretener Quellhebungen und damit verbundener teilweiser Zerstörungen der Spritzbetonaußenschale wurden in diesem Bereich eine Pufferschicht von 1,7 m bis 2,7 m Dicke sowie eine zusätzliche Sohlankerung eingebaut.

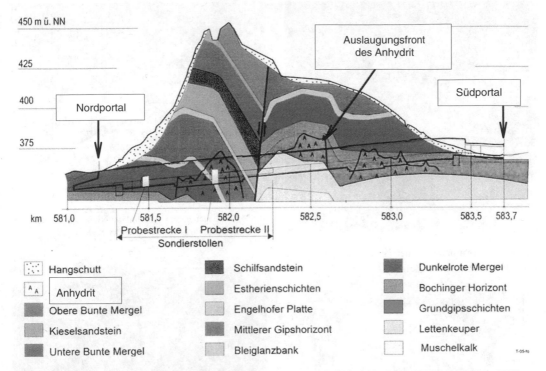

450 m ü. NN

425

400   Nordportal

375

Auslaugungsfront
des Anhydrit

Südportal

km   581,0   581,5   582,0   582,5   583,0   583,5   583,7

Probestrecke I   Probestrecke II
Sondierstollen

Hangschutt	Schilfsandstein	Dunkelrote Mergel
Anhydrit	Estherienschichten	Bochinger Horizont
Obere Bunte Mergel	Engelhofer Platte	Grundgipsschichten
Kieselsandstein	Mittlerer Gipshorizont	Lettenkeuper
Untere Bunte Mergel	Bleiglanzbank	Muschelkalk

Bild 7.   Geologischer Längsschnitt im Bereich des Engelbergbasistunnels (10fach überhöht) (Hering & Walliser 1998).

Ulmenstollen
und Kalotte

Strosse

Sohle

Bild 8.   Ausbruchquerschnitt (Hering & Walliser 1998).

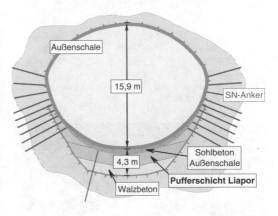

Außenschale

15,9 m

SN-Anker

4,3 m

Sohlbeton
Außenschale

Pufferschicht Liapor

Walzbeton

Bild 9.   Regelquerschnitt im Bereich des Anhydrits.

Damit wurden die Hebungen bis zum Einbau der gewölbten Innenschale, welche die Quelldrücke danach aufnehmen konnte, so gering wie möglich gehalten. Um die Pufferschicht unter der planmäßigen Sohle der Innenschale von 3 m Dicke unterbringen zu können, musste jedoch der Großquerschnitt von 265 m^2 auf über 310 m^2 vergrößert werden.

Der bergmännische Vortrieb beider Röhren dauerte insgesamt 27 Monate. Mit dem Einbau der Innenschale konnte bereits vor Beendigung der Vortriebsma ßnahmen begonnen werden, da die im Abstand von 350 m angeordneten Querschläge und Verbindungstunnel für den Schutterbetrieb verwendet werden konnten.

23

Bild 10.   Bodenaustausch im Bereich des Anhydrits (Hering & Walliser 1998).

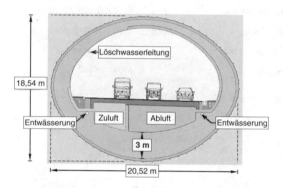

Bild 11.   Innenschale und Ausbau im Bereich des Anhydrits (Hering & Walliser 1998).

Bild 12.   Herstellung des Sohlgewölbes (Hering & Walliser 1998).

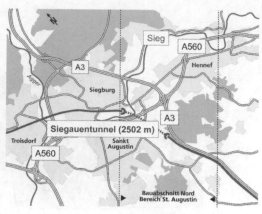

Bild 13.   Siegauentunnel als Teil der NBS Köln-Rhein/Main zwischen Köln und Frankfurt (Sänger et al. 2000).

Bild 14.   Siegauentunnel mit Unterquerung der Pfarrkirche (Blick nach Norden) (Sänger et al. 2000).

Der Engelbergbasistunnel nimmt allein aufgrund der Querschnittsfläche des Ausbruchs eine Sonderstellung unter den Tunnelbauwerken ein. Dazu kommen die besonderen Anforderungen aus dem quellenden Gebirge, welchen in diesem Fall mit einer

Kombination aus Widerstands- und Ausweichprinzip Rechnung getragen wurde.

2.2.3 Siegauentunnel (1998 bis 2002)

Der Siegauentunnel ist das Herzstück des 5,2 km langen Bauabschnittes 2.4/St. Augustin der ICE-Neubaustrecke Köln-Rhein/Main. Er unterquert von Norden kommend Siegaue, Sieg, die Autobahn A560, das Gemeindezentrum St. Martinus mit Pfarrkirche und Friedhof sowie die Pleisaue. Der geologisch

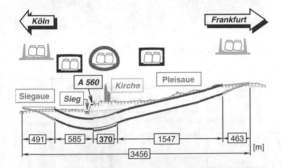

Bild 15. Längsschnitt des Bauabschnitts 2.4 der NBS Köln-Rhein/Main (Sänger et al. 2000).

gesehen schwierigste Abschnitt von 370 m Länge musste aufgrund der o.g. Bebauung bergmännisch aufgefahren werden.

Vereinfachend können im Projektgebiet drei Bodenschichten unterschieden werden. Unterhalb nichtbindiger Auffüllungen aus Sanden und Kiesen stehen quartäre Schichten aus leicht schluffigen Kiesen und Sanden mit Steinen und Blöcken an. Das darunter anstehende Tertiär besteht hauptsächlich aus Tonen und Schluffen, die jedoch unterschiedlich große Feinsandan- teile besitzen. Die Grundwasserstände imQuartär korrespondieren mit dem Siegwasserstand, welcher Schwankungen von bis zu 5 m Höhe unterworfen ist.

Beim Konzept des Tunnelvortriebs mussten u.a. folgende Randbedingungen berücksichtigt werden:

– Ausbruchsquerschnitt 152 m^2 (ohne Überhöhung)
– Setzungsempfindliche Bebauung
– Grundwasser bis maximal 6 m über Tunnelfirste
– Teilweise wenig standfeste und fließgefährdete Böden

Das daraufhin gewählte technische Konzept beinhaltete im Wesentlichen folgende Bestandteile:

– Vorpressen eines Pilotstollens (Innendurchmesser 2,8 m)

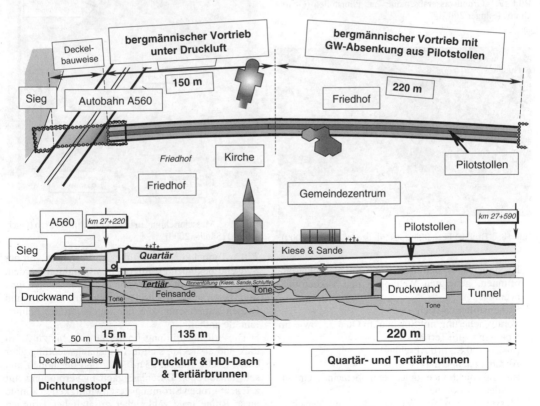

Bild 16. Bergmännischer Tunnel (Grundriss und Längsschnitt) (Sänger et al. 2000).

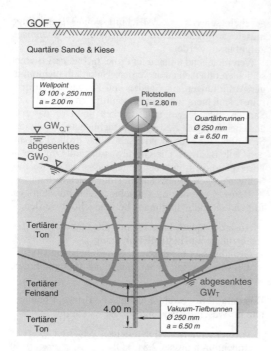

Bild 17. Grundwasserabsenkung aus Pilotstollen (Quer-schnitt) (Sänger 2000).

Bild 18. Brunnenherstellung aus Pilotstollen (Sänger 2000).

Bild 19. Tunnelanschlagwand "Süd".

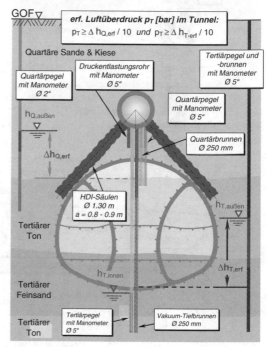

Bild 20. Messeinrichtungen beim Druckluftvortrieb (Quer-schnitt) (Sänger 2000).

- Herstellen und Betreiben von vertikalen Quartär-brunnen, geneigten Wellpoints und Vakuumtief-brunnen
- Ulmenstollenvortrieb
- Voraussicherung mit Spießen im Quartär sowie im Übergang zum Tertiär
  (bis zu 100 St. pro Abschlag)
- Abschlagslänge max. 0,8 m
- Hebungsinjektionen unter dem Gemeindezentrum und der Pfarrkirche
- Messprogramm für Wasserstände und Verschie-bungen

220 m Tunnel konnten mit diesem Verfahren von der Südseite beginnend erfolgreich aufgefahren werden. Danach wurde im Firstbereich der Ulmenstollen eine sehr heterogene Wechsellagerung aus Sanden und Kiesen aufgefahren, die nur untergeordnet Schluf-fanteile enthält.

Diese Rinnenfüllung im Tertiär war bis zu 3,5 m dick und erstreckte sich in Tunnellängsrichtung über eine Strecke von ca. 85 m. Aufgrund der hohen und sehr anisotropen Durchlässigkeit in Verbindung mit z.T. sehr großen Strömungsgeschwindigkeiten konnte diese Rinne trotz zahlreicher zusätzlicher Brunnen nicht stabilisiert werden.

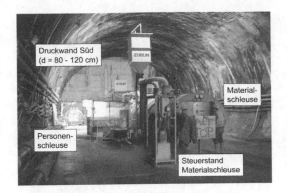

Bild 21. Druckwand mit Personen- und Materialschleuse.

Bild 22. Autobahnkreuz Frankfurt – Knotenpunkt von A3 und A5 am Frankfurter Flughafen.

Der weitere Vortrieb wurde nach einer Gegenüberstellung der in Frage kommenden Vortriebskonzepte vollständig auf Druckluft umgestellt. Um die Luftverluste im Quartär zu minimieren und die Sicherheit gegenüber plötzlichen Luftverlusten, d.h. Ausbläsern, zu erhöhen, wurde vorab aus dem Pilotstollen heraus ein abdichtendes HDI-Dach hergestellt. Die Vakuumtiefbrunnen im Tertiär wurden zur Reduzierung des erforderlichen Luftüberdruckes im Tunnel parallel betrieben.

Im Schutz der Druckluft konnten die verbleibenden ca. 150 m erfolgreich aufgefahren werden. Die Vortriebsleistung betrug dabei bis zu 1,0 m pro Tag mit einem Druckluftverbrauch zwischen 60 m³ und 400 m³ angesaugter Luft pro Minute.

Der Siegauentunnel konnte nur mit umfangreichen Zusatzmaßnahmen in der Spritzbetonbauweise aufgefahren werden. Bei den hier vorliegenden, vorher nicht vollständig bekannten geotechnischen Randbedingungen ist die Spritzbetonbauweise als grenzwertig hinsichtlich der technischen Ausführbarkeit zu bezeichnen. Für die wirtschaftliche Anwendung eines Schildvortriebs ist die Tunnellänge mit ca. 370 m jedoch zu gering.

2.2.4 Tunnel Frankfurter Kreuz (1995 bis 2001)

Im Rahmen der ICE-Unterquerung des mit 300.000 Fahrzeugen pro Tag meistbefahrenen Autobahnkreuzes Frankfurt durch die NBS Köln-Rhein/Main sowie dem gleichzeitigen Umbau des Autobahnkreuzes zur Erhöhung der Kapazität wurde neben Tunneln in offener Bauweise, diversen Brücken und Kreuzungsbauwerken sowie dem Um- und Neubau des Bahnhofes Zeppelinheim auch die Herstellung eines 281 m langen Tunnels in bergmännischer Bauweise zur Unterquerung der Autobahn A3 erforderlich.

Der Bauabschnitt PA36 der NBS liegt im nördlichen Oberrheingraben im Bereich der sog. Kelsterbacher Tiefscholle. Der Baugrund besteht aus dicht bis sehr dicht gelagerten, mittel- bis grobkörnigen Sanden mit wechselnden Kiesanteilen und sandigen Kiesen.

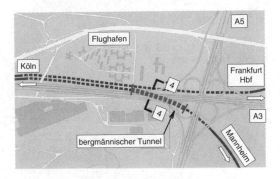

Bild 23. Lage bergmännischer Tunnel.

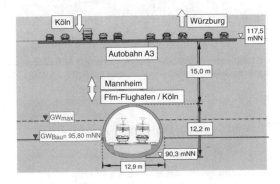

Bild 24. Querschnitt bergmännischer Tunnel (4-4).

Eing elagert sind Quarzit- und Sandstein gerölle sowie Schluff- und Tonlagen geringer Dicke. Das Grund wasser stand während der Tunnelbauar beiten ca. 6 m über der Tunnelsohle an. Eine Grundwasserabsenkung war nicht zugelassen.

Aufgrund der geringen Überdeckung zur A3 von nur 7 m bis 15 m waren bei der hier vorhandenen

Geologie umfangreiche First- und Sohlinjektionen für das Auffahren des bergmännischen Tunnels in Spritzbetonbauweise erforderlich. Der Kalottenausbruch erfolgte im Schutz eines Düsenstrahl Injektionsschirms (HDI-Schirms) mit einer Länge von jeweils 14,5 m und einer Überlappung von 3,5 m.

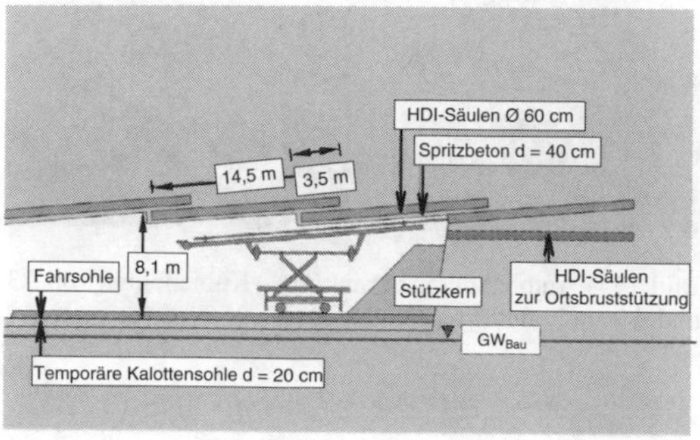

Der im Grundwasser liegende Teil des Querschnitts wurde mit einem 1,5 m dicken HDI-Gewölbe abgedichtet. Die Tunnelinnenschale wurde in wasserundurchlässigem Beton ohne eine zusätzliche Abdichtung ausgeführt.

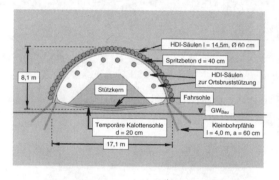

Bild 25. Herstellung des HDI-Schirms und Ausbruch der Kalotte (Längsschnitt).

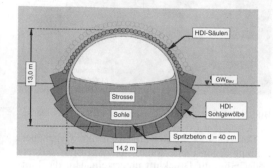

Bild 28. Herstellung des HDI-Sohlgewölbes und Ausbruch der Strosse und der Sohle (Querschnitt).

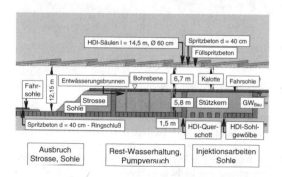

Bild 26. Herstellung des HDI-Schirms und Ausbruch der Kalotte (Querschnitt).

Bild 29. Herstellung des HDI-Schirms.

Bild 27. Herstellung des HDI-Sohlgewölbes und Ausbruch der Strosse und der Sohle (Längsschnitt).

Bild 30. Ausbruch der Kalotte im Schutz des HDI-Schirms (Hofmeister & Roth 1999).

Die Besonderheit des hier vorgestellten Tunnels unter dem Frankfurter Kreuz besteht in der geringen Überdeckung in kohäsionslosem Untergrund sowie in den strengen Vorgaben für die einzuhaltenden Grenzwerte der Senkungen an der Geländeoberfläche.

Bild 31. Bau der 4. Elbtunnelröhre westlich des bestehenden Tunnels (Lohrmann & Miemietz 2002).

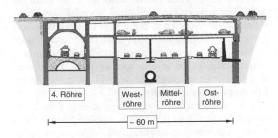

Bild 32. Tunnelquerschnitt (Lohrmann & Miemietz 2002).

Um den starken Verkehr auf der unterquerten Autobahn A3 nicht zu behindern bzw. zu unterbrechen, wurden die Verformungen mit Hilfe von digitalen, vollautomatisch gesteuerten Nivelliergeräten online kontrolliert. Die größten aufgetretenen Senkungen lagen aufgrund der gewählten Sicherungsmaßnahmen bzw. Bauabläufe mit ca. 4 cm unter dem vorgegebenen Grenzwert von 5 cm.

2.3 Schildvortriebe

Röhre Elbtunnel (1995 bis 2003)
Die Kapazität des im Jahre 1975 in Betrieb genommenen sechsspurigen Elbtunnels reicht für das Verkehrsaufkommen der A7 mit mehr als 100.000 Fahrzeugen pro Tag nicht mehr aus. Eine Verbesserung der Verkehrssituation konnte nur durch den Bau einer zusätzlichen 4. Tunnelröhre erreicht werden. Das Bauvorhaben wurde in 4 Lose aufgeteilt, wobei das Los 2 den im Schildvortrieb aufzufahrenden 2561 m langen Hauptabschnitt enthält.

Der durchfahrene Baugrund ist ausgesprochen heterogen und besteht im Wesentlichen aus Ablagerungen aus den verschiedenen Eiszeiten (Wechsellagerungen aus Sand-/Kiesschichten und Geschiebemergel mit Sandlinsen und Findlingen bis zu 2 m Durchmesser) sowie darunter aus eiszeitlich vorbelasteten, harten tertiären Glimmertonen. Der mittlere Wasserstand oberhalb des Gradientientiefpunktes beträgt ca. 40 m, wobei bei Sturmfluten auch 50 m möglich sind.

Die verwendete Schildvortriebsmaschine mit flüssigkeitsgestützter Ortsbrust besaß einen Durchmesser von 14,2 m bei einem Gewicht von ca. 2.600 t. Die Gesamtlänge mit Schildmantel (12,9 m, 2.000 t) und den beiden Nachläufern betrug ca. 60 m.

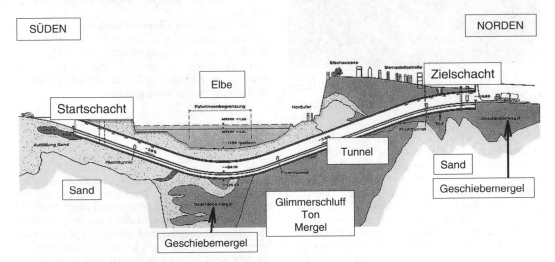

Bild 33. Geologischer Längsschnitt entlang des Elbtunnels (Lohrmann & Miemietz 2002).

Bild 34. Tunnelvortriebsmaschine mit Hydro-Schild (Lohrmann & Miemietz 2002).

Bild 36. Tübbingringausbau (Lohrmann & Miemietz 2002).

Tag. Das Schneidrad konnte um max. 80 cm in Längsrichtung verschoben werden. Das Zentrum wurde durch einen eigenständigen, ebenfalls längsverschieblichen Mix-Schild mit einem Durchmesser von 3 m entlastet. Durch in die Speichen integrierte Sender und Empfänger konnte erstmalig der Boden seismisch durch Schallwellen vorerkundet werden. Damit wurden Steine ab einer Größe von 60 cm bis zu einer Entfernung von 15 m geortet.

Die verwendeten Stahlbetontübbinge sind im Mittel 2 m breit und 0,7 m dick. Ein Ring besteht aus 8 Normalsegmenten sowie einem Schlussstein. Jeder Tübbing besitzt eine Bogenlänge von 5,2 m bei einem Gewicht von 18 t. Insgesamt wurden 11.500 Tübbinge verbaut.

Bei der 4. Röhre Elbtunnel handelt es sich um den weltweit größten Tunnel, der je im Schildvortriebsverfahren im Lockergestein unter einem Fluss hindurch aufgefahren wurde. Mit diesem Bauvorhaben wurde damit der Stand der Technik auf dem Gebiet der Schildvortriebsmaschinen gezeigt.

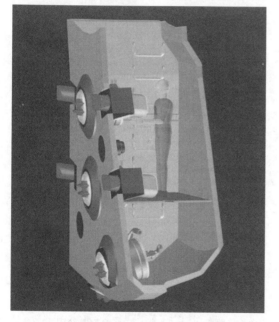

Bild 35. Werkzeugwechsel in den begehbaren Hauptspeichen (Herrenknecht 2001).

Die 5 Hauptspeichen des Schneidrades waren hohl und begehbar. Sie trugen 31 Rollenmeißel mit Doppeldisken zur Zerkleinerung der bereits angesprochenen Findlinge. An den Speichenkanten und an den 5 Hilfsspeichen waren 111 Schälmesser angeordnet. Die Abbauwerkzeuge konnten über die 5 Hauptspeichen unter atmosphärischen Druckverhältnissen ausgewechselt werden.

Der erreichte Vortrieb betrug bei ca. 1 Umdrehung pro Minute je nach Baugrund zwischen 2 mm (bei Steinen) und 25–30 mm (in Sand- und Kiesschichten). Die Vortriebsleistung betrug damit bis zu 12 m pro

3 AUSBLICK

3.1 *Allgemeines*

Nach dem vorangegangenen Überblick aktueller, bereits abgeschlossener Tunnelbauprojekte, die den Stand der Technik repräsentieren, soll nachfolgend auf einige Entwicklungsansätze der Tunnelbautechnik in Deutschland eingegangen werden. Dabei werden sowohl Entwicklungen im konventionellen Vortrieb, d.h. der Spritzbetonbauweise, als auch beim Schildvortrieb vorgestellt.

3.2 Konventioneller Vortrieb/Spritzbetonbauweise

Neben Entwicklungen in der Beton- und Maschinentechnologie ist bei der Spritzbetonbauweise in Deutschland, wie in anderen europäischen Ländern, das Bestreben zu erkennen, aus baubetrieblichen und terminlichen Gründen auch im Lockergestein Tunnel mit größeren Querschnitten im Vollausbruch aufzufahren. Der Hauptvorteil des Vollausbruchs besteht dabei in einer Verbesserung der baubetrieblichen Randbedingungen und damit einer Verringerung der Bauzeit im Vergleich zum Auffahren des Tunnels in Teilquerschnitten. Durch den bei einem Vollausbruch sehr schnell zu erzielenden Ringschluss können sowohl die Senkungen an der Geländeoberfläche als auch die Biegebeanspruchungen der Spritzbetonschale beherrschbar bleiben. Der wesentliche Nachteil eines Vollausbruchs im Lockergestein besteht jedoch im erhöhten Aufwand bei der Sicherung der Ortsbrust.

Zur Vergrößerung der Ortsbruststandsicherheit kann dabei z.B. ein vorauseilender HDI-Schirm dienen. Damit wird die Auflast reduziert, die sich aus der Überlagerung ergibt und potentiell in den Tunnelquerschnitt abrutschende Erdkeile belastet. Durch die Abschirmung des Erdkeils werden damit die treibenden Kräfte reduziert. Zur Vergrößerung der haltenden Kräfte des Erdkeils können zusätzliche Sicherungselemente durch die Tunnelortsbrust eingebracht werden. Dies können z.B. HDI-Brustsäulen sein oder auch GFK-Brustanker. Art, Durchmesser und Länge ergeben sich u.a. aus den speziellen geometrischen und geotechnischen Randbedingungen.

Die für einen wirtschaftlichen Einsatz der Sicherungsmittel erforderliche wirklichkeitsnahe Ermittlung der Ortsbruststandsicherheit erfolgt mit verschiedenen Rechenverfahren, sowohl auf analytischer als auch numerischer Grundlage (z.B. Methode der finiten Elemente). Der Verbesserung dieser Rechenverfahren dienen aktuelle Forschungsaktivitäten. In jedem speziellen Fall bleibt abzuwägen, ob der erforderliche zusätzliche Aufwand an Sicherungsmitteln zur Stützung der Ortsbrust durch den baubetrieblichen Vorteil, der sich durch den Wegfall der Teilausbrüche ergibt, wirtschaftlich gerechtfertigt ist.

3.3 Schildvortrieb

Hinsichtlich eines Vergleiches von Spritzbetonbauweise und Schildvortrieb ist ein wesentlicher Aspekt, dass Schildvortriebsmaschinen i.a. für jede Baumaßnahme speziell gebaut werden. Der zusätzliche Zeitbedarf für Herstellung, Transport und Montage der Maschine sowie die Bau- und Vorhaltekosten sind nur bei einer Mindestlänge des Tunnels bzw. in Kombination mit der wesentlich höheren Vortriebsgeschwindigkeit wirtschaftlich zu rechtfertigen. Der Kostenvergleich mit der Spritzbetonbauweise fällt dabei bei jeder Tunnelbaumaßnahme anders aus.

Zu berücksichtigen sind bei der Wahl des Vortriebsverfahrens jedoch auch Aspekte, die sich unmittelbar in Geld auszudrücken sind. Dies können z.B. eine bessere politische Akzeptanz oder die Minimierung geotechnischer Risiken sein. Auch Umweltschutzgründe können für einen Schildvortrieb sprechen, wenn z.B. eine Grundwasserabsenkung nicht zulässig ist bzw. durch eine aktive Ortsbruststützung

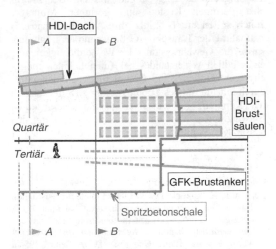

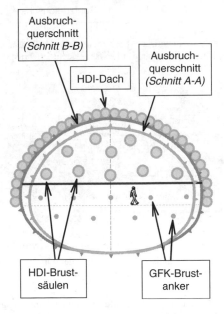

Bild 37. Möglicher Einsatz von Sicherungsmitteln beim Auffahren eines Großquerschnitts im Vollausbruch (Längsschnitt).

Bild 38. Möglicher Einsatz von Sicherungsmitteln beim Auffahren eines Großquerschnitts im Vollausbruch (Querschnitt).

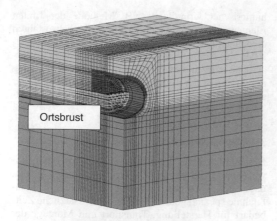

Bild 39. Numerische Berechnung der Ortsbruststandsicherheit: 3dimensionales FE-Netz.

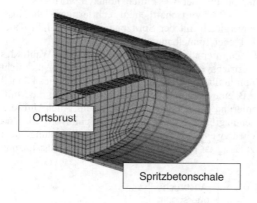

Bild 40. Numerische Berechnung der Ortsbruststandsicherheit: 3dimensionales FE-Netz – Detail Ortsbrust.

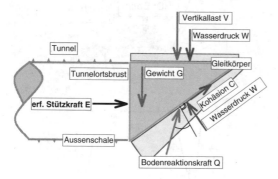

Bild 41. Analytische Berechnung der Ortsbruststandsicherheit: am Gleitkeil angreifende Kräfte.

Senkungen an der Geländeoberfläche klein gehalten werden müssen.

In innerstädtischen Bereichen wird der Schildvortrieb zunehmend dort eingesetzt, wo früher aufgrund der ungünstigen Baugrund- bzw. Grundwasserverhältnisse die offene Bauweise gewählt wurde bzw. kein Tunnelbau möglich war. Die Aufgaben, die sich speziell im innerstädtischen Schildvortrieb ergeben, sind unmittelbar mit der Sicherung des Bestandes verbunden: die mit dem Schildvortrieb verbundenen Senkungen sind zu begrenzen. Innovative Konzepte bedienen sich einer Kombination von baubegleitenden Messungen (Verschiebungen, Spannungen), baubegleitenden numerischen Simulationen und einer auf Grundlage der Mess- und Rechenergebnisse durchführbaren Maschinensteuerung.

Gemessen werden u.a. die Verschiebungen an Gebäuden bzw. der Geländeoberfläche sowie die Spannungen im Baugrund, der Stützdruck an der Ortsbrust und die Druckverteilung innerhalb des Verpressgutes im Bereich des Schildschwanzes. Durch die Schildschwanzverpressung wird sichergestellt, dass sich das umgebende Bodenmaterial nicht in den durch den Überschnitt entstehenden Ringspalt verschiebt. Dafür muss der Druck im Verpressgut bis zu dessen Abbinden aufrecht erhalten werden.

Die numerischen Berechnungsergebnisse werden mit Messergebnissen verglichen und die Berechnungen kalibriert. Auf der Grundlage der Mess- und Rechenergebnisse kann eine gezielte Steuerung der Maschine hinsichtlich der Minimierung der vortriebsbedingten Senkungen erfolgen. Dies geschieht z.B. mit einer Erhöhung des Stützdrucks an der Ortsbrust oder einer Verringerung der Vortriebsgeschwindigkeit. Die Steuerung vortriebsbegleitender Hebungsinjektionen unter bestehenden Gebäuden kann ebenfalls auf dieser Datengrundlage erfolgen.

Abschließend soll noch kurz auf die Forschungen und Entwicklungen auf dem Gebiet des einschaligen Tübbingausbaus eingegangen werden. Bei möglichst geringen Kosten und geringen Montagezeiten sollen die Tübbinge ihre Aufgaben, wie z.B Aufnahme der Transport-, Gebirgs- und Pressenlasten, sowie die Gewährleistung der Wasserdichtigkeit und eines hohen Widerstandes bei Tunnelbränden möglichst gut erfüllen. Eine Optimierung soll dabei u.a. durch ein innovatives Tübbingdesign z.B. hinsichtlich der Kopplung der einzelnen Segmente und durch Entwicklungen in der Betontechnologie oder durch den Einsatz von Stahlfaserbeton erreicht werden.

LITERATUR

Bernhardt, K. & Rock, D. 1995. Der Engelbergbasistunnel: Bau und Finanzierungskonzept eines Autobahntunnels mit außergewöhnlichen Querschnittsabmessungen in schwierigen Gebirgsformationen. Weltneuheiten im Tunnelbau: Vorträge World Tunnel Congress/STUVA-Tagung '95 in Stuttgart, Studiengesellschaft für Unterirdische Verkehrsanlagen e.V. (STUVA), Düsseldorf, Alba, (Forschung + Praxis; 36).

Bielecki,R. 1998. Besonderheiten bei der Planung und Ausführung der 4. Röhre des Elbtunnels in Hamburg. Neue Akzente im unterirdischen Bauen: '97 in Berlin, Studiengesellschaft für Unterirdische Verkehrsanlagen e.V. (STUVA), Düsseldorf, Alba, (Forschung + Praxis; 37.

Dietz, W. & Becker, C. 1995. Kriterien zur Auswahl und Bewertung von Tunnelvortriebsmaschinen – Eine Empfehlung des DAUB –. Weltneuheiten im Tunnelbau: Vorträge World Tunnel Congress/STUVA-Tagung '95 in Stuttgart, Studiengesellschaft für Unterirdische Verkehrsanlagen e.V. (STUVA), Düsseldorf, Alba, (Forschung + Praxis; 36).

Dietz, W. Spuler, B. & Härle, D. 1998. NBS Köln-Rhein/Main: Erhöhte Sicherheit bei einem Vortrieb mit vorauseilendem HDI-Schirm durch kontinuierliche Setzungsmessungen. Neue Akzente im unterirdischen Bauen: '97 in Berlin, Studiengesellschaft für Unterirdische Verkehrsanlagen e.V. (STUVA), Düsseldorf, Alba, (Forschung + Praxis; 37).

Haack, A. 1995. Unterirdisches Bauen in Deutschland. STUVA-/ITA-Tagung '95, Stuttgart, STUVA –Studiengesellschaft für Unterirdische Verkehrsanlagen e.V. und DAUB – Deutscher Ausschuß für unterirdisches Bauen e.V., Gütersloh, Bertelsmann.

Haack, A. 1999. Unterirdisches Bauen in Deutschland 2000. STUVA-Tagung '99, Frankfurt, STUVA –Studiengesellschaft für Unterirdische Verkehrsanlagen e.V. und DAUB – Deutscher Ausschuß für unterirdisches Bauen e.V., Gütersloh, Bertelsmann.

Hering, S. & Walliser, T. 1998. Ohne Stau durch Europas größten Straßentunnel. ZÜBLIN-Rundschau 30, Stuttgart.

Herrenknecht, M. 1998. Innovationen bei Tunnelvortriebsmaschinen – dargestellt am Beispiel der 4. Röhre Elbtunnel. Neue Akzente im unterirdischen Bauen: '97 in Berlin, Studiengesellschaft für Unterirdische Verkehrsanlagen e.V. (STUVA), Düsseldorf, Alba, (Forschung + Praxis; 37).

Herrenknecht, M. 2001. Praktische Beispiele zur Anpassung von Tunnelvortriebsmaschinen an die Erfordernisse beim Vortrieb. Unterirdisches Bauen 2001. Wege in die Zukunft: 2001 in München, Studiengesellschaft für Unterirdische Verkehrsanlagen e.V. (STUVA), Gütersloh, Bertelsmann, (Forschung + Praxis; 39).

Hirsch, D. & Graf, R. 2000. Besondere Verfahrenstechniken und Stabilisierungsmaßnahmen beim Auffahren des Siegtunnels der NBS Köln-Rhein/Main. Unterirdisches Bauen 2000. Herausforderungen und Entwicklungspotentiale: '99 in Frankfurt am Main, Studiengesellschaft für Unterirdische Verkehrsanlagen e.V. (STUVA), Düsseldorf, Alba, (Forschung + Praxis; 38).

Hofmeister, G. & Roth, H. 1999. 300.000 Baustellenbesucher kommen täglich mit dem Auto. ZÜBLIN-Rundschau 31, Stuttgart.

Kleffner, H.-J. & Denzer, G. 2000. Bauvorbereitung und –durchführung der Kammquerung des Thüringer Waldes im Zuge der A71/73. Unterirdisches Bauen 2000. Herausforderungen und Entwicklungspotentiale: '99 in Frankfurt am Main, Studiengesellschaft für Unterirdische Verkehrsanlagen e.V. (STUVA), Düsseldorf, Alba, (Forschung + Praxis; 38).

Lohrmann, W. & Miemietz, J. 2002. Seismischer Vortrieb steigert Leistung. ZÜBLIN-Rundschau 34, Stuttgart.

Lorscheider, W. & Kuhnhenn, K. 1998. Bautechnische Besonderheiten beim Bau des Engelbergbasistunnels. Neue Akzente im unterirdischen Bauen: '97 in Berlin, Studiengesellschaft für Unterirdische Verkehrsanlagen e.V. (STUVA), Düsseldorf, Alba, (Forschung + Praxis; 37).

Mayer, P.-M., Hartwig, U. & Schwab, C. 2003. Standsicherheitsuntersuchungen der Ortsbrust mittels Bruchkörpermodell und FEM. (zur Veröffentlichung).

Prax, K., Groten, A. & Decker, A. 2000. Deutschlands längster Autobahntunnel. ZÜBLIN-Rundschau 32, Stuttgart.

Sänger, C. 2000. Neubaustrecke Köln-Rhein/Main Los 2.4 – Siegauen-Tunnel. Vorträge der Baugrundtagung 2000 in Hannover, Deutsche Gesellschaft für Geotechnik (DGGT), Verlag Glückauf, Essen.

Sänger, C., Roth, H. & Hofmann, S. 2000. Bergmännischer Tunnelvortrieb in schwierigster Geologie. ZÜBLIN-Rundschau 32, Stuttgart.

Geotechnical Measurements and Modelling, Natau, Fecker & Pimentel (eds)
© 2003 Swets & Zeitlinger, Lisse, ISBN 90 5809 603 3

Innovations in geotechnical on-site engineering for tunnels

W. Schubert
Graz University of Technology

G.M. Vavrovsky
HL-AG, Vienna

ABSTRACT: The limits in the accuracy of the prediction of the geological architecture and rock mass characteristic require an observational approach for the safe and economical construction of tunnels. This especially applies for tunnels in weak ground and/or sensitive environments. It has become common practice to assign a geotechnical engineer to difficult sites to continuously evaluate the geotechnical situation and assist in the routine decision making process. Appropriate site organization and advanced methods for monitoring data evaluation and interpretation are a precondition for efficient work on site. The paper deals with the basic requirements to successfully implement an observational approach, the respective tools for monitoring, and monitoring data evaluations.

1   INTRODUCTION

The observational method is widely applied in underground construction. The EUROCODE 7 specifies conditions for the application of the observational method. Requirements to be met before construction are:

- Acceptable limits of the behavior shall be established
- The range of possible behaviors shall be assessed and it shall be shown, that there is an acceptable probability that the actual behavior is within the acceptable limits
- A plan of monitoring shall be devised which will reveal whether the actual behavior lies within the acceptable limits. The monitoring shall make this clear at a sufficiently early stage and with sufficiently short intervals to allow contingency actions to be undertaken successfully
- The response time of the instruments and the procedures for analyzing the results shall be sufficiently rapid in relation to the evolution of the system
- A plan of contingency actions shall be devised which may be adopted if the monitoring reveals behavior outside the acceptable limits.

In underground construction the prediction of the system behavior made during the design needs to, in most cases, be refined during construction to arrive at an economical and safe solution.

To be able to adjust criteria during construction, several conditions must be fulfilled. On the one hand the geomechanical design needs to be done in a coherent way to be able to identify the parameters, which have a dominant influence on the system behavior.

In underground engineering there are two major aspects that must be addressed during the design phase. The first and most important is developing a realistic estimate of the expected rock mass conditions and their potential behaviors as a result of the excavation. The second is to design an economic and safe excavation and support method for the determined behaviors. As the design not only depends on the behavior of the rock mass, but also on boundary conditions, regulations, and local as well as system requirements, a systematic approach is required.

Such an approach is outlined in the Guideline for the Geomechanical Design of Underground Structures (OGG 2001). The process outlined in the guideline clearly distinguishes between rock, the rock mass, influencing factors, boundary conditions, rock mass behavior, and system behavior.

Due to the fact, that in many cases the rock mass conditions cannot be defined with the required accuracy prior to construction, during construction it is necessary to continuously update the geotechnical model and adjust the excavation and support to the actual ground conditions.

The final determination of the excavation methods, as well as support type and quantity, in most cases is possible only on site. In order to guarantee the required safety, a safety management plan needs to be established and followed during construction.

The approach for the excavation and support determination on site, as outlined in the Guideline (OGG 2001) can be divided into four basic steps:

Step 1 – Determination of the encountered Rock Mass Type

To be able to determine the encountered Rock Mass Type (RMT), the geological investigation (documentation) during construction has to be targeted to collect and record the relevant parameters that have the greatest influence on the Rock Mass Behavior.

Step 2 – Determination of the actual Rock Mass Behavior Type

Observations during excavation, such as signs of excessive stress, the deformation pattern and observed failure mechanisms, and results from probing ahead are used to continuously update the geotechnical model.

The reaction of the ground to the excavation has to be observed, using appropriate geotechnical monitoring methods and layouts.

Based on observations and measurement results during construction, a short-term prediction is made, and the Rock Mass Behavior Type (BT) for the coming excavation step(s) is determined.

Step 3 – Determination of excavation and support

To determine the appropriate excavation and support the criteria laid out in the baseline construction plan have to be followed. Consequently, the actual rock mass conditions (RMT; BT) continuously have to be compared to the prediction for compliance. A continuous detailed analysis of the rock mass and system behaviors is used to update the geotechnical model. The additional data obtained during construction form the basis for the determination of the applied excavation and support methods. The goal is to achieve an economical and safe tunnel construction, considering the requirements (RQ) to be observed.

Based on the evaluated BT, and the determined excavation and support layout, the system behavior for each section has to be predicted.

Both excavation and support, to a major extent, have to be determined prior to the excavation. After the initial excavation only minor modifications, like additional bolts, are possible. This fact stresses the importance of a continuous short-term prediction.

Step 4 – Verification of System Behavior

By monitoring the behavior of the excavated and supported section the compliance with the requirements

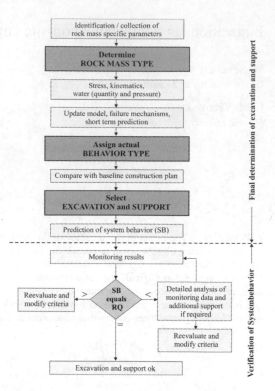

Figure 1. Flow chart of the basic procedure of determining excavation and support and verify the system behavior (OGG 2001). RQ = requirements.

and criteria defined in the geotechnical safety management plan can be checked. When differences between the observed and predicted behavior occur, the parameters and criteria used during excavation for the determination of RMT and the excavation and support have to be reviewed. When the displacements or support utilization are higher than predicted, a detailed investigation into the reasons for the different System Behavior has to be conducted, and if required improvement measures (like increase of support) ordered. In case the System Behavior is more favorable than expected, the reasons have to be analyzed as well, and the findings used to better calibrate the geotechnical model and the delimitating criteria and parameters.

2 SITE ORGANIZATION

For a successful implementation of an observational approach an appropriate site organization is required. Direct communication between the parties involved in construction allows one to maximize the value of information and to minimize time required for decision-making. To account for the responsibilities, a clear and efficient organizational structure is required.

As the available reaction time in case of deviations is a crucial factor for an observational approach, the bureaucratic procedures should be minimized as much as possible.

3 MONITORING

The procedure outlined above demonstrates the importance of efficient monitoring methods, and data evaluation and interpretation on site.

Methods for measuring displacements, as well as for evaluating the data have considerably developed during the last 15 years. The information contained in the data, especially when using spatial displacement measurements, can be evaluated in many different ways. When properly used, the information can considerably benefit a project. It is basically in the interest of the owner to promote the proper use of monitored data, as this definitely has a positive influence on the construction costs and safety.

Reviewing literature, and the common practice on site, it is apparent, that there is still considerable potential for improvement in the data collection and interpretation techniques. This starts with the collection of geological data, where in most cases standardized parameters are recorded, which may or may not have relevance for the geological conditions. Very rarely can it be found, that a routine short-term prediction of the geological conditions ahead of the face is performed continuously, which is a precondition for the prediction of rock mass and system behaviors.

It is admitted, that there are not many accepted rules as how to evaluate and interpret monitoring data. Due to the variety of geological conditions and influencing factors, it is difficult to set up generalized rules.

There are several more or less advanced tools available on the market, which make data handling, evaluation and interpretation easier and more efficient on site. There is, however, also a demand for more basic research to be able to develop "smart" tools for the site. The following chapters shall demonstrate how monitoring data can be used for the benefit of a project. The merits and limitations of each method of evaluation will be shown with examples and case histories.

3.1 Spatial displacement monitoring

The measurement of spatial displacements of targets fixed to the lining has widely replaced the traditional convergence measurements with the tape (Rabensteiner 1996). Due to the increase in information with this type of measurement, the use of additional methods as for example extensometers has decreased.

The accuracy of absolute displacement measurements is in the range of one millimeter, which is good enough for the purpose.

The observation of the transient displacements in space allows a much better evaluation of the influence of the rock mass structure, than with traditional relative measurements.

3.2 Extensometer measurements

Extensometers have not much changed during the last decades. Naturally there is now the possibility to automatically record the measured values using LVDTs instead of the dial gage.

Extensometers are used to determine the depth of the zone influenced by the excavation, or to detect or verify assumed failure modes. As mentioned, the use of absolute displacement monitoring with geodetical methods has limited the application of extensometers to special problems.

3.3 Inclinometer measurements

When inclinometers are used in connection with tunneling, commonly they are installed from the surface to either record slope movements or to get a better insight into the ground movement caused by the excavation of a tunnel.

Most recently horizontal inclinometers have been used in connection with pipe roof supports. Displacements ahead of the face can be measured efficiently, and thus the total displacement path determined (Volkmann 2003).

3.4 Strain measurements

Strain measurements are occasionally taken in shotcrete linings in order to back-calculated stresses, respectively the stress intensity factor. Although having been applied in several cases, not much has been published on the results.

4 METHODS OF EVALUATION AND DISPLAY

In the following chapters the focus is put on the absolute displacement monitoring, which is the most commonly used method.

4.1 Displacement histories

Plotting displacement versus time for one displacement component is the most common way of displaying measurement data in tunnels. The interpretation of the curve is easy for homogeneous rock mass conditions and a continuous advance rate. The condition for a satisfying stabilization, respectively the stress redistribution is a steadily decreasing displacement rate.

The displacements can be split into a component related to the face advance and a component describing

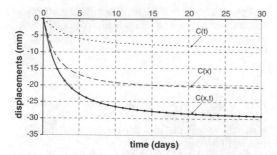

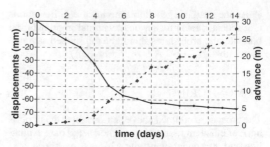

Figure 4. Displacement history and advance for a non-steady excavation rate; distance between face and measuring section plotted as dashed line.

Figure 2. Development of displacements split into the time dependent (C(t)) and advance dependent components (C(x)) for a continuous excavation rate.

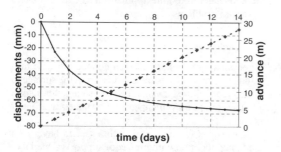

Figure 3. Displacement history and advance for a steady excavation rate; distance between face and measuring section plotted as dashed line.

the time dependent closure. Sulem et al. (1987) have formulated a relationship for the advance and time dependent closure of tunnels. Those formulations were used to produce figures 2, 3, and 4. Figure 2 shows the development of the time and advance dependent components, and the total displacement for a constant advance rate. It can be clearly seen, that the displacement rate is constantly decreasing, signaling a normal stabilization process.

Depending on the rock mass type, the influencing factors, and the type and amount of support the ratio between time dependent and advance dependent displacements change.

Figures 3 and 4 show the influence of the advance rate on the development of the displacements. The example shown in figure 3 was produced with a constant advance rate. The final displacements can easily be estimated by extrapolating the measured curve.

With a non-steady advance rate it is far more difficult to judge, if the development of the displacements is "normal". Figure 4 shows such an example with a varying advance rate. It can be seen, that a judgment of the displacement development in such a case is rather difficult. To be able to make a well founded judgment additional tools are required, like using the equations given by Sulem.

With just a visual inspection of the plot shown in figure 4, it would be hard to judge, if the stabilization process is normal, especially in the first few days. With additional headings, heterogeneous rock mass conditions, or time dependent behavior of the support it is even more difficult to properly interpret the results when only using the displacement histories.

This difficulty to visually check the normality of a displacement development was one of the reasons to develop a tool for the prediction of displacements. Sellner (2000) based on Sulem et al. and Barlow (1986), extended the functions and added new features, like the possibility to consider additional support. The extended capabilities were implemented in a code, called GeoFit© (Sellner & Grossauer 2002). He uses a set of variables, describing the time dependent and advance dependent displacements, the support, and the ratio of pre-displacements to total displacements. Curve fitting techniques are used to back calculate some of the required parameters.

The software has been used on a number of projects, and several improvements made. There is still considerable basic research required to determine the dependencies between rock mass quality, influencing factors, and support to be able to provide unique solutions. Presently still a lot of experience is required to arrive at reasonable predictions. The tool nevertheless does not want to be missed by those who used it.

With a daily update of the monitored data it can be easily detected, when the system behavior deviates from the "normal". In such a case a detailed analysis of the causes is required. If the deviation is expected to lead to an unstable condition, strengthening measures need to be done.

Figures 5 to 7 show such a process to predict the displacement magnitude and the comparison of the predicted to the measured values. Two days after the zero reading the top heading excavation was stopped for the bench excavation and a temporary top heading invert installed. After approximately two months the excavation in the top heading was resumed. Figure 5 shows the predicted displacements after two days

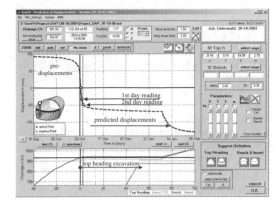

Figure 5. Back calculation of the function parameters after 2 days of measurements, and prediction of displacement development.

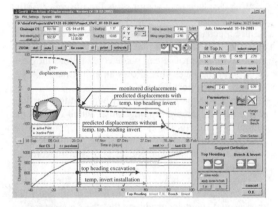

Figure 6. Simulation of the effect of a temporary top heading invert.

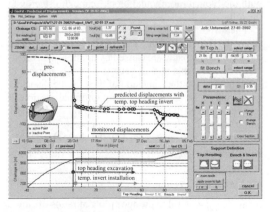

Figure 7. Comparison of the predicted displacements to the measured ones, shown as dots.

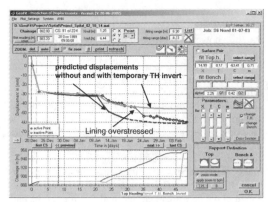

Figure 8. Deviation of system behavior from predicted behavior due to overstressing of the lining and partial loss of capacity.

without the temporary invert. Figure 6 shows the predicted system behavior with temporary invert, while in figure 7 the measured displacements are compared to the predicted ones. It can be seen, that in this case a very good compliance was obtained.

Figure 8 shows such a deviation from the predicted behavior. A measuring section was installed in the top heading immediately before the Christmas break. In addition to the primary lining and rock bolts a temporary top heading invert was used. After restart the behavior was as expected for approximately 10 days. When the face entered poor rock mass conditions, additional loads were transferred to the sections further back, leading to an overstressing of the temporary top heading invert, which lost part of its capacity. It can be seen, that after a sudden increase of the displacements, the system stabilizes again. Although a loss of lining capacity is never desirable, this case was non-critical in terms of stability.

The disadvantage of using displacement history plots is that it is difficult to obtain an overview of the processes, as each section has to be inspected separately.

4.2 Deflection curves

To be able to observe a spatial overview of the displacements deflection curves are frequently used. They are produced by connecting the measured values of one component (for example the vertical or horizontal component) at a certain time along the tunnel. By plotting these lines in regular time intervals, the influence of the progress on the sections behind the face can easily be seen. This is the reason why the deflection lines frequently are called influence lines. Details and examples of application can be found in Vavrovsky & Ayaydin (1988), Vavrovsky (1988), Vavrovsky & Schubert

(1995). Producing trend lines from the deflection lines, a certain extrapolation beyond the face is possible. Practice however shows that the extrapolation in many cases does not reveal much about the conditions ahead of the face. To be able to show comparable data from different monitoring sections on one plot, the determination of the displacements occurring prior to the zero reading is important. Zero readings of the targets are not always done at the same distance behind the face or time after excavation. This implies that besides the displacement occurring ahead of the face, an additional part of the displacements is not recorded. To make displacement measurements comparable, normalization is required. Commonly the displacements ahead of the face are neglected, and the value at the face taken to zero. Various methods to determine the missing portion of the displacements between the face and the measuring section are used. The most appropriate method is to use time- and distance-dependent functions, as described by Sellner (2000).

It is very important to accurately record the location of the face and the time of excavation to achieve comparable pre-displacement values for different measuring sections. The graph in the top of figure 9 shows the deflection lines without consideration of pre-displacements. The trend 3 m behind the face is also shown. In this example the zero reading at measuring section 320 was done when the face was at station 322. In case the face station at the time of the zero reading

would be erroneously recorded at station 321, the deflection curves and trend line would look like in the central plot in figure 9, while the bottom plot shows the curves with the exact recording of the face station. This rather simple example clearly demonstrates the importance of recording the data accurately. One can very easily arrive at wrong conclusions, when there are mistakes in the data.

With appropriate data recording and evaluation the deflection curves are a useful tool to quickly obtain an overview, and to judge the influence of the excavation at the face on the sections further back.

4.3 Displacement vectors

Displacement vectors can either be displayed in a cross section or the longitudinal section. In the first case, the radial displacements are displayed, in the second case, the combination of the vertical and longitudinal components are visualized.

With displacement vector plots the influence of the rock mass structure can easily be observed, as well as failure mechanisms detected. Using the vector plot in a cross section, structures like faults or slickensides outside the tunnel profile can be detected before they can be seen at the face. This allows an adjustment of excavation and support in time.

Figure 10 shows a situation, where the top heading approaches a steeply dipping fault, which crosses the tunnel from the right to the left side. The top plot shows a section without of influence of the fault. The fault outside the right sidewall causes higher displacements (middle).

The third plot shows a section, where the fault left the profile and is located close to the left sidewall. The increased displacements due to the stress concentration between sidewall and fault can be clearly seen.

Figure 11 shows displacement vector plots in the cross section and longitudinal section. The phyllites with the typical rhomboidal block shape dip steeply to the left.

The strike is in an acute angle to the tunnel axis (from the right to the left). The relatively high displacements on the left side are caused by dilation. The orientation of the displacement vectors is nearly normal to the foliation. On the right side of the plot the displacement vectors are shown in the longitudinal section. It can be observed, that the left sidewall moves forward, while the right sidewall shows the opposite trend. This phenomenon is clearly caused by the shear movements along the foliation.

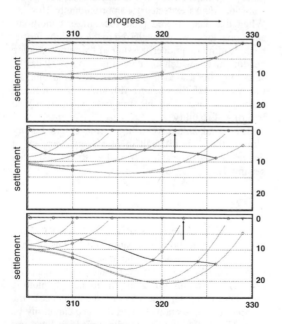

Figure 9. Deflection lines without consideration of pre-displacements (top), with pre-displacements but wrong face position (center) and with pre-displacements and correct face position (bottom) (Schubert et al. 2002).

4.4 Ratios of displacement components

It can be very useful to produce plots of ratios of single components. It can be assumed, that the ratios of displacement components remain the same if there are no

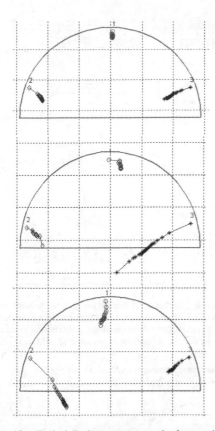

Figure 10. Typical displacement vector plot for a top building, where a steeply dipping fault crosses the profile from the right side to the left side. Top: normal vector orientation and magnitude; middle: fault close to the right sidewall causes increase of displacements; bottom: fault located close to the left side wall, showing in an increase of the displacements.

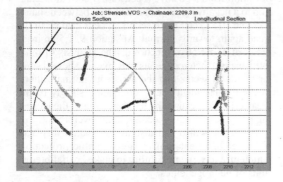

Figure 11. Top heading in foliated rock; foliation dips steeply to the left, strike is in acute angle to the tunnel axis (see Mueller flag). Left: Displacement normal to foliation indicates dilation, right: vectors show influence of the strike of foliation.

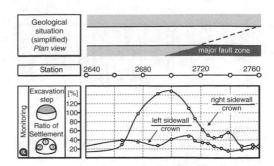

Figure 12. Trends of ratios of sidewalls and crown settlement; the relative increase of the settlements of the right sidewall indicates approach of a fault on the right side (Steindorfer 1998).

major changes in the rock mass quality in the vicinity of a tunnel. Each contrast in stiffness, or singularities, like slickensides or faults changes the stress distribution around the tunnel and this reflects in a different displacement pattern. Similar to the vectors in the cross section, those plots can give an early warning of changing geological conditions outside the visible excavation area. For example the ratio between the vertical displacement of the crown and the sidewall can be used to detect faults outside the tunnel. While in homogeneous conditions, there will be a constant ratio between the crown and sidewall displacement, in case the excavation approaches a moderately to steeply dipping fault, the displacements at the sidewall close to the fault will increase in a higher proportion than those at the crown. Routines can be incorporated into the evaluation software to automatically check on the "normality" of the respective ratios of displacement components, and to issue warnings in case certain limits are exceeded.

4.5 Longitudinal displacements

Following the idea, that the displacement pattern changes when the excavation approaches rock masses with different quality, one arrives at the evaluation of the spatial displacement vector orientation. From observations on site (Schubert 1993) it was concluded, that the longitudinal displacements are more sensitive to changes in the rock mass structure than the radial displacements. Using trends of the ratio between longitudinal and radial displacements, the capability of short term prediction increases significantly (Schubert & Budil 1995, Schubert & Steindorfer 1996).

Figure 13 shows the deflection curves of vertical and longitudinal displacements and the trend of the orientation of the displacement vector of the crown in a section of the Inntaltunnel. Nearly over the whole length of the 12.7 km long railway tunnel phyllitic rocks were encountered. The strike was nearly parallel to the tunnel axis, the dip moderately to the north.

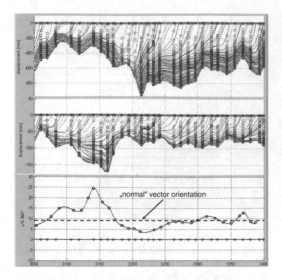

Figure 13. Settlements of the crown (top), longitudinal displacements (middle), and ratio of longitudinal and vertical displacement (bottom) at the Inntaltunnel.

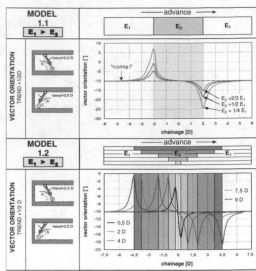

Figure 14. Influence of the length of a weak zone on the deviation of the displacement vector orientation from normal for different stiffness contrasts (top), and different extensions of the weak zone (bottom) (Grossauer et al. 2003).

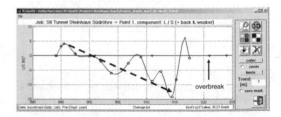

Figure 15. Trend of displacement vector orientation of crown-point prior to an overbreak.

A regional fault zone influenced the excavation for a length of more than 2.000 m. Within this fault zone a cataclastic section of approximately 100 m was encountered. In figure 13 it can be seen clearly, that the vector orientation considerably deviates from the "normal" orientation well before the excavation runs into the completely crushed rock mass. Once the excavation is in the fault zone, the vector orientation goes back to normal again. Remarkable at this project were the extremely high longitudinal displacements in some sections. It is also remarkable, that the magnitude of the longitudinal displacements is lower in the weak rock than in the stiffer block ahead of the tectonized zone.

Besides evaluating data from a number of sites, the phenomenon has been studied and confirmed with numerical simulations (Budil 1995, Steindorfer 1998, Golser 2001, Grossauer 2001). Grossauer found, that as well as the stiffness contrast, the extension of a zone with different stiffness influences the magnitude of the deviation of the vector orientation from "normal" up to a certain critical zone length (figure 14).

As can be seen from figure 15, the displacement vector orientation deviates from the normal in the opposite direction, when stiffer material is approached. It has been experienced, that in such situations there is an increased risk of overbreak. It is assumed that this is due to the comparatively low stresses in the area of the face, as stresses are concentrated in the stiffer material ahead.

Such a situation can be seen in figure 14, where a shallow tunnel is excavated in a tectonic mélange. The trend of the displacement vector orientation shows a deviation from a positive (backwards) normal orientation to a negative (forward) orientation between station 895 and 915. Then there seems to be a change in the trend again towards normal. At station 921,5 an overbreak of about 50 m³ occurred. The analysis of other overbreaks in heterogeneous rock masses showed similar trends. It seems that the apparent "normalization" of the displacement vector orientation shortly before the overbreak is caused by the beginning of the loosening process, eventually leading to the overbreak.

Some tools also can display the spatial displacement vector orientation in stereographic projection, which provides the possibility to make spatial predictions of heterogeneities.

4.6 Evaluation of stresses in linings

Once data are recorded and stored, one should make the maximum use of the information contained in the

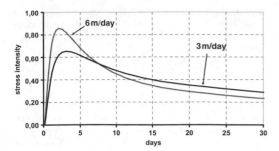

Figure 16. Development of stress intensity index in a shotcrete lining for progress rates of 3 m/day and 6 m/day.

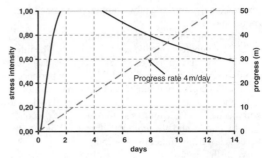

Figure 17. Development of stress intensity factor for an advance rate of 4 m/day.

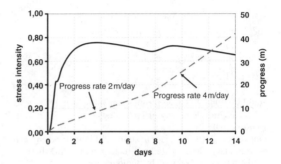

Figure 18. Utilization of the shotcrete lining due to reduction of progress rate.

data. One of the methods to increase the level of information is to analyse stresses in the lining and compare them to the strength. Rokahr and Zachow (1997) have done pioneering work in this field and the model is practically applied (Rokahr et al. 2002). Another model, simulating the complex behaviour of shotcrete is currently under development, and has been tested on one site so far (Hellmich et al. 1999, Macht 2002). Especially for tunnels with low overburden, where the shotcrete lining is the predominant support and the rock mass plays a minor role in the stress redistribution, the knowledge of the development in the stress intensity index is an important decision aid. With higher overburden the integrity of the shotcrete lining usually looses importance, because in most cases the natural "rock arch" can compensate the loss in lining capacity, under the condition that is has still reserves or is properly reinforced by rock bolts.

In both routines measured displacement data are used and strains in the lining calculated. Considering the strain development and transient properties of the shotcrete, the stresses can continuously be compared to the actual strength, leading to a stress intensity factor. Not only the magnitude of displacements, but also their timely development influences the stresses in the lining. It can be seen in figure 16 the advance rate influences the displacements, which again determine the stresses. The higher the displacement rate in the beginning, the higher is the initial utilization of the lining. Due to larger creep magnitudes the stress intensity then decreases more rapidly than with a slower progress.

Using tools to predict displacements (for example GeoFit) allows also a prediction of the stress intensity factor. The influence of different advance rates on the future stresses can be evaluated and thus the progress controlled in a way that the stresses in the lining remain within acceptable limits.

This shall be demonstrated with an example. An excavation proceeds with a rate of 4 m per day. Twelve hours after installation of the support the measured displacements in this section are available. The prediction of displacements based on the first reading is done.

Based on this the development of the stress intensity is predicted.

It shows, that the stresses would exceed the strength of the shotcrete after less than two days (figure 17). To keep the stresses in the shotcrete below its strength at all times, the advance rate is reduced to 2 m per day for about one week. Then the advance rate is increased again to the original 4 m per day. This results in a stress intensity below 0, 8 (figure 18).

This example demonstrates that measurements have to be taken in relatively short intervals, and the data immediately processed and evaluated to be able to react in time.

5 CONCLUSION

For successful tunnelling in difficult geotechnical conditions or in sensitive environments, an observational approach during construction is indispensable. To be able to meet the requirements connected to the observational method, serious preparation prior to construction and an efficient monitoring and organizational program is needed. Besides a state of the art design, professionally and socially competent engineers are required on site.

Monitoring and evaluation techniques have considerably developed during the last two decades. The information, which can be extracted from displacement monitoring data, is enormous. Still a lot of experience is required to correctly evaluate and interpret monitoring data, and draw the right conclusions. A sound rock- and soil mechanical education is an indispensable precondition to understand the complex transient and spatial processes during tunnel excavation.

With the improved tools to evaluate measurement data it is possible to study and compare case histories of various sites. This can be used to further improve monitoring and interpretation techniques. Quite some research is required to establish accepted rules, gradually building a basis for the often empirical assumptions. It is in the owner's interest to further advance monitoring and evaluation techniques.

REFERENCES

Budil, A. 1996. Längsverformung im Tunnelbau. PhD thesis at the Graz University of Technology.
EUROCODE 7.
Golser, H. 2001. The application of Finite Element and Boundary Element Methods in Tunnelling. PhD thesis at the Graz University of Technology.
Grossauer, K. 2001. Tunnelling in Heterogeneous Ground – Numerical Investigation of Stresses and Displacements. Diploma Thesis at the Graz University of Technology.
Grossauer, K., Schubert, W., Kim, C.Y. 2003. Tunnelling in heterogeneous ground – stresses and displacements. Proceedings of the 10th ISRM Congress, South Africa.
Hellmich, Ch., Macht, J., Mang, H. 1999. Ein hybrides Verfahren zur Bestimmung der Auslastung von Spritzbetonschalen. Felsbau 17, No. 5, 422–425.
Macht, J. 2002. hybrid analyses of Shotcrete tunnel linings: Assessment and online monitoring of the level of loading, Doctoral Thesis, Technical University Vienna.
Rabensteiner, K. 1996. Advanced tunnel surveying and monitoring. Felsbau 14, No. 2, 98–102.
Rokahr, R., Zachow, R. 1997. Ein neues Verfahren zur täglichen Kontrolle der Auslastung einer Spritzbetonschale. Felsbau 15, No. 6, 430–434.
Rokahr, R., Stärk, A., Zachow, R. 2002. On the art of interpreting measurement results. Felsbau 20, No. 2.
Schubert, W. 1993. Erfahrungen bei der Durchörterung einer Großstörung im Inntaltunnel. Felsbau 11, Nr. 6.
Schubert, W., Budil, A. 1995. The Importance of Longitudinal Deformation in Tunnel Excavation; Proceedings 8th Int. Congress on Rock Mechanics (ISRM), Vol. 3, 1411–1414, Balkema Rotterdam.
Schubert, W., Steindorfer, A. 1996. Selective displacement monitoring during tunnel excavation. Felsbau 14, No. 2, 93–97.
Schubert, W., Steindorfer, A., Button, E.A. 2002. Displacement monitoring in tunnels – an overview. Felsbau 20, No. 2, 7–15.
Sulem, J., Panet, M., Guenot, A. 1987. Closure analysis in deep tunnels. Int. Journal of Rock Mechanics and Mining Science, 24, 145–154.
Sellner, P. 2000. Prediction of displacements in tunnelling. pp 129. In Riedmüller, Schubert & Semprich (eds), Gruppe Geotechnik Graz, Heft 9.
Sellner, P., Grossauer, K. 2002. Prediction of displacements for tunnels. Felsbau 20, No. 2, 22–28.
Steindorfer, A. 1998. Short term prediction of rock mass behavior in tunnelling by advanced analysis of displacement monitoring data. In Riedmüller, Schubert & Semprich (eds), Gruppe Geotechnik Graz, Heft 1.
Vavrovsky, G.M., Ayaydin, N. 1988. Bedeutung der vortriebsorientierten Auswertung geotechnischer Messungen im oberflächennahen Tunnelbau. Forschung und Praxis, 32, 125–131.
Vavrovsky, G.M. 1988. Die räumliche Setzungskontrolle – ein neuer Weg in der Einschätzung der Standsicherheit oberflächennaher Tunnelvortriebe. Mayreder Zeitschrift 33.
Vavrovsky, G.M., Schubert, P. 1995. Advanced analysis of monitored displacements opens a new field to continuously understand and control the geomechanical behaviour of tunnels. Proc. 8th ISRM Congress Tokio, 1415–1419, Balkema Rotterdam.
Volkmann, G. 2003. Rock Mass – Pipe Roof Support Interaction Measured by Chain Inclinometers at the Birgltunnel. Proceedings GTMM, Karlsruhe.

Die Anwendung der Beobachtungsmethode bei Planung und Ausführung am Beispiel des Tunnels Denkendorf

R. Floss & A. Bauer
Technische Universität München, Zentrum Geotechnik

H. Quick & Th. May
Prof. Dipl.-Ing. H. Quick, Ingenieure und Geologen GmbH

ABSTRACT: Im Los Süd der Neubaustrecke Nürnberg-Ingolstadt wird der Tunnel Denkendorf mit einer Länge von 670 m und einer maximalen Baugrubentiefe von bis zu ca. 20 m in offener Bauweise gebaut. Aufgrund einer Rutschung im südlichen Voreinschnitt wurde eine Neuplanung der Baugrube des Tunnels Denkendorf erforderlich. Diese Neuplanung musste sowohl die sich gegenüber der ursprünglichen Baubeschreibung verändert darstellenden hydrogeologischen Verhältnisse als auch die als Schwächezonen wirkenden, flach einfallenden Trennflächen in den tertiären Tonen, berücksichtigen. Zur wirtschaftlichen Optimierung wurde die Baugrube mit der Beobachtungsmethode geplant und ausgeführt. Es wird über Messergebnisse und Erfahrungen bei der Planung und Durchführungen der Beobachtungsmethode berichtet.

1 EINLEITUNG

1.1 *Allgemeines zur Baumaßnahme*

Im Los Süd der Neubaustrecke Nürnberg-Ingolstadt ist der Tunnel Denkendorf mit einer Länge von 670 m und einer maximalen Baugrubentiefe von bis zu ca. 20 m in offener Bauweise geplant. Nach Süden und Norden schließen an den Tunnel geböschte Einschnitte mit Höhen von bis zu 16 m an.

Der Tunnel liegt teilweise in Festgesteinen des Weißen Juras, überwiegend aber in tertiären Decklagen, die sich aus einer inhomogenen Wechsellagerung von bindigen Sanden, Tonen und tonigen Kalken, durchzogen von dünnen Süßwasserkalkbänken zusammensetzen. Das Grundwasser wurde ausschließlich gebunden an die durchlässigeren Sandlagen und Süßwasserkalkbänke erwartet.

Darauf aufbauend wurde ein mehrfach rückverankerter Bohlträgerverbau mit offener Wasserhaltung geplant, wobei keine Wasserdruckkräfte bei den Nachweisen für die Baugrubenumschließung und die Verankerung angesetzt wurden.

1.2 *Böschungsbruch im be nachbarten südlichen Voreinschnitt*

Im südlichen Voreinschnitt des Tunnels, der ausschließlich in den tertiären Decklagen liegt, ereignete

Abbildung 1.   Böschungsrutschung im Voreinschnitt.

sich im Winter 2000 nach einem Voraushub von ca. 8 m unter Gelände ein etwa 80 m langer Böschungsbruch mit tiefen Abrisskanten. Nach dem Böschungsversagen hatten sich zwei auch gegeneinander abgescherte quasi starre Bruchkörper ergeben, die vom Gelände abgerissen und in den Einschnitt abgerutscht waren. Die jeweiligen Abrisskanten waren dabei nahezu vertikal.

Als Sofortmaßnahme wurde u.a. die Baugrube wieder um ca. 2.0 m–3. 0 m aufgefüllt. Als Schadensursache wurden Schwächezonen im Baugrund infolge von weichen Schichten und wesentlich komplexere hydrogeologische Verhältnisse vermutet.

Die Nähe des Böschungsbruchs führte zu einem Baustopp am geplanten Tunnel Denkendorf und zu umfangreichen Erkundungsmaßnahmen.

2 WEITERE BAUGRUNDERKUNDUNG

2.1 *Konzeption*

Zur weiteren Erkundung wurden von Seiten der DBProjektBau GmbH Prof. Quick und von Seiten der Arbeitgemeinschaft Nürnberg- Ingolstadt, Los Süd, Prof. Floss als Sachverständige beauftragt, im Bereich des Tunnels Denkendorf und der Vorschnitte ein Erkundungsprogramm durchzuführen und gemeinsam zu betreuen. Dabei sollte die Schadensursache geklärt und bautechnische Empfehlungen für die erforderlichen Neuplanungen der Einschnitte und des Tunnels gegeben werden.

Es wurden ca. 50 Bohrungen in der Regel bis in Tiefen von 15 m bis 20 m, vereinzelt zur Erkundung der Felsoberkante und Eigenschaften bis 35 m abgeteuft. Zusätzlich wurden die bereits hergestellten Voraushubbereiche geologisch kartiert und mehrere große Schurfaufnahmen durchgeführt.

Im Rahmen der Untersuchungen wurden fast alle Bohrungen entweder mit Piezometern zur Messung des Wasserdruckes bestückt oder zu Grundwassermessstellen ausgebaut. Die Filterstrecken der Grundwassermessstellen hatten in der Regel eine Länge von 1 m bis 2 m, so dass eine eindeutige Zuordnung der Messergebnisse möglich war. Durch die Bestückung einer Bohrung mit mehreren Piezometern sollte die Ableitung einer Wasserdruckverteilung über die Tiefe

ermöglicht werden. Die ursprünglich im Rahmen der Vorerkundung hergestellten Grundwassermessstellen hatten lange Filterstrecken, die unterschiedliche Bodenschichten miteinander hydraulisch verbanden. Eine eindeutige Zuordnung der gemessenen Grundwasserstände zu einzelnen wasserführenden Bodenschichten und deren Tiefenlagen und damit die Ableitung einer eindeutigen Wasserdruckverteilung im Tertiär waren somit nicht möglich.

Zusätzlich zu umfangreichen Pumpversuchenwurde im Bereich des Tunnels Denkendorf ein sog. Testfeld, bei dem 4 Dreifachpiezometermessstellen, eine Grundwassermessstelle und ein Brunnen dicht beieinander liegend hergestellt wurden, mit dem Ziel der Bestimmung der Absenkreichweite und der Ermittlung der Absenkkurve eingerichtet.

2.2 *Ergebnisse der Untersuchungen*

Die Untersuchungen ergaben eine gegenüber den ursprünglichen Beschreibungen komplexere Grundwassersituation. Daneben wurden harnischartige, flach einfallende große Trennflächen in den tertiären Tonen angetroffen. Die Kombination aus der Grundwassersituation sowie flach einfallenden Trennflächen in den Tonen machte eine Neuplanung des Tunnels und der Voreinschnitte erforderlich.

2.3 *Geologie*

Nach den Erkundungsergebnissen lässt sich der Baugrund in die geringmächtigen quartären Decklagen, die bis zu 30 m mächtigen tertiären Schichten und den unter diesen Lockergesteinen anstehenden weißen Jura (Kimmeridge) unterteilen. Die Untergrundverhältnisse sind insgesamt als sehr inhomogen zu bezeichnen.

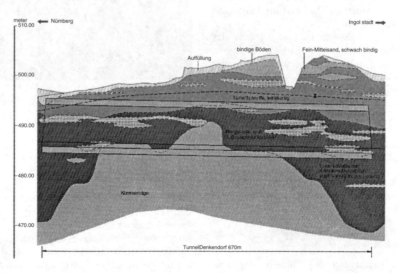

Abbildung 2.   Geologischer Längsschnitt.

Bei den quartären Decklagen (q) handelt es sich im wesentlichen um z.T. schwach sandige/kiesige Tone/ Schluffe.

Die tertiären Böden (t) wurden unterschieden in Tone/ Schluffe (t_t), nicht bis schwach bindige Sande (t_s), schwach bindige bis bindige Sande mit tonig/ schluffigen Einschaltungen($t_{s/t}$), Süßwasserkalke (t_{sk}) sowie kalkig verfestigte Tone/ Schluffe (t_k). Innerhalb der t_t- und t_k-Böden wurden Mergel- Mergelkalk- und Süßwasserkalkbänke (t_{sk}) beschrieben, die eine Mächtigkeit von bis zu 2 m und Ausdehnungen bis zu mehreren hundert Metern aufweisen können.

Unterhalb der tertiären Decklagen befindet sich das als massig bezeichnete Kimmeridge. Klüfte und Trennflächen wurden nur bei den Gesteinen des Kimmeridges und bei den tertiären Süßwasserkalken beschrieben.

2.4 Hydrogeologie

Innerhalb des vorliegenden Trassenabschnitts wurde das Grundwasser im Fels nicht aufgeschlossen. Er stellt in Folge seiner durch die Verkarstung bedingten teilweise sehr hohlraumreichen Struktur im Wesentlichen einen Karstgrundwasserleiter dar, dessen Grundwasseroberfläche in etwa bei ca. 60 bis 75 m unter GOK liegt.

Ursprünglich wurde in den den Fels lokal bis über 30 m überdeckenden, sehr heterogenen tertiären Ablagerungen kein durchgehender Grundwasserspiegel, sondern nur einzelne, als schwebende Grundwasservorkommen bezeichnete Schichtwässer erwartet.

Diese schwebenden Grundwasservorkommen sollten im Wesentlichen an die Verbreitung der Süßwasserkalkbänke (t_{sk}) und zum anderen an die sandig/kiesigen Einschaltungen (t_s) in den tonig/ schluffigen tertiären Ablagerungen (t_t) gebunden sein, wobei auch bereichsweise ergiebige, gespannte Grundwasservorkommen in den Süßwasserkalkbänken beschrieben waren.

Die in den verschiedenen Tiefenlagen und Bodenschichten verfilterten neuen Grundwassermessstellen und Mehrfachpiezometermessstellen haben gezeigt, dass entgegen der ursprünglichen Beschreibung in allen Lockergesteinsschichten Grundwasser vorhanden ist und Wasserdrücke vorliegen. Die Wasserdrücke sind in ihrer Größe und tiefenabhängigen Verteilung sehr unterschiedlich, wobei hydrostatische Drücke ebenso vorliegen wie deutliche Abweichungen davon. Genaue Angaben sind infolge der sehr inhomogenen Baugrundverhältnisse und der komplexen Strömungsverhältnisse des Grundwassers nicht möglich, da das Grundwasser sowohl in den Poren der Böden, in den Klüften der Süßwasserkalkbänke sowie in den festgestellten Trennflächen fließen kann, so dass man letztendlich für die Bemessung zunächst von einer hydrostatischen Wasserdruckverteilung ausgehen muss.

Bei den Pumpversuchen konnten keine bzw. nur sehr geringe Absenktrichter und damit auch keine eindeutigen Absenkreichweiten ermittelt werden, da auch benachbarte Grundwassermessstellen, die in der gleichen Bodenschicht und Tiefe verfiltert waren, keine bzw. nur sehr geringe Absenkung aufwiesen. Auch im Testfeld mit Abständen zwischen dem Brunnen und den Messeinrichtungen von 3 m bzw. 4 m reagierten die Messeinrichtung nicht auf die mehrwöchige, dauerhafte Absenkung im Brunnen und auch nicht auf den Versuchsbetrieb eines Brunnens mit Vakuum.

2.5 Trennflächen

Das Antreffen von Trennflächen in den tertiären Tonen im südlichen Voreinschnitt führte diesbezüglich zu einem weiteren Untersuchungsschwerpunkt im Rahmen des V-EKP.

In einem Schurf wurden Trennflächen in den t_t-Böden mit unterschiedlichen Einfallswinkeln zwischen 18° und 50° und unterschiedlichen Streichrichtungen erkundet. Ein regelmäßiges Trennflächensystem konnte hier nicht abgeleitet werden. Auch in dem ca. 4 m tiefen Voraushub des südlichen Voreinschnittes wurden in den t_t- Böden steil einfallende (70° bis 90°), teilweise auch regelmäßig wiederkehrende Trennflächen erkannt. In den tiefer gelegenen t_k-Böden wurden zwischen 20° und 55° geneigte Trennflächen festgestellt. Die Oberflächen der Trennflächen waren überwiegend rau, teilweise vorhandene Belege lassen auf Wasserführungen schließen. Teilweise waren die Oberflächen glatt und feucht und deuten auf eine Harnischbildung mit verringerter Scherfestigkeit hin. Die Trennflächen hatten Längenabmessungen im Dezimeterbereich und werden einer Mikrotrennflächenstruktur in den tertiären Tonen zugeordnet.

Neben dieser Mikrotrennflächenstruktur wurden große, weitgehend parallele Trennflächen festgestellt, die die Böschung sehr flach (ca. 3°) unterschneiden. Diese Trennflächen waren glatt, teilweise nass und hatten einen breiig – weichen Belag sowie teilweise Harnischstriemen. Sie sind augenscheinlich weit in die Böschung hinein durchhaltend und können als Makrotrennflächenstruktur bezeichnet werden.

Angaben über die Trennflächensysteme lagen im Rahmen der ursprünglichen Erkundung nicht vor. Bei den weiteren Untersuchen wurde festgestellt, dass diese Trennflächen erst im Zuge des Aushubs und nicht vorab z.B. durch Bohrungen zuverlässig erkannt werden können.

Diese Trennflächen stellen somit eine erhebliche Schwächezone im Baugrund dar, da sie ein Abgleiten eines Bodenblocks auf der Trennfläche ermöglicht. Aufgrund der geringen Scherfestigkeit auf der Trennfläche kann ein Abgleiten und damit ein Böschungsversagen schon bei sehr geringen Aushubtiefen auftreten. Beispielsweise sind solche

47

Rutschvorgänge in anderen Einschnitten des Loses Süd bereits bei 2 m bis 4 m Aushub aufgetreten.

3 NEUPLANUNG DES TUNNELS DENKENDORF MIT DER BEOBACHTUNGSMETHODE

3.1 Planungsgrundlagen und Beobachtungskonzept

Aufgrund der sich verändert darstellenden hydrogeologischen Verhältnisse und der potentiell vorhandenen Trennflächen in unterschiedlichen Tiefen der Baugrube war das ursprüngliche Baugrubenkonzept, bestehend aus einem verankerten Bohlträgerverbau nicht mehr geeignet. Die neue Planung musste sowohl an die Beanspruchungen infolge potentieller Trennflächen und an die Grundwassersituation angepasst werden. Parameterstudien unter Ansatz hydrostatischen Wasserdruckes in Kombination mit Blockgleiten auf Trennflächen zeigten, dass eine Baugrube nicht wirtschaftlich herstellbar ist. Die Erkenntnisse aus dem Testfeld und durchgeführte FEM- Berechnungen ergaben, dass eine Grundwasserabsenkung weder vom technischen Aufwand noch vom zeitlichen Rahmen sinnvoll realisierbar ist. Deswegen wurde folgendes Baugrubenkonzept vorgeschlagen:

- Die Baugrubenumschließung sollte wasserdurchlässig ausgeführt werden, so dass sich im Nahbereich der Umschließung eine Absenkung einstellt, die den Ansatz eines Restwasserdruckes als Bemessungsgrundlage rechtfertigt. Die Größenordnung des Restwasserdruckes wurde numerisch überprüft. Bei der Ermittlung der Verankerungslänge mit dem Nachweis in der tiefen Gleitfuge sollten hydrostatische Wasserdrücke angesetzt werden.
- Der Ansatz einer hydrostatischen Wasserdruckverteilung würde infolge der sehr wechselhaften Schichtenfolge und Durchlässigkeit eine Entspannung des Baugrunds unterhalb der Baugrubensohle erfordern. Da im Fels das Grundwasser weit unterhalb der Baugrube ansteht, sind diese Entspannungsbrunnen nur in den Bereichen mit tertiären Lockergesteinen erforderlich. Die Anzahl der Entspannungsbrunnen sollte sowohl durch die Erkenntnisse während der Umschließungsarbeiten und durch Messungen mit zusätzlichen Grundwassermessstellen überprüft und optimiert werden.
- Für den Nachweis des Blockgleitens auf potentiellen Trennflächen wurden reduzierte Sicherheiten angesetzt. Der Bemessung wurde die Tiefenlage und Neigung der Trennflächen zugrunde gelegt, die für den Verbau die größte Beanspruchung liefert.

Diese Vorgehensweise setzt eine möglichst genaue Beobachtung der Grundwasserverhältnisse sowie der Verformungen der Baugrubenumschließung, der Baugrubensohle und der Nachbarbebauung voraus. Für die Beurteilung waren Signalwerte festzulegen, die die Erfordernis zusätzlicher Maßnahmen aufgrund von Beobachtungen bestimmen.

3.2 Konzeption der Baugrube

Aufgrund der Erkenntnisse aus dem verdichtenden Erkundungsprogramm wurde eine aufgelöste Bohrpfahlwand mit Spritzbetonausfachung ausgeführt, die durch den Einsatz von Dränmatten wasserdurchlässig ausgebildet wurde, so dass der Bemessung zur wirtschaftlichen Optimierung ein Restwasserdruck zugrunde gelegt werden konnte. Die Umschließung wurde mehrfach verankert. Zur Vermeidung von teilweise großen Ankerlängen wurden in einigen Abschnitten der Baugrube Ortbetonsteifen bzw. Sohlaussteifungen vorgesehen. In der Abbildung 3 ist ein Zwischenaushubzustand dargestellt. Inzwischen ist bis auf einen noch erforderlichen Restaushub am nördlichen Portal die Baugrubensohle erreicht und mit der Herstellung der Tunnelblöcke begonnen worden.

3.3 Messtechnik und Beobachtungsprogramm

Es wurde ein umfangreiches Messprogramm installiert, welches die Überwachung der Verformungen der Baugrubenwände, der Hebungen der Baugrubensohle, der Ankerkräfte und der Verformungen der angrenzenden Bebauung beinhaltet.

Der Überwachung und Messung der Verformungen des Baugrundes, der Erd- und Wasserdrücke sowie der Ankerkräfte kommt insbesondere bei der Anwendung der Beobachtungsmethode große Bedeutung zu. Deswegen ist die Planung von Messprogrammen mit detaillierten Angaben von Zuständigkeiten und Verantwortlichkeiten notwendig, die jeweils an

Abbildung 3.   Foto eines Zwischenaushubzustandes.

die Besonderheiten der verschiedenen Bauwerke angepasst sind.

Die Anordnung der Messeinrichtungen sollte in Messquerschnitten erfolgen, die vorgesehene Aushubtiefe, nahegelegene Bauwerke oder Verkehrseinrichtungen sowie die Eigenschaften des Baugrundes berücksichtigen. Dies bedeutet hier vor allem im Bereich der nahe gelegenen Autobahnauffahrt zur BAB A9 bzw. der angrenzenden Industriebebauung eine Konzentration der Messeinrichtungen.

Bei der Planung der Grundwassermessstellen zur Überprüfung des Erfolges der Grundwasserentspannungsmaßnahmen und zur Kontrolle der Wasserdruckverteilungen sollten die Erkenntnisse aus den durchgeführten Untersuchungen berücksichtigt werden und zusätzliche Messstellen wieder zugeordnet zu den stratigraphischen Einheiten und Durchlässigkeitseigenschaften der Bodenschichten eingerichtet werden.

Unter Anwendung der Beobachtungsmethode als Planungs- und Bemessungsgrundlage der Bauwerke war es hier wegen der sehr wechselhaften geologischen Verhältnisse notwendig, Messquerschnitte durchschnittlich alle 50 m bis 100 m anzuordnen. Aushub, Bohrpfahl- und Ankerarbeiten waren geologisch zu dokumentieren.

Insgesamt wurden 17 Vertikalinklinometer zur Messung der horizontalen Verschiebungen im Baugrund direkt hinter der Baugrubenumschließung bzw. der sog. Ersatzankerwand, 4 Gleitdeformeter zur Messung der Hebungen der Baugrubensohle und 25 Ankerkraftmessdosen zur Überprüfung der Veränderung der Ankerkräfte eingebaut. Diese Messeinrichtungen wurden noch durch geodätische Messquerschnitte ergänzt. Zur Beurteilung der Grundwasserverhältnisse stehen insgesamt 29 Messstellen zur Verfügung.

4 BEOBACHTUNGEN UND BAUTECHNISCHE FOLGEN

4.1 *Auftriebsicherheit und Sohlentspannung*

Die speziell unter der Baugrubensohle verfilterten Messstellen ergaben, dass Entspannungsbrunnen zur Einhaltung der Auftriebssicherheit trotz der Nähe zum Fels erforderlich waren. Durch die genaue Dokumentation im Zuge der Bohrpfahlarbeiten konnte der Umfang der erforderlichen Entspannungsmaßnahmen auf zwei vergleichsweise kurze Abschnitte am nördlichen (Blöcke 1 bis 10) und südlichen (Blöcke 41 bis 52) Portal des Tunnels und auf einen kleinen Bereich bei Block 18 beschränkt werden. Die Messergebnisse konnten die Wirksamkeit der Entspannung bei einem mittleren Brunnenabstand von 20 m nachweisen.

4.2 *Ankerkräfte*

Zur Überprüfung der Ankerkräfte wurden die insgesamt 25 Kraftmessdosen in 6 Messquerschnitten angeordnet. Die Verpresskörper der Anker liegen sowohl in den tertiären Tonen t_t und t_k sowie im Fels. In der Abbildung ist exemplarisch der Verlauf eines der 6 Messquerschnitte dargestellt. Es handelt sich hier um einen Bereich ohne Steifenlage.

Die Messergebnisse zeigen, dass die Ankerkräfte bei Verpresskörpern sowohl in den tertiären Decklagen als auch im Fels weitgehend konstant blieben. Trotz des maximalen Aushubs von bis zu 20 m war kein signifikanter Anstieg der Ankerkräfte festzustellen. Andererseits war auch während der bisherigen Belastungsdauer von mehr als einem Jahr kein Kriechen der Anker in den Tonen der Decklagen festzustellen.

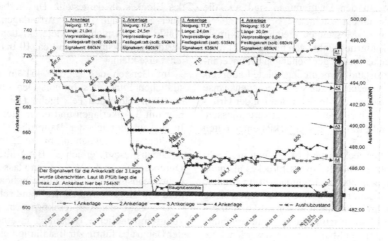

Abbildung 4.   Zeitlicher Verlauf der Ankerkräfte.

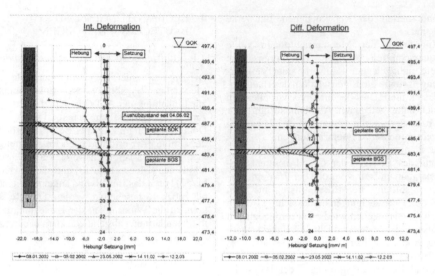

Abbildung 5.   Gleitdeformeter zur Messung der Hebungen der Baugrubensohle.

4.3 Verformungen der Baugrubenumschließung und der Baugrubensohle

Die 4 Gleitdeformeter wurden in den nördlichen und südlichen Abschnitten installiert, in denen der Tunnel im Lockergestein gegründet wird. Sie dienen u.a. zur Überprüfung des Quell- und Schwellverhaltens der Tone, das im Zusammenhang mit den Verformungsanforderungen der Festen Fahrbahn besondere Bedeutung auch bei Tunnelgründungen haben kann. In der Abbildung sind exemplarisch die Messergebnisse eines Gleitdeformeters dargestellt.

Die Abbildungen 5 bis 9 zeigen jeweils links die integrierte Darstellung der Hebungen bzw. Biegelinien und rechts die differenzielle Darstellung der Messergebnisse. In den Diagrammen sind u.a. die Schichtenprofile mit dargestellt.

Die bisherigen Messungen zeigten maximale Hebungen der Baugrubensohle von bis zu ca. 21 mm, wobei der zeitliche Verlauf der Hebungen bei unverändertem Aushubzustand zuvor durchgeführte Laborversuchsergebnisse bestätigte, die eine vergleichsweise geringe Quellneigung der Tone im Bereich der Gründungssohle des Tunnels Denkendorf erwarten lassen, sofern eine Mindestpressung von 60 kPa eingehalten wird.

Sowohl die messtechnische Überprüfung des rechnerischen Ansatzes eines Restwasserdruckes als auch die visuelle Erkennung von Trennflächen im Aushub einer verbauten Baugrube erschien nicht zuverlässig durchführbar. Durch die Messung der Reaktionen des Verbaus sollte dessen tatsächliche Beanspruchung abgeschätzt und mit der Statik verglichen werden. Dies erfordert einen besonders hohen Aufwand bei den Verformungsmessungen der Baugrubenumschließung, besonders in Hinblick auf die sich in Folge der unterschiedlichen geologischen Verhältnisse ergebenden statischen Randbedingungen. Durch die Anordnung von 17 Vertikalinklinometern in Verbindung mit den geodätischen Messquerschnitten konnte dies erreicht werden.

Die Abbildung 6 zeigt die Verformungsmessungen im Bereich des Blockes 18, bei dem die Bohrpfähle nicht im Fels einbinden. Die Verbauwand ist hier mittels 4 Ankerlagen gestützt. Die Messungen lassen auf eine Drehung der Wand um den Fußpunkt mit maximalen Verschiebungen am Wandkopf von ca. 20 mm schließen.

In der Abbildung 7 sind die Messungen im Bereich des Blockes 38 dargestellt. Die Verbauwand ist mit einer Steifenlage und 4 Ankerlagen gestützt. Die Pfähle binden mehrere Meter in den Fels ein. Die Verformungen sind hier mit ca. 10 bis 12 mm vergleichsweise klein.

Im Bereich des Blockes 51 in der Nähe des südlichen Endes des Tunnels Denkendorf erreichen die Bohrpfähle den Fels nicht. Die Verbauwand ist hier mit 3 Ankerlagen und einer Steifelage gestützt. In diesem Bereich der Baugrube stellten sich die größten Verformungen mit bis zu 25 mm ein.

Die in diesem Bereich der Baugrube festgestelltem Durchbiegungen der Verbauwand lassen auf eine sehr hohe Beanspruchung und einen hohen Ausnutzungsgrad der Biegesteifigkeit der Wand schließen.

Die Lage der Verpresskörper in der Nähe setzungsempfindlicher Bebauung in einem Teilbereich der Baugrube führte zur Installation eines Vertikalinklinometers im Bereich der sog. Ersatzankerwand

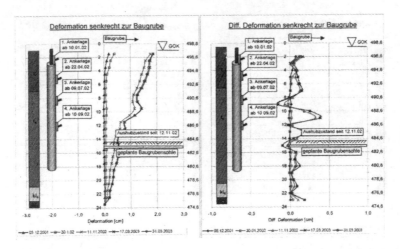

Abbildung 6.   Vertikalinklinometermessungen im Block 18.

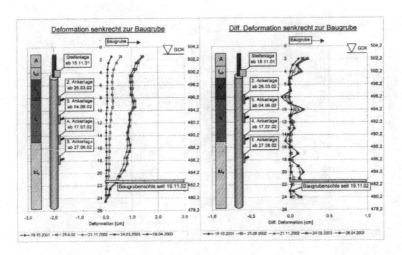

Abbildung 7.   Vertikalinklinometermessungen im Block 38.

knapp hinter den Verpresskörpern.. Aus den Messungen ist deutlich der Einfluss der Ankerkräfte auf die Verschiebungen des Baugrundes ableitbar. Die Messungen zeigen, dass sich im Bereich der Gebäudegründungen nur geringe Verschiebungen von weniger als 3 mm eingestellt haben.

Im Zuge der geologischen Dokumentation und Aushubkartierung wurde eine flach einfallende Trennfläche in einer Aushubtiefe von ca. 7 m und einer Länge von ca. 250 m durchgehend festgestellt.

Zusammenfassend haben die bisher vorliegenden Dokumentationen und Messergebnisse die Bemessungsannahmen und Prognosen bestätigt. In einigen Bereichen der Baugrube wurden die Signalwerte der Verformungen beinahe erreicht, in keinem Bereich wurden sie überschritten.

5   ZUSAMMENFASSUNG UND AUSBLICK

Eine Rutschung im südlichen Voreinschnitt führte zur Neuplanung der Baugrube des Tunnels Denkendorf. Diese Neuplanung musste sowohl die sich gegenüber der ursprünglichen Baubeschreibung verändert darstellenden hydrogeologischen Verhältnisse als auch die als Schwächezonen wirkenden, flach einfallenden Trennflächen in den tertiären Tonen, berücksichtigen.

Zur wirtschaftlichen Optimierung wurde die Baugrube mit der Beobachtungsmethode geplant und ausgeführt. Die für den Verbau vergleichsweise großen Verformungen zeigen die hohe Auslastung des Systems und damit die durch die Anwendung der Beobachtungsmethode möglichen wirtschaftlichen

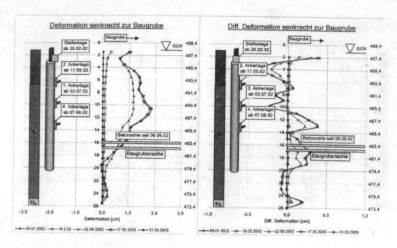

Abbildung 8. Vertikalinklinometermessungen im Block 51.

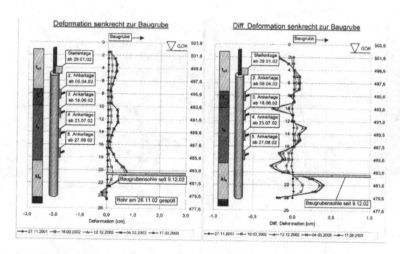

Abbildung 9. Vertikalinklinometermessungen im Block 32 im Bereich der Ersatzankerwand.

Vorteile. Diese Baugrube stellt für das Los Süd quasi ein Pilotprojekt dar. Die Erfahrungen werden inzwischen mit Erfolg bei den anderen Bauwerken und Einschnitten im Los Süd genutzt.

LITERATUR

Floss, R. & Quick, H. 2000. Gutachtliche Stellungnahme zu den Ergebnissen aus dem V-EKP einschließlich bautechnischer Empfehlungen zur Planung. Unveröffentlicht.

Floss, R. & Quick, H. 2001. Lockergesteinseinschnit te E1, E2, E7 bis E12 und Tunnel Denkendorf: Konzept Beobachtungsmethode. Unveröffentlicht.

Floss, R. & Quick, H. 2001. Verbau Tunnel Denkendorf – Erläuterungsbericht zum geotechnischen/geodätischen Messprogramm. Unveröffentlicht.

Quick, H., Belter, B. & Wegerer, P. 2002. Hochgeschwindigkeitsstrecken der Deutschen Bahn AG – Anforderungen und Grenzen der Geotechnischen.

Erkundung. In: Baugrundtagung 25–28. September 2002, Mainz.

Belastungsmodelle von Tunnelausbauten auf der Basis von Spannungsmessungen

B. Fröhlich
gbm Gesellschaft für Baugeologie und –meßtechnik mbH Baugrundinstitut, Limburg, Deutschland

ABSTRACT: Pressure cells (hard inclusion) are often used to determine the stresses in a concrete lining of tunnels. Some problems of these methods are not solved – zero point, temperature effects, contact –. Especially stresses in an existing lining are not possible to measure by this method.

The necessary of the knowledge of the loading systems in old tunnels which are to repair was reason for testing and comparison of different stress measurement systems. The compensation method – using flat jacks in a sawed slot – is the most effective method (quality and cost). It is possible to use these flat jacks also as a permanent monitoring system.

An example shows the comparison of the measured stresses with the predicted stresses based on deformation measurements and the time-depending stiffness of the shotcrete.

Examples show the loading systems as an excellent method to explain the geotechnical model based on site investigations. For repairing it is necessary to know the load to calculate construction stages and to determine the safety level.

In more than twenty analysed tunnels the results were always extremely good for planning the tunnel work and calculating the safety level.

1 EINFÜHRUNG

Die Standsicherheit eines Tunnelausbaus wird durch seine innere Tragfähigkeit und die äußere Belastung bestimmt. Beim Tunnelneubau sind beide Faktoren nicht einfach zu ermitteln. Die Berechnungen sind z. T. nur als Studie im Rahmen einer Prognose zu verstehen – auch wenn ihnen eine höhere Aussagekraft zu diktiert wird -, die durch Beobachtungen beim Bau zu überprüfen sind. Bei Tunnelneubauten sind jedoch die Materialeigenschaften des Ausbaus steuerbar. Noch schwieriger sind Belastung und Tragfähigkeit alter Tunnelausbauten zu bestimmen. Bei alten gemauerten Gewölben sind die Baugeschichte und die Veränderungen der Konstruktion und der Materialeigenschaften über die Lebensdauer zusätzlich abzuschätzen; Messinstrumente wurden beim Bau nicht installiert. Die Standsicherheit alter Tunnel interessiert nicht nur allgemein; deren Kenntnis ist insbesondere auch nötig, wenn in das System Tunnel (Ausbau und Gebirge) wegen Umbauarbeiten und Instandhaltung eingegriffen wird.

In den nachfolgend genannten Tunnelbeispielen wurde durch entsprechende Sicherungsarbeiten das erforderliche Sicherheitsniveau hergestellt.

2 ERMITTLUNG DER BELASTUNG VON TUNNELAUSBAUTEN

Die Ermittlung der Spannungen in einem Tunnelausbau setzt für die verschiedenen Spannungsmessverfahren unterschiedliche Anforderungen an das Materialverhalten. Da der Ausbau nicht stark kriechend sein darf, sind flache Druckmessdosen als hard inclusion nur in den Fällen geeignet, in denen sie in den Ausbau beim Bau einbetoniert werden. Die Abkühlung des Betons beim Erhärten führt auch zu einer Abkühlung der Druckmessdose; deshalb sind sie nach dieser Phase nachzuspannen. Dadurch ist der Ausgangspunkt der Messkurve manuell beeinflussbar. Druckänderungen werden danach i. a. zuverlässig gemessen. Der Einbau erfordert eine hohe Präzision, damit die Druckmesszellen, die den Gebirgsdruck messen sollen, auch kraftschlüssig zum Gebirge eingebaut sind.

In bestehenden Bauwerken ist die nachträgliche Messung nur mit Spannungsmessverfahren möglich, die entweder elastisches Materialverhalten (z. B.: doorstopper Methode, Triaxialzelle) und die Kenntnis des E-Moduls voraussetzen oder die Methode des Kompensationsverfahrens, das die Kenntnis des E-Moduls und linear elastisches Verhalten nicht

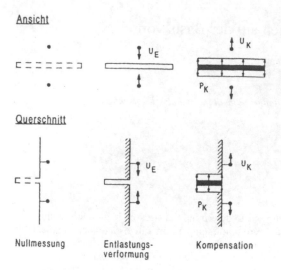

Ansicht

Querschnitt

Nullmessung   Entlastungs-   Kompensation
               verformung

U_E Entlastungsverformung

U_K Rückverformung bei Kompensation

P_K Kompensationsdruck

Abbildung 1.   Prinzip der Spannungskompensation.

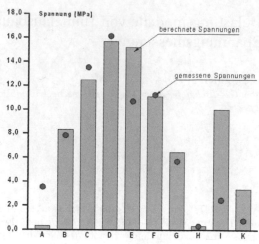

Abbildung 2.   Vergleich gemessener Spannungen im Spritzbeton mit geodätisch abgeleiteten Spannungen.

vorausetzt. Es ist lediglich die Annahme zu erfüllen, dass nach einer Entlastung die Verformungen infolge Wiederbelastung bei gleicher Spannung reversibel sind. Die Reversibilität ist bei Beton und auch bei Mauerwerk in den vorliegenden Belastungsbereichen hinreichend genau gegeben.

Bei der Durchführung der Spannungsmessungen nach der Kompensationsmethode werden zunächst an der Tunnelwandung Messpunkte gesetzt, deren Abstand mit 1 μm Ablesegenauigkeit gemessen wird. Nach der Nullmessung wird ein kreissegmentförmiger Schlitz in den Ausbau gesägt und der Abstand der Messpunkte infolge der Schlitzentlastung gemessen. Anschließend wird mit einem passgenauen Druck-kissen der Schlitz wieder stufenweise so lange belastet bis die Entlastungsverformungen infolge Sägeschlitz kompensiert sind (Abb. 1). Der Kompensationsdruck in dem Zustand des ursprünglichen Messpunktab-standes vor dem Herstellen des Sägeschlitzes entspricht dem Spannungszustand im Ausbau (Schuck & Wullschläger, 1994; gbm, 1991). Dabei werden nur die Spannungen normal zum Schlitz kompensiert; ist die Hauptspannungsrichtung – allgemein in Umfangsrich-tung und axial zum Tunnel – nicht abschätzbar, sind verschieden orientierte Schlitze erforderlich. Über den Umfang verteilt werden an ca. 7 Messstellen des Innen-randes im Ausbau die Spannungen gemessen. Mit einem Stabwerkmodell wird der in Geometrie und Steifigkeit simulierte Ausbau belastet; die Belastung

wird so oft variiert, bis in den gemessenen Punkten die Übereinstimmung mit den berechneten Spannungen erzielt wird. Somit sind die äußeren Kräfte auf den Ausbau bekannt. Sie werden für weiterführende Unter-suchungen zu Grunde gelegt. Das Messverfahren wurde in über zwanzig Tunnel mit Mauerwerks-gewölbe und in Betonausbau mit besten Erfahrungen eingesetzt. (Fröhlich, B., 1996; Fröhlich, B., Schlebusch, M., Berwanger, W. & Pape, H., 1999; Gilsdorf, A., Fröhlich, B., Schröder, K.-H. & Wullschläger, D.,1999; Rustemeier, A., Weishäupl, S., Fröhlich, B. & Schlebusch, M., 1999). Dieses Prinzip der Spannungsmessung in gemauerten Tunnelgewöl-ben wurde in die Richtlinie 853 – Modul 853.0005 Geotechnische Messungen – der DB übernommen (vgl. Schuck, W. & Fecker, E., 1998).

Die Ermittlung der Belastung kann beim Tunnel-neubau mit Spritzbeton auch erfolgen über die Mes-sung der Verschiebungen des Ausbaus an hinreichend vielen Punkten des Umfangs mit ausreichend häufi-gen Wiederholungsmessungen. Aus den Verschiebun-gen der Punkte werden die Dehnungen und Span-nungen des Beton errechnet. Da sich der E-Modul des Beton in dem Frühstadium signifikant ändert, ist für die Spannungsänderung der jeweils aktuelle E-Modul zu verwenden (Rokahr, R. & Zachow, R., 1999). Die – auf der Basis der geodätisch ermittelten Verformungen – Tangentialspannungen im Spritzbeton wurden mit den gemessenen Spannungen verglichen (Abb. 2). Die Spannungen wurden nach der Kompensations-methode bestimmt. Die Punkte geben die gemesse-nen Spannungen an. Aus dem Vergleich ergibt sich eine sehr gute Übereinstimmung, insbesondere auch unter dem Gesichtspunkt, dass bei der rechnerischen Ermittlungen der Spannungen auf der Basis der

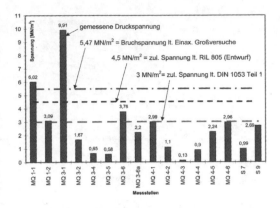

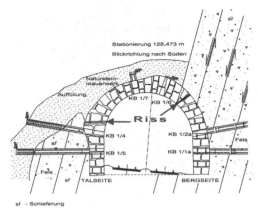

Abbildung 3. Vergleich von gemessenen Spannungen und zulässigen Spannungen (Beispiel unveröffentlicht).

Abbildung 4. Querschnitt durch den Kammereck Tunnel (offene Bauweise).

Verschiebungsmessungen nicht eingehen kann, ob der Spritzbeton auf der Baustelle immer die exakt gleiche Dosis an Erstarrungsbeschleuniger hat und somit die rechnerisch angenommene E-Modul Entwicklung im Einzelfall genau zutrifft.

3   TRAGFÄHIGKEIT VON GEMAUERTEN TUNNELGEWÖLBEN

Um die Standsicherheit eines Tunnelausbaus beurteilen zu können wird zusätzlich zur Belastung dessen Tragfähigkeit benötigt. Die Angabe der Tragfähigkeit ist bei alten gemauerten Tunnelgewölben schwierig. Die Angaben bzgl. der zulässigen Spannungen in DIN 1053 sind so niedrig, dass diese in sehr vielen Tunneln weit überschritten werden. Wegen der großen Ausbaudicke und der Bewertung der Randspannungen dürfen Ergänzungsfaktoren eingeführt werden. Die Richtlinie 805 (Entwurf) der DB AG und der UIC-Kodex 778-3 erlauben eine Ermittlung der zulässigen Spannungen unter Einbeziehung der Festigkeit von Stein und Mörtel, dem Fugenanteil und der Mauersteinart. Daraus ergeben sich höhere Werte als in DIN 1053. In vielen Fällen werden auch diese Werte deutlich überschritten (Abb. 3).

In den meisten Fällen ist es sinnvoll, die Mauerwerksfestigkeit an ausreichend großen Proben zu bestimmen. Dabei ist zu beachten, dass im Versuch die Proben orientiert in Umfangsrichtung des Tunnels belastet werden. Dies bedingt die Entnahme großer Mauerwerksproben, aus denen im Versuchlabor entsprechende Prüfkörper präpariert werden. In dem Beispiel in Abb. 3 ist erkennbar, dass die an Prüfkörpern im Labor bestimmten Bruchfestigkeiten der Einzelproben teils überschritten werden. Die setzt voraus, dass im Bereich der Messstellen mit den

hohen Spannungen das Mauerwerk fester war. Dies wird bei den Spannungsmessungen dadurch bestätigt, indem die Mauerwerksfestigkeit versuchstechnisch ermittelt wird, in dem man nach dem Spannungsmesszyklus die Belastung mit dem Druckkissen so weit steigert bis das Mauerwerk versagt. Das Verhältnis der Bruchspannungen – mit den Druckkissen ermittelt – zu den Kompensationsspannungen variierte zwischen 1,2 und 6,6 und betrug im Mittel für einen Tunnel ca. 3,8. Maßgebend für die Sicherheit sind jedoch die niedrigsten Werten.

4   KAMMERECK TUNNEL

Der Kammereck Tunnel auf der linken Rheinseite zwischen Köln und Bingen wurde 1858/59 gebaut. Er durchörtert 230 m bergmännisch Grauwacken, Grauwackenschiefer und glimmerreiche Tonschiefer (Unterdevon) des Rheinischen Schiefergebirges. Die maßgeblichen Trennflächen werden durch die Schieferung geprägt. Im südlichen Bereich schließt sich eine 59 m lange Strecke an, die in offener Bauweise errichtet wurde. Der Ausbau besteht aus Natursteinmauerwerk. Durch den Aushub wurden spitzwinklig zum Tunnel streichende, 70° steil zum Tunnel einfallende Gesteinspakete unterschnitten (Abb. 4).

Talseitig steht bereichsweise noch ein Felswiderlager an (Abb. 4). Durch die hohe Belastung und das Widerlager ist das Tunnelgewölbe gerissen; die Rissufer hatten einen Versatz. Abb. 5 zeigt die gemessenen Spannungen des Innenrandes im Tunnelgewölbe. Die lokal hohen Spannungen (bis ca. 15 MPa!) lassen sich durch ein Gewölbe mit 10 m Spannweite (Abb. 6) erklären, des oberhalb des Tunnels ein unterschnittenes Gesteinspaket unterfangen hat. Lokal werden Kräfte entsprechend einer Auflast von 33 m Höhe eingeleitet.

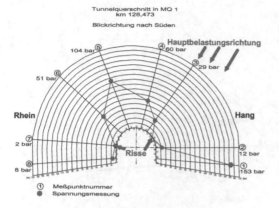

Abbildung 5. Innenrandspannungen im Tunnelgewölbe (offene Bauweise).

Abbildung 6. Ankersicherung über dem Tunnel.

Die Sicherung des Gewölbes und des Gebirges erfolgte innerhalb und oberhalb des Tunnels, um den Eisenbahnbetrieb möglichst wenig zu behindern. Die Belastungen infolge der oberhalb anstehenden, unterschnittenen Schichten wurden über lange, vorgespannte Daueranker über dem Tunnel aufgenommen (Abb. 6 und 7).

Nach der Sicherung über dem Tunnel wurde der Ausbau in Rippenbauweise instandgesetzt. Zunächst wurde das bergseitige Gewölbe mit Injektionsbohrankern (4 m bis 8 m) gesichert. Die Injektionsbohranker erlauben durch die Koppelung von kurzen

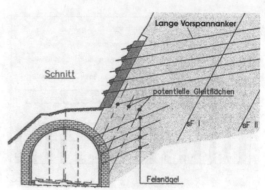

Abbildung 7. Querschnitt mit Sicherungsmittel.

Ankerstücken die Arbeiten in einem gesperrten Gleisbereich und den Fahrbetrieb auf dem Nachbargleis. Die ausgebrochenen Rippen wurden mit einem Gitterträger bewehrt und ebenfalls rückverankert. Die Felder zwischen den Rippen wurden bis zu 25 cm tief ausgebrochen und mit armiertem Spritzbeton erneuert; in den Bereichen mit ausreichendem Profil wurden mindestens 5 cm des Mauerwerks entfernt, um einen guten Verbund zwischen Spritzbeton und Mauerwerk zu erreichen. Die Spritzbetonschale wurde auf einen Fundamentbalken aufgesetzt.

Damit die Rippen vor dem Ausbruch der Felder ihre Tragfähigkeit erreichten, mussten sie auf dem gegenüber liegenden Gleisbereich ergänzt werden. Insgesamt wurden somit mehrere "Gleiswechsel" für die Arbeiten erforderlich. Zur Durchführung der Arbeiten im Firstbereich waren Sperrpausen auf beiden Gleisen erforderlich.

Die Arbeiten wurden messtechnisch überwacht.

5 TÜLLINGER TUNNEL

Der 884 m lange Tüllinger Tunnel aus dem Jahre 1888/90 zwischen Weil am Rhein und Lörrach ist zweigleisig gebaut, wird aber nur eingleisig genutzt. Die Überlagerung beträgt bis zu ca. 59 m. Der Tunnel durchquert Gesteine der Molasse (Abb. 8; Tonmergel, Kalkstein, Gipsmergel, teils ausgelaugt). Oberhalb des Tunnels sind Geländerutschungen bekannt; sie tangieren aber nicht den Tunnel. Der nachfolgend erläuterte Tunnelabschnitt ist durch Tonstein/Tonmergel unterhalb der Firste und eine 6 m bis 8 m mächtige Kalksteinbank über der Firste geprägt. Die Schichtgrenzen variieren in ihrer Höhenlage.

Der Ausbau besteht aus 0,6 m bis 0,7 m dickem Quadermauerwerk (Sandstein) im Gewölbe und aus ca. 3 m hohen, bis 1,2 m dicken Widerlagern aus großformatigen Sandsteinen. Die Sohle ist offen und hat einen Mittelkanal zur Tunnelentwässerung.

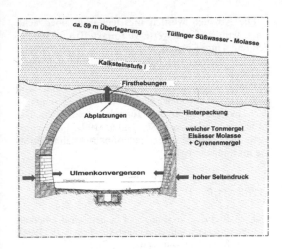

Abbildung 8.   Querschnitt durch den Tüllinger Tunnel.

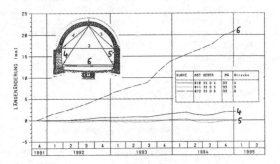

Abbildung 9.   Konvergenzmessungen im Tüllinger Tunnel.

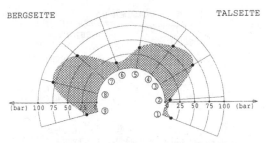

Abbildung 10.   Gemessene und berechnete Spannungen.

Abbildung 11.   Bauzustand Sohlaushub mit hydraulischer Queraussteifung zwischen den Widerlagern.

In verschiedenen Bereichen wurde in den vergangenen Jahrzehnten Instandsetzungsarbeiten ausgeführt (Fugenverpressungen im Gewölbe, Spritzbetonarbeiten). Der Tunnelausbau in dem nachfolgend erläutertem Abschnitt war visuell durch leichte Abplatzungen an den Lagerfugen des Mauerwerks in den Kämpfern geprägt. Visuell waren leichte Verdrückungen zu vermuten.

An 9 Stellen des Umfangs (Punkte in Abb. 10) wurden die Innenrandspannungen gemessen. Die Messschlitze wurden in der Mitte eines Mauersteins angeordnet. Die minimale Drücke betrugen ca. 1 MPa; die maximalen Drücke erreichten Werte von ca. 8 MPa. Die Drücke und ihre Verteilung (Abb. 10) entsprachen der örtlichen Einschätzung. Es wurden 2 Messquerschnitte untersucht. Die Rückrechnung der äußeren Belastung ergab eine wirksame Auflasthöhe von $h_{min} = 12\,m$ bei einem von der Seite wirkendem Druck von 20% bzw. einer Obergrenze von $h_{max} = 20\,m$ und 50% seitlichen Druck. Durch eine geringe Hanglage war die Belastung etwas asymmetrisch.

Die Messung der Tunnelverformungen in Konvergenzmessquerschnitten mit einem Distometer ergaben fast keine Veränderungen der Diagonalmess- trecken. Die unten liegende horizontale Messstrecke zwischen den Widerlagern zeigte deutliche Konvergenzen mit einer Rate von ca. 6 mm pro Jahr! Später bekannt gewordene Bauakten ergaben bei einem Vergleich des Planprofils mit dem heutigen Ulmenabstand eine Verringerung von bis zu 1,05 m in den Jahren von 1892 bis 1995! Auf Grund von Extensom- etermessungen werden die Verschiebungen auf beiden Widerlagerseiten etwa gleich angenommen. Nivellements ergaben, dass die Firste sich hebt, was auf ein geringe Bettung in der Firste schließen lässt. Diese Verschiebung ist plausibel bei den starken Horizontalkonvergenzen. Widerlagerabsenkungen wurden nicht gemessen.

Die Konvergenzmessungen zeigten, dass sie sich schnell beschleunigten, wenn Arbeiten im Tunnel ausgeführt wurden, z. B. Erkundungsbohrungen oder Probeanker.

Durch das Bergwasser weichte lokal der tertiäre Tonmergel unter den Widerlagern auf. Infolge der seitlichen Belastungen wurden die Widerlager nach innen geschoben. Durch die Verschiebungen bilden sich Risse im Tonmergel, der wiederum infolge des

57

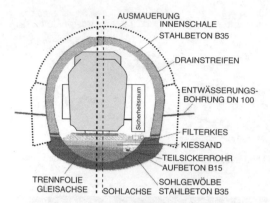

AUSMAUERUNG
INNENSCHALE
STAHLBETON B35

DRAINSTREIFEN

ENTWÄSSERUNGS-
BOHRUNG DN 100

Sicherheitsraum

FILTERKIES
KIESSAND
TEILSICKERROHR
AUFBETON B15
TRENNFOLIE
GLEISACHSE
SOHLGEWÖLBE
SOHLACHSE STAHLBETON B35

Abbildung 12.   Altes Gewölbe und neue Innenschale.

eindringendes Wassers entfestigt und stärker kriecht und die Widerlager seitlich belastet, die wiederum nachgeben und eine Verstärkung der Risse im Gebirge zur Folge haben.

Da der Tunnel zweigleisig gebaut und eingleisig genutzt wurde, war das Lichtraumprofil ausreichend, um eine Innenschale aus Stahlbeton mit einem Sohlgewölbe einzubauen (Abb. 12).

Der kritischste Bauzustand war der Aushub des Sohlgewölbes. Die gekrümmte Sohle sollte nicht nur die Widerlager des neuen Stahlbetongewölbes aufnehmen. Es musste auch für den Bauzustand bis zum Einbau des Betongewölbes die Widerlager des bestehenden Mauersteingewölbes unterfangen und deren Horizontalverschiebung verhindern. Für die Zeit des Aushubs bis zum Erhärten des Sohlbetons wurden daher die Widerlager mit einem hydraulisch verspanntem, 8 m langen Wanderverbau ausgesteift (Abb. 11). Die Aussteifungskräfte wurde auf 4 × 3.300 kN festgelegt; die Festlegelast betrug 100 kN, um den Ausbau nicht aktiv nach außen zu pressen. Der Sohlaushub erfolgte in Abschnitten von max. 3,35 m und einer Tiefe bis 2,55 m unter SO. Zur Absicherung der Arbeiten wurde in der Sohle zunächst eine 2-lagig bewehrte Spritzbetonschicht eingebaut Die Betondicke des Gewölbes beträgt im First 50 cm und im unteren Ulmenbereich 60 cm. Dehnfugen wurden in 5 m Abständen angeordnet.

Die Arbeiten wurden messtechnisch überwacht. Dazu wurden Grenzwerte von max. 10 mm Setzungen und 4 mm Konvergenz festgelegt. Die Setzungen und die Konvergenzen haben jedoch 2 mm nicht überschritten.

Diese Teilerneuerung wurde auf eine Länge von 195 m entsprechend der geologischen Gliederung und der Beobachtungen im Tunnel erforderlich. Die Arbeiten wurden von September 1997 bis August 1998 durchgeführt. Die Wiederinbetriebnahme erfolgte im Frühjahr 1999.

LITERATUR

Fröhlich, B. (1996). Belastung alter Tunnel: Schadensbilder – Untersuchungen – Interpretation. Geotechnik 19 (1996), Nr. 3, Essen: Vlg. Glückauf.

Gilsdorf, A., Fröhlich, B., Schröder, K.-H. & Wullschläger, D. (1999). Die Teilerneuerung des Tüllinger Tunnels; Gebirgsmechanische Anforderungen und bautechnische Lösungen. Felsbau 17 (1999) Nr. 6, Essen: Vlg. Glückauf.

Fröhlich, B., Schlebusch, M., Berwanger, W. & Pape, H. (1999). Tunnelinstandsetzung im unterschnitten Hangbereich des Rheintals. Felsbau 17 (1999) Nr. 6, Essen: Vlg. Glückauf.

Rustemeier, A., Weishäupl, S., Fröhlich, B. & Schlebusch, M. (1999). Vom Schadensbild bis zur Instandsetzungsmaßnahme am Beispiel Ramholz- und Ziegenberg-Tunnel. Felsbau 17 (1999) Nr. 6, Essen: Vlg. Glückauf.

Schuck, W. & Wullschläger, D. (1994). Spannungsmessungen im Mauerwerksausbau von Eisenbahntunneln. Felsbau 12 (1994) Nr. 3, Essen: Vlg. Glückauf.

Deutsche Bahn AG. (1999). Richtlinie Ril 853 – Eisenbahntunnel planen, bauen und instandhalten. Hrsg. & Vlg. DB AG (01/99).

Rokahr, R. & Zachow, R. (1999). Betonspannungsermittlung in der Spritzbetonaußenschale mit dem Programm STRESS am Beispiel Eggetunnel. Persönliche Information, Universität Hannover – Institut für Unterirdisches Bauen.

gbm – Gesellschaft für Baugeologie und –meßtechnik mbH Baugrundinstitut (1991). Studie zur Beanspruchung des Ausbaus alter Eisenbahntunnel – In situ Spannungsmessungen und rechnerische Analyse. Ettlingen 1991 (unveröffentlicht).

Schuck, W. & Fecker, E. (1998). Geotechnische Messungen in bestehenden Eisenbahntunneln. Tunnelbau-Taschenbuch (22) 1998, Essen: Vlg. Glückauf.

Geotechnical Measurements and Modelling, Natau, Fecker & Pimentel (eds)
© 2003 Swets & Zeitlinger, Lisse, ISBN 90 5809 603 3

JointMetriX3D – Measurement of structural rock mass parameters

A. Gaich
3G Software & Measurement GmbH, Austria

A. Fasching
3G Gruppe Geotechnik Graz, ZT GmbH, Austria

W. Schubert
Institute for Rock Mechanics and Tunnelling, Graz University of Technology, Austria

ABSTRACT: This contribution describes a measurement and documentation system that significantly improves present practice in collecting structural rock mass parameters. The system relies on the generation of high resolution panoramic images of a rock face and software components that enable spatial measurements of geometric features visible in the images. This approach allows indirect measurements increasing safety for the engineering geologist, enhancing data quality, and representing an objective documentation of actual rock mass conditions. The improved data quality helps to optimise excavation concerning efficiency and safety both in tunnelling and mining.

1 INTRODUCTION

Field mapping as usually performed by an engineering geologist comprises the determination of structural rock mass parameters. These geometric features are usually measured using a compass-clinometre device and a tape measure. Although measuring with a compass-clinometre device is convenient and easy to apply in field, certain drawbacks can be identified.

It requires physical contact to get an orientation measurement. If the measuring point cannot be reached safely, the measurement cannot be taken or only at risk. In order to get a measurement sometimes high efforts are required due to the need for the manual application of the compass. This often results in a rather poor number of measurements which even gets worse if the time provided for the measurements is restricted.

Using high resolution images of a rock mass the measurements can be taken indirectly on the computer by analysing the images. This solves the problem for accessing problematic or hazardous regions. The time to take the data (images) is no longer dependent from the complexity of the rock mass structure. Additionally the images represent an objective documentation of the rock mass conditions and allow also a later review and analysis of the situation even if the original conditions are no longer existing, e.g. due to excavation or erosion.

Besides, it is easily possible to change the point of observation during the analysis on the computer. A quick change between an overview and a detailed location where to measure an orientation, improves the quality of the assessment. The quality of a measurement itself is improved as well, as it is not influenced by the surrounding, e.g. a large amount of support at a tunnel site might deviate a compass needle.

The resulting data and models are ready for further algorithmic use, such as stability computations or numerical simulations.

2 RELATED WORK

Several attempts were taken in the past to measure structural rock mass parameters from images. Among those, Linkwitz (1963) and Rengers (1967) used a photo theodolite and a stereoscope to measure points from a rock surface deriving discontinuity orientations. Both showed that measurements in inaccessible regions become possible and claimed a more reliable analysis over large areas.

A more recent approach applying the same principles but using digital images and automatic procedures to assess the images is described by Roberts & Poropat (2000). It showed that orientation measurements for mapping a high rock wall can be derived from standard digital images, although a complete approach which contains a proper imaging system and a referencing mechanism for the images remained open.

The approach within this contribution relies on a special imaging device that allows the generation of high resolution panoramic images and several specially designed software components that process the images. The system thus covers ideas from aerial photogrammetry, digital image processing, or computer graphics and leads those techniques into the field of engineering geology/geotechnics.

3 WORK FLOW

The procedure to get structural rock mass parameters comprises several steps. First at least two images of the rock face from different standpoints are taken. Then an interactive procedure is used to reference the images according to visible reference points, i.e. points with known co-ordinates in some reference co-ordinate system.

Next an automatic routine computes corresponding points within the images which represent the basis for the spatial reconstruction of the surface points. Those are connected to a surface description (a mesh) which can be displayed and interactively inspected.

Interactive "drawing" on the model by a comfortable editor leads finally to geometric features from which lengths, areas, or orientation measurements are derived.

3.1 Imaging

The imaging system is easy to transport and apply. It is set up on a tripod similar to an ordinary camera and has a notebook computer that controls the camera and stores the enormous data amount that accrues during the image generation.

The camera itself uses a rotating line-scan principle which requires a still object scene as it scans the surrounding during its movement. A single scan takes several seconds up to some minutes depending on the required image resolution and the illumination conditions. Figure 1 shows a photograph of the imaging system applied at an active tunnel site.

Images taken with that camera have 75–120 Megapixels for a full panoramic sweep depending on the used focal length of the lens (confer a currently good standard digital camera has about 6 Megapixels). Considering the possibility of having 48 Bits colour-depth in the images, this leads to uncompressed image sizes of 450–720 MB. Images of these sizes require a special treatment concerning storage, display, and manipulation.

The need for such big images originates from the need to identify small structures within the image, namely those locations that are relevant for the assessment of the whole rock face. If those details are not

Figure 1. The imaging systems applied at a tunnel site.

captured by the images, a serious geotechnical analysis is hard to perform.

An example for such an image is given in figure 2. It shows a part of quarry in Austria. The overall height of the rock face is about 70 m. The image shows a spatial resolution of 35 mm/pixel. Applying the system at a tunnel face, spatial resolutions in the range of 2 to 3 mm/pixel are usual.

3.2 The spatial image

Taking an image means mapping the three-dimensional world onto a two-dimensional image plane. As this procedure implies a loss of information, spatial data can be recovered if two images of an object are taken from two different standpoints.

The recovery of the surface needs four steps and can be done with minimal user interaction.

Camera orientation: The orientation of a camera divides into the interior orientation and the exterior orientation. The interior orientation describes how the mapping of an object point onto the image plane actually happens. The exterior orientation refers to the spatial position and orientation of the camera in relation to the reference co-ordinate system.

Image matching: This is the process of identifying corresponding points within the image pair. Software components automatically search for according locations within the images. The resulting point sets are the basis for the spatial reconstruction.

Spatial reconstruction: For each pair of corresponding points in the images one spatial surface point can be reconstructed (see figure 3). Prerequisite is the knowledge of the camera orientations.

Surface generation: The reconstructed positions lead to a bunch of spatial points. Connecting the points with each other leads to a geometric surface description of the rock face. If one of the underlying images from which the surface was computed is draped on the

Figure 2. Image taken in a quarry. The spatial resolution is 35 mm/pixel leading to an image size of about 25 Megapixels for the rock face. The imaging distance is about 120 m off the rock face.

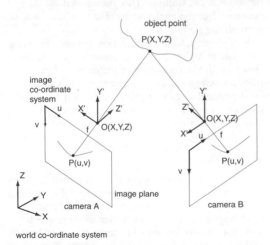

Figure 3. Surface reconstruction principle. Two corresponding points lead to one surface (object) point.

reconstruction, a result as depicted in figure 4 is derived: a surface model that can be inspected and analysed in 3D.

3.3 Parameter identification

Having a spatial reconstruction of the rock face and a proper software to inspect and analyse the model, geometric features can be measured from it. Metric distances like spacing, trace lengths, or gap lengths can be determined, just as area sizes or discontinuity orientations.

At this stage it should be clearly stressed that measurements from images cannot deliver all parameters that are commonly collected for rock mass characterization. Of course there are several magnitudes that cannot be derived from observing images, e.g. the filling material of a discontinuity.

Table 1 gives some selected parameters identifiable from images and some that are not.

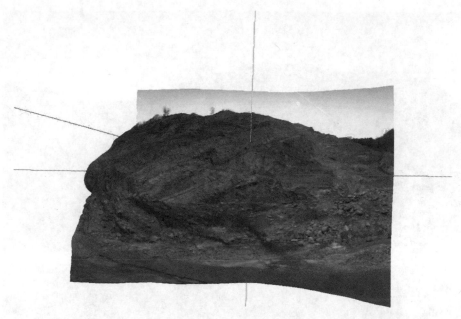

Figure 4. Reconstructed surface model. The model contains 80,000 surface points and a high resolution texture map.

Table 1. Selected parameters of a geological tunnel documentation. See Beer (2003).

Parameter		Identification from images
Rock type/	Type	Partially Yes
kind of rock mass,	Orientation	Yes
rock mass type,	Shape/curvature	Yes
discontinuity	Roughness	Yes
	Persistence	Yes
	Spacing	Yes
	Type of filling	No
	Thickness of filling	Yes
	Termination	Yes
	Aperture	Yes
Jointing of	Shape of joint bodies	Yes
rock mass	Dilation of rock mass	No
	Degree of fracturing	Yes
Degree of weathering		Yes
Water	Occurrence	Yes
	Amount	No
	Pressure	No

4 APPLICATIONS

The system was already applied in quarries as well as in subsurface environments, such as tunnel sites and mines. The application is especially useful when field work is difficult to perform, if it might get dangerous to collect measurements, or if efforts to provide manual access are high.

4.1 Tunnelling

In the case of the tunnel site, the application of the system does not require the geotechnical expert to be on-site for every recorded rock face, as certain conditions usually do not change significantly between subsequent excavations. This means that all data is captured, but the analysis can also be done later, or even off the site.

Furthermore, the documentation that one gets using the system highly surpasses conventional methods. However, the better quality of the derived data represents not only a good documentation. Daily decisions on the tunnel site concerning excavation or support measures can be made from a more objective basis which implies that they can be more cost-efficient.

4.2 Mining and quarrying

In the case of surface applications the rock faces to assess are often larger than a tunnel face. Doing field work sometimes leads to the problem that being close to the large rock face, the identification of proper locations for orientation measurements becomes harder, as the overview got lost. Using the system the point of view can be quickly changed on the computer, and the relevant features easily be identified. Having large rock faces, the advantage of measuring also inaccessible regions becomes even more obvious.

As the data is ready for further processing steps, excavation planning can be supported efficiently:

- The general layout of mine or quarry walls, especially their orientation and inclination in relation to main discontinuities can be controlled systematically.
- The acquired data can be directly used for identifying potential failure modes, and stability assessments of the face, berms, or the whole structure.
- Taking images periodically it is possible to compute a difference between reconstructed surfaces which equals to the excavated volume.
- The comprehensive and periodical documentation of quarries enables the optimisation of blasting patterns as well as the selective exploitation concerning the mining of controlled rock block sizes, or different material qualities for example.
- In underground mining images can be taken even from large chambers, where a direct access to the exposed rock is impossible and dangerous. Also in this case resulting data can be used for stability assessments and excavation planning.

5 CONCLUSION

JointMetriX3D is a measurement and documentation system that allows the determination of structural rock mass parameters. It relies on the generation of at least two high resolution panoramic images and software components that enable a spatial assessment of the recorded rock faces. The system was applied on tunnel sites, at quarries and mines and considerably improves the work of the engineering geologist.

The structural rock mass data can be measured indirectly which decouples the data acquisition from the analysis process. Beyond, the generated images are a perfect documentation of the actually encountered rock mass situation which allows a later review and analysis if needed.

The system represents a valuable supplement for current field work and might help to create a new quality standard.

REFERENCES

Beer, G. (ed.). 2003. *Numerical Simulation in Tunnelling.* Wien, New York: Springer.

Crosta, G. 1997. Evaluating Rock Mass Geometry From Photogrammetric Images. *Rock Mechanics and Rock Engineering* 30(1): 35–38.

Fasching, A. 2000. *Improvement of Data Acquisition Methods for Geological Modelling*, Phd thesis. University of Technology, Graz, Austria.

Gaich, A. 2000. *Panoramic Vision for Geotechnical Analyses in Tunnelling*, Phd thesis. University of Technology, Graz, Austria.

Hagan, T.O. 1980. A Case for Terrestrial Photogrammetry in Deep-Mine Rock Structure Studies. *International Journal of Rock Mechanics & Mining Sciences* 17: 191–198.

Linkwitz, K. 1963. Terrestrisch-photogrammetrische Kluftmessung. *Rock Mechanics and Engineering Geology* I: 152–159.

Rengers, N. 1967. Terrestrial Photogrammetry: A Valuable Tool for Engineering Geological Purposes. *Rock Mechanics and Engineering Geology* V: 150–154.

Roberts, G. & Poropat, G. 2000. Highwall mapping in 3-D at the Moura Mine using SIROJOINT. In Beeston, J.W. (ed.) *Bowen Basin Symposium 2000 Proceedings*: 371–377.

Schubert, W., Klima, K., Fasching, A., Fuchs, R. & Gaich, A. 1999. Neue Methoden der Datenerfassung und Darstellung im Tunnelbau. In: *Beiträge zum 14. Christian Veder Kolloquium* – Die Beobachtungsmethode in der Geotechnik: 13–27. Graz: University of Technology.

Slama, Ch.C. (ed.). 1980. *Manual of Photogrammetry.* Fourth Edition. Fall Church, Va.: American Society of Photogrammetry.

Uetlibergtunnel/Zürich – Geotechnische Modellierung zur Auswahl des Vortriebsverfahrens in den Lockergesteinsstrecken und deren Erfahrungen

H. Hagedorn & Th. Eppler

Amberg Engineering AG, Regensdorf-Watt, Schweiz

ABSTRACT: The western drive round of Zurich is from the view of technically traffic management of great national and international importance. The Uetlibergtunnel crosses two crests of ridge, the Uetliberg hill and the Ettenberg hill in the south western territory of Zurich. In the core of both hills you will find plane bedded deposits of the Upper Freshwater Molasse. In the portal areas and the site between the two hills there have to be crossed between 200 m and 400 m of soil coverage. The three sections with soil coverage had to be crossed with a capable heading up to the rock area. For this in the design phase and in the preparation for the tender there had been analyzed five possible methods of heading in detail: crown excavation/bench heading/jet grouting, crown excavation divided into bench heading/jet grouting arch, core construction method, crown excavation with cutting shield, crown excavation with the protection of a body of frozen ground. The assortment of the building method depended on the results of the geotechnical modellings. Tunnel driving in the soil material had been successfully carried out with the core construction method. Especially in problematic sections additional structural engineering methods have been carried out, such as pipe screens, protection shields of lances, advanced groundwater lowering, subdivisions of the face in the upper side wall tunnel and bolts in the tunnel face. All of the three tunnels in the soil coverage had been built without any special occurence. The deformation of the tunnel walls abided in the expected amplitude.

1 EINFÜHRUNG

1.1 *Allgemeines*

Der Uetlibertunnel ist das Herzstück der zur Zeit in Bau befindlichen Westumfahrung Zürich. Die zwei Röhren des mit ca. 4.4 km längsten Tunnels der gesamten Westumfahrung Zürich verbindet die Umfahrung Birmensdorf (Einzugsgebiet aus dem Westen und Norden von Zürich) mit der bestehenden Nationalstrasse Zürich-Chur im Osten.

Die beiden parallel angeordneten Tunnelröhren sind alle 300 m mit einer begehbaren und alle 900 m mit einer befahrbaren Querverbindung gegenseitig erreichbar. Der Abstand der SOS-Nischen beträgt 150 m. An den Portalstationen des West- und Ostendes des Tunnels finden sich technische Einrichtungen. Das von Westen nach Osten etwa gleichmässige Gefälle beträgt 1.6%.

In der Baugrube des Zwischenangriffes Reppischtal in Landikon wird später eine unterirdische Lüftungszentrale angeordnet, die über einer ebenfalls unterirdischen Verkehrsüberwachung zu liegen kommt.

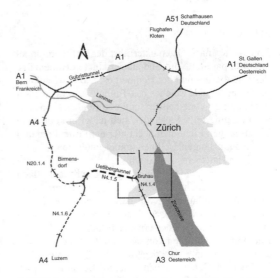

Abbildung 1. Nationalstrassen im Raum Zürich. Gestrichelt gezeichnet ist die im Bau befindliche Westumfahrung mit dem Uetlibergtunnel als zentrales Verbindungsteil.

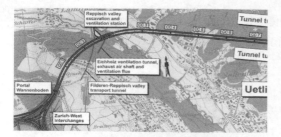

Abbildung 2. Lageplan des Uetlibergtunnels mit den wichtigsten Pro.

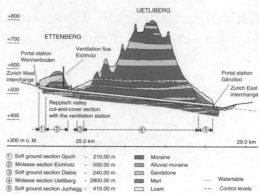

① Soft ground section Gjuch	~ 210.00 m	■ Moraine
② Molasse section Eichholz	~ 500.00 m	▦ Alluvial moraine
③ Soft ground section Diebis	~ 240.00 m	▨ Sandstone
④ Molasse section Uetliberg	~ 2800.00 m	■ Marl ---- Watertable
⑤ Soft ground section Juchegg	~ 410.00 m	□ Loam -- Control levels

Abbildung 3. Gelogisches Längenprofil mit Vortriebsverfahren.

Im Normalfall wird der Tunnel durch die natürliche Längslüftung (Kolbenwirkung der Kraftfahrzeuge) in beiden Röhren belüftet. Über die unterirdische Lüftungszentrale kann aber auch Luft aus den Röhren abgesaugt werden und über ein Verteilnetz in der Zwischendecke des Tunnels über einen seperaten Abluftstollen dem Abluftschacht Eichholz zugeführt werden und dort ins Freie geblasen werden.

Das Normalprofil in den Lockergesteinsstrecken sowie in der Molassestrecke Eichholz ist ein Hufeisenprofil mit Sohlgewölbe und weist im Ausbruch Abmessungen von 14.7 m Breite und eine Höhe von 12.7 m auf. Aufgrund der Geometrie ergibt sich eine Ausbruchsfläche von ca. 143 m^2 bis 148 m^2. Das Normalprofil der mit einer Tunnelbohrerweiterungsmaschine aufzufahrenden Molassestrecke unter dem Uetliberg ist nahezu kreisförmig. Bei einer Ausbruchsfläche von ca. 160 m^2 variiert der Radius zwischen 14.2 m und 14.4 m.

1.2 Geologie

Von Westen nach Osten unterfährt der Uetlibergtunnel die zwei parallel Nord-Süd-ausgerichteten Höhenzüge des Ettenberges und des Uetliberges. Dazwischen liegt das Reppischtal, welches das Tunnelbauwerk in zwei unabhängig voneinander aufzufahrende Tunnelabschnitte unterteilt. Es handelt sich dabei um den rund 720 m langen Eichholztunnel unter dem Ettenberg und den 3.5 km langen Uetlibergtunnel unter dem gleichnamigen Hausberg von Zürich.

Der Kern beider Höhenzüge besteht aus flach gelagerten Schichten der oberen Süsswassermolasse, einer Wechsellagerung von harten Sandsteinbänken und weichen Mergelschichten. Eingelagert finden sich hier immer wieder organisch-kohlige Schichten mit einer Mächtigkeit von wenigen Zentimetern.

Die maximale Überlagerung beträgt unter dem Uetliberg ca. 320 m. An den späteren Ost- und Westportalen sowie am Zwischenangriff im Reppischtal sind zum Erreichen der Molassestrecken drei Lockergesteinsabschnitte mit Moränenmaterial aufzufahren.

Es handelt sich dabei um die Abschnitte Gjuch, Diebis und Juchegg.

Die Lockergesteinsstrecken bestehen aus einem sehr inhomogenen Moränenmaterial und Hangschuttmaterial. Das Material reicht von sehr steinreichen Kiesen bis zu tonig-siltigen Ablagerungen. Bereichsweise liegt die Tunneltrasse unter dem Grundwasserspiegel im Lockermaterialbereich.

1.3 Projektelemente des Uetlibergtunnels

1.3.1 Lockergesteinsstrecken
Zum Erreichen der festen Molasse sind vorgängig die oberflächennahen Moränenablagerungen zu durchfahren.

1.3.2 Molassevortrieb Eichholz
Der rund 500 m lange Abschnitt unter dem Ettenberg (Molassestrecke Eichholz) wird sprengtechnisch erstellt, wobei der Ausbruch in Kalotte, Strosse und Sohle unterteilt wird.

1.3.3 Molassestrecke Uetliberg
Der Ausbruch dieser Felsstrecke erfolgt mit einer Tunnelbohrmaschine mit einem Durchmesser von 5.0 m und einer nachfolgenden Tunnelbohrerweiterungsmaschine mit Hinterschneidtechnik auf einen Durchmesser von 14.2 m/14.4 m. Damit wird der vorgängig gefräste Pilotstollen auf den vollen Querschnitt aufgeweitet. Die Sicherungsmittel, bestehend aus Seilankern, Spritzbeton, Swellex-Ankern und Netzen, werden direkt hinter dem Schneidrad eingebaut. Im Bereich des Nachläufers der Tunnelbohrerweiterungsmaschine erfolgt der Ausbau der Tunnelsohle bis Höhe der späteren Planumsoberkante der Fahrbahn. Abdichtung, Sohle, Werkleitungskanal (Fertigteile) und seitliche Hinterfüllung sind in den Nachläufer der Ausweitungsmaschine integriert.

1.3.4 Transportstollen Filderen-Reppischtal

Zum Abtransport des Ausbruchmaterials aus dem Hauptvortrieb der Molassestrecke Uetliberg sowie zur Belieferung des Betonwerkes mit Zuschlagstoffen wurde vorab ein Transportstollen mit einem Durchmesser von 3.7 m und einer Länge von 540 m erstellt.

1.3.5 Abluftschacht und Abluftstollen Eichholz

Die über die Lüftungszentrale im Reppischtal abgesaugte Luft wird später über den Abluftstollen und einen Schacht mit aufgesetztem Kamin auf dem Ettenberg ins Freie geblasen. Abluftstollen und Schacht wurden im konventionellen Sprengvortrieb erstellt.

2 LOCKERGESTEINSSTRECKEN

Aus den geologischen Untersuchungen und den ersten Überlegungen zum Bau des Uetlibergtunnels hat es sich herausgestellt, dass die Lockergesteinsstrecken Gjuch, Diebis und Juchegg nicht ohne besondere Bauhilfsmassnahmen zur Gewährleistung der Hohlraumstabilität im Bauzustand aufgefahren werden können. Erste Abschätzungen hatten gezeigt, dass im unverfestigten Zustand der Boden die Stabilität bis zum Einbringen der definitiven Verkleidung nicht oder nur ungenügend gewährleisten kann.

Für die zuverlässige Projektierung der Bauhilfsmassnahmen waren deshalb Modellsimulationen unerlässlich. Alle Bauhilfsmassnahmen, welche auf der

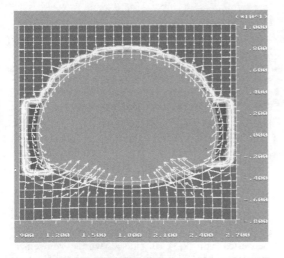

Abbildung 4. Ohne Sohlgewölbe ergeben sich grosse Hebungen. First und Parament verfestigt. Beispiel aus der numerischen Modellsimulation. Axangaben in Meter (10^1 m).

Verfestigung des natürlichen Bodens beruhen, verändern die Bodenkennwerte, welche das Elastizitäts- und Festigkeitsverhalten des Bodens prägen. Eine realistische Simulation von Bauhilfsmassnahmen ist somit nur dann gewährleistet, wenn die Materialkennwerte sowohl für den gewachsenen wie auch für den verfestigten Baugrund bekannt sind. Dort, wo keine zuverlässigen Bodenparameter angegeben werden konnten, mussten Parameterstudien ausgeführt werden.

Die Ergebnisse der Modellsimulationen waren Grundlage für die Auswahl des definitiven Ausbruchsverfahrens.

3 BODENPARAMETER DER LOCKERGESTEINSSTRECKEN

Die Lockergesteinsstrecken bestehen vorwiegend aus Grundmoräne mit teilweise überlagertem verschwemmtem Gehängeschutt. Die Kornzusammensetzung des Lockermaterials an den drei Standorten war bereits lokal sehr heterogen und konnte zwischen den drei geplanten Angriffspunkten nicht korreliert werden.

Gemäss den ermittelten bodenmechanischen Versuchen wies die Kornverteilung in der Moräne Diebis einen hohen Anteil von tonigem Kies (USCS-Klasse: GC-CL) auf, während die Moräne Juchegg einen höheren Anteil von tonigem Sand (SC-CL) enthält. Die prozentualen Anteile an kiesigem und tonigem Silt (GM-ML, SM-ML) sind in beiden Moränen etwa gleich. Die Moräne Gjuch weist einen geringen Anteil an lehmigem Moränenmaterial (Anteil CL) auf und enthält vorwiegend kiesiges und sandiges Material (GC und SM). In der Moräne Gjuch ist eine sehr heterogene Verteilung der USCS-Klassen vorherrschend.

Die Lagerungsdichten im Gehängelehm waren vorwiegend locker bis mitteldicht, in den Moränen kompakt bis sehr heterogen (Moräne Gjuch).

Die Wasserspiegel zum Zeitpunkt der geologischen Erkundungen lagen minimal an den Portalen bei ± 0.0 m und einem max. Wasserstand von ca. 15 m über Tunnelfirste.

Die Wasserdurchlässigkeiten wurden mit $k_f = 10^{-7}$ m/sec $\cdots 10^{-8}$ m/sec angegeben.

4 GEFÄHRDUNGSBILDER

Für die Lockergesteinsstrecken wurden die möglichen Gefährdungsbilder in zwei Kategorien gegliedert. Zum einen in Gefährdungsbilder, die Einfluss auf die statischen Berechnungen haben, und zum anderen in Gefährdungsbilder, die zwar relevant sind, aber in den statischen Berechnungen nicht berücksichtigt werden.

4.1 Gefährdungsbilder mit Einfluss auf die statischen Berechnungen

- *Auflockerungsdruck*: Auflockerungen im First-bereich durch Firstsenkungen. Dadurch entstehen hohe Belastungen der Ausbruchsicherung und der Verkleidung im Kalottenbereich. Die Ergebnisse sind in der statischen Bemessung der Ausbruch-sicherung und des Innengewölbes zu berücksichtigen.
- *Oberflächensetzungen*: Verformungen des Tunnels führen zu Setzungen und Verschiebungen, das Gebirge lockert sich auf. Die Auswirkungen können u. U. bis an die Oberfläche reichen, wo unzulässig tiefe Setzungsmulden eine Gefährdung von Gebäuden, Strassen, Kanalisation, Werkleitungen, etc. verursachen. Frühe Ringschlusszeiten, ausreichende Dimensionierungen der tragenden Elemente und eine Anpassung der Abschlagslängen sind hier erforderlich.
- *Ortsbrustinstabilität*: Der Tunnelvortrieb wird von einer Entlastung des Baugrundes begleitet. Die Scherfestigkeit des Bodens wird reduziert, die Standfestigkeit der Ortsbrust stark eingeschränkt. Mit einer aktiven Ortsbruststützung mittels Spritzbeton, Ankern oder Jettingpfählen sowie einer Ausbruchsunterteilung der Ortsbrust kann diesem Gefährdungsbild begegnet werden.

4.2 Gefährdungsbilder, die konstruktiv berücksichtigt werden müssen

Durch konstruktive Massnahmen wurden die folgenden Gefährdungsbilder berücksichtigt:

- Wasserdruck
- Entspannung des Porenwasserdruckes
- Schwebende Grundwasserstockwerke
- Mangelnde Festigkeit des Spritzbetons
- Deformationsverhalten des Kernes
- Variation der Stärke des Jettinggewölbes
- Versagen der Kalottenwiderlager
- Geologische Rinne

5 UNTERSUCHTE BAUVERFAHREN

In einer Vorauswahl wurden fünf Bauverfahren festgelegt, die zum Einsatz kommen könnten und die näher untersucht wurden.

Als zusätzliche Bauhilfsmassnahmen können folgende Massnahmen dabei zur Ausführung kommen:

- Grundwasserabsenkung und Drainage
- Injektionen
- Jet Grouting
- Gefrieren
- Rohrschirm

5.1 Kalottenausbruch mit Jet-Grouting und Spritzbeton mit nachträglichem Strossenabbau

Unabhängig von der Wahl der Sicherungsmittel wird die Kalotte vollständig ausgebrochen. Der Strossenabbau erfolgt nachträglich. Der Strossenabbau kann unmittelbar an den Kalottenausbruch erfolgen oder erst nach dem Durchschlag der Kalotte. Eine weitere Unterteilung in Strosse und Sohle ist möglich.

5.2 Kalottenausbruch unterteilt

Bei diesem Verfahren wird die Kalotte mindestens einmal unterteilt. Der Ausbruch erfolgt somit in einzelnen Etappen. Strosse und Sohle können wie oben beschrieben abgebaut werden.

5.3 Kernbauweise

Der Ausbruch in der Kernbauweise ist in seinen einzelnen Etappen vorgegeben: Linker und rechter Paramentstollen (oben und unten), Kalotte, Kern und Sohle. Die Abstände der einzelnen Etappen aufeinander kann stark variiert werden und hängt im Wesentlichen von der Standfestigkeit des Bodens ab.

5.4 Kalottenvortrieb mittels Messerschild

Die in kleinem Querschnitt zu erstellenden Paramentstollen werden als Widerlagerkonstruktion für das Messerschild ausgebildet. Im Schutze des Messerschildes kann die Kalotte auf der gesamten Breite aufgefahren werden.

5.5 Kalottenvortrieb im Schutz eines Gefrierkörpers

Das Verfahren gleicht dem Kalottenvortrieb, wobei vorgängig ein Gefrierkörper als Bauhilfsmassnahme zum Schutz des Kalottenausbruches angelegt wird.

5.6 Ergebnisse der Voruntersuchungen

Für die oben genannten Bauverfahren wurden eine Vordimensionierung zur Ermittlung der notwendigen bodenmechanischen Parameter ausgeführt und Überlegungen zur generellen Machbarkeit angestellt. Dabei reduzierte sich die Anzahl der möglichen Ausbruchsverfahren von fünf auf drei, da einige Bauhilfsverfahren nicht zur Anwendung kommen konnten.

Aufgrund des geringen Wasserdurchlässigkeitsbeiwertes von $k_f < 10^{-6}$ m/sec wurde von einer systematischen Grundwasserabsenkung Abstand genommen.

In dem anstehenden Moränenmaterial mit einem sehr heterogenen Anteil läge der injezierbare Anteil an Bodenmaterial bei ca. 60%. Die restlichen 40% Bodenanteil müssen als nicht injezierbar angesehen werden.

Jetting und Rohrschirm sind in allen anstehenden Lockergesteinsmaterialien anwendbar.

Im Zuge der Vorüberlegungen erwiesen sich nur der Kalottenausbruch mit Jet-Grouting und Spritzbeton, die Kernbauweise und der Kalottenvortrieb mittels Messerschild als ausführbar. Allen drei Verfahren gemeinsam ist, dass ein entspannter Porenwasserdruck vorausgesetzt wird.

6 MODELLBERECHNUNGEN

Die numerischen Modellberechungen wurden mit dem Finite-Differenz-Programm FLAC von ITASCA ausgeführt.

6.1 *Kalottenvortrieb mit Jet-Grouting*

Der zu erstellende Tragring mit einer mittleren Stärke von 70 cm führt zu sehr hohen Spannungskonzentrationen in den Kalottenfüssen. Ohne Ausbildung eines entsprechenden Widerlagers kann es hier zu einem Grundbruch kommen. Für den Strossenausbruch ist ein Parament mit einer mittleren Stärke von 1.2 m und ein Fussauflager von 2.2 m Breite erforderlich. Ausserdem ist ein Sohlgewölbe mit einer Stärke von 25 cm erforderlich. Bei diesen Ausbaudimensionierungen ergeben sich rechnerisch Deformationen in der Grössenordnung von 1 cm bis 2 cm.

6.2 *Kernbauweise mit Stahleinbau und Spritzbeton*

Bei der Kernbauweise werden die beiden seitlichen Paramentstollen in zwei Phasen ausgebrochen, jeweils unterteilt in oberen und unteren Teil. Die Ausbruchsicherung erfolgt mittels Stahlbögen und faserbewehrtem Spritzbeton.

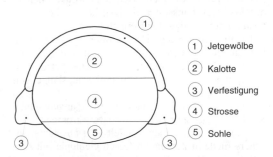

1 Jetgewölbe
2 Kalotte
3 Verfestigung
4 Strosse
5 Sohle

Abbildung 5. Kalottenausbruch mit Jetgewölbe (schematische Darstellung).

Ein Vorteil dieser Baumethode ist, dass die Paramentstollen auch erheblich vorgezogen werden können und als Drainagestollen oder als Erschliessungsstollen für Start- und/oder Zielkavernen dienen kann. Der Ausbruch von Kalotte, Kern und Sohle kann danach relativ rasch aufeinander erfolgen, sodass ein schneller Ringschluss erreicht wird.

Für die inneren Spritzbetonwände wurden Stärken von 25 cm angesetzt, die äusseren Spritzbetonwände wurden mit 30 cm angenommen.

Aus den numerischen Modellierugen ergaben sich Deformationen in der Grössenordnung von 5 cm bis 10 cm.

6.3 *Kalottenausbruch mit Messerschild*

Die Erstellung der Paramentstollen erfolgt wie bei der Kernbauweise, einschl. der einzubauenden Sicherungsmittel. Zur Aufnahme der grossen Kräfte aus dem Kalottenausbruch sind unter den Paramentstollen Jetting-Pfähle zur Widerlagerverfestigung auszuführen.

Der Ausbruch der Kalotte erfolgt im Schutz eines Messerschildes, wobei als Ausbruchsicherung ein bewehrtes Ortbetongewölbe eingebracht wird, dessen

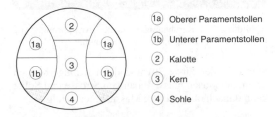

1a Oberer Paramentstollen
1b Unterer Paramentstollen
2 Kalotte
3 Kern
4 Sohle

Abbildung 6. Kernbauweise (schematische Darstellung).

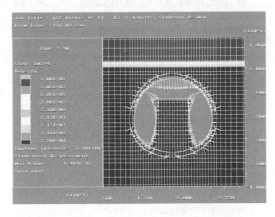

Abbildung 7. Bauzustand Kalottenaubruch bei der Kernbauweise. Ermittelte Deformatinen hier 5 cm.

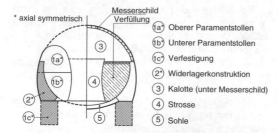

(1a*)	Oberer Paramentstollen
(1b*)	Unterer Paramentstollen
(1c*)	Verfestigung
(2*)	Widerlagerkonstruktion
(3)	Kalotte (unter Messerschild)
(4)	Strosse
(5)	Sohle

Abbildung 8. Kalottenausbruch mit Messerschild (schematiche Darstellung).

Abbildung 9. Kernbauweise in der Ausführung. Alle Bauzstände der Kernbauweise sind erkennbar.

Stärke in der Firste 30 cm und im Auflagerbereich 55 cm aufweisen muss. Abbau von Strosse und Sohle kann danach ohne weitere Massnahmen erfolgen.

7 RISIKOBETRACHTUNGEN UND WAHL DES BAUVERFAHRENS

Zur Auswahl des Bauverfahrens wurde aus geomechanischer Sicht eine Risikobetrachtung ausgeführt, wobei folgende Aspekte berücksichtigt wurden:

- Spannungsumlagerungen im Boden
- Wasserandrang
- Stabilität der Ortsbrust
- Auflockerungen im Boden
- Machbarkeit der Bauhilfsmassnahmen
- Bodenpressungen der Auflagerkonstruktionen
- Festigkeitsentwicklung des Spritzbetons
- Deformationsverhalten einzelner Elemente
- Beanspruchung der Ausbruchsicherung
- Dauerhaftigkeit der Konstruktion
- Komplexität des Bauverfahrens
- Störungsempfindlichkeit des Bauverfahrens
- Flexibilität und Anpassung des Bauverfahrens

Aus der Risikobetrachtung ergab sich, dass der Kalottenvortrieb mit Jetting die Variante mit dem grössten Risiko darstellte.

Unter Berücksichtigung der Kostenvergleiche, der Terminsituation und der geotechnischen Risikobetrachtung unter Zugrundelegung der Ergebnisse der numerischen Modellrechnungen fiel die Entscheidung für die Kernbauweise, die als Amtsvorschlag ausgeschrieben und ausgeführt wurde.

8 ERFAHRUNGEN MIT DER KERNBAUWEISE

In den drei Lockergesteinsstrecken wurde die Kernbauweise mit Erfolg angewandt. Durch die Flexibilität des Ausbruchverfahrens konnten auch kritische Strecken mit instabilen Bodenverhältnissen ohne besondere Vorkommnisse aufgefahren werden.

Gegenüber dem ursprünglichen Konzept der Kernbauweise wurden für die Ausführung einige Modifizierungen vorgenommen:

- *Rohrschirm* für das Anfahren des Tunnels aus der Baugrube
- *Firstsicherung mit Spiessen* bei nachbrüchiger Firste
- *Brustankerung* zur Stabilisierung der Ortsbrust in den Paramentvortrieben
- *Drainagerohre* in der Brust zur voreilenden Entwässerung
- *Teilausbruch* beim Ausbruch der oberen Paramentstollen in schwierigen Böden

9 ERGEBNISSE DER GEOTECHNISCHEN MESSUNGEN

Die messtechnische Überwachung der Lockergesteinsvortriebe erfolgte mit verschiedenen Messeinrichtungen wie 3D-Konvergenzmessungen, Extensometer und Stahldehnungsmessgeber.

Die aus den numerischen Modellierungen ermittelten Gesamtverformungen lagen zwischen 5 cm und 10 cm. Die effektiv gemessenen Deformationen in den Lockergesteinsstrecken lagen bei ca. 7 cm, also noch im prognostizierten Bereich.

Die mit den versetzten Extensometern gemessenen Bewegungen sind ca. 5 mal kleiner als die gemessenen Deformationsbewegungen. Ursache hierfür sind Bewegungen, die über die Verankerungslänge von 12 m hinausgehen.

Die an den Stahlträgern angebrachten Stahldehnungsgeber wiesen relativ grosse Unterschiede auf. Zum einen handelt es sich sicher um Unterschiede in der angetroffenen Geologie mit sehr heterogenen Bodeneigenschaften, zum anderen sind die unterschiedlichen Werte auf das nicht vollständige Einspritzen der Stahlbögen zurückzuführen.

10 ZUSAMMENFASSUNG

Für die Auswahl des Ausbruches der Lockergesteinsstrecken am Uetlibergtunnel waren die durchgeführten numerischen Modellsimulationen eine wichtige Entscheidungsgrundlage. Zusammen mit anderen Überlegungen konnte das optimale Ausbruchverfahren ausgewählt werden. Die Erfahrungen beim Bau bestätigten in vollem Umfang die in den numerischen Simulationen getroffenen Annahmen.

LITERATUR

Mauerhofer, S. (2002). Erfahrungen mit dem Lockergesteinsvortrieb.- Amberg – Tagung Uetliberg, 14. Seiten; Regensdorf.

Stadelmann, R. (2002): Evaluation der Lockergesteinsvortriebe – Projektierung, Geomechanik.- Amberg – Tagung Uetliber, 22 Seiten; Regensdorf.

Geotechnical Measurements and Modelling, Natau, Fecker & Pimentel (eds)
© 2003 Swets & Zeitlinger, Lisse, ISBN 90 5809 603 3

The role of time in NATM

D. Kolymbas

Institute of Geotechnical and Tunnel Engineering, University of Innsbruck

ABSTRACT: A simple but reasonable assumption for stress relaxation in rock leads to a satisfactory description of squeezing behaviour. It is shown how field tests in boreholes can be used towards optimization of support. The role of rock anisotropy is demonstrated by two simple examples.

1 INTRODUCTION

Tunnelling is a sort of underpinning: We remove a bearing element and we replace it with another one before the whole structure collapses. Such an operation is only possible within statically undeterminate systems, i.e. systems with redundance in bearing capacity. Failure of a part does not necessarily imply collapse of the whole. 3D continua, such as a rock mass, are statically indeterminate. One of the basics in civil engineering is that statical indeterminate systems can only be analyzed on the basis of their deformation behaviour (i.e. stress–strain relations, constitutive equation). This is why constitutive relations play (or, better: should play) a paramount role in tunnelling. Underpinning is always related with deformation. In general, this deformation implies that the subsequently installed bearing element receives less load than the initial one (see Fig. 1 a,b). This insight appears trivial in civil engineering, but was not introduced in tunnelling until the NATM came. Of course, the principle 'the more deformation, the less load' is to be applied cautiously. Exaggerated deformations can become contraproductive (Fig. 1 c) leading to a strong increase of load upon the bearing construction. To point on this was also a merit of NATM. The collapse which NATM had in mind was the softening (and the related loosening) of geomaterials. It should be emphasized, however, that it was not addressed the gentle stress reduction subsequent to the peak, as this is obtained with soil mechanics laboratory tests with remoulded or reconstituted soil samples. Much more was meant the very drastic strength reduction observed in poor rock due to loss of structural cohesion.

It is an (odd) tradition in civil engineering to distinguish between deformation and failure (collapse) of a structure. It is, however, impossible to find a

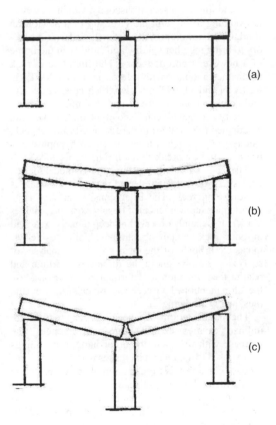

Figure 1.   Stages of underpinning.

genuine difference between these two notions. Virtually, failure is nothing but an overtly large deformation. At any rate, large deformations are to be avoided. How to achieve this in underpinning/tunnelling?

There are two ways: Either early and rigid support (which is not economic) or by keeping the size of the excavated cavities small. The latter option is pursued in tunnelling. There are two ways to do this:

- partial excavation instead of full face excavation.
- small advance steps.

2 TIME EFFECTS

Of course, too small excavation steps wouldn't be economic. So, tunnellers try to make the excavation steps as large as possible. In doing so, they exploit the rock strength as far as possible. This means that the peak of the stress–strain curve is approached as close as possible. There, however, rock (we are speaking of poor rock, since good rock does not matter) exhibits a very remarkable property: rate dependence and time dependence. This notion includes creep and relaxation and implies a sort of viscosity, which, however, has nothing to do with the viscosity of a newtonean fluid. Up to now, it proves extremely difficult for theory to bring together solidity and fluidity in the frame of a however rigorous theory. The literature in rock mechanics abounds with references to MAXWELL, KELVIN and VOIGT models which consists of several combinations of strings and dashpots, i.e. linear elasticity and newtonean viscosity. None of them can be adapted (calibrated) to the real behaviour of rock. The recently launched theory of visco-hypoplasticity promises to be a break-through (Niemunis 2003).

The lack of a satisfactory constitutive model for viscoplasticity of rock are reflected in the present analysis of squeezing rock. The latter consists in very large deformation (convergence) of freshly excavated cavities in rock. It is mainly observed in deep tunnels excavated in poor rock containing laminated silicates (clays, micas etc.). Some of the most prominent analyses of squeezing simply ignore the dimension of time and restrict their attention to deformation. Other ones use the aforementioned viscoelastic models and remain, thus, purely academic.

The manifestations of squeezing rock are two fold: unlined cavities exhibit large convergences that increase with time. On the other hand, rigid linings are exposed to a load that increases with time. The latter case corresponds to relaxation, since deformation is inhibited. However, relaxation usually means that stress *decreases* under constant deformation, whereas the pressure exerted by squeezing rock upon rigid linings *increases* with time!

3 ANALYSIS OF SQUEEZING

To analyze the phenomenon let us first consider the deformation around a circular tunnel (radius r_0)

excavated within a frictionless rock ($\varphi = 0$, $c > 0$) with constant hydrostatic primary stress ($\sigma_x = \sigma_y = \sigma_\infty$, $\gamma = 0$), as this is obtained in the conventional way, i.e. considering rate-independent elastic and plastic behaviour. In the elastic regime ($r > r_e$) the stresses follow from LAMÉ's solution as

$$\sigma_r = \sigma_\infty - (\sigma_\infty - \sigma_e)(r_e/r)^2$$
$$\sigma_\vartheta = \sigma_\infty + (\sigma_\infty - \sigma_e)(r_e/r)^2$$
$$\sigma_{r\vartheta} = 0.$$

In the plastified regime ($r_0 < r < r_e$) the stress field can be obtained from the equilibrium equation $d\sigma_r/d_r + (\sigma_r - \sigma_\vartheta)/r = 0$ and the limit condition $\sigma_\vartheta - \sigma_r = 2c$ as

$$\sigma_r = 2c \ln(r/r_0) + p$$
$$\sigma_\vartheta = \sigma_r + 2c$$

p is the support pressure. We furthermore obtain

$$r_e = r_0 \exp\left(\frac{\sigma_\infty - c - p}{2c}\right). \tag{1}$$

In the elastic regime the radial displacement reads

$$u = (c/2G) \cdot r_e^2/r. \tag{2}$$

Under the simplifying assumption of isochoric plastic deformation, equ. 2 prevails also in the plastified regime.

We may now assume that the above stress and deformation fields appear simultaneously with excavation and, subsequently, creep sets on, i.e. the deformation increases under constant stress. However, this assumption does not cover the second manifestation of squeezing, i.e. the continuous increase of pressure exerted upon a rigid support (e.g. a shield). Therefore, it appears more appropriate to assume that the stress deviator in the plastified regime (i.e. the cohesion c) decays with time. Experiments point to the fact that stress relaxation is proportional to the logarithm of time: $c = c_0 - \alpha \ln t$.

It should be added that experimental evidence on stress relaxation is very scarce. Thus, we do not know what occurs with cohesion after very long time lapses. Does it decay to zero? Or to some stationary value $c_\infty > 0$? Of course, $c = 0$ means that the material behaves as a fluid, in the long range. In this case all stresses will eventually become hydrostatic, and the full overburden pressure γh will act upon the lining of a tunnel. However, such large pressures have not been observed, and we may assume that a non-vanishing cohesion c_∞ persists with time. In most cases we have to admit that neither c_∞ nor the speed of decay of c are known (at least, not a priori). However, stress relaxation in the plastified rock explains both, the large convergence of unlined tunnels and the pressure increase upon rigid linings. To show this we assume that relaxation proceeds along with the function $c(t)$, i.e. $c = c(t)$, $c(0) = c_0$, $c(\infty) = c_\infty$, $\dot{c}(t) < 0$, $\dot{c}_\infty = 0$. Apart from these conditions, the function $c(t)$ is not further specified here.

Denoting the convergence $u(r = r_0)$ as w we obtain from (1) and (2):

$$w = \frac{c}{2G} \cdot r_0 \exp\left(\frac{\sigma_\infty - c - p}{2c}\right) \qquad (3)$$

Differentiating equ. (3) with respect to time t and noting that $\dot{r}_0 = -\dot{w}$ we obtain

$$\dot{w}\left(1 + \frac{c}{2G} \cdot \exp\cdots\right) = \frac{\dot{c}r_0}{2G} \cdot \exp\cdots + \frac{cr_0}{2G}$$
$$\cdot \frac{2c(-\dot{c} - \dot{p}) - 2(\sigma_\infty - c - p)\dot{c}}{4c^2} \cdot \exp\cdots$$

$w = $ const, i.e. $\dot{w} = 0$, (rigid lining) means that the expression in brackets vanishes. Hence,

$$\dot{p} - \frac{\dot{c}}{c}p = \frac{\dot{c}}{c}(2c - \sigma_\infty). \qquad (4)$$

Equ. (4) is a differential equation for p. Its solution reads:

$$p(t) = \frac{1}{M(t)}\left(\int g(t)\, M(t)\, dt + C\right)$$

with

$$f(t) := -\dot{c}/c, \quad g(t) := (\dot{c}/c)\,(2c - \sigma_\infty),$$
$$M := \exp\left(\int f(t)\, dt\right)$$
$$= \exp\left(-\int (\dot{c}/c)\, dt\right) = \exp(-\ln c) = 1/c$$

Thus we obtain

$$p(t) = c\left(\int \frac{\dot{c}}{c}(2c - \sigma_\infty) \cdot \frac{1}{c}\, dt + C\right)$$
$$= c(2\ln c + \sigma_\infty/c + C).$$

The integration constant C can be specified if we consider the initial condition:
At $t = 0$: $p = p_0$, $c = c_0$. Hence

$$p(t) = c\left(2 \ln\left(\frac{c}{c_0}\right) + \frac{\sigma_\infty}{c} + \frac{p_0 - \sigma_\infty}{c_0}\right).$$

The final pressure upon the lining will be:

$$p_\infty = c_\infty\left(2 \ln\left(\frac{c_\infty}{c_0}\right) + \frac{\sigma_\infty}{c_\infty} + \frac{p_0 - \sigma_\infty}{c_0}\right) \qquad (5)$$
$$= \sigma_\infty - c_\infty \ln\left(\frac{c_0}{c_\infty}\right)^2 - \frac{c_\infty}{c_0}(\sigma_\infty - p_0)$$

Clearly, $p_\infty \to \sigma_\infty$ for $c_\infty \to 0$. Note that this result, i.e. equ. (5), is independent of the speed of decay of c.

For unsupported tunnel ($p = 0$) it follows from (3):

$$w = \frac{c}{2G} r_0 \exp\left(\frac{\sigma_\infty - c}{2c}\right) \qquad (6)$$

$$\dot{w} = \frac{\dfrac{r_0 \dot{c}}{2G}\left[1 - \dfrac{\sigma_\infty}{c}\right] \cdot \exp\left(\dfrac{\sigma_\infty - c}{2c}\right)}{1 + \dfrac{c}{2G} \cdot \exp\left(\dfrac{\sigma_\infty - c}{2c}\right)}$$

For $c \ll 2G$ we may write

$$\dot{w} = \frac{r_0 \dot{c}}{2G}\left(1 - \frac{\sigma_\infty}{c}\right) \cdot \exp\left(\frac{\sigma_\infty - c}{2c}\right).$$

Since, $\dot{c} < 0$ high $\dot{w}$-values (i.e. rates of convergence) are obtained if σ_∞ considerably exceeds, c, i.e. if the oberburden load $\sigma_\infty \equiv \gamma h$ is considerably higher than the rock strength. This is why squeezing is observed in rock of low strength and deep tunnels. In fact, HOEK reports that minor squeezing problems are expected for $c/\sigma_\infty \approx 0.11,\ldots,0.18$ and extreme ones for $c/\sigma_\infty < 0.05$ (Barla 2002).

4 INTERACTION WITH SUPPORT

Let us now consider the interaction of the squeezing rock with the support, which is here assumed as a lining (shell) of shotcrete. We neglect curing and assume that shotcrete attains immediately its final stiffness and strength. Usually, the rock reaction line (i.e. the curve p vs. w) and the support reaction lines are plotted in the same diagramm. Their intersection determines the convergence and the pressure acting upon the support (see Fig. 2).

For squeezing rock, the rock reaction line is not unique. There are infinitely many rock reaction lines. Each of them corresponds to a particular-value (Fig. 3). In this case the intersection point of rock and support lines moves along the support reaction line and crosses the various rock reaction lines. The evolution in time of this process can be obtained

75

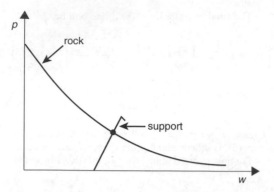

Figure 2. The intersection of the reaction lines determines the pressure p and the convergence w of the support (non-squeezing case).

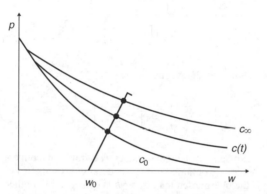

Figure 3. Fan of rock reaction lines for squeezing rock.

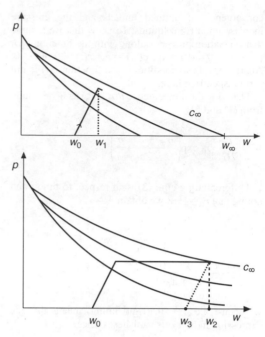

Figure 4. A brittle support collapses at $w = w_1$ (left) and the convergence continues until the ultimate value $w = w_3$. A ductile (or yielding) support remains active until the ultimate convergence $w = w_2$ is obtained.

if we introduce into equ. 3 the support reaction line $p = p_s(w)$, e.g. $p = k \cdot (w-w_0)$, $k = $ const, for $w > w_0$:

$$w = \frac{c}{2G} r_0 \exp\left(\frac{\sigma_\infty - c - p_s(w)}{2c}\right) \quad (7)$$

With, $c = c(t)$, equation (7) determines the relation $w(t)$.

No matter how the relation $w(t)$ looks like, after a sufficiently long time lapse the ultimate rock reaction line, i.e. the one corresponding to c_∞, will be reached. It may happen, however, that the support cannot support the corresponding load. This can be the case with a brittle support, whereas a ductile support will yield until an intersection with the ultimate rock reaction line is obtained (Fig. 4).

Examples of yielding support show Figs. 5, 6, 7, that refer to the Strengen tunnel (Austria).

A yielding ('elastic-ideal plastic') support can be achieved by interrupting the lining with arrays of steel tubes, which are designed to buckle under a

Figure 5. In the upper left part is visible the array of pipes.

specific load (Fig. 8). The position of the tube arrays within an idealized circular lining is shown in Fig. 8. The circumferential length of the lining is $L = 2\pi r$. The total length of the embedded steel tubes is ml_s ($m = $ number of tube arrays). With

E_s: Young's modulus of shotcrete
d: shotcrete thickness
n: number of steel tube arrays per unit length of tunnel

Figure 6. Anchor headplates and pipe array.

Figure 7. Squeezed anchor headplate. The two steel tubes had initially circular cross sections. Their squeezing indicates that this anchor has been overloaded.

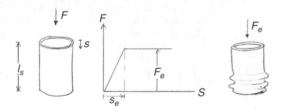

Figure 8. Force–displacement characteristic of embedded tube.

we obtain the following relation between the thrust N and the shortening ΔL of the lining

$$\Delta L = \left[(2\pi r - ml_s) \cdot \frac{1}{d \cdot E_s} + \frac{1}{n} \frac{s_e}{F_e} \right] N$$

Figure 9. Steel tubes embedded in shotcrete lining.

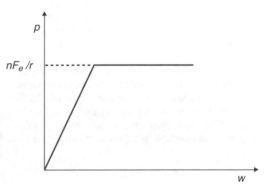

Figure 10. Reaction line of yielding support.

for $N/n < F_e$. With $w = \Delta L/(2\pi)$ and $p = N/r$ we obtain the support reaction line $p = p(w)$ (Fig. 10) as

$$p = w \frac{2\pi}{r[(2\pi r - ml_s)/(d \cdot E_s) + s_e/(n \cdot F_e)]}$$

for $p < nF_e/r$.

5 ROLE OF MEASUREMENTS

The question arises whether a yielding support should be installed, which is costly. The alternative would be to install a usual support at a *later* time, when the rock has converged by the amount w_3 (see dotted line in Fig. 4). Of course, one has to determine the value of w_3. This can be achieved with borehole tests, i.e. the borehole represents the tunnel and the convergence is measured within the borehole. Equation 6 gives a relation between w and c for an uncased borehole. Measuring w-values at various times yields (with numerical elimination of c from this equation) the corresponding c-values. Thus, the function $c(t)$ can be determined (for some discrete values of time t). With knowledge of $c(t)$, the complete rock reaction line and its interaction with the support can be determined as shown before.

6 ANISOTROPIC ROCK

As already mentioned, squeezing rocks have low strength and are characterized by layered silicates that predominate in schists. The orientation of schistosity (or foliation) imposes a mechanical anisotropy to such rocks. It appears reasonable to assume that stress relaxation affects only shear stresses acting upon planes of schistosity. As a consequence, tunnels that perpendicular cross the planes of schistosity (Fig. 11a) are not affected by squeezing, even at high depths. In contrast, tunnels whose axes have the same strike as the schistosity planes (Fig. 11b) can be considerably affected by squeezing.

Typical examples for the two cases are the Landeck and Strengen tunnels in western Austria. Both tunnels have been headed within the same type of phyllitic rock. The Landeck tunnel was oriented as in Fig. 11 a and encountered, therefore, no squeezing problems despite a considerable overburden of up to 1,300 m.

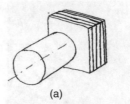 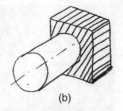

(a)   (b)

Figure 11. Two different orientations of schistosity relative to tunnel axis.

In contrast, the Strengen tunnel was oriented as in Fig. 11 b and had considerable problems with squeezing. Its overburden was up to 600 m.

REFERENCES

Barla, G. 2002. Tunnelling and squeezing rock conditions. In: Tunnelling Mechanics, Kolymbas, D. (ed.) Berlin: Logos.

Niemunis, A. 2003. Extended hypoplastic models for soils, Habilitation thesis, Gdansk.

Geotechnical Measurements and Modelling, Natau, Fecker & Pimentel (eds)
© 2003 Swets & Zeitlinger, Lisse, ISBN 90 5809 603 3

Mechanism of swelling tunnels driven through mudstone

R. Nakano

Formerly professor of Gifu University, Gifu, Japan

ABSTRACT: The mechanism of so-called swelling tunnels is analytically elucidated, taking advantage of the concept of Critical or Fully-softened State Soil Mechanics and Kastner's classical approach. Case histories are of a railway tunnel and drainage tunnels as countermeasures against landslides driven through mudstone of Shiiya and Teradomari Neogene formations in Niigata Prefecture, typical problem materials in Japan. It is shown that Kastner's approach can be successfully applied for calculating this kind of pressure on a circular tunnel driven through mudstone severely faulted or folded into clay, if shear strength parameters in terms of total and effective stress at fully-softened state are used for short term and long term problems respectively, as has so far been successfully applied in practice for stability analyses of clay slopes and that swelling test is of no use for design purpose.

1 INTRODUCTION

Very heavy so-called swelling pressure has often been encountered so far when tunneling through mudstone of Tertiary type landslide area in Japan.

Therefore it is quite reasonable to assume that there must be something in common with the fundamental mechanisms of landslides and swelling tunnels. However, the important new findings obtained for the past several decades by the researchers of soil mechanics, especially those in the field of landslides, have not been fully taken into consideration in elucidating the mechanism of swelling tunnels and there are still many tunnel engineers who believe in Terzaghi's statement (Terzaghi, 1969) "The term swelling rock refers to rocks the squeeze of which is chiefly due to swelling ....The pressure on the support in tunnels through swelling rock depends primarily on the swelling capacity of the rock which is analogous to the swelling capacity of clays. If the volume of samples of the freshly exposed rock or of samples from a freshly recovered core does not increase by at least 2 percent during immersion in water, the rock can not be classified as swelling rock." Thus, there are still not a few practicing tunnel engineers who observe the specification on "Swelling test of clay for tunnels" in the tunnel-design manual published from the Japan Society of Civil Engineers (JSCE 1964). The purpose of this paper is to elucidate the real cause of swelling tunnel by presenting experimental data, theoretical analyses and case histories (Nakano 1974, 1996).

2 LOCATIONS OF TUNNELS AND SAMPLING

2.1 *Tertiary type landslides and so-called swelling tunnels*

Among the typical three types of landslides in Japan (Nakano 2002), the number of Tertiary type is predominant, prevailing in tuffaceous mudstone in Neogene formations in Green-tuff region which extends from south-west Hokkaido down to Japan-Sea side of south-west Japan as shown by screen-toned area in Figure 1. Figure 2 shows the intensely folded structure of Niigata sedimentary basin (area ① in Figure 1) in Northern Fossa Magna (a great ruptured zone) situated between two tectonic lines, i.e. Itoigawa and Shibata-Koide lines. The area ⑪ shown in Figure 2 relevant to the sites referred to in this paper is located within this basin which develops in NE-SW direction, having many folding axes running in the same direction. This basin is composed of Neogene mudstone where Tertiary type landslides are most densely distributed due to the softening of mudstone caused by complex folding, faulting and by artesian pressure with methane gas pressure.

Figure 3 shows a geological map of the area ⑪ in Figure 2 which is predominantly composed of thick deposit of Shiiya and (continued to page 3) Teradomari formations which are known to be the most notorious problem materials in Japan.

In this Figure, ①,②,③ are the locations of recent well-recorded typical huge landslides in this area. Nabetachiyama railway tunnel (upper right of this

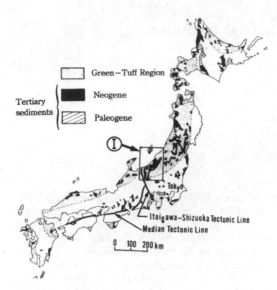

Figure 1. Distribution of Tertiary sedimentary rocks and sampling locations.

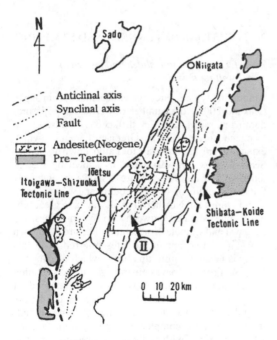

Figure 2. Enlarged map of the area ① in Figure 1.

Figure) and the drainage tunnels driven for counter-measures against Itakura landslides (Landslide ③, lower left of this Figure) suffered a very heavy so-called swelling pressure. Iiyama tunnel now under construction for New Hokuriku Trunk Line (near Itakura landslide, lower left of this Figure) is also

reported to be suffering from a very heavy pressure. Test samples of mudstone and clay were taken from the working face of Nabetachiyama tunnel and the drainage tunnels of Itakura landslide.

2.2 Physical and Mechanical properties of samples

The typical physical properties of Tertiary type land-slide clay of Shiiya and Teradomari formations are $w_L = 50 \sim 150\%$, $w_p = 30\%$, $CF = 70\%$. The unit weight γ_t of undisturbed sample is $20\,\text{kN/m}^3$. The liquid limit is very large because of the abundant clay minerals of smectite group (montmorilonite). To measure the mechanical properties of this kind of clay in a Critical or Fully-softened state (Atkinson & Bransby 1978, Skempton 1970), triaxial consolidated drained (CD) tests were carried out on reconstituted samples normally consolidated from slurry prepared by mixing water with the powder of wet mudstone or fault clay. One dimensional preliminary consolidation from the slurry was carried out in a cylindrical oedometer with top and bottom drainage only, using a pressure of $100 \sim 200\,\text{kPa}$. Cylindrical soil samples with a diameter of 35 mm and a height of 70 mm were then cut out from this normally consolidated clay and consolidated further in a triaxial cell under pressures higher than the preliminary consolidation pressure. Triaxial CD tests were then performed on these samples using top and bottom drainage only, with a slow axial strain rate ($\varepsilon_a = 1 \times 10^{-3}\%/\text{min}$) to minimize the non-uniform distribution of water content within the samples (Nakano 1996). Figure 4 shows the relationship between the triaxial CD compressive strength $q = (\sigma_1 - \sigma_3)_{max}$ and the water content w_s at fully-softened state. The compressive strength q and Young's Moduli (E_t and E_{50}) obtained from CU test for the reconstituted samples without radial drainage and those obtained from UU test for the undisturbed specimens sampled from the face in the faulted zone of Nabetachiyama tunnel are also plotted in this Figure, indicating that these strengths fall in the same shaded solid line i.e. fully softened or critical state line. Triaxial compressive CD strength q of reconstituted samples with radial drainage, incorrect data due to the non-uniform water content distribution within the sample, are also plotted with ▲ in this Figure for comparison. The relationship between w_s and the average effective confining pressure $p' = (\sigma_1' + 2\sigma_3')_{max}/3 = (q/3) + \sigma_3'$ at fully-softened state is shown in Figure 5. CD direct shear tests were also carried out for the normally consolidated samples of this clay and its failure line is shown in Figure 6 together with the failure envelope and Mohr's circles obtained from CD triaxial tests. Direct shear tests give a little higher shear strength but there is no appreciable difference between the two failure lines. It is worthy of note that both lines are a little curved

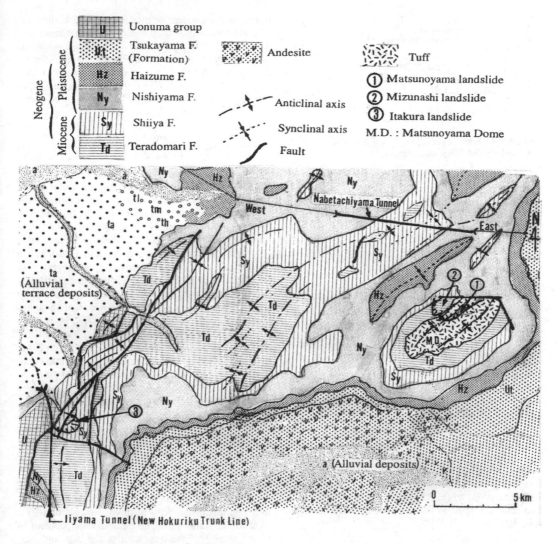

Figure 3.   Geological map of the area ⑩ in Figure 2.

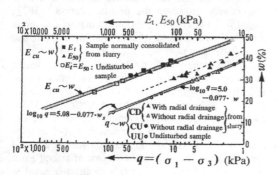

Figure 4.   Relationship between compressive strength q, Young's modulus E_{cu} and water content w.

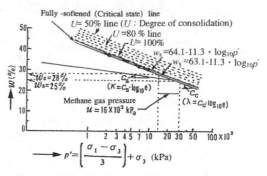

Figure 5.   Relationship between average effective confining stress p' and water content.

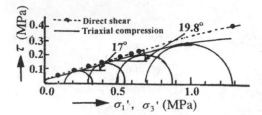

Figure 6. Failure lines at fully softened state obtained from direct and triaxial compression tests.

i.e. non-linear. The relationship between q and w_s and that of w_s and p' can be expressed from Figures 4 and 5 as follows,

$$\log_{10} q = 5.04 \pm 0.04 - 0.077 \cdot w_s \tag{1}$$

$$w_s = 63.60 \pm 0.50 - 11.3 \cdot \log_{10} p' \tag{2}$$

3 FORMULAE FOR ANALYSES

3.1 *Strength parameters as a function of p'*

Substituting the equation (2) into (1) and eliminating w_s, the following equation is obtained.

$$q = 1.39 \cdot p'^{0.87} \tag{3}$$

This indicates that the relationship between q and p' is non-linear which is consistent with the fact that the Mohr-Coulomb's failure lines are curved as mentioned above. This also means that the fully-softened strength parameters c'_s (kPa) and ϕ'_s can be expressed as a function of p' as follows (For details of derivation, ref. Nakano 2002),

$$c_s = \frac{0.54 \cdot p'}{2\sqrt{\left(3 \cdot p'^{0.13} - 1.21\right)\left(2.42 + 3 \cdot p'^{0.13}\right)}} \tag{4}$$

$$\phi_s = \sin^{-1} \frac{3.63}{6 \cdot p'^{0.13} + 1.21} \tag{5}$$

3.2 *Formulae for analyses of a circular tunnel*

Various analytical formulae for stress and strain analysis around a circular tunnel have been published so far, assuming a plastic behavior of soil, according to Mohr-Coulomb's failure criterion. To carry out a simple analytical approach without losing physical insight into the problem, Kastner's classical approach (Kastner 1962) was adopted with some modifications

based on the concept of modern soil mechanics, i.e. $\phi = \phi_u = 0$ and $c = c_u = q_u/2$ for short-term and $\phi = \phi_s \neq 0$, $c = c_s$ for long-term tunnel problems respectively . Kastner derived the following differential equation for the plastic (broken or sheared) zone based on Mohr-Coulomb's failure criterion and equilibrium of forces, assuming the coefficient of lateral earth pressure as $K = 1$,

$$r \cdot \frac{d\sigma_r}{dr} + \sigma_r \left(1 - \frac{1 + \sin \phi}{1 - \sin \phi}\right) - \frac{2c \cdot \cos \phi}{1 - \sin \phi} = 0 \tag{6}$$

In solving this equation, strength parameters mentioned above have been introduced, differentiating the problems into two, short-term and long-term. The final results are summarized in Table 1. For the meaning of notations, refer to Notes underneath this Table (For details of derivation, ref. Nakano 1996). When unconfined compressive strength q_u is known either from sample or in-situ test, the competence factor for short-term $F_c = q_u/(\gamma_t \cdot H)$ can be obtained and when strength parameters c_s, ϕ_s are known, that for long-term $F_c = q_s/(\gamma_t \cdot H)$ can be obtained since q_s is given by the formula given at the bottom of Notes. Thus, R_a/R_1 is obtained as a function of β as is given either by equation (9) or (14). Substituting these into equation (10) or (15) and further putting these into equation (17), the relationship between $\beta = \sigma_i/\sigma_0$ and convergence U_a (i.e. characteristic line) is obtained.

4 CASE HISTORIES AND DISCUSSIONS

4.1 *Nabetachiyama tunnel*

Nabetachiyama tunnel, a 9,117 m long railway tunnel, was constructed by JRCPC (Japan Railway Construction Public Corporation) in three work sections, i.e. west, middle and east. The natural water content of the mudstone of the mountain in the east and west sections were much less than w_s (i.e. competent) and tunneling was normally finished in 1978 and 1982 respectively. However the middle section, 3387 m long (St. 31 km324 m ~ 34 km711 m), was severely disturbed, suffered a very heavy swelling pressure and the construction work was discontinued in 1982, leaving the especially difficult ground of 645 m length (between 32 km404 m ~ 33 km049 m) unexcavated (For details of geological profile, ref. Nakano 1996). Just west side of this abandoned section (at 33 km 142 m), the natural water content w_n was 25%. The pressure σ_i exerted on the support just after excavation was 4×10^2 kPa, distributed hydrostatically around the circular drift, i.e. the coefficient of lateral earth pressure $K = 1$. Thus, p' at this location is equal to the pressure of the overburden $H = 150$ m i.e. $\sigma_v = \gamma_t \cdot H = 30 \times 10^2$ kPa. As shown in Figure 5, w_s

Table 1. Formulae for stresses, radius of broken zone and convergence of a squeezing tunnel.

Short-term after tunnel excavation (in terms of total stress)	Long-term after tunnel excavation (in terms of effective stress)

$$\sigma_r = \sigma_i + q \cdot \ln\left(\frac{r}{R_1}\right) \qquad (7)$$

$$\sigma_r' = \frac{q_u}{\zeta - 1} \cdot \left\{\left(\frac{r}{R_1}\right)^{\zeta-1} - 1\right\} + \sigma_i \cdot \left(\frac{r}{R_1}\right)^{\zeta-1} \qquad (12)$$

$$\sigma_\theta = \sigma_r + q \qquad (8)$$

$$\sigma_\theta' = \zeta \cdot \sigma_r' + q_u \qquad (13)$$

$$\frac{R_a}{R_1} = 0.61 \cdot \exp\left(\frac{1-\beta}{F_c}\right) \qquad (9)$$

$$\frac{R_a}{R_1} = \left\{\frac{2}{1+\zeta} \cdot \frac{\zeta - 1 + F_c}{\beta \cdot (\zeta - 1) + F_c}\right\}^{1/(\zeta-1)} \qquad (14)$$

$$\frac{U^* R_a}{R_1} = \gamma_t \cdot H \cdot \left(1 - \beta - F_c \cdot \ln\frac{R_a}{R_1}\right) \times \frac{1+v}{E_{cu}} \cdot \left(\frac{R_a}{R_1}\right) \qquad (10)$$

$$\frac{U^* R_a}{R_1} = \gamma_t \cdot H \cdot \left[1 - \frac{F_c}{\zeta - 1} \cdot \left\{\left(\frac{R_a}{R_1}\right)^{\zeta-1} - 1\right\} - \beta \cdot \left(\frac{R_a}{R_1}\right)^{\zeta-1}\right] \cdot \frac{1+v}{E_{cu}} \cdot \left(\frac{R_a}{R_1}\right)^{\zeta-1} \qquad (15)$$

$$p = \frac{(\sigma_r + \sigma_\theta)}{2} \qquad (11)$$

$$p' = \frac{(\sigma_r' + \sigma_\theta')}{2} \qquad (16)$$

$$\frac{U_a}{R_1} = 1 - \sqrt{1 - \left(\frac{R_a}{R_1}\right)^2 + \left(\frac{R_a}{R_1} - \frac{U^* R_a}{R_1}\right)^2} \qquad (17)$$

Notes: $\zeta = (1 + \sin\phi)/(1 - \sin\phi) \cdot \cdot \phi = 0$ for short-term and $\phi = \phi_s$ for long-term, E_{CU}: Young's modulus from consolidated CU test, σ_r: Radial stress, σ_θ: Tangential stress in polar coordinate. U_{Ra}^*: Elastic radial deformation at the elastic–plastic boundary, $\beta = \sigma_i/\sigma_0$: Ratio of pressure σ_i on tunnel support to initial overburden stress $\sigma_0 = \gamma_t \cdot H$, H: Overburden depth, $F_c = q_u/\sigma_0$: Competence factor for short-term, $F_c = q_s/\sigma_0$: Competence factor for long-term, here, $q_s = 2c_s \cos\phi_s/(1 - \sin\phi_s)$.

corresponding to this pressure is 25% i.e. this section was tectonically disturbed to fully-softened state. The convergence $2U_a$ measured shortly after the installation was about $45 \sim 50$ cm, i.e. the absolute convergence $2u_a$ is about 95 cm. The compressive strength q_u corresponding to $w_s = 25\%$ is 13×10^2 kPa as obtained either from equation (1) or Figure 4, hence the competence factor F_c for short-term is 0.43. Thus the ratio of mountain pressure to the overburden $\beta = \sigma_i/(\gamma_t \cdot H)$ is 0.13. Referring to the bottom solid characteristic line, arrowed $F_c = 0.43$ in Figure 7, prepared by the procedures explained above, it is clear that these points fall just on this characteristic line. For the long-term problem, fully-softened strength parameters $c_s = 60$ kPa, $\phi_s = 12°$ corresponding to the average effective confining stress p_{av}' in the plastic zone, are obtained either from the equations (4), (5) or from the straight line tangent to the curved line of $q < p'$ at $(p')_{av} = 2000$ kPa as shown in Figure 8. Thus, $q_s = 2 \cdot 60 \cdot \cos 12°/(1 - \sin 12°) = 150$ kPa and $F_c = 0.05$ are obtained, corresponding to the dashed line with an arrow $F_c = 0.05$, suggesting that the pressure increases in long-term from the point S to the point L_2 where $\beta = 0.28$, $\sigma_i = 840$ kPa. If we adopt the strength parameters $c_s = 0$, $\phi_s = 17°$ obtained at low stress range of p' neglecting the non-linearity of $q \sim p'$ relationship, we over-estimate q as shown in Figure 8 and hence under-estimate the pressure as $\sigma_i = 690$ kPa since $\beta = 0.23$ at the point L_1 on the dashed line arrowed $F_c = 0$. This is an unsafe assumption in tunneling. The construction of the once-abandoned section of 645 m was resumed in August 1985 and

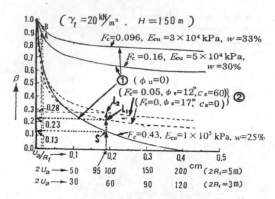

Figure 7. Relationship between mountain pressure ratio β and convergence $2U_a$ (Characteristic line) for Nabetachiyama tunnel, Note: ① for short-term, ② for long-term.

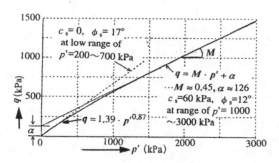

Figure 8. Relationship between q' and p'.

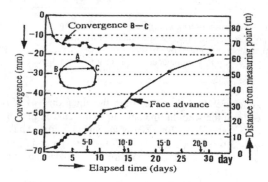

Figure 9. Convergence and elapsed time (days).

a central circular drift with a diameter $2R_1 = 3$ m was driven through in October 1992, overcoming an extraordinarily heavy so-called swelling pressure. During the construction, at the section between 32 km 865 and 32 km 874 m, methane gas with a pressure of 16×10^2 kPa was encountered. The water content w_n

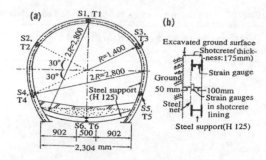

Figure 10. Layout (a) and locations of gauges in plan (b).

here was 28% which is equal to w_s corresponding to $p' = (30 - 16) \times 10^2 = 14 \times 10^2$ kPa as is clear from Figure 5, meaning that the gas pressure u was acting as a pore-pressure. At the location 32 km 770 ~ 32 km 780 m, w_n was 30 ~ 33%, due probably to even higher gas pressure, and the pressure on the support (rigid enough to allow the absolute convergence only 3 ~ 5 cm) reached as high as $24 \times 10^2 \sim 30 \times 10^2$ kPa i.e. 80 ~ 100% of the total overburden pressure. This can also be explained by the points A,B on the upper two solid characteristic lines in Figure 7.

4.2 Drainage tunnel No. 3 in Itakura landslide

Three drainage tunnels were driven as countermeasures against slowly moving huge Itakura landslide, approximately 1 km wide, 1 km long and 150 m deep, occurring in the same Neogene mudstone. All three tunnels suffered a very heavy swelling pressure. At 681 m from the portal of the drainage tunnel No. 3, in-situ measurement of convergence (Figure 9), that of stresses induced in steel support (H $-125 \times 125 \times 6.5 \times 9$ mm, 0.6 m pitch) and in steel-fiber-shotcrete lining (0.175 m thick) were carried out using strain gauges. Figure 10 shows the layout of the strain gauges in the steel support and lining. Figures 11 and 12 show the relationship between the axial forces N_{st}, N_c (supported by steel support and shotcrete respectively, calculated by the formulae shown above Figures 11,12) and time. Judging from the distribution of combined axial forces $N_{comp} = N_{st} + N_c$ at six points, the pressure was approximately hydrostatic and its average value was 1.2 and 1.4 MN after 5 and 140 days respectively. Assuming that the tunnel is approximately circular, the swelling pressure was $\sigma_i = N_{comp}/0.6R_1 = 1.2 \sim 1.4$ MN/m² ($R_1 = 1.65$ m). As the overburden is 82 m, $\beta = \sigma_i/(\gamma_t H) = 0.73 \sim 0.85$. As shown by Figure 9, the convergence was 18 mm (absolute $2u_a$ is 36 mm), the convergence ratio U_a/R_1 is 0.02. F_c obtained from in-situ soft-rock cone-penetrometer test was about 0.4. These points fall on the bottom solid characteristic line in Figure 7,

84

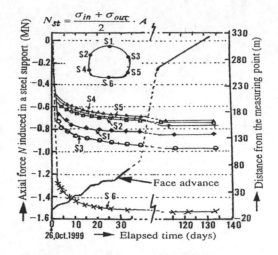

$$N_{st} = \frac{\sigma_{in} + \sigma_{out}}{2} \cdot A$$

Figure 11. Axial force N_{st} in steel support and elapsed time (σ_{out}, σ_{in}: stress at ground side and inside respectively).

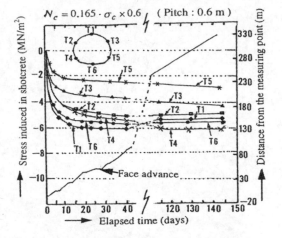

$N_c = 0.165 \cdot \sigma_c \times 0.6$ (Pitch : 0.6 m)

Figure 12. Axial force N_c in shotcrete lining and elapsed time.

re-confirming the validity of this approach. It is to be emphasized that no data on swelling test have been used at all so far. Although more rigorous approach should be made taking into account the progressive failure mechanisms, the present author's approach may be taken as a compromise between theory anpractice at the present state of knowledge.

5 CONCLUSION

The so-called swelling pressure is essentially a squeezing pressure and can be explained only using q_u and fully-softened strength parameters c_s, ϕ_s. Swelling test is of no use for design purpose.

REFERENCES

Atkinson J.H. & Bransby P.L. (1978): *The Mechanics of Soils – An Introduction to Critical State Soil Mechanics*, McGraw Hill.

Kastner H. (1962): Statik des Tunnel und Stollenbaues, Springer Verlag, Berlin.

Nakano R. (1974): On the design of water tunnels, *Bull. Natl. Res.Inst. Agr.Eng.Minisry of Agriculture & Fisheries Japan,* No.112, 89–142.

Nakano R. (1996): Cause of swelling phenomena in tunneling and a proposal for a design procedure for a swelling tunnel, *Soils and Foundations* 36(4), 101–112.

Nakano R. (2002): On the validity of Japanese conventional method of taking cohesion as a function of depth in back analyses of landslides, *Proc. 1st European. Conf.*, Prague, Czech Republic, 661–666.

Skempton A.W. (1970): First-time slides in overconsolidated clays, *Geotechnique* 20(3), 320–324.

Tezaghi K. (1964): Rock tunneling with steel supports, Commercial Shearing and Stamping Co. Youngstown, Ohio.

Geotechnical Measurements and Modelling, Natau, Fecker & Pimentel (eds)
© 2003 Swets & Zeitlinger, Lisse, ISBN 90 5809 603 3

Approach of field tests and adjacent site investigations to improve the geotechnical site model and design of dams and underground structures

V. Schenk
Consulting Engineering Geologist

ABSTRACT: Examples of three case histories dealing with geotechnical site investigations specificly tailored for the relevant site geology and proper design and construction of large dams and underground structures are reported on.

Soil and rock mechanics field tests and adjacent geological/geotechnical investigations are appropriate tools to yield realistic parameters of strength, deformation and stress properties of a rock mass. In view of Consulting Engineering, selected projects from Germany and abroad, realized in the meantime or presently under construction, were dealt with.

1   INTRODUCTION

Beginning in the sixtieth and especially in the seventieth of the last century large scale field tests and supplementary measurements were commonly applied to obtain realistic geomechanical parameters for design and construction of large dams and underground structures, mostly based on the ideas of Leopold Müller and Manuel Rocha.

An early example is the introduction of the so called TIWAG radial-press, which was installed in pressure shafts to determine the deformation behavior of the rock mass/support (Lauffer, 1960). Presumably Leopold Müller was the first who consequently performed a large in-situ field test programme for the Kurobe arch dam in Japon (Müller, 1963). It was then largely believed that the scale effect to be fitted, describing the geomechanical properties of rock mass realistically, best can be determined by appropriate field tests, carried out at representative test locations, preferably to be selected both by the geologist and the engineer. Moreover the geotechnical site model due to the results of the site investigations and geomechanics lab and field tests could be refined accordingly. Then bearing in mind the philosophy, that the foundation rock or rock mass of large underground works is to be integrated as essential construction elements into the project, these tests including structural geological investigations and other supporting geotechnical measurements, hence, shall deliver realistic rock mass parameters to optimize both, design technically and economically.

In this context a review of the author's experience (then with Lahmeyer International Consulting Engineers, Germany), conducting complex geotechnical site investigations for large dams and underground plants, is presented including some remarks on construction.

In the seventieth as a young site geologist already the author has had the opportunity to be acquainted with an extensive rock mechanics site investigation program, which was carried out for an arch dam of a drinking water project in the Rheinische Schiefergebirge (Slate Mountains in the Rhine region) in Germany (Schenk & Köngeter, 1982). According to the distinct strength and deformation anisotropy of the sensitive foundation rock, i.e. particular emphasis on the interaction of dam and rock mass was laid and called for an extraordinary programme of in-situ and adjacent tests, mainly as follows.

Many large scale flat-jack, field shear tests and numerous measurements of primary state of stress (triaxial cell, Doorstopper and mini flat-jack tests), also with respect to the intense micro seismicity in the vicinity of the seismic active Rhine Graben structure, were carried out. These tests were performed in four exploration adits driven into the abutments, which were mapped engineering geologically in detail for proper selection of test locations. During the field campaign the author has had the opportunity to meet Manuel Rocha, one of the developers of such field tests and a renown expert of rock mechanics at that times. Unfortunately the project, which was investigated for more than 10 years was cancelled because of shrinkage of drinking water demand in the region. In this period which was a fruitful and exciting climate for the development and application of large scale tests the author met Otfried Natau 1979 with a big tunnel project of the New German High Speed Railway, where he suggested the multi-step triaxial test technology of his Karlsruhe based institute, performed on large scale drilled cores (diam. 0.6 m). These tests proved as a very appropriate tool in particular of soft rock and hard soil-testing. It should be mentioned here that there is a comprehensive overview of geotechnical measurement devices and field tests in rock (Fecker, 1997).

In view of Consultant Engineering the following projects from Germany and abroad realized in the meantime or presently under construction shall be reported on.

2 PUMP STORAGE PLANT HERDECKE, GERMANY

The 140 MW Koepchenwerk erected in 1930 near Dortmund, was the first major plant of this type. Because of some operational problems, the owner RWE decided to replace the four old pump sets by a single 150 MW unit to be housed in a 46 m – deep and 19 m wide power shaft (Schenk et al., 1986). The upper reservoir and existing open – air powerhouse, connected by a four – lane penstock, maintained in operation until the end of construction in 1988. The new constructed 400 m long pressure tunnel and the shaft, partly extended to the lower reservoir, are protected by special cut-off measures from the upper and lower reservoir respectively. The plant is predominantly located in moderately to slightly jointed massive siltstone of medium strength, with minor sandstone intercalations of Upper Carboniferous.

2.1 Geotechnical investigation programme

To investigate the rock mass conditions, i.e. to establish a realistic geotechnical site model and to determine the rock mechanical parameters for the shaft, tunnel and slope design the programme mainly comprised:

- Some 800 m core drillings and Lugeon tests in selected locations of structures;
- Sinking of an approx. 30 m deep exploration shaft in the powerhouse shaft area to map the rock mass conditions, i.e. joint systems, faults and to perform field tests and test groutings as well;
- 20 dilatometer tests with the LNEC probe;
- 2 triaxial lab tests on large scale drilled core samples (diam. 0.6 m, length 1.0 m);
- Detailed structural geological mapping of the access road slopes at site for joint plane statistics (FE calculations), Fig. 1.

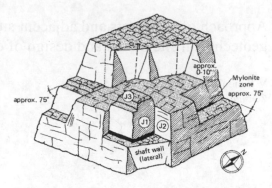

Figure 1.   Koepchenwerk, geomechanical site model.

The existing rock slopes up to 30 m height, enabled a realistic assessment of the geometry/distribution of planes of separation and stability conditions of the open-pit, with a total height of some 75 m above powerhouse shaft floor without further investigations.

2.2 Rock mechanics tests and deformability and shear strength characteristics

The rock mass deformability was determined mainly by dilatometer tests and large scale lab triaxial-compression tests by the multi-stage technique (Natau et al., 1983). Tests were evaluated using the total deformations in the elastic stress range. Moreover, the first loading and unloading stress – strain curves were used to obtain the lower boundary results. It was found:
– increase of moduli with increasing depth, proving the decreasing influence of joint planes; and, – moderate to slight anisotropy of deformability perpendicular to bedding and consequently small influence of joints in sandstone intercalations.

Test sample 1 for large scale triaxial test was taken from a weathered surface zone and sample 2 is representative of a moderately jointed siltstone at a depth of 20 m, both for loading perpendicular to bedding. The resulting moduli were 1500 and 2500 MPa, respectively, for the medium stress range related to failure stress. In the higher stress range of dilatometer tests (>5 MPa) some creep deformation was observed, but could be disregarded because of the moderate effective primary and secondary stresses.

Concerning the shear strength of the rock mass in weathered and moderately jointed rock mass by the large scale shear tests were determined: $\varphi' = 43.7°$ and $\varphi' = 45°$ for the angle of friction and $c' = 1.45$ MPa and $c' = 4.0$ MPa as peak values. The residual shear strength for test No. 2 can be estimated at $\varphi' = 40°$ and $c' = 2.5$ MPa and is appropriate for a homogeneous model of the rock mass. For a model consideration the actual joint shear strength, lower values were determined by small scale tests of which results were

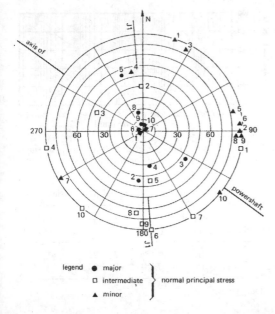

legend ● major ⎫
        □ intermediate ⎬ normal principal stress
        ▲ minor ⎭

Figure 2. Results from overcoring in-situ stress measurements.

considered in some special sliding stability analyses only (slopes of 30–75 m height).

2.3 Primary state of stress

In a vertical borehole, sunk from the floor of the trial shaft, 10 overcoring tests were carried out. The majority of the tests resulted in nearly vertical major and N–S and E–W directed horizontal intermediate and minor principal stresses of which values are shown in Fig. 2.

2.4 Rock mechanics analysis parameters

The results of the investigations discussed were assumed to be representative for the project area because of the small scattering of results and the other sources of detailed information on the rock mass behavior. The rock mechanics parameters listed in Table 1 are derived for stability analyses by the finite element method.

An extensive rock mechanics monitoring programme executed while construction, consisting of extensometer, inclinometer and anchor load measurements as well, and a horizontal and vertical precise geodetic survey. The redundant, widely distributed measurements largely confirmed the rock mechanics consumptions made and helped in the selection of support measures which were appropriate to actual load needs.

Table 1. Rock mechanical parameters for FE analyses.

(a) Rock mass

Mat.	E_1 (MPa)	E_2 (MPa)	ν_1 (–)	ν_2 (–)	G_2 (MPa)
(1)*	250	250	0.40	0.40	–
(2)*	3000	2000	0.25	0.17	1000
(3)*	5000	3300	0.25	0.17	1600

Mat.	c (MPa)**	φ (°)**	σ_t (MPa)**
(1)*	0	32.5	0
(2)*	1.0	38	0.5
(3)*	3.5	42	1.5

(b) Joint planes (large scale strength)

Mat.	J1/J2 c (MPa)	φ (°)	σ_t (MPa)	J3 c (MPa)	φ (°)	σ_t (MPa)
(1)*	–	–	–	–	–	–
(2)*	0	30–35	0	0	10***	0
					–20	
(3)*	0.05–0.1	35–40	0	0	20–25	0

* (1) Weathering zone (2) jointed rock (3) massive rock;
** Intact rock strength; *** Mylonite zone.

3 HEILSBERG MOTORWAY TUNNEL, GERMANY

In 1988 the 450 m long double tube motorway tunnel was comissioned for the A 81 Singen (SW Germany) to the swiss border (Schenk, 1993). The tunnel (each tube diam. 10.0 m, max. cover approx. 25 m) lies in sandy to silty, gravelly and glacially preloaded Würm moraine with some boulders, was the first tunnel of that size and soil of that type in Germany.

Due to its high content of coarse pebbles embedded in a matrix of varying cohesion it was not possible to obtain drilled undisturbed core samples for lab tests. Thus it was decided to sink a 28 m deep trial shaft (diam. 3.0 m), between the two tubes to gain realistic soil mechanics design parameters, i.e. to clarify the overall soil conditions and refining the site model.

The results of the geological mapping of the shaft is shown in Fig. 3. It can be seen that the tunnel lies in coarse gravelly marl with some boulders (diam. max. 0.8 m) well above the bedrock – clay marls-, Tertiary. Local sand lenses of poor cohesion and density were encountered. The global permeability of the series was determined very low (k =10^{-9} m/s), i.e. technically watertight.

To obtain realistic strength and deformation parameters at both the roof and floor level of the designed tunnel, cores (length 1.0 m, diam. 0.6 m) were extracted to

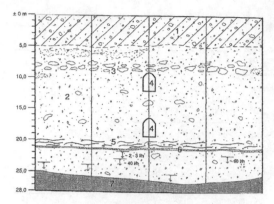

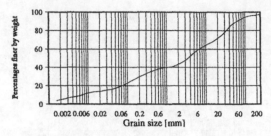

Figure 5. Typical grain size distribution, silty to sandy marl, pebbles, moraine, shaft depth 10 m.

Figure 3. Heilsberg motorway tunnel, geological mapping of trial shaft. 1: weathering zone, 2: ground moraine, gravelly, marly silt, boulders(diam. max. 0.8 m) not to scale, 3: layer of boulders, 4: adits at tunnel roof and floor level, 5: bands of Fe- oxide 6: marl layer, 7: top of Molasse (Tertiary), hard clayey marl.

Table 2. Results of large scale triaxial compression tests.

Sample no.	Depth (m)	c (MPa)	φ (°)	γ (kn/m³)	CaCO₃ (%)
1	10	0.186	42	23.7	*
2	19	0.181	40	23.3	21

* Only fines.

3.1 Soil mechanics test results

The strength and deformation behavior of the ground moraine was determined by the improved multi-step technique as triaxial compression test.

The grain size distribution (Fig. 5) shows a curve ranging from silt to the size of pebbles (<20 cm), which was considered to be representative both for the samples (shaft results) and the overall tunnelling soil. Pebbles diam. >10 cm were disregarded. The triaxial tests were run using 3 different loading steps. Due to the stresses of cover, loading steps of 500, 700, 900 kN/m² were selected. The test results are listed in Table 2. Obviously the results show the same magnitude.

The high angles of friction and surprisingly high cohesion are judged to be the result of the glacially compacted moraine, loaded by an ice shield a few 100 m thick, strengthened by its high lime content and favourable grain size distribution, as well.

However, according to the drilling results along the tunnel alignment the silt content varies so that the shear strength in that combination could not be representative for the whole tunnel section. More-over, it has to be mentioned that the parameters are also influenced by changing lime contents.

Consequently, for the stability analysis the following conservative parameters were selected:

cal φ' = 35–38°   cal c' = 70–100 kN/m²
cal γ = 24 kN/m³   cal E def. = 150–200 MPa

Convergency measurements during shaft sinking revealed negligible deformations and creeping of soil could also disregarded as confirmed by the triaxial test results. Finally it can be stated that the shotcrete

Figure 4. Extraction of core samples (diam. 0.6 m) in gravelly moraine for large scale triaxial compression test.

be tested on the large scale triaxial machine of the Soil Mechanics Institute of the Technical University of Karlsruhe.

The extraction of large scale drilled cores was not possible because of the high content of pebbles of the soil and limited space in the shaft. Therefore for the first time in such a coarse soil a steel tube, loaded by 50 kg was continually driven into the soil to permit manual excavation around the tube (Fig. 4).

When a core height of 1 m was reached the space between the core and tube was filled by gypsum to achieve proper contact. After break off from the ground, top and bottom plane of the sample were also sealed for safe transportation to the lab.

method was applied successfully during construction by crown and bench heading without major tunnelling difficulties.

4 HYDROELECTRIC POWER PLANT BAKUN, SARAWAK, EAST MALAYSIA

4.1 Introduction

Since 1981 the 2400 MW HEPP Bakun is under investigation and presently under construction (Schenk & Lee, 1987), (Failer & Abong, 1995). It is the most challenging geotechnical task, of which the author and his team were engaged with the Int. SAMA Consortium under the leadership of Lahmeyer International. Until 1987 an extensive site investigation programme was carried out for feasibility and tender design stage. Client during this design phase was the German Agency of Technical Cooperation (GTZ). In a remote area of central Sarawak (West Borneo), accessible by boat or helicopter only in a gorge of the Balui River up to 700 m deep, about 37 km upstream of Belaga Town, a 210 m high arch dam was designed originally. By later design modifications it was changed to an concrete face

rockfill dam. The main components of the project are (Fig. 6):

- Concrete face rockfill dam, approx. $28 \, mio \, m^3$, mainly fresh sandstone, zones with varying mudstone contents, 30 and up to 50% respectively;
- Power conduit system of 8 pressure tunnels;
- Open air powerhouse, housing eight 300 MW units;
- 3 diversion tunnels, length about 1.6 km each; excavation diam. 14.0 m;
- Gated spillway, chute length 750 m;
- Cut slopes at forebay > 100 m height;
- Submarine power transmission cable link to peninsular Malaysia, some 700 km length.

Because of the decreasing Malaysian economy in the second half of the eighties the realisation of the project was postponed. However after extraordinary economic growth in the nineties the project was transformed into privatisation, thus Ekran Berhad, a construction and timber company was requested to proceed with the development of the Bakun project (Failer & Abong, 1995). Construction commenced at the end of 1995 with the excavation of the river diversion tunnels (Fig. 7), while the river was diverted successfully

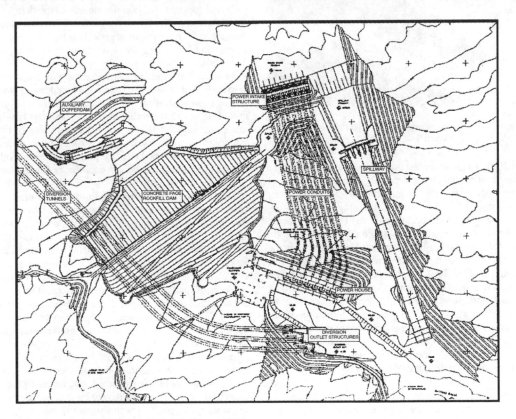

Figure 6.   Layout of HEPP Bakun (Failer & Abong, 1995).

Figure 7. Construction of diversion tunnels, diam. 14.0 m, upstream inlets, situation 1997.

Figure 8. Overstreaming of auxiliary cofferdam, May 2002, Foto LI "Aktuell", Oct. 2002.

in 2001. To minimize the owner's risk regarding design, construction and cost and to safeguard the financing of this huge plant, the entire project (including power transmission) is to be constructed under a single contract on a fixed price, turnkey basis (Failer & Abong, 1995). When completed it will be the largest HEPP in SE Asia and the biggest privatised hydropower scheme in the world. After construction of the auxiliary cofferdam (Fig. 8) in 2002 the project is now to proceed with the main project components. According to an arch dam foundation, planned originally, on a rock mass characterised by a distinct strength and deformation anisotropy of a greywacke/shale (mudstone) alternation an extensive geotechnical site investigation programme was performed during feasibility study, tender design stages and final design modifications:

– Some 7000 m of core drillings, at site predominantly with water pressure tests, exploration of borrow areas for rockfill and aggregates;

– Geological survey at the site and borrow areas;
– Three adits (totalling 504 m) for abutment exploration, such as detailed engineering/structural geological mapping and rock mechanical tests;
– Three large flat jack tests, two in-situ shear tests and stress measurements by compensation method;
– Tests on suitability of construction materials;
– Geological reconnaissance survey and investigations to assess the potential of reservoir induced seismicity (RIS) of the project area;
– Extensive ecological and environmental studies.

4.2 Geotechnical setting

At the site the Balui River cuts the mountain range favourably nearly at right angle thus a very regularly NE–SW striking greywacke/mudstone succession with argilaceous or silicous matrix dipping upstream (60°) is developed. The rock mass belongs to the slightly metamorphic Belaga Formation of Upper Cretaceous to Lower Tertiary, which in places resembles to Flysch sequences. At the dam site, massive, strong and moderately jointed greywacke (sandstone) beds alternate with mudstone/shale intercalations and greywacke layers of medium strength. Sandstones are the most predominant rock type (approx. 70%) in the vicinity of the dam site, showing up to 100 m thick massive, beds, of which quartz and feldspar grains are poorly sorted. The shales comprising 30% of the series, are indurated claystones, mostly silty.

Bedding is geomechanically seen the most important plane of separation. The spacing of bedding planes ranges from several metres in massive greywacke and in shale to only a few milimetres in the laminated shale/ mudstone sequences. The joint system consists of two steeply inclined transversal sets, striking more or less parallel to the riverbed (release joints) and a relatively flat dipping joint set, which cuts the bedding plane almost perpendicularly.

Due to the distinct strength anisotropy of the rock mass, minor shearzones, i.e. mylonitic, are obvious at the interfaces of strong sandstones and relative lower strength mudstones formed by tertiary compressional tectonics (spacing some 10 m to 30 m, a few cm thick, in rare cases 50 cm). Major faults running through either abutment were not detected. The massive greywacke beds tend to have deep-seated spheroid weathering of erratic distribution of decomposed sections, thus in places, penetrating much deeper (up to 60 m in depth!) into the abutments than in the comparatively softer mudstone/shale layers (up to 10 m in depth only). Consequently those sections of higher permeabilities require proper treatment to be considered when the grout curtain will be constructed. The reservoir basin proved to be built up exclusively by greywacke/shale sequences, of excellent water tightness and will prevent waterleakage to neighbouring river basins.

4.3 Seismicity

The natural seismicity of the tectonically calm Sunda Shield (Island of Borneo), including the South China Sea is low, i.e. forms the relatively stable SE lobe of the Chinese Plate and located on the inner side of the active volcanic belt of Indonesia (Hamilton, 1981). Earthquakes of small to medium magnitude have been reported along Sarawaks's coast from Kuching to Miri since 1870. The most intense tremor observed in Sarawak on 12. February 1994, with a magnitude of about 5.1 on the Richter scale and an epicentre located some 100 km east of Sibu. However no major structural damage was reported (Failer & Abong, 1995). According to the reservoir depth of about 200 m and a size of approx. 600 km^2 possibly occurence of RIS when in operation has to have considered due to ICOLD recommendations.

Regarding the insufficient number of tremors, the deterministic approach was selected to derive realistic seismic design parameters. The highest peak ground acceleration (PGA) was derived by applying a rupture-length-magnitude – relationship developed for the Philippines and surrounding areas (Failer & Abong, 1995). A seismic potential of M = 6.3 was calculated resulting in a horizontal PGA of about a_h =0.12 g at the site. Taking into account admissible reduction and importance, size and risk category of the Bakun project (according ICOLD Bull. 72, 1989) horizontal and vertical accelerations of a_h =0.1 g and a_v =0.05 g were used as basic design parameters also covering RIS.

4.4 Some further results

Concerning rock mass classifications for the tunnels it can be noted that a geologically site specific procedure was applied sucessfully. Derived from the engineering geological mapping of the adits and surface mapping, different rock mass types were established, to which the typical rock mass parameters were referred. It was obvious that due to the sandwich like alternation of comparative soft mudstones and strong sandstones, classifications such as NGI or BIENIAWSKY were not helpful. These methods would deliver unrealistic heavy support in the mudstone/ shale series. As examples the following geomechanical parameters of greywacke and mudstone were selected (Failer & Abong, 1995):

	Compressive strength (MPa)	Moduli of deformation (MPa)
Greywacke (SW)*	100–150	6000–8000
Greywacke (MW)**	50–100	800–2000
Mudstone (SW)*	15–40	1000–3000
Mudstone (MW)**	10–20	3000–6000

* SW: slightly weathered;
** MW: moderately weathered.

According to the low tensile strength of the rock mass, in places, particularly at the interfaces (bedding planes) of sandstone and laminated shales hydraulic fracturing occurred frequently when tested by Lugeon tests. Therefore its critical stress behavior has to be considered when the subsoil sealing will be carried out.

5 CONCLUSION

In order to determine realistic geomechanical parameters in particular for the design and construction of large or difficult dam and underground structures it is indispensable to perform appropriate geotechnical field tests embedded in the geotechnical site investigations. They are properly to be selected according to the results of detailed engineering/structural geological studies, regarding the relevant interaction of structure and rock mass.

Such an approach may optimize both construction, costs and the life time of the plant as well. Nevertheless saving costs at the wrong side during site investigations and design stage often fails, i.e. frequently cause higher expenditures when later design modifications or during construction were required, as many case histories show.

REFERENCES

Failer, E. & Abong, M.D. 1995. The 2400 MW Bakun Hydroelectric Project. *Int. Water Power and Dam Construction, Nov. 1995.*

Fecker, E. 1997. Geotechnische Meßgeräte und Feldversuche im Fels.-pp. 204. Stuttgart: Enke.

Hamilton, W. 1981. Tectonic Map of the Indonesian Region. *Dept. of Interior, U.S. Geol. Survey, Washington D.C.*

Müller, L. 1963. *Der Felsbau, Vol. 1, Felsbau über Tage.* Stuttgart: Enke.

Natau, O., Fröhlich, B. & Mutschler, T. 1983. Recent Developments of Large-Scale Triaxial Tests. *Proc. 5th. Int. Congress of Rock Mechanics, Melbourne.* Rotterdam: Balkema.

Schenk, V. & Köngeter, J. 1982. The Influence of anisotropic Foundation Rock on Design of an Arch Dam in the German Mountains. *Proc.,14th ICOLD Congress, Rio de Janeiro.* Paris.

Schenk, V. & Lee, S.-S. 1987. Geotechnical Aspects of the Bakun Dam in Sarawak Malaysia. *Proc. 9th. SE- Asian Geotechnical Conference, Bangkok, Dec. 1987.*

Schenk, V. 1993. Engineering Geological Aspects of Hard Soil Tunnelling in Germany. *Proc. Int. Symposium of Geotechnical Engineering of Hard Soils – Soft Rocks, Sept. 1993, Athens.* Rotterdam: Balkema.

Schenk, V., Hönisch, K. & Scheibe, H.-J. 1986. Geotechnical Investigations for Koepchenwerk. *Water Power and Dam Construction, July 1986.*

Geotechnical Measurements and Modelling, Natau, Fecker & Pimentel (eds)
© 2003 Swets & Zeitlinger, Lisse, ISBN 90 5809 603 3

Monitoring of Metro Line C construction in Prague

V. Veselý, K. Kolesa & R. Bucek
SG-Geotechnika a.s. Prague, Czech Republic

ABSTRACT: High density of population in the north parts of Prague evoked extending of Metro Line C. Geological conditions of this area are very difficult, designer (Metroprojekt), decided to build new part of metro line using NATM. Monitoring is an integral part of this method. Our article describes design, process and results of monitoring. We also focus on summarization of advantages and disadvantages of used monitoring methods and instrumentations and their context to both analytical and numerical methods of calculation.

1 GEOLOGICAL CONDITIONS

The new Line C, part IV.1, crosses Ordovician beds (shale stones and quartz-sandstones in vertical position, with clay fill on contact) on the route, which are covered by cretaceous sandstone. Contact of Ordovician beds and upper sandstone plate is situated right above the roof of top heading, unfortunately. This contact is heavy saturated by underground water. The new line C underpass mostly inhabitant area, overburden achieves up to 30 m.

2 TECHNOLOGY OF EXCAVATION

2.1 Basic dates

Total length of all new line C, part IV.1, exceeds 3.9 km, tunneled part is 2.6 km long. Two stations were designed, drilled Kobylisy station in the middle of the route, Ladvi station excavated in the open pit in the end of the route. Platform is situated in the middle of both stations. The most of the new line C (IV.1) consists of double track tube (75 m^2), single track tubes (54 m^2) were drilled in short parts ahead and after the stations.

The most challenging part of all route was 150 m long one-space Kobylisy station with 220 m^2 in cross section. NATM was used as a technology for excavation both on the double and single-track tube and both on Kobylisy station.

2.2 Excavation steps

Single and double track tubes were divided onto top heading and invert. Advance of top heading exceeded up to 50–70 m due to the results of monitoring.

Kobylisy station was excavated in several steps. Basic idea of the design (Metroprojekt) was typical vertical dividing of the face. Left and right side tunnels closed by inverts, top heading, bench and invert. Rock mass in good condition appeared in the bottom during excavation of side tunnels. Basic design was changed; cross section was divided into top heading (3 parts) and invert (2 parts). Left and right top heading and theirs temporary inverts were drilled first. Middle top heading and bench followed, invert was excavated in two parts, left and right (fig. 1). Weathered and saturated contact zone were 3 m above the roof of station. Rock bolting in this area was minimized, not to let underground water flow into the tunnel. That's why the

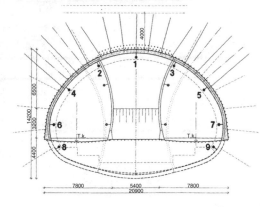

Figure 1.   Kobylisy station – typical cross section – final type.

last change was done; space of side top headings were enlarged, the span of middle top part could be shorter.

3 MONITORING

3.1 Introduction dates

The basic idea was to build up the complex system of monitoring. The system should provide overview about safety and economic factor of construction and its influent on area with high density of population and, last but not least, specify prediction of influence in further difficult part of the rout. Construction of the new metro line C (part IV.1) was cover by various type of monitoring instrumentations; this article is focused on monitoring of Kobylisy station.

3.2 Associate profile

Main attention was given to associate profiles. Full Profile consisted of measurement of convergences, pressure cells on contact of rock mass and primary lining, extensometers bored from the surface and leveling of surface in cross profile (fig. 2). Pressure cells are product of Glotzl; all another instrumentations are the products of SG-Geotechnika. All measurements were provided by corporation SG-Geotechnika and Inset, except leveling of surface and buildings was provided by client itself (IDS). There were four full profiles installed on double track route and two full profiles were instrumented on Kobylisy station.

Convergence measurement, leveling of surface and buildings, extensometers and inclinometers in various intensity covered all 2.6 km route, of course. Tilt bases instrumented buildings in 1st zone of influence;

cracks were monitored by dilatometers, as usually. Seismic impacts were watched periodically, blasting depended on building's conditions.

3.3 Results

Associate profile in km 15.480 is selected to demonstrate typical development of deformation during Kobylisy station's excavation.

Results of **convergence** measurement are presented in main steps: 1. side tunnels excavated 2. middle top heading and bench finished and internal walls destroyed 3. invert closed. Table 1 shows deformation of point nr. 3 (see fig. 1).

Extensometers bored from surface enabled to measure hidden convergence ahead the face of the tunnel. Table 2 presents deformation of lowest point of extensometer nr. 21.017, bored 3 m left of the tunnel axis. This point is 2 m above the tunnel lining situated in geological contact zone between Ordovician beds and upper sand stones. This point developed highest settlement, another upper levels follows settlement of the surface. The results of extensometer measurement led us to conclusion that sand stone table above the tunnel decrease deformation of the overburden on affordable level. Due to these results

Table 1. Deformation of tunnel primary lining in main steps of excavation – Kobylisy station – point nr. 3 (see fig. 1).

	Settlement	Hor. def. (mm)	Vector
L & R side tunnels	6	6	9
Top heading & bench	30	6	32
Invert	54	11	**56**

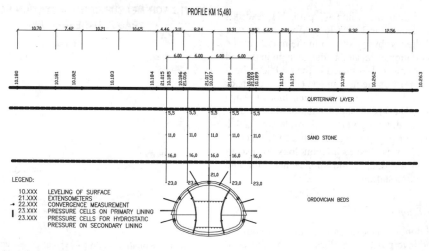

Figure 2. Associate profile – Kobylisy station.

96

Table 2. Extensometer nr. 21.017. Total (real) settlement of the surface and lowest point (2 m above the tunnel lining) in main moment of excavation.

Settlement time	Surface (mm)	Lowest level (mm)
L & R side tunnels	28	32
Top heading crossing	33	39
After top heading	59	66
Total (after invert)	**75**	**98**

the technology of excavation could change with minimum influence on the traffic and buildings on the surface.

Leveling of surface contains both cross profiles (including associate profiles) both longitudinal profiles. As written above, sand stone table above the tunnel absorbed tunneling influence. After all the depressed area appeared wider than was expected. Width of this zone increased up to 130 m, maximum settlement in tunnel axis achieved 75 mm.

Tilt bases and dilatometers monitored buildings. Tilt of houses correspondence with declination of surface and achieved up to 1:330.

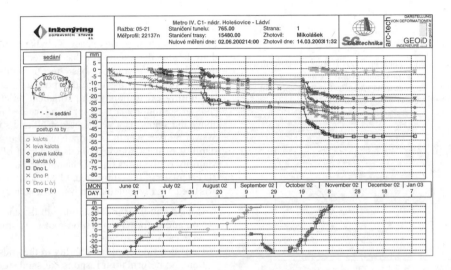

Figure 3. Settlement of tunnel primary lining. Kobylisy station, associate profile.

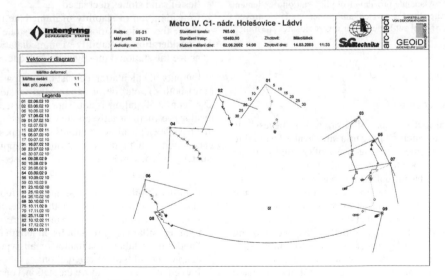

Figure 4. Vectors of deformation. Kobylisy station, associate profile.

97

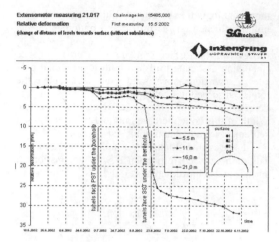

Figure 5. Extensometer nr. 21.017. Relative deformation; distance between surface and several levels. Vertical lines in comments means moments when left side, resp. right side, resp. top heading is crossing extensometer.

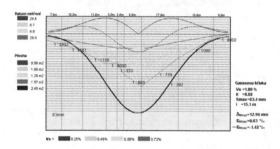

Figure 6. Associate profile – Kobylisy station. Settlement after top heading excavated and inner walls destroyed. Prognosis of total settlement after invert closure – 63 mm.

4 MODELLING

4.1 *Loss of ground*

Analytical method Loss of Ground was successfully applied. This method gave us a simple and fast tool to forecast settlement; we used Loss of Ground software (Bucek). Associate profiles in double track tube served the basic data into basic back analysis. Comparing input data from those profiles and first results of measurements on Kobylisy station led us to estimate further deformation on the surface and buildings. Reliability of this method had been moving about 85%. Next figures (figs 6, 7) show forecast of settlement development in several excavation steps. Light lines means measured settlement on associated profile; stiff line means prognosis.

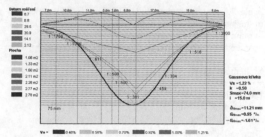

Figure 7. Associate profile – Kobylisy station. Total settlement of the surface – 75 mm.

4.2 *FEM calculation*

FEM calculation of one-space Kobylisy station was done before the excavation has started. The aim of modelling was to approve basic presumptions of the design. The design assumed that both temporary pillars during excavation, both permanent pillars in the ends of station are stiff enough and sand stone table above the station is able to carry itself and surcharge of over burden over the station. Next question was the stability of the Ordovician beds as a roof of station. Majority of the tasks had 3D character; possibility of 2D calculation was eliminated. FEM Plaxis – 3D code was used, finally. Calculation was divided into 4 steps:

1. Consideration of pillar stability in the ends of station and deformation response of sand stone layer on excavation process
2. Consideration of pillar stability in the ends of station and deformation response of sand stone layer on excavation process – deformation characteristics of sand stones decreased
3. Consideration of stability of immediate rock roof above the station – mostly in quartz stone
4. Consideration of stability of immediate rock roof above the station – mostly in shale stone

Outputs of calculation consist both of partial (running) both of total deformations and tension states. The length of station is too short to present all the results; we focus strictly on answering the questions that have been put above. Outputs show appearance of plastic points that indicate tensions overlapping shear capacity of rock mass that could collapse. Answers led to following recommendations:

1. Stability of pillar between single-track tubes in the end of station is suitable. If the convergences of single-track tube achieve 36 mm, we recommend tying up pillar using rock bolts, in length of 25 m.
2. Sand stones table is suitable for all type of surcharge and all type of calculations
3. Immediate rock roof above the station, consists both of quartz stone both of shale stone, does not form

the nature vault. Primary lining will carry out full surcharge of this layer. The total surcharge on lining will consist not only of slim part of overburden straight above the roof but of wide area of rock mass bounded by inclined planes crossing sides of tunnel. Surcharge could be double than weight of pillar above the tunnel profile. Lining could be surcharge eccentrically by block of rock 50 resp. 1200 t. of weight. We consider that to use only structural capacity of shotcrete lining is too risky and we recommend increasing the stiffness of overburden using rock bolts or similar technician improvement.

5 CONCLUSIONS

Monitoring is an undetachable part of NATM technology of tunneling. The aim of monitoring is not only estimation of safety and economic factor during excavation process, last but not least the aim is the possibility to create exact prognosis of further tunneling. Complex monitoring that combine various instrumentation in associate profiles goes hand in hand with back analysis needed for precision of analytical and FEM modelling.

Loss of Ground as an analytical method of modelling is simple and very fast tool for estimation the settlement of surface. Method is very useful in areas with high density of population, reliability of this method in Prague's metro conditions had been moving about satisfactory 85%.

FEM modelling had been developed as precision tool during last 20 years. There are cases in technical praxis where 2D code still does fit well. Situation around complicated underground metro stations with theirs connection into single-track tubes is typical example. Development of 3D code is one of the ways to calculate these situations properly. Calculation of one-space Kobylisy station in Prague's new metro line C (part IV.1) shows possibility of 3D modelling.

REFERENCES

Bucek, R. 1996. *Loss of Ground software*, Prague: SG-Geotechnika.
Kolesa, K & Veselý, V. 2002. Monitoring on Construction of Metro Line IVC.1. Prague: SG-Geotechnika.
Metroprojekt a.s. 2001. Design of Kobylisy metro station, Metro line IVC.1 Prague.

Geotechnical Measurements and Modelling, Natau, Fecker & Pimentel (eds)
© 2003 Swets & Zeitlinger, Lisse, ISBN 90 5809 603 3

Influence of the zero reading time and position on geodetical measurements

G.M. Volkmann, E.A. Button & W. Schubert
Institute for Rock Mechanics and Tunneling, Graz University of Technology, Austria

ABSTRACT: Geodetic measurements have become an accepted method to record tunnel deformations. With this increase in data quality different evaluation methods have been developed to assess the utilization of the support, as well as make short term predictions of the rock mass quality ahead of the excavation. The results from these evaluations are directly related to the quality of the input data. Settlement data from continuously measured chain inclinometers are used to assess the relative magnitude of the settlements occurring both ahead of the excavation and before the zero reading of the geodetic survey. It is shown that delays in this measurement can have a significant effect on the magnitude of the measured data and therefore on the quality of the evaluations made with this data.

1 INTRODUCTION

Geotechnical engineers often rely on observing behaviors and evaluating measured data to optimize construction projects in difficult ground conditions. The most important aspect when using the observational method is that the right information is obtained and that this information is sufficiently accurate to evaluate the problem at hand. If the engineer evaluating the data does not understand the potential errors in the utilized measurement systems and how they can influence the evaluations, the information may become more detrimental then beneficial.

Over the past 15 years, geodetic measurement systems have increasingly been utilized during underground construction to monitor the absolute 3-D displacements during excavations, largely replacing relative convergence measurements. The accuracy of these systems is typically within 1 mm which is adequate for most applications. With any monitoring program there are potential inaccuracies, caused by inadequate program planning or oversight, environmental conditions, equipment problems, or operator error. During underground construction environmental conditions as well as equipment or operator errors can have an influence on data quality. Additionally, large errors are possible if the initial measurement readings are delayed. These errors affect all subsequent analyses that are based on the measured data and should be avoided.

As part of an ongoing research project focused on the performance and design of pipe roof tunnel support, horizontal inclinometers were installed over three consecutive pipe roof fields at a tunnel project in Austria (Volkmann, 2003). The settlements measured during this investigation are used here to demonstrate the influence of the time and position of the zero reading on the non-measured displacements quantitatively. The results confirm previous experience and theoretical considerations as discussed by Schubert et al. (Schubert et al., 2002).

2 SITE AND MONITORING SYSTEM

2.1 *Project and Geology*

The measurements discussed in the following sections were acquired over a 44 m section of a double track rail tunnel constructed as part of the modernization of the "Tauernachse" between Salzberg and Villach, Austria.

The rock mass conditions consisted of weak highly sheared cataclastic fault gouge surrounding blocks of fractured stiff blocks associated with the "Tauernnordrandstörung" (3G & BGG, 2001). Typical rock mass behavior for this type mass ranges from nearly isotropic in matrix dominated zones, as encountered in this measurement section, to highly anisotropic in block influenced zones, further discussions about rock mass behavior in these rock mass conditions can be found in the following articles (Button et al., 2003).

2.2 *Excavation sequence and support*

The excavation; with a cross section of approximately 130 m^2, consisted of 1m top heading advances

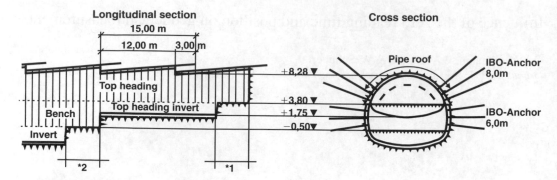

Figure 1. This figure shows on the left side the longitudinal section and on the right side the cross section of the excavation steps. *1: The maximum distance between the top heading face and the temporary top heading invert closure was 5 m. *2: The maximum distance between the face of the bench and the invert closure was 6 m.

followed by a bench-invert sequence as shown in Figure 1. Additionally, the top heading was subdivided into three sections and a temporary invert.

Support consisted of an initial shotcrete lining combined with a 15 m long pipe roof system with a overlapping distance of 3 m and additional rock bolts on the sidewall. The tunnel face was reinforced with shotcrete and up to 5 face bolts, support was installed immediately after opening each section.

2.3 Data acquisition system

Two methods were used to measure the tunnel deformations during the excavation. Geodetic surveying was used to record the absolute tunnel deformations at three points per monitoring section, the crown, and both side walls. Monitoring sections were spaced approximately every four meters in this evaluation section. Additionally, 20 m long chain inclinometers were installed above the crown to measure the settlements continuously both in front of and behind the face. Additionally, a geodetic point was installed at the beginning of each inclinometer string to locate its absolute vertical position with time. Two inclinometers were used to provide an overlapping distance of 8 m over three consecutive measurement fields. Volkmann (Volkmann, 2003) gives a detailed discussion of both the monitoring system and its performance.

The measurements from the combined monitoring system allow the magnitude and trends of the pre-displacements to be compared with the settlements occurring in the tunnel. Figure 2 shows both the inclinometer measurements and the geodetic measurements for section 59.75. When the geodetic measurements are adjusted to the corresponding time in the inclinometer measurements there is an excellent agreement in the post excavation settlements. However, the inclinometers provide additional information concerning the settlements associated with each excavation

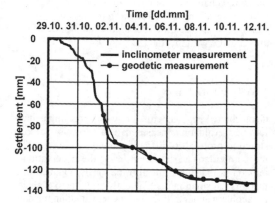

Figure 2. Time settlement curve for the top heading excavation at station 59.75.

step that occur before the zero reading is made for the geodetic survey.

3 GEODETIC VERSUS INCLINOMETER MEASUREMENTS

3.1 Pre-settlements ahead of the face

One of the limitations to geodetic surveys in the tunnel is that they are not able to measure the entire displacement path. In some situations this information is not critical for the construction project. In shallow tunneling where limited surface settlements are required this information is valuable for the assessment of the support adequacy.

Figure 2 shows the settlement path measured with the inclinometer at station 59.75 for the top heading excavation. It can be seen that approximately half of the settlements associated with this portion of the

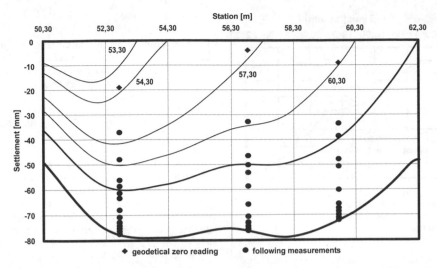

Figure 3.   Inclinometer deflection curves and the geodetic measurements for the settlements measured in one pipe roof field.
The zero reading for the geodetic measurements are shown with a diamond, following measurements with a dot.

excavation sequence are not measured with the geodetic survey.

3.2   *Settlements behind the face*

The settlements that occur after the excavation can be
divided into two parts, those occurring before the
geodetic zero reading and those after. Figure 3 shows
the settlements for one pipe roof field. Inclinometer
measurements are made every two meters and shown
as deflection curves, while the geodetic measurements are made every 4 meters and are shown as
points. The zero reading is represented as a diamond
while subsequent measurements are dots.

It can be seen that the magnitude of the settlements
occurring before the zero reading vary. There are two
major influences affecting this value. One is the
displacement rate in the heading area, which is
related to the interaction between the rock mass and
the excavation-support system. The other is the delay
between the excavation and the time of the geodetic
zero reading.

4   INFLUENCES OF THE ZERO READING
TIME AND POSITION

4.1   *Best measuring time*

In order to provide consistent results from the geodetic
survey the zero reading for each measurement section
should be made as soon as possible during the same
part of the excavation and support sequence. Figure 4
shows a detail of two top heading excavation steps and

the appropriate time to make the geodetic measurement. The excavation was divided into three sections,
each of which results in a settlement increment.

In Figure 3 the zero readings for the three geodetic
measurement sections are shown. The readings for
sections 56.82 and 59.75 are made within the time
recommended in Figure 4. The non-measured settlements in these two cases are less than 1 cm.

4.2   *Delayed zero readings*

The magnitude of the non measured displacements
depends on the delay between the excavation and the
zero reading. The zero reading of the geodetic measurement for section 52.73 was measured after the next
excavation step. In this case the non-measured settlements include the settlements resulting from two excavation steps. Compared to sections 56.82 and 59.75 the
non-measured settlements are approximately twice as
large.

In some cases a measurement point is destroyed
during the excavation process. In this case a new zero
reading is required and the preceding settlement increments can not be measured. If this occurs between the
zero reading and the next measurement the non-
measured displacements are significant. In the examples
shown in Figure 3 this would result in up to 40% of the
measured settlements behind the face.

4.3   *Face distance influence*

The distance to the face position can also have an effect
on the magnitude of the non-measured displacements.

103

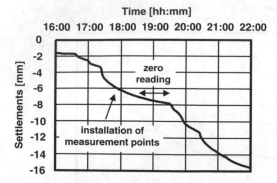

Figure 4. Detail of a time settlement diagram showing two excavation sequences highlighting the appropriate time for the zero reading.

Both the sections 56.82 and 59.75 shown in Figure 3 were made before the next excavation step. However, the distance to the actual face position at station 59.75 was 0.8 m farther from the measurement section then at station 56.82. This could be the reason for the difference in the non-measured displacements for these two points.

5 DISCUSSION

Geodetic measurements are increasing being used to evaluate the utilization of the shotcrete lining (Rokahr et al., 2002; Rokahr & Zachow, 1997) in addition to making short term predictions of the rock mass conditions ahead of the tunnel excavation (Steindorfer, 1998; Sellner, 2000). Both of these methods rely on an assessment of the settlement or displacement magnitudes and their trends in time at individual, as well as consecutive measurement sections. It was shown that significant displacements can occur before the zero reading of the geodetic measurements is made. This can effect the prediction of the rock mass conditions ahead of the tunnel face (Schubert et al., 2002). When the zero reading delay occurs after the installation of the initial shotcrete support any evaluations of the lining stress intensity will be inaccurate.

6 CONCLUSION

In order to maximize the information obtained from geodetical surveys the zero reading should be made as soon as possible after the excavation, as well as at a consistent time in the excavation and support sequence. Additionally, the relative magnitude associated with the different zero reading times and positions were discussed using continuously measured settlements during a tunnel excavation. If the geodetic measurements are to be used for evaluations during the excavation the errors resulting from the delay in the zero reading should be avoided to prevent misinterpretations of the evaluated data. In order to assure the quality of the geodetical survey the contractual documents should specify when the zero reading needs to be made during the excavation and support sequence.

REFERENCES

3G & BGG, 2001. Gutachten zur Geologie, Geomechanik and Hydrologie, unpubl.
Button, Schubert & Riedmüller, 2003. The Use of Monitoring Data and Geological Documentation for Defining Rock Mass Behavior Types. Proceedings New Methods of Geotechnical Engineering, Bratislava, Juni 23–24. in press.
Rokahr, Stärk & Zachow, 2002. On the Art of Interpreting Measurement Results, Felsbau 20 No. 2.
Rokahr & Zachow, 1997. Ein neues Verfahren zur täglichen Kontrolle der Auslastung einer Spritzbetonschale, Felsbau 15 No. 6
Schubert, Steindorfer & Button, 2002. Displacement Monitoring in Tunnels – an Overview, Felsbau 20 No. 2.
Sellner, 2000. Prediction of Displacements in Tunnelling, Ph.D. Thesis Graz, University of Technology, Gruppe Geotechnik Graz Heft 9.
Steindorfer, 1998. Short Term Prediction of Rock Mass Behaviour in Tunnelling by Advanced Analysis of Displacement Monitoring Data. Ph.D. Thesis Graz, University of Technology, Gruppe Geotechnik Graz Heft 1.
Volkmann, 2003. Rock Mass – Pipe Roof Support Interaction Measured by Chain Inclinometers at the Birgltunnel, Proceedings Geotechnical Measurements and Modelling – Karlsruhe.

Geotechnical Measurements and Modelling, Natau, Fecker & Pimentel (eds)
© 2003 Swets & Zeitlinger, Lisse, ISBN 90 5809 603 3

Rock mass – Pipe roof support interaction measured by chain inclinometers at the Birgltunnel

G.M. Volkmann

Institute for Rock Mechanics and Tunneling, Graz University of Technology, Austria

ABSTRACT: The increased use of pipe roof support systems in tunneling has not been followed by an increased understanding of the system behavior of this support system. In order to address this problem detailed investigations on the rock mass–pipe roof support interaction have been initiated. The knowledge about this system was increased by an on-site monitoring system which included continuously recorded chain inclinometer measurements combined with geodetic measurements. Characteristics of the measured settlements indicate that the design of the overlapping length should consider the bearing capacity of the rock mass.

1 INTRODUCTION

Increasingly forepoling methods and/or soil improvements are utilized ahead of the face in tunnels with shallow cover. These should improve the behavior of the surrounding rock mass with respect to stability and deformation during the excavation process. To improve the stability of the area near the face, face bolts, spiles, and tubes often combined with grouting are in use.

Technological advances in pipe roof drilling systems have reduced costs and the installation time over the last years. This has resulted in the increased application of pipe roofs as a normal support system during shallow tunneling instead of being limited to special geotechnical problems.

This rapid increase in use has not been followed by an increased understanding of the interactions between the support system and the surrounding rock mass. Therefore, there are only numerical simplifications or conservative rules for design. Currently during tunnel design, as well as on site, parameters for a pipe roof system e.g. tube length, overlapping length, distance between the tube axes and the strength of one tube are fixed by experience or an empirical approach. In order to improve design methods for these systems a detailed study has been initiated.

Numerical programs can solve a given geotechnical model very well, but an appropriate model can only be selected after calibrating it by measured settlement data from a tunnel site. That data set should have the following specifications: On one hand it should be as exact as possible and on the other hand it should describe the area ahead of and behind the face as well as on the surface. In order to understand and model the behavior of a pipe roof support system a monitoring program was developed to collect the necessary data to calibrate the geotechnical model.

2 PROJECT AND GEOLOGY

The 950 m long Birgltunnel (Austria) is a double track rail tunnel constructed as a part of the upgrading of the "Tauernachse" between Salzburg and Villach (Austria) with an excavation area of around 130 m².

A majority of the tunnel is situated north of the "Tauernnordrandstörung" in the "Grauwackenzone". The west portal and an approximately 80 m long section of the tunnel are situated within the fault zone. This section was constructed by using the NATM (New Austrian Tunneling Method) utilizing a pipe roof support system. In the evaluated zone the overburden raises form 30 m up to 50 m.

This part of the "Tauernnordrandstörung" consists of a clayey, cataclastic fault zone material with shear lenses composed of more competent blocks (3G & BGG, 2001). Due to the potential for encountering blocks the rock mass behavior was described as

Table 1. Strength parameters for "Tauernnordrandstörung" in the area of Birgltunnel (3G & BGG, 2001).

	Parameter	Value
Matrix	Rock mass strength	0,3–0,8 MPa
	Friction angle	20°
	Cohesion	up to 0,03 MPa
Blocks	Rock mass strength	up to 100 MPa

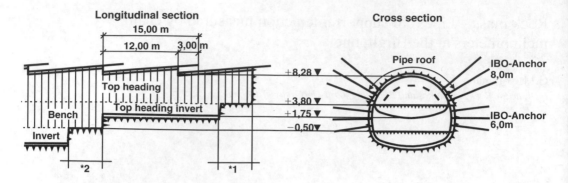

Figure 1. This figure shows on the left side the longitudinal section and on the right side the cross section of the excavation steps. *1 The maximum distance between the top heading face an the temporary top heading invert closure was 5 m. *2 The maximum distance between the face of the bench and the invert closure was 6 m.

ranging from isotropic to highly anisotropic. The design rock mass parameters are shown in Table 1.

The design idea was to pass this weak rock zone with a stiff support system. To achieve this, a temporary top heading invert and a short bench were applied. In the section discussed here, the top heading face was opened in three parts, with a 1 m advance supported with a pipe roof and additional rock bolts at the sidewalls. The stability of the face was ensured by a combination of shotcrete and up to five face bolts. After the top heading, a top heading invert was installed for a temporary ring closure at a maximum distance of 5 m. The ring closure followed the bench excavated with a maximum distance of 6 m (fig. 1). Using that system the weak rock section was passed without any problems.

3 DATA ACQUISITION SYSTEM

Two methods were used to acquire the settlement data used in this evaluation. Chain inclinometers were used to measure the settlements in the crown both in front of and behind the face. Geodetic measurements were taken as part of the normal monitoring program and used to describe the absolute position of the inclinometer.

3.1 Geodetical measurements

Each measurement section of the top heading consists of three points (one in the crown, one on each sidewall) with a longitudinal distance between two sections of approximately four meters. The zero reading of the points was made within 12 hours of the excavation passing that section and then once a day.

Additional geodetical points were situated at the beginning of each chain inclinometer string.

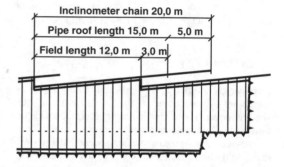

Figure 2. Position of the inclinometer chain relative to the pipe roof.

3.2 Inclinometer measurement system

The system which was used at Birgltunnel is an in-place 20 m long chain inclinometer consisting of ten links (Boart Longyear Interfels, 2002). Three pipe roof fields in a row were equipped as shown in figure 2. The tilt meters measured the inclination continuously and transferred data every minute to the data acquisition system.

As neither the start point nor the end point of the inclinometer chain can be considered to be fixed points, the geodetic measurements are used to determine the spatial position of the beginning of the inclinometer.

3.3 Problems during the data evaluation

During a study of technical literature it was assessed that no comparable measurements were published. That raises the problem of how to fix the measured values in between the time of two geodetical measurements.

A comparison between a linear and an inclination dependent approach showed that using the inclination of the first or the last inclinometer string multiplied by a linear factor gives better results for the vertical settlements over the measured time period. The data quality of the chosen system is shown in the time settlement line of figure 3. The errors between the geodetical measurements and the inclinometer measurements can be minimized to a value lower than 1 mm. But the absolute accuracy of this inclinometer measuring system can

never be higher than that of the geodetical system, because as discussed before the vertical position was given by that system.

3.4 Advantages of the inclinometer chain measurements

In figure 3 it can also be seen that the settlement path of the inclinometer starts nearly 4 days earlier than the geodetical measurements. This advantage exists because the inclinometer measurement starts immediately after the installation and measures the inclination continuously for 20 m in front of the face position where it was installed (fig. 4). Due to the overlapping length of the inclinometer chains the settlements were measured for a minimum length of 8 m ahead of the face.

This arrangement allows the evaluation of the settlements ahead of the face in their absolute value. Additionally, the given settlement rates and their relationship to the face can be evaluated for the designation of the stress transfer length during tunneling. The measured rates were also used to get information about the system behavior of the pipe roof system during single excavation steps.

Besides the advantage that the pre-displacements can be measured, the precision of the dataset in time creates the possibility to observe the time dependent rock mass behavior. Another evaluation possibility is

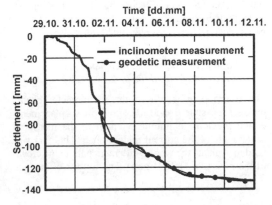

Figure 3. The inclinometer chain and geodetical measurements of the section 59.75 compared to each other shows that the results are equal with respect to the accuracy.

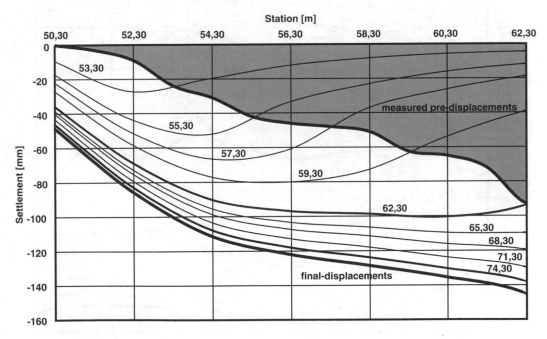

Figure 4. The deflection curve diagram shows the data of the inclinometer chain from station 50.3 to 70.3 in the discussed pipe roof field.

107

that every construction step can be analyzed with regard to its influence on the settlement increments (fig. 5).

Because of these advantages this monitoring program was chosen for the evaluation of the pipe roof support influences in tunneling.

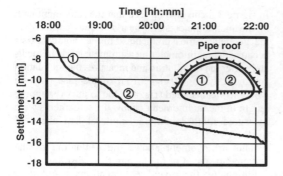

Figure 5. The time settlement line in that diagram shows the settlements two meters ahead of the new face position with a face area divided into two parts during one meter of excavation.

4 RESULTS OF THE MEASURING CAMPAIGN

The evaluation of the measured dataset leads to a few significant results which are discussed in the following paragraphs.

4.1 *Time dependent settlements*

As shown in Figures 3 and 5 the stress transfer requires a given period of time until the settlement rate indicates stability. As this process occurs, the shotcrete's support strength increases. Therefore, the influence of each phenomenon cannot be determined uniquely.

4.2 *Ratio between pre-displacements and settlements behind the face*

The overlapping length of the inclinometer chains makes it possible to add measured pre-displacements of the previous inclinometer chain to the measured settlements of the current inclinometer. This fact was used to evaluate the whole settlement path in the crown of the pipe roof field (fig. 6). The top three lines are

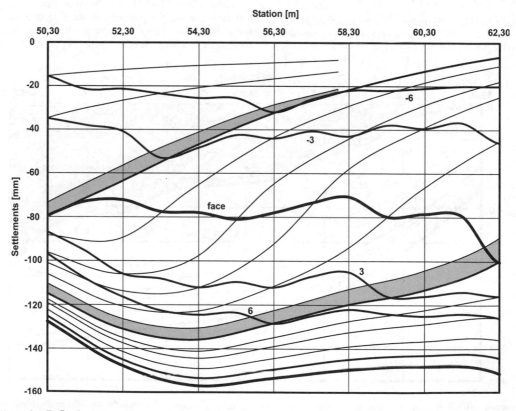

Figure 6. Deflection curves diagram with the whole settlement path in the pipe roof field between station 50.3 and 62.3.

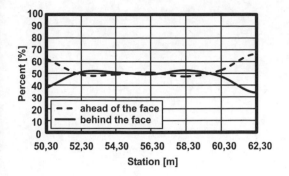

Figure 7. The percentage between the settlements ahead of and behind the face in one pipe roof field.

measurements from the first inclinometer and the remaining lines are from the second inclinometer. The grey shaded area in that figure shows the time dependent settlements during the installation time of the pipe roof.

The "face" line divides the pre-displacements from the settlements behind the face. Over two third of the pipe roof field, the settlements ahead of the face are equal to those occurring behind the face (fig. 7).

4.3 Stress transfer area

The first measured settlements appear in a region around 15 to 18 m in front of the face. That is three times the height and around one and a half times the width of the top heading area. The line " − 6" in Figure 7 displays the settlements which occur 6 m in front of the advancing face. These settlements are up to 15% of the whole measured settlements. The settlements occurring behind the face are nearly the same.

4.4 Characteristics in a pipe roof field

As described before, the percentage of the settlements ahead of and behind the face in the middle of the pipe roof field are nearly the same. At the beginning and at the end of every measured pipe roof field the pre-displacements are up to 65% of the total settlements.

The analysis illustrates that the pre-displacements increase in between the face and three meters ahead of the face. The foundation length of the pipes ahead of the face is also 3 m. This results from an increase in the loading level directly ahead of the face because of the installed pipes.

Relative to the high amount of settlements ahead of the face, the stiff shotcrete arch reduces the settlement amount for the first three excavations steps. On this account the total settlements in the pipe roof field are nearly the same at each position.

5 CONCLUSION

The increasing use of pipe roofs in tunneling calls for a scientific based design method. Therefore, a monitoring campaign was designed to evaluate the system behavior of the pipe roof support system during the excavation process. The measured dataset fulfills the chosen requirements for further theoretical investigations and shows significant characteristics for the settlement behavior in a pipe roof field.

In the evaluated section of the Birgltunnel (Austria) the excavation dependent vertical displacements in the crown start around 15 m ahead of the face and by the time the face passes around one half of the total measured settlements have occurred. At the beginning and at the end of the pipe roof fields the amount of the pre-displacements raise up to 65% of the total settlements but the stiff arch at the drilling position of the pipe roof counters this effect, leading to similar total displacements. The evaluation of the settlement data indicates that the bearing capacity of the rock mass should be included in the design of the overlapping length of the pipe roof fields.

REFERENCES

3G & BGG, 2001. Gutachten zur Geologie, Geomechanik und Hydrologie, unpubl.

Boart Longyear Interfels, 2002. Geotechnical Instrumentation catalogue, http://www.interfels.com

Geotechnical Measurements and Modelling, Natau, Fecker & Pimentel (eds)
© 2003 Swets & Zeitlinger, Lisse, ISBN 90 5809 603 3

3D numerical analysis of deep tunnel through short weak zone

W. Wu, S. Malla, J. Kaelin & D. Fellner
Electrowatt Infra Ltd. & Electrowatt-Ekono Ltd, Zürich, Switzerland

ABSTRACT: A 3D FE-analysis of a deep tunnel through a short weak zone is presented. The weak zone is sandwiched between competent rocks. The tunnel is to be driven by drill-and-blast. The support consists mainly of shotcrete and rock bolts. The analysis takes into account the nonlinear ground properties, changing tunnel geometry and excavation sequence. Emphasis is placed on the arching effect due to the competent rock on both sides of the weak zone.

1 INTRODUCTION

Despite the rapid development in design and construction of tunnels, tunnelling at great depth remains a challenging task for designers and contractors alike. The reason for this is twofold. First, there has been scant experience, since there are few deep tunnels constructed in the past. Second, there is limited information on the ground conditions at great depth due to the high exploration cost. As a consequence, the geological uncertainty (geological formations, mechanical properties and in-situ stresses) is comparatively high for tunnels at great depth. In general, rock mass at great depth has gone through a complex tectonic and metamorphic history, which gives rise to weak zones of various extensions (shear zones, faults and foliation). The strength of the material in the weak zones is substantially reduced. These weak zones are particularly relevant for the tunnelling activities.

Frequently major weak zones with large extensions can be ascertained at the outset of a project by a detailed study of the regional geology. The major weak zones should be avoided through a proper choice of tunnel alignment. However, it is not always possible to avoid all weak zones. In particular minor weak zones with small to medium extensions have to be circumvented by appropriate tunnelling techniques. In general, some improvement of rock quality with depth can be expected. This applies also to the weak zones at great depth owing to the confinement exerted by the high in-situ stresses. However, the response of weak rock at great depth, i.e. under the combination of high stress level and poor strength/deformation properties, to tunnelling is not yet fully understood.

Basically there are three approaches in the design of deep tunnels – empirical, analytical and numerical. A successful application of the empirical approach requires a broad database in similar ground conditions, which is usually not available for tunnels at great depth. The analytical approach, e.g. the confinement-convergence method based on the ground response curve (Pacher 1964, Kovari & Anagnostou 1995), is based on the following assumptions: circular tunnel cross section, full-face excavation, homogeneous ground and plane strain/axisymmetric conditions. These assumptions severely limit its applicability. For tunnels with changing geometry, sequential excavation/support and complex ground conditions, the numerical approach remains the last resort.

The present paper considers a deep tunnel through a weak zone sandwiched in competent rocks, which give rise to an arching effect. The arching effect greatly reduces the convergence and the load on the support.

2 GEOLOGICAL SETTING

We are concerned with a tunnel with an overburden of about 1250 m. The north-south tunnel alignment in the investigated section runs through a series of nearly vertically structured geological units (crystalline massif), which are composed mainly of different gneisses, in particular normal gneiss and schistose gneiss. The crystalline massif has been subjected to large tectonic deformation in the past. The intensive shearing has given rise to weak zones of various extensions. The rock in the weak zone consists mainly of schist with clayey and silty fills. The stiffness and strength of the schist in the weak zones are largely reduced compared with the surrounding gneiss. The geological units plunge down to the south and strike nearly perpendicular to the tunnel alignment.

Figure 1 shows the ground model of the investigated section together with the tunnel crossover (see next section). A 20 m thick weak zone is sandwiched by normal gneiss (Rock 1) and schistose gneiss (Rock 2). The weak zone dips vertically and strikes perpendicular to the tunnel axis. The difference between the weak zone and the competent rocks can be appreciated by comparing the uniaxial compressive strength. The uniaxial compressive strength is about 77 MPa for Rock 1; 70 MPa for Rock 2 and 10 MPa for the weak zone. According to the classification suggested by the ISRM (1981) Rock 1 and Rock 2 can be regarded as *strong rock*, whereas the rock in the weak zone can be classified as *weak rock*.

Several exploration campaigns were carried out including geological mapping, drilling and geophysical sounding. The investigated area was investigated by a 1750 m long inclined wire cable drilling. By taking advantage of the steep dip of the strata, the ground conditions at the tunnel level can be inferred from the boreholes by projecting the bore-log onto the tunnel alignment. Meanwhile an extensive testing program (in situ and in lab) was carried out. The design parameters are given in Table 1.

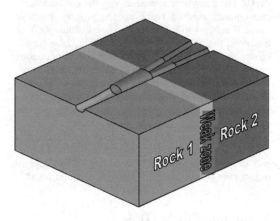

Figure 1. Ground model with weak zone sandwiched between competent rocks.

Table 1. Design parameters for the competent rocks and the weak zone.

Rock type	E [GPa]	ν [–]	φ [°]	c [MPa]
Rock 1	29	0.25	37.0	4.25
Rock 2	19	0.25	33.5	3.25
Weak zone	6	0.30	29.8	1.10

The specific weight is about 27 kN/m³. For an overburden of about 1250 m a vertical stress of about 34 MPa is obtained. An isotropic stress state is assumed, although some variation of the horizontal stress is possible.

3 TUNNEL DESIGN

3.1 Crossover

The railway tunnel consists of two parallel single-track tunnels with a spacing of about 50 m. The cross-section of the single-track tunnels is dictated by the loading gauge and the primary stress state. A circular cross-section is chosen with a clearance diameter of about 7.7 m. In case of an accident and for maintenance it is necessary for the trains to change from one tube to the other. For this purpose several crossovers are provided along the alignment.

3.2 Excavation scheme

The crossovers will be excavated in the following way. The running tunnel is widened gradually until the bifurcation point, where the cross-section is large enough to accommodate the running tunnel and the connection tunnel (see Figure 1). The tunnel cross-section at the bifurcation point is shown in Figure 2. The maximum span is about 22 m and the height about 14 m. From the bifurcation point, the running tunnel will be excavated followed by the connection tunnel. In view of the changing tunnel geometry, the squeezing potential, the crossover will be excavated by drill-and-blast.

In view of the large excavation area in the enlarged section, the excavation is divided into top heading, bench and invert. The ring closure, i.e. the distance between bench and invert, is limited to 15 m. The top heading with its large span is further divided into two side galleries and a centre part. Figure 2 shows the

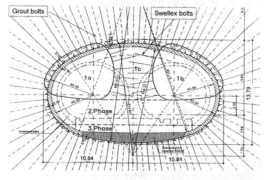

Figure 2. Tunnel cross-section in the enlarged section together with the support.

excavation scheme together with the support elements. The side galleries with their small excavation area can be handled more easily. Besides, the side galleries provide also information on the ground conditions and allow some ground improvement through drainage.

3.3 Support measure

The support in the enlarged section consists mainly of a 50 cm strong shotcrete lining in combination with 9 m long radial grouted bolts and 12 m long horizontal grouted anchors in front of the face. The design bearing capacity of the bolts is about 32 t. The bolt density is one bolt every 1.25 m². Some Swellex anchors are provided in the centre part between the side galleries.

The running tunnel and the connection tunnel will be driven by full-face excavation. The tunnels will be supported by a 25 cm shotcrete lining and 6 m long radial rock bolts with a bolt density of one bolt every 1.5 m². The excavation length in the weak zone is assumed to be 1 m.

4 NUMERICAL MODEL

The common practice in tunnel design is based on plane strain model neglecting the arching effect. For a short weak zone, however, a plane strain model leads to an over-conservative design. The problem calls for a 3D analysis.

4.1 Finite element code

The analysis was carried out with the finite element code ADINA Version 7.5. The rock mass was discretised by 20 node elements and the shotcrete lining by 8 node shell elements. An ideally plastic model with the failure criterion of Drucker-Prager was used for the rock mass. The intermediate failure cone between triaxial compression and triaxial extension was adopted (Chen & Mizuno 1990). The associated flow rule was used, since non-associated flow rule was not available in ADINA. The associated flow rule overestimates the tunnel convergence to some extent.

The behaviour of shotcrete is simulated by an elastic model with an apparent modulus of 8 GPa. This is a reduced modulus with reference to the modulus of about 30 GPa according to the structural code. The reason for the reduced modulus lies mainly in the time dependent behaviour of green shotcrete (Pöttler 1992).

4.2 Finite element mesh

A rectangular block of 200 m × 200 m × 200 m is descretised. The bifurcation point is placed in the middle of the weak zone. The quality isoparametric elements in ADINA enable reasonable accuracy with a

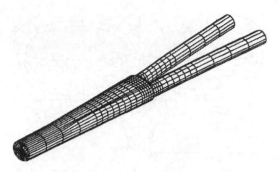

Figure 3. Perspective view of the mesh of the crossover (only the excavated volume is shown).

relatively coarse mesh. A finer mesh is used in the vicinity of the weak zone and the excavation area. Although our primary concern is the weak zone and the immediate vicinity, a larger block has been chosen in order to minimise the influence of the boundary conditions and generate an initial stress state for the excavation of the weak zone.

A perspective view of the mesh of the crossover is shown in Figure 3. The mesh is generated by interpolating among a number of cross-sections. The spacing between the cross-sections corresponds to the excavation length. A large spacing of about 15 m is adopted for the 4 excavation rounds near the block boundaries. For the adjacent excavations until the edge of the weak zone a spacing of about 5 m is used. The spacing in the weak zone corresponds to the design excavation length of 1 m.

4.3 Excavation simulation

Excavation and support can be conveniently simulated by death and birth of elements in ADINA. During excavation, the elements within the excavation boundary are removed. The out-of-balance forces on the boundary are calculated und serve as load for the subsequent iterations. A stress release factor characterising the stress release prior to support installation, which is required in a 2D analysis, need not be specified.

Installation of the shotcrete lining is modelled by the birth of shell elements. After their birth the shell elements are attached to the solids elements along the excavation boundary. Both elements share the same nodes and displacement along the excavation boundary. The shotcrete lining is installed two excavation rounds behind the tunnel face. Figure 4 shows the excavation sequence in the enlarged tunnel section.

4.4 Simulation of rock bolts

The radial rock bolts as shown in Figure 2 are treated as an equivalent cohesion. Recent experimental investigation shows that the bolts can be adequately considered

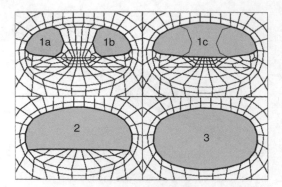

Figure 4. Excavation sequence in the enlarged tunnel section.

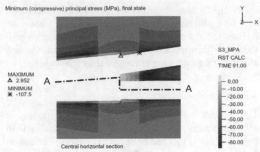

Figure 5. Contours of major principal stress in the vicinity of the bifurcation point.

as reinforcement of the rock mass by increasing the cohesion while leaving the friction angle virtually unchanged (Spang 1988). The equivalent cohesion c_a can be calculated as follows:

$$c_a = \frac{T_0}{A_a} \tag{1}$$

where T_0 is the equivalent bolt force by considering the dowel effect; A_a is the average anchored area per bolt. According to Spang (1988), the equivalent bolt force can be expressed by the following empirical relationship:

$$T_0 = 1.55 \, P_z \, \sigma_c^{-0.14} (0.85 + 0.45 \tan \varphi) \tag{2}$$

where P_z is the bolt capacity, σ_c is the uniaxial compressive strength of rock and φ is the friction angle of rock mass. The same procedure applies also to the horizontal anchors installed at the tunnel face. The horizontal anchors are also treated as an equivalent cohesion.

5 NUMERICAL RESULTS

The numerical results are presented for a 60 m long section around the bifurcation point (30 m before and 30 m behind the bifurcation point). Following the sign convention in continuum mechanics, compressive stress and force are negative.

5.1 Arching effect

Figure 5 shows the contours of the major principal stress on a horizontal plane through the tunnel axis. The displacement of the weak zone towards the excavation area reduces the stresses in the weak zone. The drag of

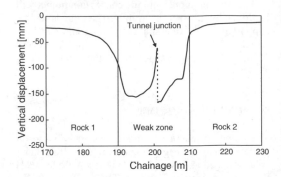

Figure 6. Final vertical displacement along the crown of the enlarged tunnel section and of the running tunnel (A-A in Figure 5).

the weak zone is sustained by the competent rocks bounding the weak zone. The stress concentration on the edge of the competent rock indicates arching in the weak zone, which resembles the trap-door problem in soil mechanics. Moreover, the pillar between the running tunnel and connection tunnel is highly stressed in the competent rock.

The arching effect can also be observed in Figure 6, where the settlement along A-A in Figure 5 is depicted over the chainage. The settlement trough in the weak zone shows a discontinuity at the bifurcation point. This is ascribed to the change of tunnel geometry at the bifurcation point. The settlement in the middle of the weak zone amounts to about 17 cm, which can be compared to the 54 cm settlement from 2D analysis. Note that no slip is allowed between competent rock and weak zone. Allowing slip would give rise to localized deformation near the edges of the weak zone.

5.2 Forces in shotcrete lining

Unlike shallow tunnels, overstressing of shotcrete lining in deep tunnels does not necessarily mean an impending collapse (Wu & Saraiva 1999). However

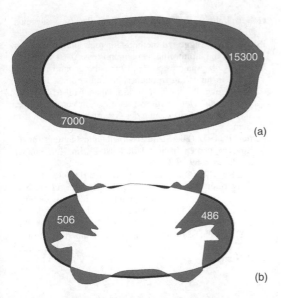

(a)

(b)

Figure 7. Thrust [kN/m] (a) and moment [kNm/m] (b) in a cross-section immediately before the bifurcation point.

excessive overstressing may cause shear cracks and spalling undermining the integrity of support. Besides, a damaged lining poses a safety hazard for the workers.

The forces (thrust and moment) in the shotcrete lining are shown in Figure 7. The forces are obtained in a cross-section immediately before the bifurcation point. The maximum thrust is about 15.3 MN/m and the maximum moment 0.5 MNm/m. For a lining thickness of 50 cm, a maximum axial stress of about 30 MPa is obtained, which exceeds the uniaxial compressive strength of 25 MPa. In order to mitigate the overstressing, some longitudinal slits are prescribed in the shotcrete lining (Golser 1999). The slits realised by leaving recesses during spraying. Both the thrust and the moment will be largely reduced by the slits.

5.3 *3D Phenomena*

The 3D analysis reveals some phenomena, which cannot be observed in a 2D analysis. Figure 8 shows the contours of the longitudinal force in the shotcrete lining near the bifurcation point. The longitudinal force in the lining is negligibly small outside the weak zone. In the first half of the weak zone a compressive longitudinal force of about 13 MN/m is obtained. This is mainly due to the large inward deformation of the weak zone at the excavation face. In the second half of the weak zone a tensile force of about 8 MN/m is obtained. The low deformability of the competent rock is responsible for the tensile force.

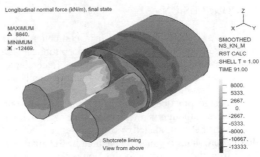

Figure 8. Contours of longitudinal force [kN/m] in the vicinity of the bifurcation point.

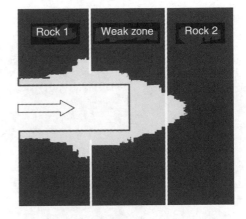

Figure 9. Plastic zone in a vertical plane through the axis of the enlarged tunnel section.

Another interesting observation can be made on the plastic zone plotted in Figure 9. The plastic zone is shown on a vertical plane through the axis of the enlarged tunnel section. It can be seen from Figure 9 that the plastic zone in the competent rock is larger than in the weak zone. At the first sight this seems quite strange, since the competent rock is stronger than the weak zone. The mechanism behind this phenomenon can be explained as follows. Since there is no slip between the competent rock and the weak zone, both rocks will experience the same strain at the edge. The higher stiffness of the competent rock will give rise to higher stresses, which may lead to a larger plastic zone.

A perusal of Table 1 suggests that the difference between Rock 1 and weak zone in terms of E-modulus is more pronounced than the difference in terms of strength parameters. This is probably the reason for the larger plastic zone in the competent rock. The plastic zone beyond the bifurcation point shows that the pillar between the running tunnel and the connections tunnel is highly stressed. The plastic zone in the pillar extends into the competent rock.

115

6 CONCLUSIONS

Short weak zones sandwiched in a competent rock matrix are frequently encountered in deep tunnels. 2D analyses without considering the arching effect tend to an over-conservative design. Such a problem calls for 3D analyses, which enable an economic design.

In design practice, however, 3D analyses are but exceptions, although many problems are of 3D nature and can be better modelled by 3D analyses. We believe that 3D problems ought to be handled by 3D analyses. User friendly and cost effective computer codes are becoming increasingly available. Let us hope to see more 3D analyses in future tunnelling projects.

REFERENCES

Chen, W.F. & Mizuno, E. 1990. Nonlinear analysis in soil mechanics, theory and implementation. Elsevier Science Publisher.

Golser, J. 1999. Innovationen im Tunnelbau in Österreich. *Tunnel*. 34–40.

ISRM. 1981. Suggested methods for rock characterisation, testing and monitoring. Pergamon Press.

Kovari, K. & Anagnostou, G. 1995. The ground response curve in tunneling through short fault zones. In: Proc. Inter. Cong. Rock Mech., Tokio, Japan. 611–614.

Pacher, F. 1964. Deformationsmessungen im Versuchsstollen als Mittel zur Erforschung des Gebirgsverhaltens und zur Bemessung des Ausbaus. In: *Felsmechanik*.

Pöttler, R. 1992. Die Standsicherheitsuntersuchung für die Kaverne der englischen Überleitstelle im Kanaltunnel. *Bautechnik*, **69**, 602–617.

Spang, K. 1988. Beitrag zur rechnerischen Berücksichtigung vollvermörtelter Anker bei der Sicherung von Felsbauwerken in geschichtetem oder geklüftetem Gebirge. Dissertation, EPFL, Lausanne, Switzerland.

Wu, W. & Saraiva, E. 1999. Strategy for dimensioning shallow NATM-tunnels. In: Proc. Int. Symp. Numerical Methods in Geomechanics. Graz, Austria. 367–372. Rotterdam: Balkema.

Geotechnical Measurements and Modelling, Natau, Fecker & Pimentel (eds)
© 2003 Swets & Zeitlinger, Lisse, ISBN 90 5809 603 3

Ensuring behavioural requirements of tunnelling through natural and man-made ground by employing the Observational Method

M.K. Zacas & V.O. Vandolas
Pangaea Consulting Engineers Ltd, Athens, Greece

N.A. Rahaniotis
Engatia Odos SA, Thessaloniki, Greece

K. Tzima
Civil & Construction Department, University of Umist, United Kingdom

ABSTRACT: Tunnelling a twin tube slope tunnel in the ring road of the city of Patras in northern Peloponnesse encountered unique geotechnical challenges in planning, design and construction. The tunnel was driven through natural (neogen formation) and man-made ground. The latter was reinforced by means of a series of stabilisation and improvement techniques such as cement stabilisation, piling and jet grouting. Because of great uncertainties in the estimation of the geotechnical parameters of the geomaterials and in the design modelling, the Observational Method was employed for effective risk management. The overall evaluation of the displacements concluded that the maximum measured values were comparable to the predicted ones. In certain cases, measurements dictated changes in the excavation and support program of the contractor to avoid undesirable consequences. The use of the Observational Method resulted in the successful and economical outcome of the project.

1 INTRODUCTION

For many years heavy traffic along the northern coast of Peloponnesse in Greece was channelled through the city of Patras creating major economical, environmental and social problems.

During last decade the Greek Ministry of Environment, Planning and Public Works proceeded in the construction of a ring road around the city as part of a new highway alignment in the area. Because of regional geomorphology a number of tunnels and bridges exist over a major or part of the ring road. The planning, design and construction of one of these tunnels, namely tunnel Sc, presented unique geotechnical challenges.

Sc tunnel is a twin tube tunnel, 160 m long, driven through a difficult terrain of neogen sediments. It was constructed by a combination of a "cover and cut" technique and tunnelling in natural and treated by jet grouting ground. Its northern end is connected with a cut and cover tunnel while its southern portal almost coincides with the abutment of a bridge.

Because of the nature and the complexity of the project, the Observational Method was employed for an effective risk management.

2 THE OBSERVATIONAL METHOD

In geotechnical engineering, the Observational Method, is a process that recognises the inherent limitations of our information and knowledge and manages the associated risk involved. Deviations from expected behaviour are carefully monitored, observed and evaluated and when appropriate, a course of action, for which provisions have already been made, is adopted. As a result, by controlling and reducing unforeseen risks, materials saving and reduced construction time can be achieved.

A key factor in the application of the method is instrumentation and monitoring.

The parameters to be selected for observation should reflect the phenomena actually influencing/governing the behaviour of the construction work. Limit values for these parameters, usually referred to as trigger criteria, should be determined from the results of analyses, based on the most probable and most unfavourable conditions. Trigger criteria or alert levels, must allow sufficient time for the implementation of the planned modifications or predefined contingency plans.

3 GROUND CONDITIONS

The geology of the area is comprised of well compacted, slightly cemented conglomerates and stiff silty – clay marls; the first are prevailing at higher altitudes, while the latter are dominant at greater depths. The ground surface is covered by a thin mantle of sandy gravel and clay, products of the weathering process in the area.

The geotechnical behaviour of marls is greatly influenced by factors such as: spatial variation of stiffness (originated to sedimentation processes), tectonic disturbance, existence, orientation, thickness and characteristics of sandy or silty – clayey sand interlayer and changes in moisture content.

The tunnel was constructed above the water table. However minor local ingress of water has been encountered while excavating the invert.

The mechanical parameters of the geomaterials were determined from the results of in situ and laboratory tests and field observations. These were

complemented with experience gained from the design, construction and performance of a variety of geotechnical structures in the wider area as well as engineering judgement.

4 SOLVING A PROBLEM

The left tube with the highest overburden was constructed by underground excavation (see Fig. 1).

The right tube has been initially designed to be constructed partially as a cut and cover tunnel (ch. 4 + 751 – ch. 4 + 803.5) and partially as a slope tunnel (ch. 4 + 803.5 – ch. 4 + 883.5). Unfortunately the Greek Board of Archaeology forbade any slope excavations in the area; a factor that mandated rational design and construction modifications. Consequently, an innovative design and construction scheme was devised. It included a series of stabilization and improvement techniques in order to provide adequate lateral confinement and stability for the right tube.

The main components of the new construction scheme were:

– Construction of a fill embankment resting on the hill slope and stabilised by cement

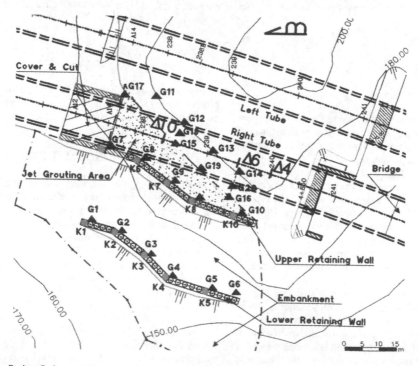

Figure 1. Project Sc layout (Gi: surface settlement point, Ki: inclinometer casing, Δi: convergence measurement station).

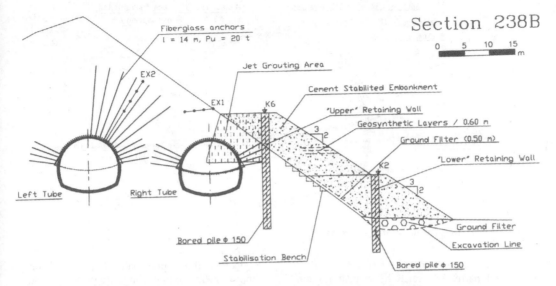

Figure 2. Typical cross section in stabilised ground area and cement stabilized side embankment.

– Construction of two retaining piles walls. In the first one (upper retaining wall) the piles, Ø150 cm in diameter, were bored (at axial distances of 150 cm), from the crest of the embankment.

In the second (lower retaining wall) the piles, Ø150 cm in diameter, were bored every 180 cm at an intermediate level between the crest and the foot of the embankment slope.

– Improvement of the embankment section between the natural hill slope and the upper retaining pile wall by jet grouting. The jet grouting columns were constructed vertically or inclined on a triangular grid (0.70 cm center to center distance).

– Construction of a cover and cut section where the right tube of the tunnel was excavated in man made ground through a "corridor" of two pile walls (Ø150 cm piles) braced with transverse concrete beams on the capping beam level (ch. 4 + 780 – ch. 4 + 796).

As a result the right tube was finally constructed as a cut and cover structure in the section ch. 4 + 711.5 – ch. 4 + 780, as a cover and cut structure from ch. 4 + 780 to ch. 4 + 832 and with underground excavation in natural and treated by jet grouting ground from ch. 4 + 796 to ch. 4 + 832 and in natural ground from ch. 4 + 832 to ch. 4 + 861.3. A plan of the Sc project layout is shown in Figure 1.

The present article deals with the last three sections of the right tube. A typical cross section in the area of tunnelling through natural and treated by jet grouting ground is presented in Figure 2.

5 TUNNEL DESIGN

Tunnel design and construction was based on the NATM principles.

Excavation was performed in two stages, i.e. top heading and bench, using a Liebher 932 tunnel excavator with a properly modified cutting head.

The primary support included shotcrete 30 cm thick reinforced with T131 wire mesh, lattice girders (spaced at 0.80–1.0 m centers for the left tube and 0.60–0.80 m for the right tube), four rock bolts, 6 m long, in every top heading sidewall at every advancing step, forepoling over the crown and face support by soil nailing.

Forepoling consisted of Ø25 mm bars, l = 5–6 m long, for the left tunnel tube and Ø76 mm tube, l = 6–8 m long, for the right tunnel tube (Fig. 3). Fiberglass nails, 12 m long, were used for the face stabilisation of the top heading of the right tube.

Additional reinforcement using Fiberglass nails, l = 11–16 m long, was installed through the right side – wall of the left tube towards the pillar and the area over the crown of the right tube.

The primary support section was closed with a 30 cm thick invert of reinforced shotcrete.

For the right tube a temporary invert 20 cm thick was foreseen to enhance stability conditions during top heading construction (Fig. 3).

Excavation and support of the top heading was followed by excavation and support of the bench with a closed invert.

The distance between top heading and bench was specified as 17.0 m for both tubes for sections bored

119

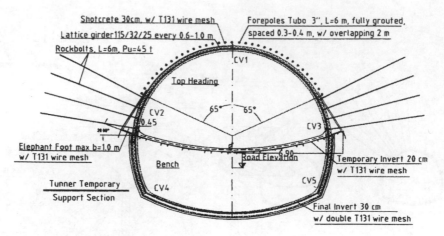

Figure 3. Right tube excavation profile and primary support measures.

in natural ground and 10.0 m for sections partly in natural and partly in treated by cement grouting ground respectively.

Both tubes were designed with a water proofing system and a permanent cast in situ reinforced concrete lining.

6 INSTRUMENTATION

The design and construction of Sc tunnel faced up to many problems and difficulties because of geometrical restrictions, restrains for surface excavations, geological and geotechnical conditions, uncertainties in design parameters and most of all the cover and cut tunnelling through ground reinforced/stabilised by jet grouting. Therefore, increased demands for quality control and monitoring of ground–structure interaction were postulated.

The purpose of monitoring the constructions was to confirm the adequacy of the tunnel and associated structures design with regard to overall stability as well as for the controlling of the ground deformation, hence ensuring the safety of the people and the project. The monitoring program was mainly directed in measuring tunnel convergence, displacements of the natural and treated by jet grouting ground as well as ground deformation around the right tube. The instrumentation used for monitoring these quantities included:

– Convergence measuring stations located every 4–5 m in each tube. Each station included five measuring points (three in the top heading and two in the bench).
– Inclinometer casings (9 pieces) installed in the bored piles of the upper and lower retaining walls of the embankment for monitoring the behaviour/

deflection of the pile walls during the ground improvement and tunnelling activities.
– Surface settlement points (20 pieces) installed on the natural and artificial (embankment) slopes, over the right tube of the tunnel. Their readings were correlated with those from the inclinometers and convergence measurements for assessing overall stability.
– Three point extensometers, EX1 (3 m, 6 m, 9 m) and EX2 (6 m, 9 m, 15 m) were installed in sections Δ238 and Δ239 for monitoring ground deformation around the right tube during its construction.

The instrumentation scheme layout is shown in Figure 1.

7 CONSTRUCTION AND MONITORING

Construction of the left tube was initiated by excavating the southern portal during 2001. Excavation of the left tube including concreting of the inner lining was completed within 2001. Construction of the side embankment, upper and lower retaining walls, was completed in the period from Jan. 2001 to Dec. 2001. Jet grouting treatment/improvement of the upper part of the embankment was carried out in the period from Jan. 2002 to April 2002.

Top heading excavation of the right tube started on Feb. 2002 from the south portal and was completed on May 2002.

Right tube excavation was lagging at least 10 m behind the jet – grouting column – construction front to allow for the hardening and strengthening of the ground – cement mixture.

To meet the requirements for an early project delivery, the program target construction rates and the top-heading bench sequence of excavation were reconsidered.

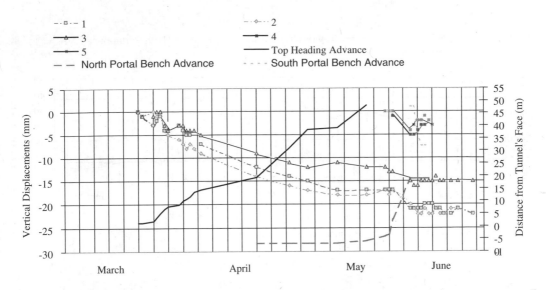

Figure 4. Vertical displacement vs construction activities & time (station Δ4).

Increased excavation rates were accepted as long as tunnel convergence measurements were within the predictions of the design and the trend of deformation rate was not continually accelerating; otherwise the provisions of the design were applied.

Speed of invert closure is an important factor in limiting deformation. Accordingly, design for the right tube specified a temporary (for the top heading), as well as a final invert (for the full section). Nevertheless site observations, continuous evaluation of the results of geotechnical measurements and experience from the construction of the south end entrance collar (closed ring section), suggested that a through going top heading excavation (not closely followed by bench excavation) was feasible without impacting the stability of the tube. This approach was finally adopted successfully. The results of the extensive monitoring of tunnel deformation and surface settlement showed that:

(i) top-heading convergence was within the expected range of 20 mm,
(ii) surface settlements above the tunnel axis were in the range of 15–20 mm and in the range of 5–15 mm on the natural slope.

Inclinometer deflections varied between 7 mm and 12 mm for the upper and between 2 mm and 4 mm for the lower retaining wall. Extensometer values were in the range of 1–2 mm.

Top heading construction was completed without any problems.

Bench excavation and support started on May 2002, initially from the south end and shortly after,

from the north end. Temporary introduced high excavation rates caused increased convergence rates, and enforced a slower progress rate under strict surveillance of construction quality. Bench construction was completed on June 2002 without any problems. Convergence values amounted to 25–30 mm. Surface settlements on the embankment reached a maximum of 35 mm above the tube and 22 mm on the natural slope. Inclinometer deflections varied between 9 mm and 16 mm for the upper and between 4 mm and 6 mm for the lower retaining wall. Extensometer values remained smaller than 3 mm. Characteristic time vertical displacement and time – construction activity diagrams for measuring points of Δ6, Δ9, Δ10 stations are presented in Figures 4, 5 and 6 respectively. Figure 7, presents typical program of the ground surface settlement in relation to time for surface settlement points G11, G12 and G13. Figure 8 shows characteristic deflections of K6 inclinometer casing (section Δ238B) in the upper retaining pile wall.

Additional measurements of tunnel and ground deformations were continued for 15 days after the completion of excavation and support works in the right tube. Equilibrium was reached within the above time period and the contractor proceeded in the construction of the final lining.

8 CONCLUSIONS

To deal with uncertainties associated with design and construction requirements for stability and safety of the Sc tunnel in northern Peloponnesse, Greece,

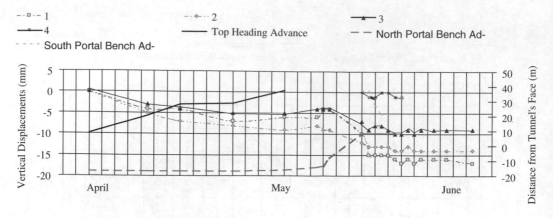

Figure 5.   Vertical displacement vs construction activities & time (station Δ6).

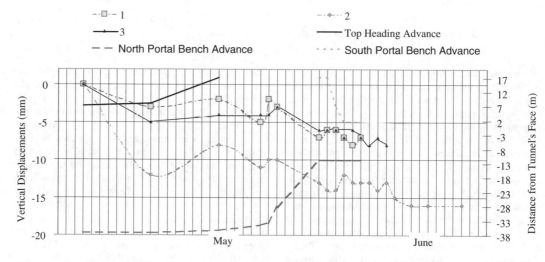

Figure 6.   Vertical displacement vs construction activities & time (station Δ10).

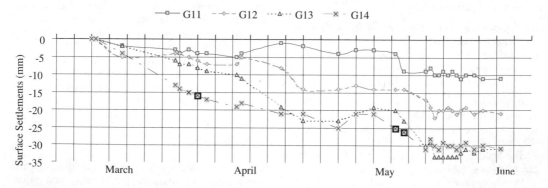

Figure 7.   Ground surface settlements (mm) vs construction period (surface settlement points G11, G12, G13, G14).

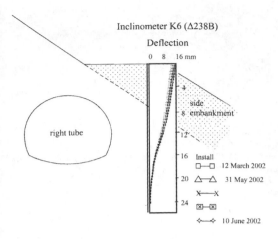

Figure 8. Pile wall deflections at Δ238B section (inclinometer casing K6).

the Observational Method was employed. An extensive observation program was implemented for monitoring the behaviour of the structures, during construction.

Optical measurements for measuring tunnel convergence and ground surface displacements as well as inclinometer and extensometer readings provided valuable information about the behaviour of the tunnel and dictated application of contingency plans.

The most significant of the contingency plans were:

– Alteration/adjustment of the ring closure distance
– Change of the length of advance for top heading and bench excavation and support.

In addition the timely evaluation of the collected measurements and the presence of experienced crew personnel were key factors in achieving significant cost and time saving.

REFERENCES

Eurocode 7 (1994). Geotechnical Design-Part 1, General Rules, *ENV 1977–1/CEN.*

Nicholson, D., Tse, C.M. & Penny, C. (1999). The Observational Method in ground engineering; principles and applications, *CIRIA Report R185*, London.

Sofianos, A.I., Katsaris, D., Giannakos, Ch. & Kalkantzi, A. (1999). Variability in the construction of the soft rock Patra highway deviation tunnels, *Challenges for the 21st century*, Allen et al. (eds), pp. 433–439.

Zacas, M. & Rahaniotis, N. (2000). Tunnelling with NATM in Neogen Marls, *Proc. of the Eurock 2000 Symposium in Aachen*, Verlag Gluekauf GmbH, pp. 535–540.

Geotechnical measurements and modelling in underground waste disposals

Geotechnical Measurements and Modelling, Natau, Fecker & Pimentel (eds)
© 2003 Swets & Zeitlinger, Lisse, ISBN 90 5809 603 3

Development of an excavation disturbed zone in claystone (Opalinus Clay)

P. Blümling
Nagra, Wettingen

H. Konietzky
ITASCA, Gelsenkirchen

ABSTRACT: Nagra has conducted a feasibility study for a potential deep geological repository for radioactive waste in the Opalinus Clay, a Jurassic claystone in Northern Switzerland. As part of these studies, the rock mechanical behaviour of the Opalinus Clay was investigated. The Opalinus Clay is a transversely isotropic soft rock (or stiff clay) and shows a remarkable dependence of the rock mechanical parameters on water content. A modified Mohr–Coulomb model was calibrated with data from laboratory tests and verified by comparing modelling results with data from a mine-by-test carried out at the rock laboratory at Mont Terri. The modelling results demonstrate that the excavation disturbed zone (EDZ) is mainly influenced by the anisotropy of the rock. The geometry of the EDZ is controlled by the instantaneous ground reaction (undrained elasto-plastic behavior) during excavation. Studies of long-term development showed the influence of compaction/de-compaction, creep processes and swelling of the host rock and demonstrated the self-sealing capacity of the Opalinus Clay.

1 INTRODUCTION

Nagra, the Swiss National Cooperative for the Disposal of Radioactive Waste, has conducted a feasibility study for a potential deep geological repository for radioactive waste in the Opalinus Clay, a Jurassic claystone in Northern Switzerland. The feasibility study consists of three major elements – siting feasibility, safety assessment and construction and operation feasibility. As part of these studies, the rock mechanical behavior of the Opalinus Clay was investigated. Data are available from testing on core samples as well as from in situ tests.

2 CHARACTERISATION OF THE OPALINUS CLAY

2.1 Geology

The Opalinus Clay is a marine claystone which was deposited about 180 million years ago and has undergone a complicated burial history. The maximum burial depth of the Opalinus Clay in the Benken area is about 1700 m. Although several different facies can be distinguished, the claystone is relatively homogeneous – in a lateral as well as a vertical direction. The main mineralogical components of the rock are – 54% clay, 20% quartz and 26% carbonate and accessory minerals (Nagra 2002). The clay consists mainly of illite, kaolinite, chlorite and mixed layers. The porosity and the water content of the Opalinus Clay at the Benken site in Northern Switzerland at a depth of about 600 m are 12% and 3.5–4.5% respectively.

2.2 Rock mechanics

The Opalinus Clay is a transversely isotropic soft rock (or stiff clay) and shows a remarkable dependence of the rock mechanical parameters on water content. The core samples taken in the Benken borehole in Northern Switzerland showed a strong tendency to disking during unloading after recovery from the core barrel. This disking process was not the classical process observed in highly stressed hard rock where the samples fail immediately after drilling, but occurred a few hours after the recovery of the core, probably due to the high porewater pressure and the very low hydraulic conductivity (10^{-13}–10^{-14} m/s) of the rock (Nagra 2002). A special core recovery procedure and a pressure container were used to avoid the problem of core disking.

Table 1. Petrophysical data for Opalinus Clay (Benken site).

	No. of data	Mean and standard deviation	
		Core data	Logging data
V_p (m/s)			
Parallel to bedding	71	4030 ± 70	–
Normal to bedding	68	3030 ± 260	3190 ± 370
45° to bedding	14	3660 ± 300	–
V_s (m/s)			
Parallel to bedding	71	2280 ± 100	–
Normal to bedding	68	1710 ± 150	1670 ± 240
45° to bedding	14	2030 ± 160	–
E_{dyn} (GPa)			
Parallel to bedding	71	33 ± 3	–
Normal to bedding	68	19 ± 3	18.9 ± 6
45° to bedding	14	26 ± 5	–
ν_{dyn} (−)			
Parallel to bedding	71	0.27 ± 0.01	–
Normal to bedding	68	0.27 ± 0.01	0.31 ± 0.03
45° to bedding	14	0.27 ± 0.01	–

The physical and mineralogical parameters of the rock samples used for rock mechanical testing were determined to allow an unbiased classification of the samples. The following parameters were measured:

- Clay, quartz and carbonate content
- Density and water content
- Ultrasonic p- and s-wave velocity
- Dynamic elastic parameters

The petrophysical parameters given in Table 1 clearly show the anisotropy (transversely isotropic material) of the rock. The dynamic elastic parameters are quite high and reach values up to 33 GPa. The consolidation due to the deep burial of the Opalinus Clay is reflected in the relatively high wet bulk density of the rock (2523 ± 40 kg/m³) and the low porosity (12%).

Laboratory tests on core samples included uniaxial and triaxial testing, Brazilian tests, creep tests, slake durability tests and swelling tests. The tests were conducted using different loading directions with respect to the bedding planes to account for the significant anisotropy of the Opalinus Clay. The main features detected during these tests are:

- Transversely isotropic strength, deformation and swelling behavior (caused by the bedding)
- Strong dependence of material strength on water content (Figure 1)
- Pronounced hysteresis during loading/unloading cycles
- Strain hardening in the pre-failure region
- Strain softening in the post-failure region

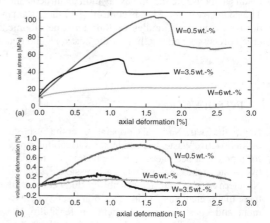

Figure 1. Stress–strain curve (upper figure) from undrained triaxial tests of Opalinus Clay (confining pressure 10 MPa) with different water content (w) – sample with intermediate water content = in situ water content, low water content = dried sample, high water content = saturated at low confining pressure. The lower figure shows the development of the volumetric strain during testing (positive number indicates volume reduction).

- Development of shear bands and distinct fracture planes
- Minor dilatancy
- Time-dependent deformation

Important strength parameters of the Opalinus Clay from the Benken site are given in Table 2.

Table 2. Static rock mechanical parameters.

Parameter	Orientation relative to bedding – mean and standard deviation – water content (w) about 4 wt.%		
	Parallel [P]	Normal [S]	45° [Z]
E_{stat} (GPa)	11.4 ± 3.7	5.5 ± 2.3	6.8 ± 4.2
ν_{stat} (−)	0.27 ± 0.10	0.27 ± 0.09	0.25 ± 0.11
UCS (MPa)	28.0 ± 5.7	30.3 ± 6.6	6.1 ± 2.6
$\sigma_{tensile}$ (MPa)	2.7 ± 0.8	1.2 ± 0.3	

Table 3. Final parameters of the modified Mohr–Coulomb model. The numbers in brackets (e.g. II) refer to Figure 2.

Parameter	Matrix	Bedding
Young's modulus	7 GPa	7 GPa
Poisson's ratio	0.27	0.27
Tensile strength ($\sigma_{tensile}$)	2.5 MPa (V)	1.2 MPa (VI)
Peak cohesion	8.7 MPa (I)	1.3 MPa (III)
	17.2 MPa (II)	10.1 MPa (IV)
Friction angle	30° (I)	34° (III)
	14° (II)	12° (IV)
Residual cohesion	4.3 MPa (I)	0.7 MPa (III)
	16 MPa (II)	5 MPa (IV)
Residual friction angle	29° (I)	33° (III)
	13° (II)	11° (IV)

bi-linear Mohr-Coulomb strength criterion with tension cut-off

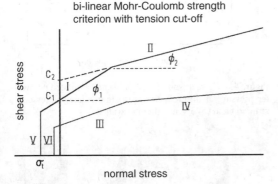

Figure 2. General representation of the Mohr–Coulomb model used for the numerical modelling of the tunnel deformation.

3 NUMERICAL MODELLING

3.1 Model set-up and constitutive law

A modified Mohr–Coulomb model is used for numerical predictions with the FLAC code (Itasca 2000) and includes multiple couplings:

- Anisotropic Darcy's law for the hydraulic behavior
- Transversal-isotropic, bi-linear, strain hardening and softening law including a tensile strength cut-off
- Creep law

Two basic models were used during this study. One was based on a linear stress–strain relationship in the pre-failure region and one assumed strain hardening. Tests showed that the general behavior was not influenced by the choice of these assumptions. Therefore, the main set of modelling runs were carried out using a linear stress–strain relationship before failure. In the post-failure region strain softening was taken into account for all modelling cases.

Figure 2 shows the basic Mohr–Coulomb model used to describe the strength of the material. A transversely isotropic bi-linear Mohr–Coulomb model with

tension cut-off adequately represented the strength of the material as observed in the laboratory tests.

Essentially, the basic set of parameters was extracted from laboratory tests on core samples from the Benken borehole (Northern Switzerland) was used to calibrate the model. The final model parameters are given in Table 3.

The time-dependent deformation was calculated by simulating the consolidation of the material during loading and 'creep' (or cataclastic flow as described in Nüesch (1991)). The creep behavior was modeled using the Salzer creep law (Salzer et al. 1998) which accounts for strain hardening. The creep law was further modified so that a threshold for the deviatoric stress could be specified. Below this threshold no creep was allowed. This feature had to be implemented because of the in situ stress field at the Benken site in the Opalinus Clay. The observation of highly deviatoric stresses could otherwise not be explained.

3.2 Model verification

Nagra is a partner in the international research project at Mont Terri, where a system of exploration tunnels and niches are used to conduct in situ experiments in Opalinus Clay. The Opalinus Clay at Mont Terri differs from the rock at the Benken site in that the maximum burial and actual depth are shallower in the Mont Terri case. Therefore, the rock at Mont Terri is less compacted and shows a porosity of 15–16% (Nagra 2002) and a higher water content (6–7%).

As explained above, the rock mechanical parameters are directly dependent on water content and the model had to be re-calibrated for the Mont Terri case. It should be noted that the calibration for the Mont Terri dataset followed the same approach as in the Benken case. In addition, the general structure of the constitutive law and the couplings were unchanged.

The calibration was carried out using a dataset gathered during a mine-by-test at the rock laboratory.

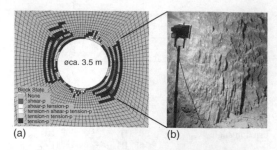

(a)   (b)

Figure 3. The numerical modelling of the excavation disturbed zone around a tunnel at Mont Terri (a) clearly shows the extensional features as observed in the sidewalls at Mont Terri (b, photo B. Niederberger) at the intersection of two tunnels.

The available dataset is described by Martin & Lanyon (2002) and Bossart et al. (2002). The maximum convergence detected during excavation was about 25 mm in a horizontal direction and 10 mm in a vertical direction in a tunnel with a diameter of about 3.5 m (overburden 250–300 m). The failure in the excavation disturbed zone was mainly extensional (Figure 3b) in the sidewalls of the tunnels, while combined tensional and shear failure along the bedding was observed in the roof and invert. The porewater pressure measured in the sidewalls showed an increase immediately after tunnel construction and a subsequent decrease due to drainage into the tunnel.

The calculated response to tunneling in Figure 3a clearly indicates the creation of extensional features.

The comparison of the calculated and observed porewater pressures immediately after tunnel construction and 3 years later also indicate a good match (Figure 4).

3.3 Predictive modelling for the Benken site

Numerous modelling runs have been carried out to investigate the development of the EDZ along emplacement tunnels for radioactive waste.

The results demonstrate that the excavation disturbed zone (EDZ) is mainly influenced by the anisotropy of the rock and the in situ stress field. In the roof and invert, the EDZ is controlled by tensile and shear fractures of the bedding, whereas in the sidewalls there will be predominately tensile fractures through intact rock (Figure 5). The geometry of the EDZ is mainly controlled by the instantaneous ground reaction (undrained elasto-plastic behavior) during excavation.

The effect of parameter variations of about 30% and variations of the stress tensor (30%) were investigated. The results in Figure 5b–h indicate that such variations do not change the overall geometry of the EDZ compared to the reference model. Investigations of the

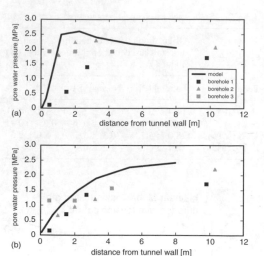

Figure 4. Comparison of measured and calculated porewater pressure at different times. (a) Immediately after construction and (b) 3 years after excavation.

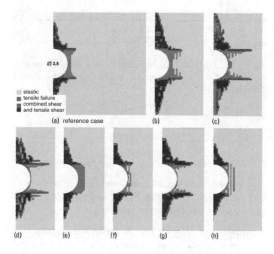

Figure 5. Predicted development of the EDZ for the Benken site. The EDZ in the roof and invert is caused by bedding plane failure while the sidewalls show tensile failure of the intact rock. Reference (a) and alternative cases (b–h), see text for explanation.

deformation and failure mechanisms show that the EDZ in the roof and invert is bounded by bands of enhanced shear deformation (Figure 6). Inside the area bounded by these shear zones, mainly tensile failure of the bedding plane is observed. This leads to deformation patterns which are well known from investigations of failure patterns in anisotropic rocks (Jacobi 1981, Ortlepp 1997, Økland & Cook 1998). The observed

130

(a) deformed and undeformed grid

displacements magnified 30 times

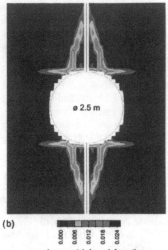

(b)

ø 2.5 m

0.000 0.006 0.012 0.018 0.024

max. incremental shear deformation

Figure 6.   (a) Modeled displacements and (b) localized shear deformations in the vicinity of the tunnels. The asymmetry of the EDZ is caused by the anisotropy in the material strength; the horizontal and vertical stresses are nearly identical (15.1 and 15.9 MPa resp.).

failure pattern is caused by the anisotropy of the material parameters rather than the stress field, as the stress field was assumed to be nearly isotropic (horizontal stress 15.1 MPa, vertical stress 15.9 MPa).

It should be noted that none of the modelling cases resulted in total failure of the tunnel and it can be stated that the construction of the emplacement tunnels can be carried out without any major support.

3.4 *Long-term development*

Numerical investigations have been carried out to evaluate the long-term development of the EDZ. The boundary conditions for this evaluation are the following: The waste is emplaced in the tunnels immediately after construction. Part of the waste emplacement procedure is the backfilling of the gaps between waste container and tunnel wall with buffer material (highly compacted bentonite). Therefore, the emplacement tunnels are open for 2 years at the maximum.

Due to this approach, tunnel convergence will be limited due to the mechanical behavior of the buffer material. The saturation of the buffer and the host rock in the immediate vicinity of the tunnel will cause high swelling pressures in the buffer on one hand and additional closure of the tunnel cross section on the other hand (Konietzky et al. 2002). This convergence will continue to compact the buffer and therefore increase swelling pressure until swelling pressure and far field stress equal out. The swelling of the host rock in the vicinity of the tunnel, together with micro-fracturing and disintegration processes, will lead to a homogenization

of pore space distribution and a reduction of the hydraulic conductivity of the EDZ, as has been demonstrated with an in situ self-sealing experiment at Mont Terri (Nagra 2002).

4   CONCLUSIONS

A comprehensive investigation of the mechanical and hydro-mechanical behavior of the Opalinus Clay has been carried out. Based on laboratory tests and in situ experiments, a constitutive model has been evaluated and a numerical model set up, calibrated and validated with in situ experiments. The convergence, deformation pattern and the porewater pressure distributions observed in a mine-by-test at the underground rock laboratory at Mont Terri (Switzerland) were calculated adequately with the numerical model. Therefore, it was assumed that the model can be used for further predictive modelling for emplacement tunnels at a potential site for radioactive waste disposal.

The conditions at the potential disposal site (Zürcher Weinland), which were explored by 3D seismic investigations and the Benken borehole (Nagra 2002), are different to those at Mont Terri. Due to deeper burial and a different tectonical setting, the stresses are significantly higher and the orientation of bedding is less favorable (parallel to tunnel axis) in the Zürcher Weinland. Nevertheless, it could be shown that tunneling is feasible under the given conditions and either no or only minor support for the emplacement tunnels is necessary. The geometry of the excavation disturbed zone (EDZ) of the emplacement tunnels is mainly controlled by the anisotropy of the Opalinus Clay. The stress anisotropy is less important for these tunnels as their axis will be aligned in the direction of the maximum horizontal principal stress. Therefore, stresses acting at the tunnel cross-section (minimum horizontal principal stress (15.1 MPa) and the vertical principal stress (15.9 MPa) are nearly identical.

REFERENCES

Bossart, P., Meier, P.M., Moeri, A., Trick, T. & Mayor, J.-C. (2002): Geological and hydraulic characterisation of the excavation disturbed zone in the Opalinus Clay of the Mont Terri Rock Laboratory, *Engin. Geol. 2040.*

Itasca (2000): FLAC-Manuals, (Vers. 4.0): Itasca Consulting Group, Minneapolis, MN, USA.

Jacobi, O. (1981): *Praxis der Gebirgsbeherrschung*, Verlag Glückauf, Essen.

Konietzky, H., Blümling, P. & te Kamp, L. (2002): Opalinuston: Felsmechanische Untersuchungen. *Unpubl. Nagra Int. Rep.*

Martin, C.D. & Lanyon, G.W. (2002): EDZ in Clay Shale: Mont Terri, *Mont Terri Techn. Rep. 2001-01.*

Nagra (2002): Projekt Opalinuston: Synthese der geowissenschaftlichen Untersuchungsergebnisse. *Nagra Tech. Ber. NTB 02–03.*

Nüesch, R. (1991): Das mechanische Verhalten von Opalinuston. *Diss. ETH Zürich Nr. 9349.*

Økland, D. & Cook, J.M. (1998): Bedding-related borehole instability in high-angle wells. Eurock '98 (Trondheim, Norway, 8–10 July 1998) *Proc. SPE/ISRM 47285.*

Ortlepp, W.D. (1997): *Rock Fracture and Rockburst* – An Illustrative Study. The South African Inst. of Mining and Metallurgy, Monograph Ser. M9, Johannesburg.

Salzer, K., Konietzky, H. & Günther, R.-M. (1998): A new creep law to describe the transient and secondary creep phase. *Proc. NUMEG 98, Springer, Wien.*

Geotechnical Measurements and Modelling, Natau, Fecker & Pimentel (eds)
© 2003 Swets & Zeitlinger, Lisse, ISBN 90 5809 603 3

Partial backfilling effect in underground mines

T. Collet & F. Masrouri
Laboratoire Environnement, Géomécanique & Ouvrages, Ecole Nationale Supérieure de Géologie, Vandœuvre-lès-Nancy, France

C. Didier
Institut National de l'Environnement Industriel et des Risques, Verneuil-en-Halatte, France

ABSTRACT: A series of instruments (borehole extensometers and total pressure cells) was used to study the behaviour of three pillars confined by partial backfilling in Livry-Gargan Mine, France. Understanding the influence of partial backfilling on the pillar stability is necessary to evaluate the interest of using this method to reduce the surface implications of underground exploitations. This article describes an on site instrumentation carried out in this mine. The specifications and the location of these instruments were chosen to complete data from a previous investigation. Numerical modelling using the finite-elements program Plaxis confirmed the backfill influence on the pillars stability: partial backfilling seems to generate a contraction of the lower part of the pillars.

1 INTRODUCTION

Pillar failure in abandoned room-and-pillar shallow mines can generate surface subsidence with possible severe public safety and environmental implications. In this context, partial backfilling of the cavities is proposed to reduce and to control the hazards linked to the underground exploitations.

This method consists of the optimisation of backfill height regarding to safety and economical reasons. Partial backfilling presents three major interests. Firstly, backfilling reduces the volume of voids, and therefore the amount of subsidence. Secondly, some direct support is provided to the floor, thus it can limit heave. Finally, backfill generates lateral support to the mine pillars.

A study based on finite-element model and on site investigations was carried out in an underground gypsum mine for a better understanding of pillars and backfill combined behaviour.

Certainly regarding to stability, it is better to fill up the underground voids as much as possible. However, the benefits of increasing backfill height must be weighed against its cost. If security conditions are conform to legislation for subsidence effects, partial backfilling of underground openings is more cost-effective than total backfilling.

2 LIVRY-GARGAN MINE

This underground gypsum mine is located at Livry-Gargan, 20 kilometers north-east of Paris, France. It has a maximum width of 500 meters and a maximum length of 1 kilometer, the depth is about 30 meters.

2.1 Mining method

The exploitation creates pillars and rooms of 17 meters high and 8 meters wide, there is a 75 % extraction ratio. The excavation is executed in three stages: one drifting of 7 meters high and two benches of 5 meters high. Gypsum in the lower 2 meters of the pillars is softer than the rest of pillar, its degradation can be faster (Fig. 1).

To prevent pillar failure, partial backfilling of 8 meters high is realized during first six months after excavation. This partial backfilling is executed as following: after a trimming of the pillars, materials are carried using trucks and bulldozers. There is no backfill compaction, except the equipment traffic effect.

Total backfilling, depending on the material supply, is realized 3 to 5 years later.

Figure 2 shows the plan view of the studied shallow mine, with total or partially backfilled areas and height of the excavation.

Figure 1. Softer layer in the lower part of the pillar.

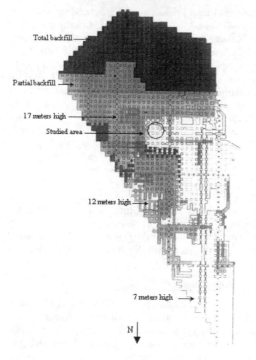

Figure 2. Livry-Gargan Mine plan view.

2.2 Why backfilling?

In this shallow mine, the pillars slenderness ratio is important. To evaluate the stability from a strictly mechanical way, the following equation was used to calculate the Rankine critical load:

$$\frac{1}{P_r} = \frac{1}{P_E} + \frac{1}{P_C} = \frac{1 + \dfrac{\sigma_c \cdot \lambda^2}{4 \cdot \pi \lambda^2 \cdot E}}{\sigma_c \cdot S} \quad (1)$$

where P_E is the critical Euler load (kN), P_C the critical uniaxial compressive load [kN], P_r the critical Rankine load [kN], σ_C the overburden pressure [kPa], λ the slenderness ratio [–], E Young's modulus [kPa] and S the column section [m^2].

The critical Rankine load is, in this case, about 2.4 MPa. Constitutive materials have a long-term uniaxial compressive strength of 2.6 MPa and a short-term strength of 5.8 MPa (Lucas 2000).

The critical Rankine load is similar to the long-term uniaxial compressive strength. So, in this context, if a degradation takes place, the dimension of the pillars could be insufficient, the overburden load could carry a pillar failure.

3 INSTRUMENTATION 1

During 2000, two pillars were instrumented with pressure cells and single extensometers. Earth pressure cells were placed in the pillar-backfill interface and borehole extensometers were installed in gypsum pillars to measure displacements due to backfill pressure.

3.1 Pressure cells

It was considered that a backfill near a pillar generates the same pressure as a soil against a retaining wall. The magnitude of the lateral earth pressure depends on the movements of the pillar relative to the backfill. The lateral pressure is minimum when the soil immediately behind the pillar is at the point of failure due to a lateral expansion and it is maximum when the failure is due to a lateral compression of the soil: the soil, at these two limiting failure conditions, is said to be in an active or passive state of plastic equilibrium.

If we consider a friction angle of 30°, the value of the coefficient K, ratio between horizontal and vertical stresses is:

– Coefficient of earth pressure at rest $K_0 = 0.5$
– Coefficient of active earth pressure $K_A = 0.33$
– Coefficient of passive earth pressure $K_P = 3$

Figure 3 shows the measured variations of the coefficient K in two cases. The cell C1, at a depth of 6 meters below backfill surface, gave values of K between 0.3 and 0.5, corresponding to an active state of plastic equilibrium. The cell C6, at a depth of 1 meter, indicated that the soil is in a passive state.

We can consider that backfilling generates a pillar contraction in its lower part and seems to lead to an extension on the level of the backfill surface.

3.2 Single extensometers

Single extensometers were installed 2 meters above the wall strata. Initially, we placed these extensometers to

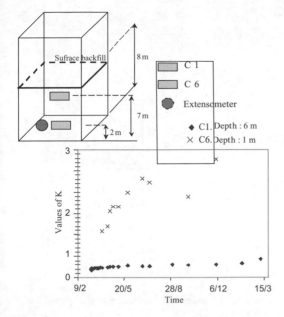

Figure 3. Variations of K for two different earth pressure cells.

Table 1. Laboratory tests results.

	Material N°1	Material N°2
W_{nat} (%)	14 ± 0.5	16 ± 3.43
γd_{opt} (kN·m^{-3})	19.2	16.6
W_{opt} (%)	13	19
% of particle size		
>50 mm	0	3
20 mm ≪ 50 mm	1.3	8.7
400 µm ≪ 20 mm	32.1	15.3
80 µm ≪ 400 µm	35.4	28.5
<80 µm	31.2	44.5
VBS	2.02 ± 0.2	2.53 ± 0.6
Atterberg limits		
w_L (%)	26.5 ± 0.5	47.5 ± 4
w_p (%)	20	35.5 ± 4
I_P	6.5 ± 0.5	12 ± 2.4
UU triaxial tests		
C_{uu} (kPa)	57	88
φ_{uu} (°)	24	35
E (MPa)	10	17

show an extension of the pillars. The measures indicated their contraction that reached 25 mm. After that, it was not possible to continue the measurements.

3.3 Site and laboratory backfill characterizations

Pressiometric and dynamic penetrometric tests were performed in the backfill. A compact layer is located at its surface due to the traffic effect and a dense layer is at a depth of 4 meters, which was probably a ramp for the backfilling.

Laboratory tests indicated the heterogeneity of the materials used as shown in the Table 1 that compares laboratory test results for two materials.

4 NUMERICAL MODELLING

A simple model using Plaxis software was performed to estimate the pillars behaviour during the partial backfilling. The results of these numerical calculations were compared to the measured data.

The geometry, shown in Figure 4, simulated the exploitation.

The interface between the backfill and the pillars was calculated by applying a 60% reduction of the soil properties.

The Mohr-Coulomb model was used as a first approximation of soil behaviour. This model involved five parameters: Young's modulus, Poisson's ratio, cohesion, friction angle, and dilatancy angle. The

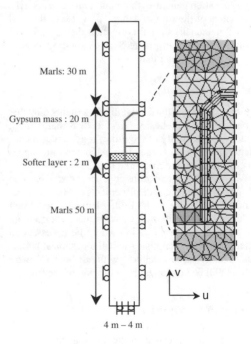

Figure 4. Geometry and boundary conditions.

values of the gypsum and marls parameters came from a bibliographic synthesis (Jardin 1975, Lucas 2000) and the backfill parameters were determined by laboratory tests.

135

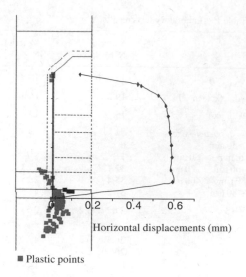

Horizontal displacements (mm)

■ Plastic points

Figure 5. Plastic points and horizontal displacements after the excavation.

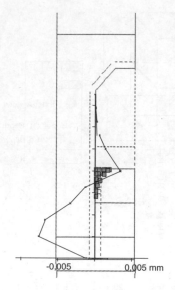

-0.005   0.005 mm

Figure 6. Plasticity and displacements after partial backfilling.

4.1 *Excavation*

The excavation generates three main movements: (i) a subsidence of the roof of 4 mm, (ii) a rise of the wall of 10 mm and (iii) a pillar horizontal displacement of 0.6 mm, as shown in Figure 5. A plasticity occurs in the lower part of the pillars.

4.2 *Partial backfilling*

Figure 6 shows the horizontal displacements only due to backfilling. A cohesive backfill of 8 meters high reduces the plasticity of the pillar, it generates a contraction in its lower part and an extension in its upper part, similarly to the measured data.

We noticed a great difference between measured (25 mm) and calculated displacements (0.005 mm), this could be partly due to the finite elements model that neglects cracks in pillars. It was considered that a more complete on site investigation would be necessary to better understand the pillar-backfill mechanical behaviour. Numerical modelling with distinct element method will be hold on in the following of the study.

5 INSTRUMENTATION 2

From July 2002, a second on site investigation was carried out in the same mine. The objective were to quantify the pillars displacement near the backfill surface (extension or contraction) and to monitor cracks in the lower part of the pillar.

The studied area of three pillars is shown in Figure 2. These pillars were totally excavated in November 2002. Four months later, mechanical cracks have been noticed in the lower part of these pillars, principally in their corners. After then, the partial backfilling was performed in April 2003.

5.1 *Earth pressure cells*

The Glötzl total pressure cells (EEKO 20/30 K5) consist of a sealed distribution pad filled with de-aired oil. The pad is connected to a pressure transducer. Variations in oil pressure resulting from changes of load acting on the pad are sensed by the electrical 4–20 mA transducer.

These total pressure cells were placed at a depth of about 2, 4 and 6 meters from the surface of the backfill. The measuring range of these cells is 0–500 kPa. The cells were surrounded by a fine soil for a best distribution of the stresses.

5.2 *Single boreholes extensometers*

Single extensometers (Model 8709, INERIS) were used near the backfill surface (in the middle of the pillars) to study the pillar displacements at this point. The displacements were measured remotely using a linear potentiometer.

To install these instruments, boreholes were drilled in the pillars. The collar of the hole was enlarged to facilitate the transducer installation to protect the head assembly from physical damage. The extensometer rods and anchors were assembled and inserted in the borehole, and fixed to a reference rod head (Fig. 7). The anchor bolting was used at a depth of 4 meters.

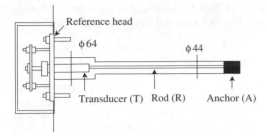

Figure 7. Single extensometers (Model 8709, INERIS).

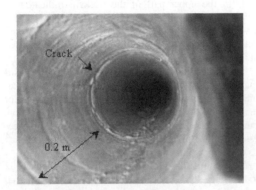

Figure 8. Crack in a borehole.

5.3 Multiple-point extensometers

In the lower part of the pillar, Telemac extensometers with three anchors (SAM-E/M borehole extensometer) were installed to know if the displacement is due to a cracking in the border or to the pillar mass deformation.

The head assembly is composed of a watertight module containing 3 electrical transducers, used to monitor the displacement between the rod and the head of individual extensometers.

To determine anchor spacing, geological factors, mainly the cracking of the gypsum mass were determined by visual inspection and endoscopy. Figure 8 shows a crack in a borehole. For each extensometer, the depth of the anchors is 0.2, 1 and 3 meters. 150 mm micrometers with a resolution of 0.01 mm were used to have a large measuring range.

5.4 Instruments installation

Figure 9 shows the position of the instruments installed in three pillars. Two faces of a pillar were instrumented with two single extensometers (E5–E6), two three-points extensometers (E8–E9) and six total pressure cells (C11 to C16) to monitor its behaviour during the backfilling.

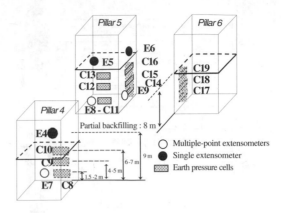

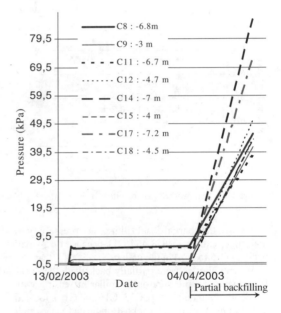

Figure 9. Position of the instruments.

Figure 10. Variations of the pressure.

Two faces of pillars were instrumented to measure the distribution of the partial backfilling stresses and the displacements in two galleries, one main gallery with major traffic and one secondary with a lower traffic.

5.5 First results

This instrumentation was installed from July 2002 to February 2003, and the partial backfilling was realized in April 2003. Here, the first results concerning pressure cells, single extensometers and one three-points extensometer are discussed.

Figure 10 shows measured pressures at different depths.

137

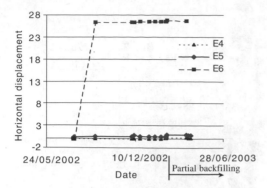

Figure 11. Single extensometers results.

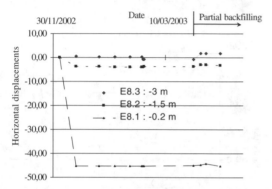

Figure 12. Extensometer E8 results.

Before the partial backfilling, no pressure was measured, except for C8 and C11 because of a small backfill (1.5 to 2 m) in the gallery.

When galleries were partially backfilled, the backfill generated a pressure on the pillar increasing with the depth. Four cells (C10, C13, C16 and C19), located at the upper part of the backfill, indicated very small values.

Single extensometers E4, E5 and E6 were installed in the middle of the pillars (9 m). Before the backfilling, an extension of the pillar 5 was indicated by E6. This could be due to a crack opening in this pillar. E4 and E5 gave very small values of extension (Fig. 11). After backfilling, there is no changes in the displacements yet.

Figure 12 shows the displacements of the three anchors for the extensometer E8. Before the backfilling, a measured displacement of about 45 mm between anchors located at 0.2 and 1.5 m could be due to a weathering of the outer part of the pillar.

After backfilling, the anchors seemed to indicate a beginning of contraction, but more values are necessary

to better study the behaviour of the pillar. For this reason, measurements will be continued until 2005.

6 CONCLUSION

To study the influence of the partial backfilling on the pillar stability, on site instrumentations with extensometers and earth pressure cells were carried out in the Livry-Gargan gypsum mine. This mine is partially backfilled (8 meters high for galleries of 17 meters high) during the first six months after excavation.

A first installation was realized in 2000. Pressure cells in the upper part of the backfill indicated an extension of the pillar, because the soil was in a passive state of plastic equilibrium. However, there was no extensometers to study the displacements of the pillar at this level.

Finite-element calculations were performed to estimate the effect of the partial backfill on the pillar stability. The backfill generates the same displacements of the pillar as those measured by on site instrumentation, but with a lower amplitude. Consequently, it seems necessary to take pillar cracking into account to approach its real behaviour.

The second installation was realized from July 2002 to February 2003. Pressure cells showed that the partial backfill generated a pressure on the pillars. Single horizontal extensometers indicated an extension of the middle of the pillars before the backfilling. Multiple-point extensometers, installed two meters above the wall strata, indicated an extension, probably due to the opening of a crack.

REFERENCES

Jardin J. 1975. Fondations sur le gypse. Expériences acquises en région parisienne. *Bulletin de Liaison des Ponts et Chaussées*, 78.
Zhou Y. 1990. Controlling subsidence effects using partial backfilling. 9th *international conference on ground control in mining*: 193–197.
Hassani F.P., Fotoohi K. & Doucet C. 1998. Instrumentation and backfill performance in a narrow vein gold mine. *International Journal of Rock Mechanics and Mining Sciences* 35 (4–5).
Lucas Y. 2000. Etude de mise en sécurité des terrains de surface à l'aplomb de carrières souterraines, INERIS report.
Collet T., Masrouri F. & Didier C. 2003. *Mise en sécurité des carrières souterraines. Post-Mining 2003, Impacts and risk management, International Symposium.* Nancy, France.
Tesarik D.R., Seymour J.B. & Yanske T.R. 2003. Post-failure behavior of two mine pillars confined with backfill. *International Journal of Rock Mechanics and Mining Sciences* 40: 221–232.

Geotechnical Measurements and Modelling, Natau, Fecker & Pimentel (eds)
© 2003 Swets & Zeitlinger, Lisse, ISBN 90 5809 603 3

Initial rock stress in the Gorleben salt dome measured during shaft sinking

S. Heusermann, R. Eickemeier, K.-H. Sprado & F.-J. Hoppe
Federal Institute for Geosciences and Natural Resources (BGR), Hannover, Germany

ABSTRACT: The Gorleben salt dome, Germany, has been investigated for its suitability as a site for the final disposal of radioactive wastes. During sinking of the two shafts of the exploration mine a comprehensive geotechnical exploration program has been performed including the measurement of rock temperature, rock deformation, shaft closure, initial rock stress, long-term stress change, and sample collection for laboratory testing. To determine initial rock stress, overcoring tests at three levels in each shaft were carried out. Evaluation of the test data included the determination of necessary material parameters in the laboratory and finite-element modelling of the overcoring method to provide a proper conversion of measured stress-release deformations to stress taking into account the nonlinear time-dependent creep behaviour of salt. The results indicate a more or less isotropic initial stress state in the Gorleben salt dome. The measured stress components agree sufficiently to the theoretical overburden pressure.

1 INTRODUCTION

To prove the suitability of the Gorleben salt dome as a permanent radioactive waste repository site, extensive geoscientific investigations have been made from above ground since 1978 and from underground since 1986. The underground exploration started with the sinking of two shafts with a total depth of 933 m (shaft GS1) and 843 m (shaft GS2). Parallel to shaft sinking and to following excavation of the exploration mine, extensive geological and geotechnical in-situ investigations have been performed to provide the basic data for the suitability statement of the final waste disposal (Wallner et al. 1998).

The geological investigations comprise the characterization of the geological structure. The geotechnical measurements in both shafts include:

- measurement of rock temperature,
- measurement of rock deformation using extensometers, inclinometers, and shaft closure measuring devices,
- measurement of initial rock stress using the overcoring technique,
- measurement of rock stress change using stationary stress monitoring probes each containing several hydraulic pressure cells,
- drilling of salt core samples for geomechanical laboratory tests to determine stiffness, strength, and creep parameters.

Drilling of measurement boreholes and installation of stationary measuring devices was made during shaft

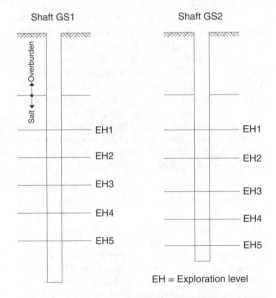

Figure 1. Schematic view of the shafts GS1 and GS2, Gorleben salt dome, with selected exploration levels EH1 to EH5.

sinking at several exploration levels to observe the salt rock behaviour at different depth. Figure 1 shows a schematic view of both shafts and the selected exploration levels EH1 to EH5. As an example, Figure 2 shows a horizontal cross section of exploration level EH2, shaft 2, at a depth of approximately 481 m

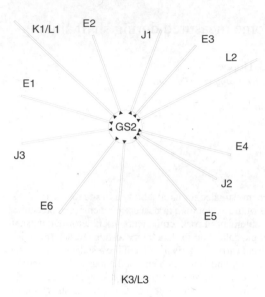

Figure 2. Cross section of the exploration level EH2, shaft GS2, with measurement boreholes (E = extensometers, J = inclinometers, K = overcoring tests, L = stress monitoring probes).

including all geotechnical exploration and measurement boreholes and shaft closure measuring points.

Special effort was spent on the measurement of initial rock stress as a fundamental parameter for geomechanical models to calculate and to assess the long-term stability of the repository and the integrity of the salt barrier. To this end, BGR carried out numerous overcoring stress measurements at different exploration levels to determine amount and direction of stresses in the virgin salt dome as well as the variation of stress with increasing depth.

2 STRESS MEASUREMENT TECHNIQUE

Over the last two decades, several overcoring probes have been developed and improved by BGR for various rock types and applications (Pahl & Heusermann 1991). Most frequently used by BGR are two-dimensional measuring probes which contain four LVDT gauges arranged in radial direction at an angle of 45° to each other. In shallow boreholes up to a depth of max. 50 m, an installation rod is used to emplace the probe at the desired position in the borehole. The probe is fixed in the pilot borehole by spring-loaded steel strips (Fig. 3). For installation in deeper vertical boreholes up to a depth of max. 200 m, the probe is lowered to the proper depth on a cable that also provides the electrical connection to the transducers. A further development of the 2D-Probe is the BGR 3D-overcoring probe which comprises the basic model as described

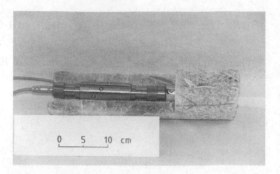

Figure 3. BGR 2D-overcoring probe in an overcored salt rock sample.

above with an additional LVDT transducer oriented along the borehole axis to detect the axial stress-release deformation of the overcored pilot borehole.

The data obtained during overcoring are transmitted via a wireline to the data amplifier and can be recorded digitally by an AD converter and a computer. Alternatively, data acquisition can be done with a special downhole computer incorporated in the overcoring probe.

BGR overcoring tests require a pilot borehole with a diameter of 46 mm which is drilled in the center of the bottom of a 146 mm-diameter borehole. The probe is inserted through the drill stem into the pilot borehole. The data is recorded continually during overcoring, permitting immediate control of the test and observation of the complete deformation behaviour of the overcored pilot borehole. With respect to the distinct time-dependent behaviour of salt and to avoid different test conditions, the drilling rate and advance of overcoring in salt rock should be approximately 2 cm/min and the time elapse between drilling the pilot borehole and performing the overcoring test should be about 1 hour.

3 OVERCORING DATA

From October 1994 until July 1995, a total of 93 overcoring tests were carried out during sinking of the Gorleben shafts GS1 and GS2. Table 1 comprises the data of the several test locations including exploration level, total depth, maximum borehole depth, borehole direction, and number of tests in each borehole. Examples of the test results are shown in Figure 4, where the change in diameter of the pilot borehole is plotted against the overcoring time. It can be seen that, initially, the pilot borehole undergoes more or less considerable convergence. This convergence is caused by the temporary stress concentration occurring at the front of overcoring and includes not only a reversible elastic component but also sizeable irreversible inelastic components. This is followed by a

Table 1.   Data of overcoring test locations in the Gorleben shafts GS1 and GS2.

Shaft/exploration level	Total depth (m)	Bore hole no.	Borehole length (m)	Borehole direction (gon)	Number of tests
GS1/EH2	479.4	K1	18.60	50.0	6
GS1/EH2	479.4	K2	25.80	169.7	10
GS1/EH2	479.4	K3	23.90	298.8	6
GS1/EH4	677.2	K1	21.92	50.4	7
GS1/EH4	677.2	K2	30.65	170.1	6
GS1/EH5	744.2	K1	24.20	49.6	5
GS1/EH5	744.2	K3	24.30	290.2	7
GS2/EH2	478.6	K1	23.52	349.9	10
GS2/EH2	478.6	K3	27.00	204.9	6
GS2/EH3	601.6	K1	26.94	350.1	9
GS2/EH3	601.6	K2	20.15	69.9	6
GS2/EH5	791.2	K2	28.13	70.0	7
GS2/EH5	791.2	K3	30.90	204.9	8

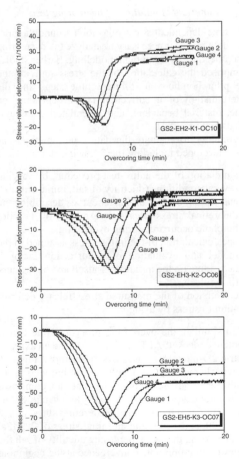

Figure 4.   Stress-release deformations measured in overcoring tests at different depth (top: shallow depth, middle: medium depth, bottom: larger depth).

phase of rapid stress release, as shown by an increase in the pilot borehole diameter. After the overcoring procedure has advanced well, more or less constant values are obtained. The observed deformation behaviour differs significantly from that behaviour measured in overcoring tests in elastic rock types and is characteristic for tests in nonelastic rock types like salt.

Comparing the three diagrams plotted in Figure 4, different types of overcoring curves can be observed depending on the depth of the several exploration levels. The upper diagram represents test data obtained at shallow depth, e.g. exploration level EH2 at a depth of about 479 m, and yields lower values of the pilot borehole convergence (up to $-18\,\mu$m). The middle diagram is characteristic for tests performed at medium depth, e.g. exploration level EH3 at a depth of about 602 m, and yields medium values of the pilot borehole convergence (up to $-32\,\mu$m). The lower diagram represents test data obtained at larger depth, e.g. exploration level EH5 at a depth of about 791 m, and yields higher values of the pilot borehole convergence (up to $-75\,\mu$m). These examples indicate that, in principle, the deformation behaviour, especially the pilot borehole convergence occurring temporarily at the beginning of the overcoring, is strongly influenced by the nonlinear time-dependent behaviour of salt rock as well as by the initial rock stress itself and, in opposite to overcoring tests in elastic rock types, must be considered during evaluation of the data.

Earlier numerical model calculations of the overcoring procedure show that, assuming nonlinear inelastic material behaviour and using a multilinear elastic−plastic constitutive model, the characteristic data of overcoring tests in salt rock can be reproduced (Heusermann 1993). The diagram plotted in Figure 5 comprises three overcoring curves calculated with different initial stress ($p = 10$, 15, and 20 MPa).

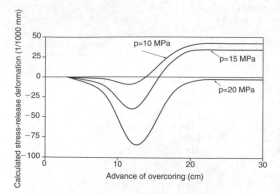

Figure 5. Variation of theoretical stress-release deformation with initial stress p obtained from finite-element modelling (Heusermann 1993).

All curves are very similar to the measured data and verify that the variation of the stress-release deformation with overcoring progress, especially the inelastic pilot borehole convergence at the beginning of overcoring, is strongly influenced by the nonlinear inelastic behaviour of rock salt.

4 ANALYSIS OF OVERCORING DATA

4.1 Elastic rock behaviour

To convert measured stress-release deformation to rock stress, the following equation is commonly used in the case of elastic rock behaviour:

$$\sigma = \frac{E}{2(1 - \nu^2)} \cdot \frac{\Delta D_e}{d} \tag{1}$$

where σ is the initial stress, E is the modulus of elasticity, ν is the Poisson's ratio, d is the pilot borehole diameter, and ΔD_e is the measured stress-release deformation (change in pilot borehole diameter determined as difference between the initial and the final pilot borehole diameter).

4.2 Nonlinear creep behaviour

Evaluation of overcoring tests in rock salt must take into account the characteristic time-dependent stress relaxation around the pilot borehole as well as the inelastic pilot borehole convergence as shown in Figure 4. To consider the nonlinear creep behaviour of rock salt, the following constitutive model can be used assuming steady-state creep (Hunsche & Schulze 1994):

$$\dot{\varepsilon} = A \cdot exp(-Q/RT) \cdot (\sigma/\sigma^*)^n \tag{2}$$

where $\dot{\varepsilon}$ is the deviatoric creep rate $(1/d)$, A a scale factor $(0{,}18\ 1/d)$, Q the activation energy (54 kJ/mol), R the universal gas constant $(8.3143 \cdot 10^{-3}\,kJ/K/mol)$, T the temperature (K), σ the deviatoric stress (MPa), σ^* the reference stress (1 MPa), and n the stress exponent (5.0).

Equation 2 is only valid if a long time elapses between drilling and overcoring of the pilot borehole. With respect to the short period of the overcoring procedure (1–2 h), transient creep effects have to be considered. Alternatively, the large creep rates occurring after 1–2 h can be taken into account assuming increased steady-state creep rates with

$$A = A^* \cdot 0.18\ (1/d) \tag{3}$$

and $A^* = 500 \div 1000$, corresponding to creep rates measured in creep tests on rock salt after 1 h.

4.3 Numerical simulation of overcoring tests

The time-dependent stress relaxation around the pilot borehole makes it necessary to analyze the overcoring test procedure by numerical modelling. Earlier studies comprised the calculation of the stress state around the pilot borehole at different time steps and the determination of a correction factor depending on time, material behaviour, and several test conditions (Heusermann 1984).

Actually, to evaluate the data a new method has been developed including the following steps:

- simulation of the entire test procedure by varying several conditions (ductility of salt, initial stress),
- calculation of a set of master curves describing the theoretical stress-release deformation of the pilot borehole occurring during overcoring,
- back calculation of the initial stress assumed in the model and evaluation of a correction factor κ depending on the initial stress itself and on the ductility of salt,
- conversion of the measured stress-release deformation to stress by using κ.

To simulate the overcoring procedure, the finite-element code ANSALT I (Nipp 1991) and an axisymmetric 2D-model was used. As an example, Figure 6 shows characteristical results assuming elastic or nonlinear creep behaviour. The diagram comprises the theoretical pilot borehole diameter change versus overcoring depth assuming different values of A^* (100, 500, and 1000) and an initial stress of 22 MPa. It can be seen, that with increasing ductility of salt the inelastic pilot borehole convergence at the beginning of overcoring increases and due to stress relaxation around the pilot borehole the total stress-release deformation ΔD decreases.

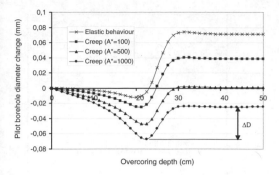

Figure 6. Theoretical overcoring curves for different creep rates of the salt rock.

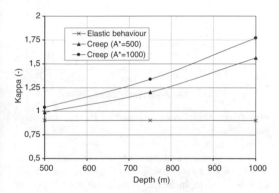

Figure 7. Variation of correction factor κ with depth (resp. initial stress).

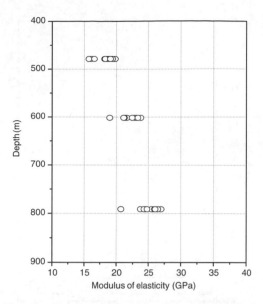

Figure 8. Variation of the modulus of elasticity with depth (shaft GS2).

The correction factor κ is determined from these master curves. The diagram plotted in Figure 7 shows κ depending on the overburden pressure respectively on the depth of the several exploration levels in the Gorleben salt dome for different values of A^*. To convert measured data to stress, κ is determined from Figure 7. The corrected stress-release deformation ΔD_e is calculated by

$$\Delta D_e = \Delta D \cdot \kappa(\sigma, A) \qquad (4)$$

and is converted to stress using Equation 1.

5 INITIAL STRESS STATE IN THE GORLEBEN SALT DOME

A basic requirement of performing and evaluating overcoring tests is the determination of the modulus of elasticity E to convert measured stress-release deformation to stress (Eq. 1). To this aim, the hollow cylinder cores drilled in situ in overcoring tests were investigated in laboratory in load-deformation tests using a Robertson biaxial cell and a measuring device very similar to the overcoring measurement equipment. For each core, an individual modulus of elasticity was determined from several unloading stages at different load levels.

As an example, Figure 8 shows the variation of the modulus of elasticity with depth in the Gorleben shaft GS2. Due to differences in the stratigraphy and in the material behaviour of several salt layers of the Gorleben salt dome, lowest values of E were obtained at the exploration level EH2 at a depth of 479 m (about 15 to 20 GPa), highest values occur at the exploration level EH5 at a depth of 791 m (about 21 to 27 GPa). The large variety of E makes it necessary to use individual values for the evaluation of each overcoring test instead of a constant average value for all tests.

The results of overcoring tests carried out in the Gorleben salt dome in the shafts GS1 and GS2 at several exploration levels are plotted in Figures 9 and 10. The diagrams show the maximum and minimum stress components S_1 and S_2 evaluated from the test data. For each exploration level, the average values of S_1 and S_2 were determined assuming a medium ductility of salt ($A^* = 750$). To consider lower and higher creep rates, a bandwidth of S_1 and S_2 is shown according to lower and higher values of κ ($A^* = 500$ and 1000).

The experimental results are compared to the theoretical overburden pressure obtained with an assumed density of rock salt of $\rho = 2.1$ to 2.3 kg/dm^3. It can be seen that only a slight anisotropy of stress occurs

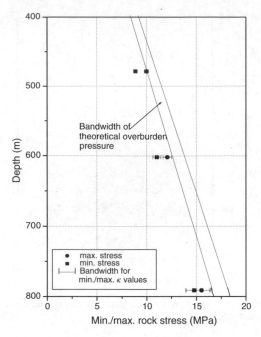

Figure 9. Variation of measured rock stress with depth (shaft GS1).

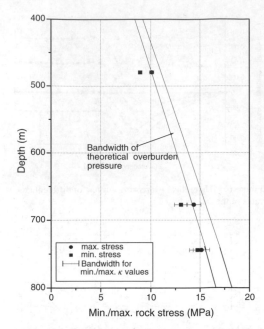

Figure 10. Variation of measured rock stress with depth (shaft GS2).

indicating a more or less isotropic stress state and that the vertical distribution is very similar to the theoretical overburden pressure ($\rho = 2.1\,\mathrm{kg/dm^3}$).

Mainly, the measured stress is slightly lower than the theoretical overburden pressure.

6 CONCLUSIONS

A total of 93 overcoring tests were made at several exploration levels during sinking of the two Gorleben shafts GS1 and GS2 to determine the initial rock stress in the virgin salt dome. To enable a proper evaluation of the test data and to take the distinct nonlinear creep behaviour of salt into account, a new method has been developed to convert the measured stress-release deformation to stress considering the typical effects of stress relaxation and inelastic pilot borehole convergence in salt rock. This method includes the calculation of several sets of master curves by finite-element modelling and the determination of correction factors κ depending on the initial rock stress itself and the ductility of salt.

The results of the overcoring tests show that the stress state in the Gorleben salt dome is more or less isotropic and that the vertical distribution is very similar to the theoretical overburden pressure.

Future work on evaluation of stress measurements in nonelastic rock types using the overcoring method will include more sophisticated constitutive models (e.g. to describe the transient and steady-state creep of salt) to enable a more precise evaluation of data.

REFERENCES

Heusermann, S. 1984. Aspects of Overcoring Stress Measurements in Rock Salt. In H.R. Hardy & M. Langer (eds), *Proc. 2nd Conf. Mech. Behavior of Salt*, 272–289, Clausthal-Zellerfeld: Trans Tech Publ.

Heusermann, S. 1993. Measurement of Initial Rock Stress at the Asse Salt Mine. In M. Ghoreychi, P. Berest, H.R. Hardy & M. Langer (eds), *Proc. 3rd Conf. on the Mech. Behavior of Salt*, 101–114, Clausthal-Zellerfeld: Trans Tech Publ.

Hunsche, U. & Schulze, O. 1994. Das Kriechverhalten von Steinsalz. *Kali und Steinsalz, 11(8/9)*, 238–255, Essen.

Nipp, H.-K. 1991. Testbericht und Freigabemitteilung für das Programmsystem ANSALT I (Release 1991-1). Bericht, Archiv-Nr. 108586, BGR, Hannover.

Pahl, A. & Heusermann, S. 1991. Determination of Stress in Rock Salt Taking Time-Dependent Behavior into Consideration. In W. Wittke (ed.), *Proc. 7th Int. Congress on Rock Mechanics*, Vol. 3, 1713–1718, Rotterdam: Balkema.

Wallner, M., Bräuer, V. & Bornemann, O. 1998. Geoscientific Charactcrization and Evaluation of the Suitability of the Gorleben Repository Site. *Proc. Distec '98*, 75–81, Hamburg.

Geotechnical Measurements and Modelling, Natau, Fecker & Pimentel (eds)
© 2003 Swets & Zeitlinger, Lisse, ISBN 90 5809 603 3

The geomechanical influence of insolubles in solution mining for mining layout in Alpine salt rock

G. Pittino
Department of Geomechanics, Mining University, Leoben, Austria

ABSTRACT: For salt production brine is won by solution mining with the borehole well method. The Alpine salt rock consists of a high share of insolubles, and therefore leached caverns are filled with clay residue, as so-called Laist, a natural backfill, according to salt content. The influence of cavern-fill-pressure on the long-term behavior of the cavern system is demonstrated based on numerical calculations, whereby the creep behavior of the salt rock is described by an elasto-viscous power law and the clay residue by the modified Cam-clay model. The research method for registering relevant parameters concerning short, as well as long-term behavior and furthermore, the obtained results are presented. Additionally, it will be demonstrated that the creep behavior of a material, which can be described via the Norton power law, can additionally be obtained from the convergence rates of drifts in different depths.

1   INTRODUCTION

Austrian salt mining has utilized the solution ability of water (solution mining) to extract salt from the Alpine salt rock (Haselgebirge) in the form of brine (Figure 1) since the Middle Ages. The Haselgebirge is composed of mixed rock similar to a conglomerate formed of salt, clay and anhydrite, thus the leached caverns are filled with clay residue (Laist) according to the salt content. With the development of the leaching solution technique that nowadays enables caverns with 100 m diameter and up to 300 m height (Gaisbauer 2000) and with the endeavor to increase the amount of brine that can be deployed per solid and cropped-out deposit volume, a geomechanical perspective of the cavern system involving the clay residue resulted as necessary.

Salt mining is performed by the borehole well method through leaching of thin slices (Scheibensolung).

After the sinking of the production bore, cementation of the standpipe, installation of the well head, tubing of the borehole with an outer and inner pipe, running of water and brine conduits and after installation of measuring and regulating devices the production process can be assumed. In the developing stage a fast cavity formation is targeted in the width through "tail-water" injection, during the central stage the cavern is supposed to develop cylindrically upward through the injection of "headwater" and in the ultimate stage the final cavern roof is created. During "tail water"

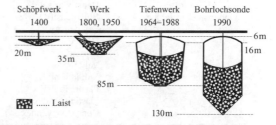

Figure 1.   History of leached caverns.

injection water is conducted through the inner pipe (tubing) to the deepest point of the borehole and finally drained through the outer pipe (casing), in case of "headwater" conduction vice-versa. Insertion of leaching liquid into the borehole is prevented through an inhibiting medium (blanket). Through the optimal combination of water insertion, specification of a measure for cauterization (Ätzmaßvorgabe) and blanket the leaching process can be regulated horizontally as well as vertically.

2   PROPERTIES OF LAIST AND HASELGEBIRGE

The method of Laist-sampling was performed with a telescopic core drilling with a single core barrel (EK 180, 160, 146, 131, 116 und 101 mm) without drilling

Table 1. Results of the field tests in MPa for Laist.

Pocket-penetrometer	Uniaxial compressive strength	S_{FD}–radial	0.24
		S_{FD}–axial	0.33
Vane shear test	Undrained shear strength	t_{FS}	0.05
Dilatometer test	Deformation modulus	G	3.7
	Un-/reloading modulus	G_{UR}	37

Table 2. Soil physical properties of Laist refer to water content ($w = m_w/m_{d+salt}$).

Drying oven	Water content	w	21.5 ± 3.4	%
Capillary-pycnometer	Density of solids	ρ_s	2.78 ± 0.05	g/cm^3
Underwater weighing	Bulk density	ρ	2.08	g/cm^3
	Dry density	ρ_d	1.71	g/cm^3
	Degree of saturation	S_r	96	%
	Porosity	n	38	%
	Void ratio	e	0.63	1
Consistency	Liquid limit	W_{LL}	30	%
	Plastic limit	W_{PL}	21	%
	Plasticity index	I_P	9	%
	Consistency	I_C	1	1

Table 3. Soil physical properties of Laist concerning brine content ($w_{brine} = m_{brine}/m_d$).

Drying oven	Brine content	w_{brine}	32.7 ± 5.7	%
Capillary pycnometer	Density of brine	ρ_{brine}	1.2	g/cm^3
	Density of solids	$\rho_{s,brine}$	2.87 ± 0.05	g/cm^3
Underwater weighing	Bulk density	ρ	2.08	g/cm^3
	Dry density	$\rho_{d,brine}$	1.57	g/cm^3
	Degree of saturation	$S_{r,brine}$	94	%
	Porosity	n_{brine}	45	%
	Void ratio	e_{brine}	0.83	1
Consistency	Liquid limit	$W_{LL,brine}$	47	%
	Plastic limit	$W_{PL,brine}$	32	%
	Plasticity index	$I_{P,brine}$	15	%
	Consistency index	$I_{C,brine}$	0.93	1

fluid 55 m in the Laist of a cavern with a drilling performance of approx. 2.5 m per day. Distributed across the borehole dilatometer tests were carried out. The core analysis was supplemented by pocket penetrometer and vane shear tests. Table 1 illustrates the results of the field tests.

The material behavior of cohesive soil depends generally on the content of pore fluid. Determining of water content via a drying oven, however, only yields small values due to the crystallization of salt during the drying phase and thus becoming part of the dry mass. Therefore, the soil physical properties were determined in consideration of the brine content in the Laist (index brine in Table 3). The saturated brine has a salt content of 28 Mass.-% ($f_{salt} = 0.28$), thus salt mass m_{salt} und mass of the brine m_{brine} can be assessed by measuring the mass of the water: $m_{salt} = f_{salt} \, m_{brine}$; $m_{brine} = m_w/(1 - f_{salt})$. With $m_d = m_{d+salt} - m_{salt}$ and

with the determined water content $w = m_w/m_{d+salt}$ the brine content can be calculated:

$$w_{brine} = \frac{m_{brine}}{m_d} = \frac{1.39w}{1 - 0.39w} \qquad (1)$$

The average water content of the Laist was determined at $21.5 \pm 3.4\%$ and results according to (1) in a brine content of $32.7 \pm 5.7\%$. The soil physical properties are indicated in Table 2 concerning the water content and in Table 3 in regard to the brine. According to sieve and elutriation analysis with brine Laist can be labeled as silt-clay-mixture.

The objective of the geotechnical tests was to determine the material behavior of Laist and to describe it through parameters based Mohr-Coulomb and modified Cam-clay constitutive models. The results are illus-

Table 4. Soil mechanical parameters of Laist.

Friction angle	$\phi' = 27 \pm 5°$
Cohesion	$c' = 0.051 \pm 0.043$ MPa
Oedometer test at σ' up to 0.6 MPa	
Loading:	$E_s = 18\ s'$ MPa
$s' = 0.25$ MPa $\rightarrow E_s = 4.5$ MPa	
Unloading:	$E_s = 176\ s'^{1.33}$ MPa
$E_s (s' = 0.25$ MPa$) = 28$ MPa	
Consolidation coefficient:	$C_v = 1.75e{-}7$ m^2/s
Hydraulic conductivity:	$k = 2.2e{-}9\ s'^{-1.5}$ cm/s
$k(0.25$ MPa$) = 1.8e{-}8$ cm/s	
Initial state	$e_{a,m} = 0.67,\ e_{a,brine,m} = 0.91$
Void ratio for $\sigma'_v = 600$ kPa	$e_{0.6,m} = 0.49,\ e_{0.6,brine,m} = 0.70$
Void ratio for NCL at p = 1 kPa	$e_{\lambda\ brine}^{1\,kPa} = 1.183$
Void ratio for NCL at p = 1 Pa	$e_{\lambda\ brine}^{1\,Pa} = 1.743$
Void ratio for CSL at p = 1 kPa	$e_{\Gamma\ brine}^{1\,kPa} = 1.136$
Void ratio for CSL at p = 1 Pa	$e_{\Gamma\ brine}^{1\,Pa} = 1.696$
Void ratio for CSL at p = 1 lb/in^2	$e_{\Gamma\ brine}^{1\,lb/in2} = 0.980$
Compression index	$C_{c,m} = 0.162,\ C_{c,brine,m} = 0.186$
Slope of NCL	$\lambda_m = 0.071,\ \lambda_{brine,m} = 0.081$
Swelling index	$C_{s,m} = 0.026,\ C_{sbrine,m} = 0.030$
Slope of elastic swelling line	$\kappa_m = 0.011,\ \kappa_{brine,m} = 0.013$
Oedometer tests at σ' up to 150 MPa	
Slope of NCL	$\lambda = 0.0905,\ \lambda_{brine} = 0.102$
Slope of elastic swelling line	$\kappa = 0.0274,\ \kappa_{brine} = 0.0309$
Void ratio of NCL at p = 1 Pa	$e_\lambda^{1\,Pa} = 1.702,\ e_{\lambda\ brine}^{1\,Pa} = 2.046$
Void ratio of CSL at p = 1 Pa	$e_\Gamma^{1\,Pa} = 1.658,\ e_{\Gamma\ brine}^{1\,Pa} = 1.997$

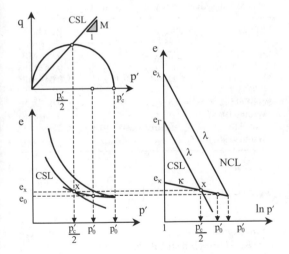

Figure 2. Modified Cam-clay failure criterion. Normal consolidation (NCL), swelling (slope κ) and critical state line (CSL).

trated in Table 4 according to the denominations in Figure 2. In the yield function

$$F = q^2 + M^2 p(p - p_c) = 0 \qquad (2)$$

of the modified Cam-clay Model M is the stress ratio q/p for critical state, thus corresponding with the CSL

(critical state line) and p_c being the preconsolidation pressure.

This material model is corresponding to the CSM (critical state model) and describes against the volumetric plastic strain for normally, as well as slightly overconsolidated soil an isotropic strain hardening and for highly overconsolidated soil a softening occurring from the maximum shear strain.

In addition to oedometer tests with pneumatic loading up to a maximum 0.6 MPa (approx. 65 m Laist superposition) with Laist specimens from different depths, the same oedometer was installed into the MTS testing system for displacement and load controlled testing up to 150 MPa. The stress history of a test based on displacement control mode at 5 to 10 µm/min displacement rate is demonstrated in Figure 3 time-dependent and in Figure 4 dependent upon the specific volume.

Table 5 illustrates the undrained shear strength s_u, the slope of NCL and the specific volume of CSL at 1 kPa, derived from index tests for a mean effective stress of $\sigma'_z = 0.25$ MPa.

Table 6 shows Laist in comparison to known soil types.

The exploration of the salt deposit by core drillings yielded specimens from Haselgebirge. Parameters in Table 7 describe the short-term behavior of Haselgebirge.

147

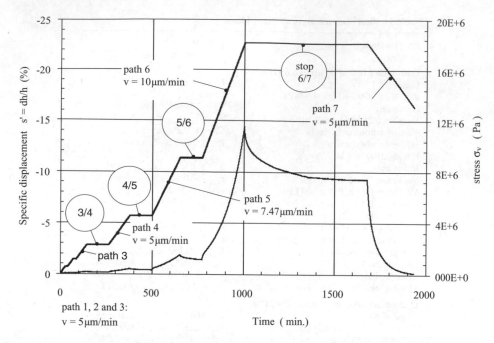

Figure 3. Oedometer test on Laist-specimen – histories of specific displacements and vertical stresses.

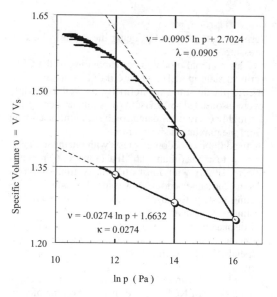

Figure 4. Oedometer test – history of specific volume versus stresses.

The creep deformations observed in Haselgebirge correspond particularly with a deformation free of fraction. Thus, long-term tests are evaluated based on an elasto-viscous power law describing the secondary creep process.

Table 5. Based on index tests: s_u, λ and Γ.

Reference	Equation	Result
Mesri (1975)	$s_u = 0.22\ \sigma_z'$	0.055 MPa
Budhu (2000)	$s_u = 0.25\ \sigma_z'$	0.063 MPa
Skempton (1957)	$s_u = \sigma_z'\ (0.11 + 0.37\ I_p)$	0.041 MPa
Schofield & Wroth (1968)	$\lambda = 0.585\ I_p$	0.088
Schofield & Wroth (1968)	$\Gamma^{1\,kPa} = \lambda \ln(2 \cdot s_u/M \cdot p_1) + \upsilon_{cr}$	2.20

$$\dot{\varepsilon}_s = A \cdot \overline{\sigma}^n = 0.27 \cdot \overline{\sigma}^{1.14} \left(\frac{\mu m}{m \cdot h} \right) \tag{3}$$

In Figure 5 steady state creep-rates, that were assessed through tests are described as a function of Mises's equivalent stress applying power law and furthermore, compared to exponential law.

3 CREEP PROPERTIES, CALCULATES FROM CONVERGENCE-MEASUREMENTS

Numerical calculations with the Finite Difference code FLAC 3.4 (Itasca 1999) and the elasto-viscous constitutive model "power law" (Equation 3) demonstrate how the stress-dependent (stress exponent n) creep

Table 6. Laist in comparison to known soil types (according to Schofield & Wroth 1968).

Parameter	Laist	Klein Belt Ton	Wiener Tegel V	London clay	Weald clay	Kaolin
λ	0.081	0.356	0.122	0.161	0.093	0.26
Γ^1 lb/in2	1.980	3.990	2.130	2.448	1.880	3.265
υ for p = 100 lb/in^2 = 0.69 MPa	1.607	2.350	1.558	1.700	1.480	2.065
M	1.07	0.845	1.01	0.888	0.95	1.02
ϕ (°)	27	21.75	25.75	22.5	24.25	26
κ_m	0.013	0.184	0.026	0.062	0.035	0.05
$\Lambda = 1 - \kappa/\lambda$	0.778	0.483	0.788	0.614	0.628	0.807
W_L	0.47	1.27	0.47	0.78	0.43	0.74
υ_L	–	4.520	2.300	3.144	2.180	2.930
W_P	0.32	0.36	0.22	0.26	0.18	0.42
υ_p	–	2.00	1.607	1.715	1.495	2.108
I_p	0.15	0.91	0.25	0.52	0.25	0.32
$\Delta\upsilon_p$	–	2.52	0.693	1.429	0.685	0.822
ρ_s	2.87	2.77	2.76	2.75	2.75	2.61
Source	Pittino	Hvorslev		Parry		Loudon

Table 7. Properties of Haselgebirge.

E (GPa) = 20	V (GPa) = 4	ν (1) = 0.2	ϕ (°) = 44	c (MPa) = 3.3

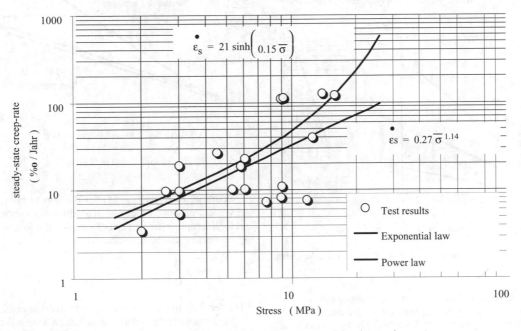

Figure 5. Comparison of steady state creep-rates for Haselgebirge – exponential and power law.

behavior can be determined by measuring convergence of different drifts in various depths. According to Dreyer (1974) depth pressure has an effect on the volume-convergence rate based on power law.

In the example depicted in Figure 6 the E-Modulus = 20 GPa, Poisson's ratio ν = 0.2, the structural parameter A = 10^{-20} (Pa$^{-2.5}$a^{-1}) as well as the stress exponent n = 2.5 are preassigned. The cavern is excavated in one step, subsequently the secondary stress state is calculated, as well as the vertical and horizontal convergence for the duration of 10 years respectively for a depth of 250, 350 und 450 m

149

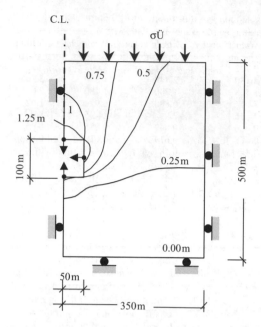

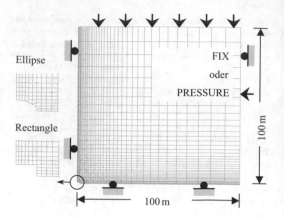

Figure 8.   Model geometry with rectangular/elliptical drift.

Figure 6.   Model geometry and displacement contours.

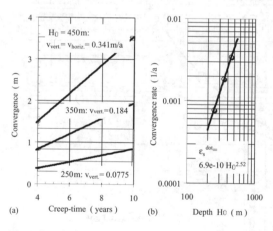

(a)   Creep-time (years)   (b)   Depth H0 (m)

Figure 7.   a) Histories of convergences. b) Stress dependent creep rate.

(Figure 7a). The convergence rates result, outlined double logarithmic as function of the depth, into the stress exponent n = 2.5 which is the preassigned creep parameter in the calculation, as the slope of the best-fit line (Figure 7b).

According to this method a stress exponent of n = 1.18 could be obtained for the cavern system Altaussee. This however, only with the inaccuracy of stress states in regions of measuring cross-section that are dependent upon the constellation of the cavern system to a large extent.

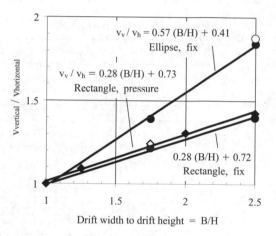

Figure 9. v_v to v_h ratio versus B to H ratio.

According to Dreyer (1974) for the horizontal v_v and vertical v_h convergence rate of a drift cross-section and for the rupture-free creep deformation the following connection can be obtained:

$$\frac{v_v}{v_h} = 1 + 0.2\frac{B}{H} \tag{4}$$

B/H refers here to the ratio of drift width to drift height.

In the example demonstrated in Figure 8 with the creep parameters $A = 2 \cdot 10^{-14}$ $(Pa^{-1.5}a^{-1})$ and $n = 1.5$ at variation of the boundary conditions (RB) and Poisson's ratio (RB "FIX" for $\nu = 0.5$ or $\nu = 0.25$; RB "PRESSURE" for $\nu = 0.5$) for drift ratios B/H with low scatter (Figure 9):

$$\frac{v_v}{v_h} = 0.72 + 0.28\frac{B}{H} \tag{5}$$

Calculations based on an elliptic cross-section and B/H ratios of semi-axes result in:

$$\frac{v_v}{v_h} = 0.41 + 0.57\frac{B}{H} \qquad (6)$$

The symbols white on the inside in Figure 9 are according to calculations B/H-ratios below 1, with registered inverse values that are utilized here for evaluating symmetrical proportions.

4 VOLUME CONVERGENCES OF CAVERNS DUE TO CREEP

An elasto-viscous power law that only characterizes the steady-state creep describes the Haselgebirge. The creep deformations of the Haselgebirge mobilize the passive fill-pressure in Laist that is described as elastoplastic with isotropic hardening through the modified Cam-clay Model. In comparison calculations are carried out as well based on the Mohr-Coulomb-Model or respectively on heavy liquids as contents of caverns. 60% salt content, as well as a corresponding swell factor lead in turn to a cavern filled with 65%. Subsequently, the cavern-filled pressure is estimated by FLAC for a time-dependent, as well as a time-independent leaching simulation.

4.1 Numerical analysis of time-dependent leaching

Die leaching process of a cavern containing 2 Mio · m³ lasts at 2-cm/d Ätzmaßvorgabe around 40 years and is simulated through 10 leaching sequences. The content of the cavern is modulated through Laist as modified Cam-clay and brine a unit weight at 12 kN/m³. The brine level is at 10 m above the final cavern roof at the pertaining drift (Figure 10). Due to the low

convergence rate of the cavern as compared to the permeability of Laist and the slowly increasing Laist level (0.12 kPa/d) a steady-state brine-pressure distribution is assumed. Thus, a coupled hydro-mechanical analysis can be disregarded. Effective stresses and brine pressure are taken into consideration by utilization of FLAC-configurations such as "config groundwater" and "flow off". For every modified-Cam-clay-element the pre-consolidation pressure and the total stresses are initiated through FISH-Routine.

Figure 10 illustrates that creep deformations of time-dependent modelling of the leaching are ascribed less importance. Therefore, further analysis will be performed by means of time-independent leaching.

4.2 Numerical analysis of time-independent leaching

In the time-independent leaching simulation a cavern as well filled with approx. 65% Laist is investigated and compared to a cavern filled with heavy liquid (Figure 13). The modelling of the cavern content is performed in case of the heavy liquid according to Figure 12 by the brine pressure corresponding to its unit weight at 12 kN/m³ acting on the cavern roof and sidewall up to the level of the Laist surface. From this height pressure of the heavy liquid (19 kN/m³) is exerted on the remaining cavern sidewall and the leaching area.

Laist as modified Cam-clay or respectively as Mohr-Coulomb material is modeled according to the description in section 4.1, with the variation that the termination of the production process is equated to the beginning of

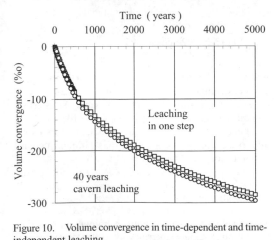

Figure 10. Volume convergence in time-dependent and time-independent leaching.

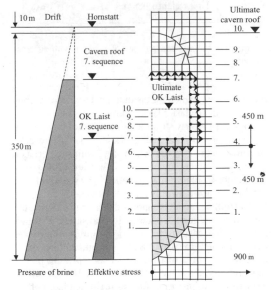

Figure 11. Time-dependent leaching model.

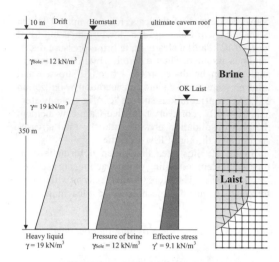

Figure 12.   Time-independent leaching models.

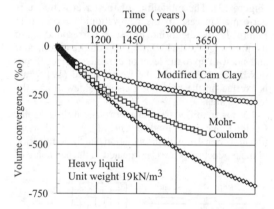

Figure 13.   Volume convergence of a cavern with different content.

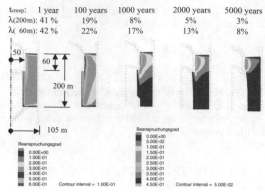

Figure 14.   Degree of utilization λ of pillar.

Table 8.   Compression modulus (GPa) and creep time (years).

t_{creep}	0	500	2000
K	0.8	1	1.5

the creep time, which means that the designed cavern is "leached" in an elastoplastic analysis.

Figure 13 clearly shows a lower volume convergence and volume converge rate of the modified Cam-clay. In comparison, caverns with heavy liquids experience a volume convergence of −250‰ after 1200 years, with Mohr-Coulomb-material after 1450 years and with modified Cam-clay only after 3650 years.

The mining layout is supposed to help in utilizing the limited salt deposit economically and safely. Through a hexagonal cavern constellation mining claims are evenly loaded. In case of a concurrent leaching of all caverns the limits of tributary areas represent planes of symmetry. Thus, the 3D-problem can be approached axisymmetric, in form of a vertically loaded, hollow cylinder. On the outer radius transverse strain is

prevented, the inner one, corresponding to the cavern radius r. Through the tributary area of a cavern for a borehole distance of s = 2 r the outer radius is calculated at R = 0.525s and the percentage of the remaining area at f_{Rest} = 0.773. The numerical analysis exhibits the same distribution concerning volume convergence as a single cavern.

The average degree of utilization of "pillar", weighted according to the axisymmetric through the volume, is calculated for the 200 m high hollow cylinder (λ_{200}) and for the unsupported 60 m high pillar range (λ_{60}). The Laist is mobilized corresponding to the fill-pressure through the creep behavior and the resulting lateral deformation of the pillar:

$$\lambda = \frac{\sigma_1 - \sigma_3}{2\sigma_3 \sin \varphi - 2c \cos \varphi} (1 - \sin \varphi) \quad (7)$$

where λ = degree of utilization; φ = friction angle; c = cohesion; σ_1 and σ_3 = principal stresses.

λ decreases from 45% in the secondary stress state e.g. after 1 year to approx. 40% (Figure 14).

The compression modulus increases proportionally to the hydrostatic stress in Laist. In intermediate depth it increases according to Table 8.

5   CONCLUSIONS

The development of leaching techniques in Austrian salt mining enables caverns with 100 m diameter and a height of several hundred meters. Along with the effort of increasing the extractable brine amount per outcrop and exposed deposit volume, a geomechanical perspective of the cavern system involving the clay residue resulted as necessary.

The research program for assessing relevant parameters in respect to short, as well as long-term behavior was presented for the Haselgebirge and additionally for the Laist. Part of it were the results of the field tests performed with help of a dilatometer, penetrometer, as well as a vane shear test and soil mechanical laboratory analyses. Soil physical parameters were determined with the inclusion of brine contained in Laist. The creep deformations of the Haselgebirge mainly correspondent with a rupture-free flow throughout centuries were calculated by means of an elasto-viscous power law, solely characterizing steady-state creep behavior. These deformations mobilize the passive fill-pressure in Laist that was described elastoplastic with isotropic hardening by means of modified Cam-clay Model.

Relatively few long-term investigations concerning the four salt rock types, the five salt "stone" types, as well as in respect to the general salt concentration of the Haselgebirge exist and thus were supported through calculations of the creep parameter via measured convergences of drifts at various depths.

The cavern-fill-pressure of Laist in regard to the long-term behavior of the cavern system could be assessed via Finite Difference Method calculations, the volume convergence, as well as the pillar stress level.

ACKNOWLEDGEMENTS

The author would like to thank E. Gaisbauer from the Österreichischen Salinen AG for entrusting with results of convergence measurements.

REFERENCES

Budhu, M. 2000. Soil Mechanics & Foundations. John Wiley & Sons, Inc. New York.

Dreyer, W. 1974. Gebirgsmechanik im Salz, Struktur und Gebirgsbewegungen. Ferdinand Enke Verlag, Stuttgart.

Gaisbauer, E. 2000. Experiences with blanket level measurement in solution mining caverns of Salinen Austria. 8th World Salt Symp. The Hague, The Netherlands.

Itasca Consulting Group 1999. FLAC Fast Lagrangian Analysis of Continua User's Guide.

Pittino, G. & Golser, J. 1997. Gebirgsmechanisches Dimensionierungsmodell – Projekt Laist. Institut für Geomechanik, Tunnelbau u. Konstruktiven Tiefbau. Montanuniversität Leoben. unpublished report.

Pittino, G. & Klade, M. 1997. Gebirgsmechanisches Dimensionierungsmodell für Lösungskavernen im Salz unter Berücksichtigung der Stützwirkung von Laugungsrückständen. Österr. Bergbautag. Bad Aussee. proceedings.

Pittino, G. 2002. Tragverhalten des Gesamtsystems Alpines Salzgebirge – Grubengebäude – Laugungsrückstand. Dissertation, Montanuniversität Leoben.

Schofield, A. & Wroth, P. 1968. Critical State Soil Mechanics. Mc Graw Hill London.

Modelling of flow and transport phenomena in a granitic fracture

A. Pudewills
Forschungszentrum Karlsruhe GmbH, Institut für Nukleare Entsorgung, Karlsruhe, Germany

ABSTRACT: The numerical simulation of the groundwater flow and solute transport in a dipole experiment in a granitic fracture is described. The objectives of the analysis are to calibrate the model and determine its parameters from a series of conservative tracer tests and to illustrate the practical application of the model for the simulation of radionuclides migration. The groundwater flow through fractured media is described by Darcy's law assuming an equivalent anisotropic porous model. The solute transport model in the saturated fracture filling material is governed by the advection-dispersion equation. A comparison of the field tests and the numerical results is presented.

1. INTRODUCTION

In the past decades, crystalline rocks have been extensively investigated as a potential host rock for nuclear waste disposal. Crystalline rocks such as granite are nearly impermeable and groundwater flows predominantly through discrete features such as fractures or joints. Such features are thus likely to provide the primary pathway for the migration of radionuclides from an underground repository to the biosphere. In order to predict the movement of radionuclides, the processes involved must be understood and quantified. For this purpose, laboratory tests, field experiments and adequate numerical models are needed.

The Colloid and Radionuclide Retardation Experiment (CRR) is carried out in the frame of Phase V of the Grimsel Test Site. The aim of this international, multidisciplinary project is to study the migration of radionuclides and colloids through a fractured shear-zone in crystalline rock under in-situ conditions in the underground facility at the Grimsel Test Site, Switzerland [Möri, 2001, Fierz et al., 2001]. This experiment included a series of tracer tests to determine the hydraulic and transport properties of the shear zone and to study the colloid and colloid-facilitated radionuclide migration at the experimental location [Hauser et al., 2002] . In the final field tests, a tracer cocktail containing representative radionuclides in absence and presence of bentonite colloids was injected in the Grimsel groundwater and the migration through the shear zone was observed.

The present paper focuses on the numerical simulation of the flow fields and transport processes in a

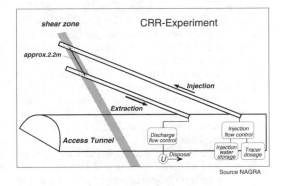

Figure 1. Layout of the dipole experiments.

dipole experiment with a distance of about 2.2 m between injection/extraction wells. Figure 1 shows the location of the access gallery and the boreholes within the shear zone plane used as for injection and extraction of water and tracers.

For numerical simulation, it is assumed that the flow and solute transport takes place in a fracture filled with fault gouge, a fine-grain material [Bossart & Mazurek, 1990]. Furthermore, it is assumed that this shear zone at the experimental location is plane which allows a two dimensional (2D) approach for the analysis. The investigations were performed with the ADINA-F finite element code [ADINA R&D, 2002]. The mathematical model is based on the Darcy's law for groundwater flow and the advection-dispersion equation for solute transport with a linear sorption of radionuclides in the fracture material. The objectives of these numerical analyses are: (1) to describe the groundwater flow in the dipole experiments, (2) to calibrate the model and determine its parameters from a series of conservative tracer tests, and (3) to illustrate the practical application of the calibrated model for the simulation of radionuclides migration in the presence of bentonite colloids. This modelling marks a milestone in our model development effort in the CRR project.

The comparison of calculational results and experimental data is presented as breakthrough curves. The agreement between the modelled and measured curves is quite good for conservative tracers and radionuclides. The modelling exercise should lead to improved understanding of flow and solute transport through a fractured shear zone of crystalline rock.

2 MODELLING

2.1 Geometry and boundary conditions

Figure 2 shows the schematic setup of the experiment in the plane of the fracture and the dimensions of the model, including the boundary conditions used. The problem studied is symmetric about the y-axis at the mid-plane. Therefore, a simplified half domain (i.e. the domain above the symmetry axis in Figure 2) was discretised in the numerical simulation.

This water-conducting zone (i.e. the fault gouge filled fracture) with a thickness of 1 cm corresponding to the value given in [Heer & Hadermann, 1996] was treated as a saturated porous medium with anisotropic, homogeneous permeability. The access tunnel was not considered.

Flow in the model domain was calculated on the assumption of impermeable top and bottom boundaries and a uniform, regional water flow from left to right, with a Darcy velocity of 10 m/year. After steady-state flow conditions were established, the injection of about 20 mg/l uranine solution was performed. The uranine concentration was measured on line in order to determine the input function. In the final field test, run #31, a tracer cocktail containing radionuclides relevant to high-level radioactive waste such Am, Np, Pu, U, Sr and I are injected in Grimsel groundwater

and the migration through the shear zone was observed. In the second test, run #32, nearly the same tracer cocktail together with a dispersion of 20 mg/l bentonite colloids was injected.

2.2 Flow and transport equations

Assuming that fluid flow in the fracture is laminar and isothermal, the momentum balance equation for fluid flow can be simplified to the well-known Darcy's law [Bear, 1979]. For the 2D incompressible flow in a porous medium the governing equation is given by

$$\mathbf{V} = -1/\eta\, \mathbf{K} \cdot \nabla p \tag{1}$$

where $\mathbf{V} = [Vy, Vz]^T$ is the vector of the components of averaged groundwater velocity in the y, z directions, η is the dynamic viscosity coefficient, $\mathbf{K}$ the absolute permeability tensor and ∇p a prescribed pressure gradient. The equation for mass transport of a single species in the y–z plane of the porous media, neglecting radioactive decay, is given by

$$\partial c/\partial t = 1/R(\nabla \cdot (\mathbf{D} \cdot \nabla c) - \mathbf{V} \cdot \nabla c) \tag{2}$$

where c is the concentration of the species in the pore fluid and R is the retardation factor for linear sorption in the porous material. $\mathbf{D}$ is the tensor of hydrodynamic dispersion. The coefficients of hydrodynamic dispersion are evaluated using the equation:

$$D_{kl} = (D_o + \alpha_T V)\, \delta_{kl} + (\alpha_L - \alpha_T)(V_k V_l) / V, \tag{3}$$

where D_o is the coefficient of molecular diffusion, α_L, α_T are the longitudinal and transverse dispersion lengths, δ_{kl} is Kronecker's delta, V_k, V_l are the components of the average groundwater velocity, V is the magnitude of the average groundwater velocity and the set (k, l) range over the indices y, z. The retardation in the porous materials, when it is assumed to be the result of linear reversible sorption, can be calculated as:

$$R = 1 + (\rho_o/n)K_d \tag{4}$$

with ρ_o the dry density of the medium, n the porosity and K_d is the sorption coefficient.

2.3 Hydraulic and transport parameters

The flow and transport properties of the fracture material obtained from model fitting to experimental breakthrough curves of conservative tracers such as uranine are summarised in Table 1. The porosity and the permeability of the fault gouge layer correspond nearly to the values determined in laboratory [Chen and Kinzelbach, 2002]. With regard to the values of

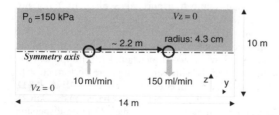

Figure 2. Schematic diagram for the 2D model of the dipole experiment.

the dispersion length, there are no measurements available. The α_L and α_T are just used as fitting parameters. The diffusion coefficients of both uranine and colloids were determined in the laboratory [Geckeis, 2002].

2.4 Finite element model

The entire mesh and a detail of the mesh near the boreholes are shown in the Figure 3. The computational domains of $14\,m \times 5\,m$ is large enough to ensure that the dipole does not significantly perturb conditions at the model boundaries. The boreholes are explicitly modelled.

3. NUMERICAL RESULTS

The numerical analysis starts with the modelling of the steady-state groundwater flow field in the dipole experiments. In Figure 4 the distribution of the fluid velocity vectors around the boreholes, the stream lines and the pore-water pressure for the uranine run

Table 1. Flow and transport parameters used for model calculation.

Porosity	n [%]	8.5–8.25
Permeability (horiz.)	κ_y [m^2]	$8.5 \cdot 10^{-12}$
Permeability (vertical)	κ_z [m^2]	$7.0 \cdot 10^{-12}$
Long. dispersion length	α_L [m]	0.04
Trans. dispersion length	α_T [m]	0.02
Diffusion coefficient	D [m^2/s]	$2 \cdot 10^{-11}$

#29 are illustrated. From the velocity vectors and the stream-line plots, it can be seen that the flow field is mainly influenced by the high water extraction rate it can be seen that the flow field is mainly influenced by the high water extraction rate compared to the injection rate (i.e. the ratio between extraction and injection rates was about 15). The strong pressure gradient around the extraction well is also caused by this high rate.

3.1 Uranine breakthrough curves

Simulations of the uranine runs #29 and #34 have been used for the calibration of the numerical models. Figure 5 shows the measured uranine input function used in the calculations. A comparison of the numerical fits and experimental data is presented as breakthrough curves where the uranine concentration is plotted versus time (Figure 6 and 7). As can be seen from this figure, the simulation results match the experimental data fairly well. However, the comparison of calculation results for run #34 in the tailing part of the curves shows some differences. Nevertheless, the overall agreement of model and experiment is reasonable. The recovery of the uranine was about 95%.

3.2 Radionuclide migration

In run #31 a cocktail containing Am, Np, Pu, U, Sr and I was injected in the Grimsel groundwater and the concentration breakthrough was studied. In the following test (run #32) bentonite colloids were added.

This chapter presents the preliminary results of the numerical modelling of the breakthrough of ^{233}U and ^{237}Np which were used in both runs. In our modelling

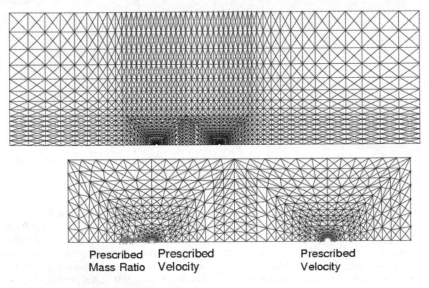

Prescribed   Prescribed   Prescribed
Mass Ratio   Velocity   Velocity

Figure 3.   Finite element mesh and detail discretisation near the boreholes.

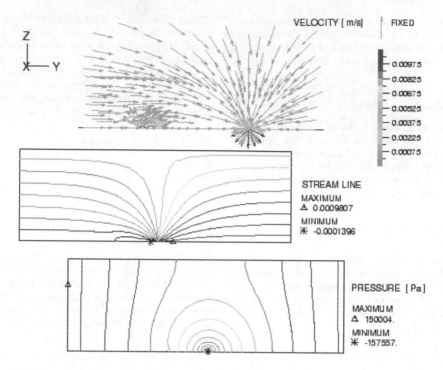

Figure 4. Distribution calculations of the velocity vectors, the stream lines and the pore pressure in the shear zone plane.

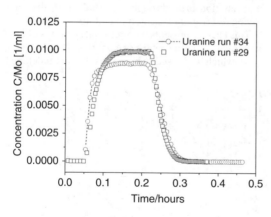

Figure 5. Measured uranine concentrations at the injection borehole.

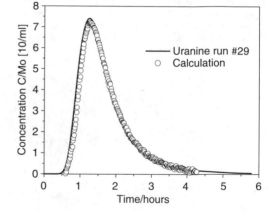

Figure 6. Comparison of measured and calculated uranine breakthrough curves.

the migration of ^{237}Np and ^{233}U are of main interest because of the particular shape of the breakthrough curves. The observed concentration peak-time corresponds to the conservative tracer curves (i.e. about 90 min) but the tailing suggests that a rather large part of the nuclides is retarded in the same manner as the Sr-curve (see also Fig. 9). Taking into account this fact,

it was assumed that 50% of the injected solution was transported like the conservative tracers or associated with colloids and the other 50% is retarded in the fracture infill with a retardation factor R of about 6 as was determined in laboratory for ^{233}U [Garcia Gutiérrez et al., 2002]. A K_d-value of about $0.21 \cdot 10^{-3}$ m^3/kg was calculated if the density of the porous

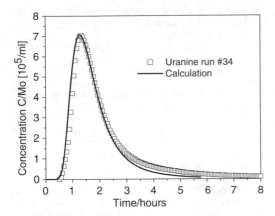

Figure 7. Comparison of measured and calculated uranine breakthrough.

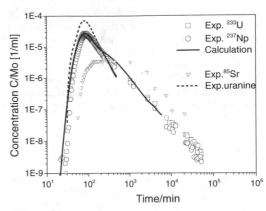

Figure 9. Calculated and measured breakthrough curves for ^{233}U in comparison with the measured uranine, ^{237}Np and ^{85}Sr curves.

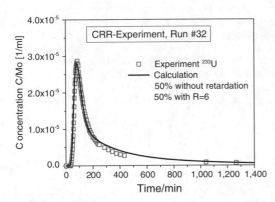

Figure 8. Simulated and measured breakthrough curves for ^{233}U.

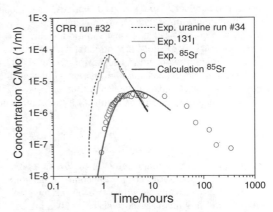

Figure 10. Calculated and measured breakthrough curves for ^{85}Sr in comparison with the measured uranine and iodine curves.

medium (i.e. fault gouge material) is 2000 kg/m³ and the porosity 8.5%. The comparison of calculated and measured breakthrough concentration of ^{233}U in run #32 is given in Figure 8. A logarithmic plot of these results together with the measured ^{237}Np, ^{85}Sr and uranine run #29 breakthrough curves is given in Figure 9. As can be seen from both figures, the agreement between the calculated and experimental curves of ^{233}U and ^{237}Np is satisfactory and the shape of the curves is quite well reproduced.

In the next step, the simulation of the strontium breakthrough curve has been performed using the retardation factor of 6 (i.e. $K_d = 0.21 \cdot 10^{-3} m^3/kg$). The results are shown in Figure 10 together with the measured breakthrough curves of uranine and iodine used as conservative tracers. The modelled and observed breakthrough curves show a relatively good agreement regarding the peak position and peak

height but the tail end properties are underestimated. A fuller discussion of this result is not possible because the assumed retardation factor for Sr is not yet supported by reliable laboratory experiments and the effect of matrix diffusion was not considered.

4. CONCLUSIONS

The numerical modelling of the dipole experiments presented in this paper demonstrates that the migration of the conservative tracers through a fractured shear zone of crystalline rock can be successfully simulated with the ADINA-F code. Finally, this calibrated model was applied to simulate the migration of selected radionuclides in the final experiment. This preliminary modelling of the radionuclide breakthrough curves shows also the limitation of the transport model used. For example, model without matrix diffusion is unable to describe correctly the tailing of breakthrough curves such as of strontium. Further development of the model will be carried out to include the matrix diffusion and the discrete fractures effects. Laboratory and field data are currently available that should allow the refinement of the model.

REFERENCES

Adina R & D Inc., 2001. ADINA (Automatic Dynamic Incremental Nonlinear Analysis), Report ARD 01–9, Watertown, MA, US.

Bear, J., 1979. Hydraulics of groundwater, McGraw-Hill, New York.

Bossart, P. & Mazurek, P., 1991. Grimsel test site: Structural geology and water flow paths in the migration shear zone. Nagra Tech. Rep. 91–12., Wettingen, Switzerland.

Chen, Q. & Kinzelbach, W., 2002. An NMR study of single- and two-phase flow in fault gouge filled fractures. *J. of Hydrology:* 259, 236–245.

Fierz, Th., Geckeis, H., Götz, R., Geyer, F.W., Möri, A., 2001. GTS V/CRR Tracer Tests #1–#16 (Sep. 1999–Jan. 2001), Report NAGRA, Wettingen, CH.

Garcia Gutiérrez, M., Missana, T., Alonso, U., 2002. Transport in fractures and porous rocks: radionuclides and colloids., CIEMAT, DIAE/54431/3/2002.

Geckeis, H., 2002. Personal communication.

Hauser, W., Geckeis, H., Kim, J.I. & Fierz, Th., 2002. A mobile laser-induced breakdown detection system and its application for the in situ-monitoring of colloid migration. *Colloids and Surfaces*, 203: 37–47.

Heer, W. & Hadermann, J., 1996. Grimsel Test Site, Modelling Radionuclide Migration Field Experiments, Technical Report 94–18, PSI, Würenlingen and Villingen, CH.

Möri, A., 2001. CRR Experiment Phase 1. NAGRA Internal Report (NIB 00–37), Wettingen, CH, in print.

Geotechnical Measurements and Modelling, Natau, Fecker & Pimentel (eds)
© 2003 Swets & Zeitlinger, Lisse, ISBN 90 5809 603 3

Geophysical methods as one way to detect and assess sources of danger in engineering and mining

H. Thoma, U. Lindner, O. Klippel & Th. Schicht
Kali- Umwelttechnik GmbH, Sondershausen, Germany

ABSTRACT: The application of geophysical methods in civil engineering and mining becomes more and more significant. The combination of geophysical data with classical geotechnical measurements can greatly improve the quality and reliability of investigations e.g., regarding stability of mines, dams or buildings because of the high data density provided by geophysical methods. In that way geophysical applications can be an important and in some cases essential source to early detect potentially dangerous conditions both in mining and engineering. A number of case examples to demonstrate the application of geophysical monitoring and exploration techniques such as engineering seismology, sonar and ground penetrating radar respectively for the assessment of the condition of mining fields, the assessment of the geological barrier of mining fields and for investigations of stability issues of dams and dikes.

1 INTRODUCTION

The application of geophysical methods in civil engineering and mining becomes more and more significant. Reasons for that are for example the nondestructive determination of properties of the investigated object as well as the possibility to describe an investigation area by geophysical methods without placing sensors directly in that area.

Independent of what geophysical methods are applied for, i.e. building ground investigations, inspection of dams and dikes or exploration and monitoring in mines, the aim is in the first place the exploration of geological structures and the determination of physical parameters of the rock formations to help solving geotechnical problems.

In engineering geophysics the question about the quality of building foundations, dams and others is frequently asked when damages are either observed or are to be expected after external stress like for example the inspection of dams and dikes after a flood.

In case of mining, geophysics can be applied both for exploration purposes and for monitoring of the geomechanical conditions. In this regard geophysical measurements can provide important information to early recognize potentially dangerous conditions. From the intensified exploitation of mines as well as from the aftertreatment phase or secondary use of the chambers after the mining result several sources of danger especially in potash and rock salt mining:

– destabilization of mining fields or parts of mining fields and resulting inadmissible dynamic stress and strain at the surface
– inadequate thickness of the protective layer or geological barrier for the protection against gases and fluids during mining and the use of the mines afterwards
– weakening and disintegration of the contour of the mine workings with a possible reduction of safety at work

The sources of danger in the field of engineering result essentially from:

– undetected near-surface structures, especially cavities
– material inhomogeneities in construction

The following case examples demonstrate the application of geophysical monitoring and exploration techniques such as engineering seismology, sonar and ground penetrating radar respectively for the assessment of the condition of mining fields, the assessment of the geological barrier of mining fields and for investigations of stability issues of dams and dikes.

2 CASE EXAMPLES

2.1 Stabilization of a carnallite mining field

The field was mined between 1982 and 1991 and started to show signs of a rapidly increasing destabilization in form of an increasing seismic activity with magnitudes larger than 1.5 and high deformation velocities already during the last years of active mining. By 1991 the geomechanical condition of the field had reached a dangerous level of weakening and in parts almost disintegration of pillars. To avoid the collapse of the entire field, it was to be stabilized by quickly refilling the remaining chambers with rock salt. The backfilling and stabilization of the field was carried out between 1991 and 1996.

Figure 1 shows the development of the seismicity over the years from the beginning of the stabilization measures. From underground deformation measurements but also especially from the distribution of the seismic events in the field a priority list of which areas in the field had to be treated first could be derived. Moreover, the combination of deformation and stress measurements and seismic monitoring made it possible to identify especially weakened zones where special precautions for safety at work had to be considered. In that way the deformation processes were significantly slowed down resulting in a notably reduction of strong seismic events almost immediately after the backfilling had started as can be seen in Figure 2.

2.2 Stabilization of a mixed salt field

This mining field was in production between 1978 and 1991 and in parts mined in up to three levels. After the closure of the mine in 1991, the field was left open but still in stable conditions, thus no backfilling of the field was planned. The field was abandoned while other parts of the mine were used as an underground waste disposal site. The situation of the mining field changed, when the field showed a sudden appearance of seismic activity in 1996. A series of strong seismic events with magnitudes of up to 2.0 were located in the beginning preferably in the overlaying horizons of the mining field.

Figure 2 shows the depth distribution of seismic events on a cross section on an East-West profile through the mining field. The distribution of the epicenters within the field is shown in Figure 3. The vast majority of the seismic events is concentrated on lines which can be related to local tectonics.

According to the distribution and strength of the seismic events a risk of the opening of fluid migration paths into the salt complex and thus the loss of the integrity of the geological barrier had to be considered. For that reason, it was decided to reopen the field and stabilize it by backfilling of rock salt analog to the previous case example. At that time the information from the seismic monitoring were the only indications of a possible dangerous development. Only limited other geotechnical data could be obtained from the field because of it's inaccessibility.

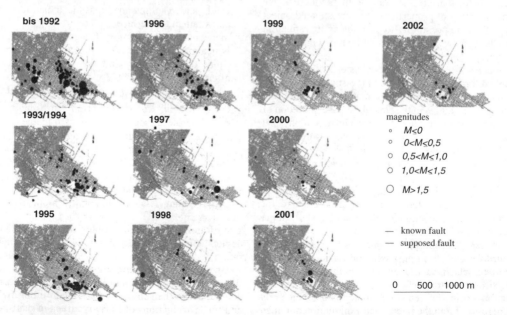

Figure 1.   Development of the seismicity of a Carnallitite mining field during (1991–1996) and after stabilization measures.

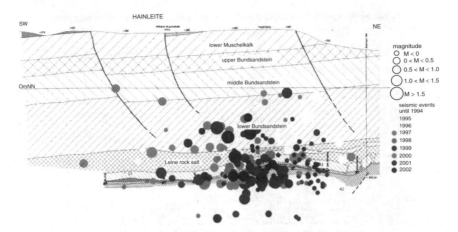

Figure 2.   Cross section through the mining field and depth distribution of seismic events.

Figure 3.   Distribution of seismic events in a mixed salt field. The majority of the events are located at structures related to local tectonics.

The stabilization measures started in 1996 are still ongoing.

Beside the alignment and grouping of the seismic epicenters along local faults and ridges as can be seen in Figure 3, the calculated focal parameters such as stress drop, dislocation, fault plane size and released seismic energy can be directly used for further geomechanical assessments.

An other way of characterizing the development over time of the field's condition by the observed seismicity is shown in Figure 4. There, the cumulative released seismic energy or respectively the square root of the energy is plotted over time.

The slope of this curve is related to the amount of stress released by the seismic events according to Benioff (1951). The shape of the curve gives

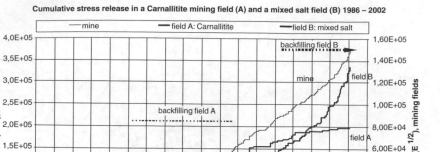

Figure 4. Cumulative stress release of a carnallitite mining field (A), a mixed salt field (B) and the entire mine computed from seismic events according to Benioff (1951).

information of how the stress is released. A high number of small events will result in a rather smooth curve while a small number of strong events would give a blocky shape. A change in the mean slope of the curve indicates a change in the characteristics of the deformation processes. If a long observation series of the seismicity in a specific area is available and under the assumption that the general deformation processes do not change in that time, the Benioffcurve also allows for an estimate of the maximum magnitude to be expected in that area (Benioff, 1951).

Figure 4 shows the cumulative stress release for the mixed salt mining field and in comparison the stress release for the previous discussed carnallitite mining field as well as for the entire mine. The sudden activation of seismicity in the mixed salt field (field B) on a high level in 1996 can clearly be identified in the curve. Since then, the seismicity remained on a high level and is lately even further increasing (increasing slope of the curve), which suggests that the ongoing stabilization measures have not yet taken any effect. Moreover, the deformation and disintegration of the field seems to accelerate. The Benioff-curve of the carnallitite field (field A) on the other hand shows a notable reduction of the cumulative released stress in 1993, already 2 years after the beginning of the stabilization measures. The seismic activity is reactivated in 1995 (see also Figure 1), but with smaller magnitudes of the single event than before (smooth curve). From 1998 on the stress release is slightly reduced. Since 2000 a seismic activity on a constant low level can be observed indicating that the deformation processes of the field have nearly come to an end – more than 3 years after finishing of the backfilling. In this regard and especially also because of the multi level mining of

the mixed salt field (field B), the deformation processes of the mixed salt field can be expected to remain on that high level for the next few years.

2.3 Application of georadar and sonar in mining

Beside passive seismic monitoring, active geophysical exploration methods can also provide valuable information for the geomechanical assessment of a mining field. Especially for salt mining both radar and sonar are well suited to easily cover large investigation areas.

Both methods are applied from e.g. roadways within the mine for:

– investigation of thickness of protective layers, both below and above the mining horizon
– exploration of geological structures
– investigation of the contours of the mine workings for safety reasons
– estimation of thickness and integrity of horizontal pillars in case of multi-level mining.

Figure 5 presents a radargram of an exploration of the underlying main anhydrite in the same mixed salt mining field from the previous case example. The main anhydrite in the area is gas bearing (methane) and a minimum distance of 15 m has to be kept for safety reasons (i.e. gas leakage, outbursts).

Because of the high contrast of the electric properties between the anhydrite and the neighboring salt as well as the high attenuation of electromagnetic waves in the anhydrite, virtually no reflections are received from beneath the main anhydrite. Thus, the salt-anhydrite boundary can be identified as the last detected reflection. In Figure 5 the top boundary of the main anhydrite is found at depths greater than 30 m from the

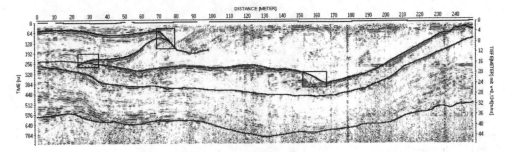

Figure 5. Radar investigation in a mine to investigate the thickness of the protective layer to the main anhydrite.

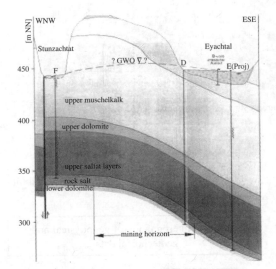

Figure 6. Cross section of the geology of the mine.

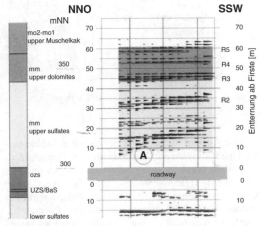

Figure 7. Results from sonar measurements in correlation with a nearby borehole.

floor of the mine workings hence the safety requirements are fulfilled. The radargram also reveals internal structures of the salt complex such as a tilted fold, indicating a much more complex geology than previously assumed.

A different situation but with similar exploration tasks can be found in a rock salt mine situated in the triasic (Muschelkalk) complex. The mining horizon is only about 130 m beneath the surface and the thickness is rather small with just around 10 m. Because of the shallow depths and thus the close vicinity of the mining horizon to the groundwater bearing formations special care has to be taken to maintain an appropriate thickness of the geological barrier.

Figure 6 shows a cross section of the main geological situation. The task for the geophysical exploration was to detect the layer boundaries of:

− rock salt/upper sulfates
− upper sulfates/upper dolomites
− upper dolomites/upper Muschelkalk

The upper dolomites and the upper Muschelkalk layers are already counted to the groundwater bearing formations. The investigations were problematic because the fine layering of the formations, especially the large number of thin shale and anhydrite layers attenuated and scattered most of the energy of the transmitted signals. The application of radar was in that case limited for only shallow investigations since the depth penetration was strongly reduced to just a few meters because of the fine layering. Although the sonar investigation was also effected by the layering, the penetration depth was still sufficient to complete the exploration tasks. The boundary between Upper Sulfates and Upper Dolomites, which was of the main interest, could be clearly detected by a dense series of reflections (Figure 7). The estimated layer boundary could be verified by a nearby borehole.

2.4 Investigations of dams and dikes

The last case example demonstrates the application of geophysical methods in the engineering sector.

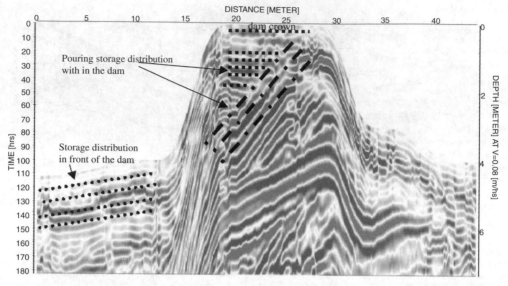

Figure 8.   Cross section through a dam with georadar, measured with 100 MHz antenna.

The August 2002 flooding of the river Mulde created a 235 m wide and 7 m deep erosion channel, leading to the breakthrough of the river into an open-pit coal mining field and washing away the street that ran in between. A dam was immediately constructed already during the flood to limit the inflow into the coal pit. The dam was built up with porphyr gravel of mixed grain size. Because of the difficult situation during construction, the new dam had to be inspected by direct and indirect means with respect to stability and tightness after the flooding.

Geoelectric and Georadar measurements were carried out both along and perpendicular to the dam axis. The geoelectric profiling enabled the differentiation of the dam body from its foundation and underlying formations (e.g. sand, gravel, clay and coal). In addition, the dam and the area in front and behind was probed by boreholes, which were also used to calibrate the results from the electric measurements.

The radar measurements were used to determine internal structures of the dam. A radargram of a profile crossing the dam is shown in Figure 8. A number of dipping reflectors can be identified within the dam, which result from the construction of the dam (sideways pouring of the gravel) and in that case imply the risk of future sliding planes.

3   CONCLUSION

The data geophysical methods can be an essential contribution to classical geotechnical measurements

for the detection of sources of danger in case of short term changes of the geotechnical situation as well as for the assessment of long term stability issues:

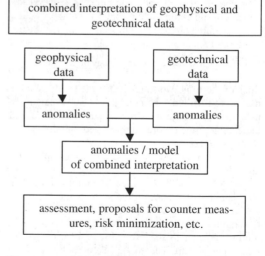

REFERENCE

Benioff, H. 1951. Colloquium on plastic flow and deformation within the earth. Trans Am. Geophys. Union. 331–514.

Geotechnical measurements and modelling
in slope engineering and retaining structures

Visco-hypoplasticity and observational method for soft ground

G. Gudehus
Institute of Soil and Rock Mechanics, University of Karlsruhe, Germany

In memoriam *Laurits Bjerrum*

ABSTRACT: For soils with soft particles, in particular clays, a visco-hypoplastic relation is introduced by numerical element tests. With a single set of equations for effective stress rates depending on strain rates, the known behaviour is obtained for oedometric compression and creep, undrained and drained shearing. OCR = p_e/p' with an equivalent pressure p_e depending on e and stress ratio is decisive. Only five key parameters are needed and can be easily determined. The natural consolidation of a soft layer can thus be followed up, and its subsequent consolidation and creep under a fill, including spreading and localization. The method is also realistic for a braced excavation in soft clay. This enables an enhanced observational method.

1 INTRODUCTION

Bjerrum (1973) gave an overview of *soft soil* behaviour. The dependence of stiffness and strength on void ratio, overconsolidation ratio, strain rate and stress component ratio was outlined in detail, also volumetric and deviatoric creep. Encouraged by his personal visit 30 years ago, we in Karlsruhe started to develop constitutive relations for soft soils. Leinenkugel (1976) introduced the *viscosity index* I_v and proposed a unified concept for rate-dependence, creep and relaxation, using Hvorslev's (1937) equivalent stress σ_e. Kolymbas (1978) proposed a rate-type constitutive relation which included viscous effects. This was the origin of *hypoplasticity*.

For the following ca 25 years our work was focussed on hard-grained soils, neglecting viscous effects. After the unification of density and pressure effects (Kolymbas 1991) hypoplasticity became successful and was validated repeatedly (Gudehus 2003a and b). Rate dependence and creep could be included by a rate-dependent granulate hardness with I_v (Gudehus 1996), but not relaxation. Niemunis (1996, 2003) could overcome this default by incorporating a modified Cam Clay concept. In my variant of his theory (Gudehus 2003c) the Cam Clay part is replaced by the three attractors of hypoplasticity, and asymptotes for compressive creep and relaxation are built in.

The *visco-hypoplastic* behaviour is presented here in a partly novel graphical manner (Sec. 2). The properties outlined by Bjerrum (1973) are covered by one or two rate-type equations. An overconsolidation ratio OCR is decisive for viscoplastic effects. Only five key parameters are needed, typical values are given for a lowly and a highly plastic clay, they serve also for classification. The validity has been shown by means of element tests (Niemunis 2003, Gudehus 2003c). The initial state field of σ' and e has to be specified, changes of state and position can then be calculated with the aid of boundary conditions. This has been validated by model tests and field data (Gudehus 2003a and b, Cudmani et al. 2003).

Peck (1969) has outlined the *observational method* developed by Terzaghi and himself. He used qualitative or conventional quantitative prediction models. His demanding general rules can be better satisfied with hypoplasticity in a number of cases. This is shown first in Sec. 3 for fills upon soft ground. The initial state field can consistently be specified by a back-analysis with visco-hypoplasticity. Creep rupture and long-term displacements are calculated with the same theory and judged by means of OCR and pore pressure. This leads to a partly novel observational method.

Excavations in soft clay with braced sheet pile walls are treated in Sec. 4. Except for neighboured buildings, the initial state is the same as in the case of Sec. 3. Displacements and strut forces are obtained as known e.g. from Chicago. Peck (1969) refers to such a case, we can now make more of it. OCR is again the key quantity in judging stability and deformations. The novel observational method can be transferred to foundations, tunnels and slopes, but limitations of the underlying theory have to be kept in mind (Sec. 5).

2 VISCO-HYPOPLASTIC BEHAVIOUR

Numerical element tests calculated with visco-hypoplasticity are outlined here for a highly plastic (H) and a lowly plastic (L) saturated clay. An *oedometric compression* (Fig. 1) with a constant strain rate, starting from a low e, can be approximated by

$$e \approx e_r - C_c \ln(\sigma'/\sigma_r) \qquad (1)$$

with a compression index C_c (e.g. 0.35 for H and 0.06 for L), a reference void ratio e_r (e.g. 2.6 for H, 0.75 for L) and a reference pressure σ_r. Different from the classical approach σ_r depends on the strain rate,

$$\sigma_r = \sigma_{rr}[1 + I_v \ln(|\dot{\varepsilon}|/D_r)] \approx \sigma_{rr}(|\dot{\varepsilon}|/D_r)^{I_v} \qquad (2)$$

with the *viscosity index* I_v (e.g. 0.05 for H and 0.02 for L), a reference strain rate, e.g. $D_r = 10^{-6}\,\mathrm{s}^{-1}$, and a reference pressure σ_{rr} (e.g. 100 kPa). The rate effect is marked for H if $\dot{\varepsilon}$ varies by several orders of magnitude, and small for L.

An *overconsolidation ratio* is defined as

$$\mathrm{OCR} = \sigma_e/\sigma' \approx \sigma_{rr} \exp\left(\frac{e_r - e}{C_c}\right)/\sigma' \qquad (3)$$

with an equivalent pressure σ_e from (1) as proposed by Hvorslev (1937), but now with σ_r by (2) for $\dot{\varepsilon} = D_r$. Thus the NCL-line for $\dot{\varepsilon} = D_r$ has OCR = 1, otherwise

OCR $\approx (D_r/\dot{\varepsilon})^{I_v}$ holds. A state 0, e.g. with $e = 2.3$ and $\sigma' = 6\,\mathrm{kPa}$ for H, has thus OCR = 4. Compression with $\sigma = 40\,\mathrm{kPa}$ imposed within 10 Minutes and pore pressure equalization with sample thickness $d = 2\,\mathrm{cm}$ and permeability $k = 10^{-9}\,\mathrm{m\,s^{-1}}$ leads to the familiar decrease of e with log t. A primary compression line associated with an $\dot{\varepsilon}$ due to filtration is reached, e.g. up to a state 1 with $e = 2.1$, $\sigma' = 35\,\mathrm{kPa}$ and OCR = 0.8. After reaching a constant σ', volumetric creep takes place which can be approximated by the known relation

$$e \approx e_0 - C_\alpha \ln(1 + t/t_d) \qquad (4)$$

with

$$C_\alpha \approx C_C I_V \qquad (5)$$

and the primary consolidation time

$$t_d \approx 10 d^2 \, C_c \gamma_w / \sigma'(1+e)k \qquad (6)$$

It leads e.g. to a state 2 after ca 10^8 hours with $e = 1.1$ and OCR = 2.2. (4), (5) and (6) hold only for a certain interval and a suitable initial OCR.

Imposing similarly now $\sigma = 105\,\mathrm{kPa}$, after pore pressure equalization a state 3 with OCR ≈ 0.9 can be reached. Subsequent creep can again be approximated by (4) with newly adapted e_0 and t_d, and the same C_α. For L this kind of creep leads to the plotted state 2 for an extremely long time because of the low C_α.

Partial un- and reloading from and back to an NC line (e.g. 5-6-5 in Fig. 1a) with a constant $|\dot{\varepsilon}|$ can be approximated with (1), replacing C_c by C_s. e_r is taken from the NC-line, σ_r is again given by (2).

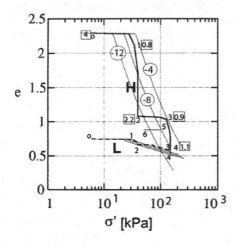

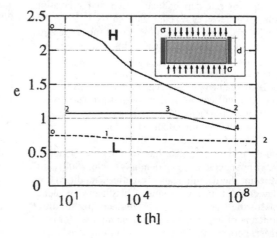

Figure 1. Evolution of void ratio with effective pressure and time for a highly (H) and a lowly (L) plastic clay in an oedometer, calculated with visco-hypoplasticity. OCR in squares, logarithm of $\dot{\varepsilon}$ in s^{-1} in circles, time steps in numbers.

These results are obtained with a *single* constitutive relation, which can be written

$$\dot{\sigma}' = \frac{\sigma'(1+e)}{2}\left[\dot{\varepsilon}\left(\frac{1}{C_S} - \frac{1}{C_C}\right)\right.$$
$$\left. - D_r\left(\frac{\sigma'}{\sigma_e}\right)^{1/I_v}\left(\frac{1}{C_S} - \frac{1}{C_C}\right)\right] \quad (7)$$

(7) leads to the familiar stiffness modulus

$$E_s = \dot{\sigma}'/\dot{\varepsilon} = \dot{\sigma}'(1+e)/C_c \quad (8)$$

for $\dot{\varepsilon} > 0$ and $\sigma'/\sigma_e = (|\dot{\varepsilon}|/D_r)^{I_v}$ by (2), and to (8) with C_s instead of C_c for $\dot{\varepsilon} < 0$. E_s is rate-dependent alongside with σ' via (2). For overconsolidated states, i.e.

$OCR > (D_r/|\dot{\varepsilon}|)^{I_v}$, (8) gets invalid, and E_s is misleading. Writing (7) for $\sigma' = 0$ and $OCR = (D_r/|\dot{\varepsilon}|)^{I_v}$ initially, (4) and (5) can be derived. (7) holds for $\sigma'_2 = K_0\sigma'_1$ with $K_0 \approx 1 - \sin\varphi_c$, otherwise a second evolution equation for σ'_2 has to be used (Niemunis 2003).

Undrained shearing shows a rather ductile behaviour when starting e.g. from OCR = 2 (Fig. 2). It leads eventually to critical states with shear stress

$$\tau_c = \sigma' \tan\varphi_c \quad (9)$$

and a constant friction angle φ_c (e.g. 10° for H and 28° for L). The critical void ratio decreases with σ' via

$$e_c \approx e_{cr} - C_c \ln(\sigma'/\sigma_r) \quad (10)$$

with C_c and e_{cr} slightly below the ones for (1). σ_r is again rate-dependent by (2), now with $|\dot{\gamma}|/2$ instead of

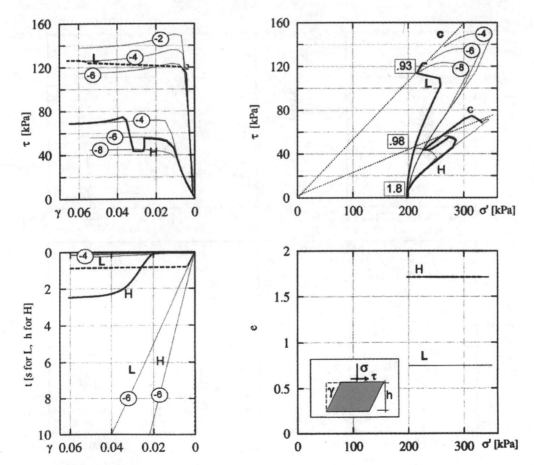

Figure 2. Evolution of shear stress, effective normal stress and void ratio, calculated with visco-hypoplasticity for undrained simple shearing from states with OCR ≈ 2. Creep under nearly constant τ is plotted with dashed lines, there are different time scales for H and L. Notations otherwise as in Fig. 1.

171

$|\dot{\epsilon}|$. τ_c depends therefore on $\dot{\gamma}$ via (2) and (9), with e via (10). Evolutions up to critical states are plotted for the shearing rates 10^{-4}, 10^{-6} and $10^{-8}\,\mathrm{s}^{-1}$. Jumps of $\dot{\gamma}$ cause transitions to the lines for the new $\dot{\gamma}$, this is plotted for H and can be used to determine I_v as proposed by Leinenkugel (1976). The garlands observed in such tests are obtained when taking the actually softer $\dot{\gamma}$-transitions for the calculation. With this remedy the calculated stress paths also have a smaller intermediate σ'-increase and are thus more realistic. All that shows that the actual 'loading program' with different and changing strain rates should be allowed for.

For the same initial σ' and e (or OCR) L is stiffer and stronger than H. Due to the rate-dependence by (2), stiffness and strength increase with $\dot{\gamma}$. OCR = 1 holds for critical states with $\dot{\gamma}/2 = D_r$, otherwise OCR $\approx$ $(2D_r/|\dot{\gamma}|)^{I_v}$. This is achieved as σ_e depends also on the stress ratio τ/σ', as was suggested already by Hvorslev

(1937). The response to a constant $\dot{\gamma}$ is ductile for H and slightly brittle for L, the calculated stress paths are also realistic, σ' is more reduced with a lower $\dot{\gamma}$.

If τ and σ are kept constant after having reached a state 2 (e.g. with $10^{-6}\mathrm{s}^{-1}$), deviatoric *creep* takes place. $\dot{\gamma}$ decreases until the $\dot{\gamma}$ for a critical state with $\tau = \tau_c$ is reached, then the creep is stationary. This is a kind of stabilization, but the equilibrium reached is only static, not thermodynamic. Creep is much faster for L than for H, and accelerated after a delay for L due to brittleness.

Starting after normal consolidation, i.e. with OCR $\approx$ 1, undrained shearing with constant $\dot{\gamma}$ causes a similar response (Fig. 3). Critical states are reached asymptotically as before. The τ-peak is again more marked with harder particles, and higher with a bigger strain rate. The peaks could be described with conventional φ' and c', the latter is nearly proportional to σ_e as by Hvorslev (1937). Such φ' and c' are arbitrary and non-physical,

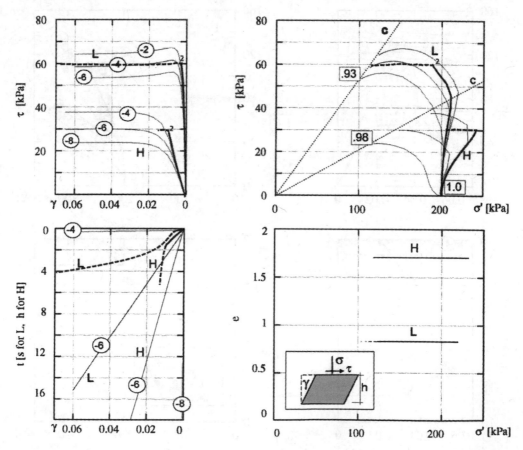

Figure 3. Evolution of shear stress, effective normal stress and void ratio, calculated with visco-hypoplasticity for undrained simple shearing from states with OCR $\approx$ 1. Creep under constant τ is plotted with dashed lines and different time scales for H and L. Notations otherwise as in Fig. 1.

the real material is cohesionless (Schofield 2002) and rate-dependent.

If τ is kept constant after having reached, say, a state 2 with $\dot{\gamma} = 10^{-4}\,\mathrm{s}^{-1}$ just prior to a peak, $\dot{\gamma}$ decreases first to the one associated with the τ-peak for a lower $\dot{\gamma}$. Afterwards it increases towards the $\dot{\gamma}$-value associated with the given $\tau = \tau_c$ and e. This kind of *creep rupture* is far more dramatic for L than for H and requires numerical caution. The critical strain associated with it depends strongly on the initial state.

Drained shearing after normal consolidation leads to a stronger stabilisation than without drainage, which is caused by densification (Fig. 4). (2) holds with $D = \sqrt{\dot{\varepsilon}^2 + \dot{\gamma}^2/2}$ instead of $|\dot{\varepsilon}|$. Critical states are reached with a lower e for a smaller $\dot{\gamma}$, but with the same τ_c for a given σ'. In the transition the soil is stiffer for a bigger $\dot{\gamma}$. Creep under constant $\tau < \sigma'$ tan φ_c and σ' takes place with a decreasing $\dot{\gamma}$. The volumetric portion is nearly the one by (4), the dilatancy ratio $\dot{\varepsilon}/\dot{\gamma}$ is determined by τ/σ'.

Drained shearing after overconsolidation causes a more brittle behaviour than in Fig. 3 (Fig. 5). With a constant $\dot{\gamma}$ a small initial contraction is followed by a dilation up to a critical state with τ_c by (9) and e_c by (10) with (2). The τ-peaks are higher with bigger $\dot{\gamma}$ and initial OCR. They may be fitted with φ' and c', but this is not physically justified and not needed in applications. Drained creep from a state 2 reached by shearing with constant $\dot{\gamma}$, say, starts with a decrease of $\dot{\gamma}$ and turns then into a *dilatant rupture*. $\dot{\gamma}$ cannot become stationary if τ exceeds τ_c. The time to rupture increases with the initial OCR and with I_v. As the dramatic increase of $\dot{\gamma}$ prevents drainage the pore pressure gets negative until the sample is cracked by cavitation.

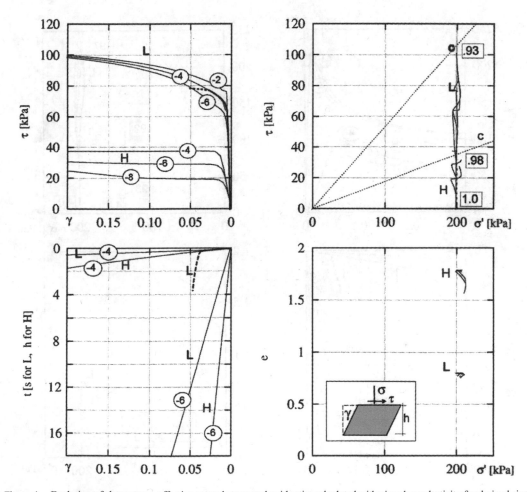

Figure 4. Evolution of shear stress, effective normal stress and void ratio, calculated with visco-hypoplasticity for drained simple shearing from states with OCR $\approx$ 1. Creep under constant τ is plotted with dashed lines and different time scales for H and L.

The behaviour plotted in Figs. 2 to 5 is described by two equations which can be written

$$\dot{\sigma}' = \sigma'(1+e)\left[L_{11}\dot{\varepsilon} + L_{22}\dot{\gamma} - N_1 D_r \mathrm{OCR}^{-1/I_v}\right] \quad (11)$$

$$\dot{\tau} = \sigma'(1+e)\left[L_{21}\dot{\varepsilon} + L_{22}\dot{\gamma} - N_2 D_r \mathrm{OCR}^{-1/I_v}\right] \quad (12)$$

The coefficients L_{11} etc. depend on τ/σ' and φ_c. (11) reduces to (7) for $\tau = 0$ with $\mathrm{OCR} = \sigma_e/\sigma'$. In general $\mathrm{OCR} = p_e/p'$ holds with the mean effective pressure p' and an equivalent pressure p_e that depends on e and τ/σ'. For critical states with $\dot{e} = 0$ and $\dot{\gamma} = \mathrm{const}$, $\dot{\tau} = 0$ and $\dot{\sigma}' = 0$, $\tau = \sigma' \tan\varphi_c$ and $\mathrm{OCR} = (2D_r/|\dot{\gamma}|)^{I_v}$ is obtained. For undrained shearing in general $\dot{\sigma}' \neq 0$ is obtained from (11). Peak stress ratios can be calcu-

lated from (11) and (12) with $\dot{\tau} = 0$ and $\dot{\varepsilon} = 0$ that depend on OCR and on $\dot{\gamma}/D_r$. For drained shearing with $\sigma' = \mathrm{const}$ the dilatancy ratio $\dot{\varepsilon}/\dot{\gamma}$ is obtained from (11) and (12). Peak stress ratios and dilatancy ratios can be calculated from (11) with $\dot{\sigma}' = 0$ and (12) with $\dot{\tau} = 0$ that depend on OCR and on $\dot{\gamma}/D_r$.

With other initial stress component ratios at the onset, including the ones not mentioned above (e.g. $K_0 = \sigma_h'/\sigma_v'$), similar curves are obtained with the same asymptotes. Other kinds of compression (e.g. isotropic) and shearing (e.g. triaxial) lead to "similar curves, but with somewhat different asymptotes. The resistance for undrained triaxial compression and extension is obtained realistically, also for field vane shearing. Compressive creep comes to an end after extremely long times so that (4) is no more valid. Relaxation and the response to alternating loading are also covered by the same theory. The described

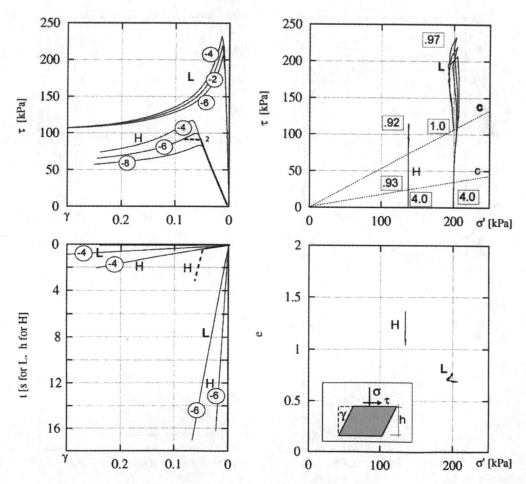

Figure 5. Evolution of shear stress, effective normal stress and void ratio, calculated with visco-hypoplasticity for drained simple shearing from states with OCR ≈ 4. Creep under constant τ is plotted with dashed lines and different time scales for H and L.

174

behaviour is close to the real one (e.g. Bjerrum 1973). Instead of peacemeal approaches this is achieved by a single constitutive equation for the evolution of the effective stress components. It is tensorial, thus objective and apt for finite element applications.

There are only five *key material parameters*, for the examples given above.

I_v, e_r, C_c and C_s are bigger for softer soils and correlated with each other and with $w_L \cdot \varphi_c$ is correlated with them for clays, but for filthy soils it can be far higher, e.g. over 60° for a diatomaeceous mud (Krieg 2000). These parameters can be used for classification. Further parameters can be used for fine-tuning, but they may be fixed empirically in many cases.

For small deformations just after path reversals, in particular for alternating loading, an intergranular strain can be introduced for modelling the initial higher stiffness (Niemunis 2003). This requires one further key parameter, the upper bound R of intergranular strain. R ranges from ca 10^{-3} to 10^{-1} for L to H clays and can be estimated by correlation.

3 FILL UPON SOFT GROUND

Consider a free field with a saturated soft layer of highly plastic clay upon dense sand with a loose sand layer on top, all under water (Fig. 6). The sand is modelled by non-viscous hypoplasticity as its $I_v \approx 0.01$ allows to neglect creep for the time intervals considered here. The H-clay may have the data of Table 1 and may have been deposited 5000 years ago together with the sand on top. It is assumed that the clay had $e = 2.8$ just after sedimentation. A void-ratio independent permeability $k \approx 10^{-9} \mathrm{m\,s^{-1}}$ is assumed for simplicity.

The plotted *evolution of state variables e and σ'* (vertical) is calculated as follows. Gravity is imposed in a first step, and the total stress σ does not change afterwards in material points. The pore water pressure $p_w = \sigma - \sigma'$ is close to σ as σ' is low by (1) for the assumed initial e. The hydraulic gradient is related with the relative velocity

$$v_w - v_s = - k \left(\frac{\partial p_w}{\gamma_w \partial z} - 1 \right) \tag{13}$$

of pore water with respect to solid. The conservation laws read

Table 1.

	I_v	e_r	C_c	C_s	φ_c
H	0.05	2.60	0.35	0.18	18°
L	0.02	0.75	0.06	0.01	28°

$$\frac{\partial(\alpha_w v_w)}{\partial z} + \frac{\partial \alpha_w}{\partial t} = 0 \tag{14}$$

for the water with its volume fraction $\alpha_w = e/(1 + e)$ for $S_r = 1$, and

$$\frac{\partial \sigma'}{\partial z} = \gamma' - \frac{\partial p_w}{\partial z} \tag{15}$$

for the linear momentum.

The effective stress rate $\dot{\sigma}'$ is given by the constitutive relation, which can be written as

$$\dot{\sigma}' = \frac{\partial \sigma'}{\partial t} + v_s \sigma' \tag{16}$$

with $\dot{\sigma}'$ by (7) in the one-dimensional case. The *initial condition* is the one outlined above for e and σ', plus $v_s = 0$. The *boundary conditions* are $p_w = p_{wh}$ (hydrostatic) and $\sigma' = \sigma - p_{wh}$ at the contact clay/sand, and $v_s = 0$ at the base. The system of equations can be solved with finite elements. The calculated evolution of e and $p_w - p_{wh}$ is plotted in Fig. 6. The decay time t_d of $p_w - p_{wh}$ can be estimated by (6) with layer thickness d and average k, e and σ'. For $t > t_d$ the void ratio decreases by (4). If t_d exceeds the resting time, in particular with non-linear permeability and very low k, the clay retains an excess pore pressure, whereas (4) can still hold. OCR tends to

$$\mathrm{OCR} \approx (t/t_d)^{I_v} \tag{17}$$

for $t > t_d$. Note that our OCR is not the conventional one, but it comes close to it in this case for $t > t_d$. The sedimentation time t_s is not relevant for OCR in case of $t_s \ll t$.

All these findings agree with field observations. The settlement s can be calculated from e, and σ' from σ and p_w. Non-homogeneous soft layers can be allowed for if $k(e)$ and the visco-hypoplastic parameters are given. A sandwich of many fine layers can be substituted by a few ones with the same overall volume fractions (Fig. 7). An equivalent permeability

$$k_e \approx k(d_s/d_f)^2 \tag{18}$$

is needed if sand bands drain laterally. The horizontal and vertical water velocities v_{wh} and v_{wv} are needed, whereas $v_{sh} = 0$ suffices for the solid if the surface is horizontal. As the sedimentation history is not precisely known an adaption by means of field data for e, p_w and c_u (cf. Figs. 2 and 3) is required.

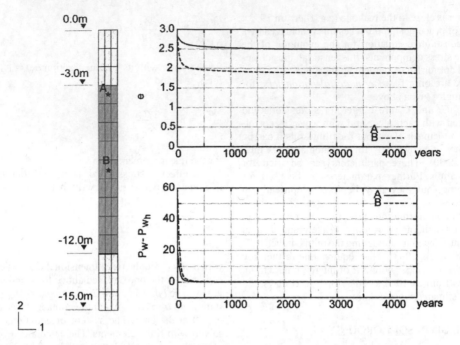

Figure 6.   Calculated evolution of void ratio and excess pore pressure of a soft layer after sedimentation.

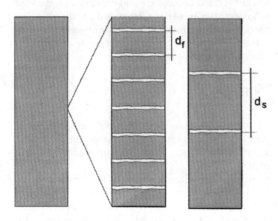

Figure 7.   Sandwich and substitute.

The state in Fig. 6 for 5000 years was used as initial state for calculating changes due to a *dam fill* (Fig. 8). The placement of layers is simulated by imposing their gravity in realistic intervals. The two-dimensional counterparts of (13) to (16) are needed if plane strain is assumed. Changes of position and state are obtained for any desired time. For instance, Fig. 8 shows displacements, in particular the settlement u_2 of the top layer and the horizontal displacement u_1 under the rim just after having placed the dam in 15 days. Afterwards

the displacements increase with a constant fill. There is a spontaneous increase of displacement rates and pore pressures, i.e. a *creep rupture* is obtained. This is accompanied by a shear localization roughly along a slip circle. It cannot lead to an extremely thin shear zone as the gradient of $p_w - p_{wh}$ keeps it wide. This is a physically based regularization.

All that is well known, but cannot be properly judged with conventional methods (e.g. Hunter and Fell 2003). Plasticity with c_u can at best give a crude estimate of failure, but it tells nothing on an autogenous increase of displacements. Creep settlements are underestimated when spreading is neglected. With visco-hypoplasticity we do not depend on conventional simplifications. The higher stiffness in the lab due to viscosity is allowed for, also lateral spreading well below limit states. The effect of a temporary excess fill can also be predicted. Other than widely assumed it is forgotten in the course of time, this 'extended swept-out of memory' (ESOM) is embedded in the constitutive relation (Gudehus 2003c). Loss of stability is obtained as spontaneous increase of velocities and pore pressures after a pre-dictable time. Limit plasticity methods are of use only for crude estimates.

Changes of state can be judged by calculated contour plots of $p_w - p_{wh}$ and OCR. A creep rupture is characterized by a spontaneous increase of both. Thus

$$d(\mathrm{OCR}_{av})/dt > 0 \qquad (19)$$

176

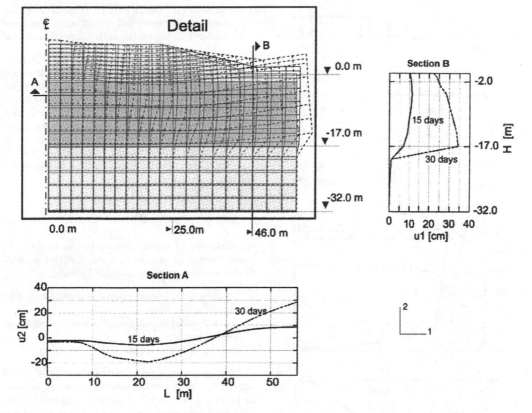

Figure 8. Calculated displacements of the soft ground of Fig. 6 during and after filling a dam. The far-field part of the mesh is not shown.

for OCR, averaged over the region of interest, is a sufficient collapse condition, and $<$ instead of $>$ in (19) holds for spontaneous stabilisation.

An *observational method* can therefore proceed as follows:

1. specify the initial state as outlined e.g. with Fig. 6 & 8, allowing also for $S_r < 1$ and inevitable scattering;
2. substitute vertical drains and granular columns similarly as a sandwich with Fig. 7;
3. calculate evolutions of position and state for design variants with different waiting times, including a temporary excess fill;
4. select variants avoiding creep rupture and too big long-term displacements;
5. select most indicative monitoring quantities;
6. execute construction and monitoring, adapt predictions and further field operations.

4 BRACED EXCAVATIONS IN SOFT CLAY

Fig. 9 shows one half of a braced excavation in soft clay. For our calculation, a moderately plastic homogeneous clay is assumed with parameters as in Table 1. With a surface pressure p_b instead of a neighbour building, the initial state is generated as for Fig. 6, without an initial $p_w - p_{wh}$. OCR ≈ 1.2 is obtained, and

$$\sigma_h / \sigma'_v \approx \sqrt{\text{OCR}} \ (1 - \sin \varphi_c) \qquad (20)$$

as known empirically (Schmidt 1966). A sheet pile wall is installed numerically, changes of state in the soil nearby due to this installation are neglected. The option for sliding of soil along the wall is built in.

The excavation in three steps is simulated by reducing p' (and thus stiffness and strength) there to zero in realistic intervals. Struts are installed and fixed simultaneously. The calculated evolution of strut forces and of some horizontal displacements is plotted in Fig. 9. There are significant autogenous changes.

The results are essentially confirmed by observations in Chicago (Wu and Berman 1953). The wall top moved backwards, the strut forces suggest a trapezoidal earth pressure distribution with a resultant that can be estimated with an adapted c_u, but it

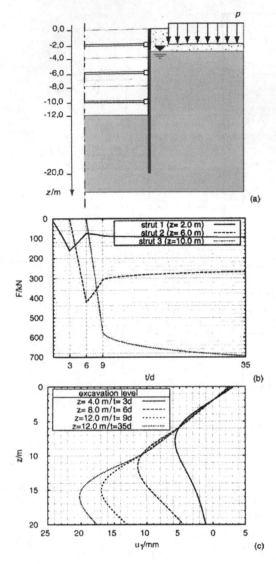

Figure 9. Calculated evolution of displacements and forces for an excavation in soft ground with braced sheet pile walls.

3. predictions for variants in order to avoid creep rupture with bottom heave and/or buckling of struts, and excessive deformations of walls and neighbourhood.

Steps 4, 5 and 6 hold as written in Sec. 3. Terzaghi and Peck (1947) mention a critical wall displacement of ca 2.5‰ beyond which creep becomes rapid. Bottom heave as described by Bjerrum (1973) can be treated similarly. This important experience is supported now by element tests (Figs. 3 and 5) and can be embedded into the analysis with visco-hypoplasticity. Destabilization can be judged with (20) and avoided with the converse of it, which is not possible with a conventional method.

5 CONCLUSIONS AND OUTLOOK

As illustrated by some numerical element tests, the known behaviour of soils with soft particles is realistically modelled with visco-hypoplasticity:

- rate-dependent compression and oedometric creep (Fig. 1);
- rate-dependent shearing and creep without drainage (Figs. 2 and 3);
- shearing with free drainage (Figs. 4 and 5).

If the overconsolidation ratio OCR $= p_e/p'$ with a stress ratio dependent equivalent pressure p_e is below ca 1.4 rate-dependence and creep are dominant, otherwise the behaviour is hypoelastic. Five key parameters are needed, some typical values are given in Table 1.

For applications the geogeneous composition and initial state have first to be specified. This is more consistent by means of a back-analysis with visco-hypoplasticity (e.g. Fig. 6). Sandwich-type formations can and have to be simplified (Fig. 7). A crust with shrinkage cracks has an estimated higher OCR and a far higher permeability k. Subsequent filling or braced excavation can be followed up realistically (Figs. 8 and 9). Changes of position and state can be predicted for design and monitoring. A spontaneous loss of stability can be judged from an autogeneous decrease of the average OCR if there are no changes at the boundaries. The converse is evident.

This enables a physically sound and already validated substitute of the observational method with conventional predictions. Conventional plasticity models cannot predict a spontaneous loss of stability, and merely empirical substitutes are insufficient. The transfer to other cases is straightforward:

- buildings upon soft ground, also with compensation grouting or underexcavation;
- any kind of excavation with fluid or solid support, e.g. shield tunnelling;
- cuts with drainage, anchors or dowels.

increases after the end of excavation. This case and similar ones are referred to by Peck (1969). He used qualitative or conventional models in his observational method, we can do better now with visco-hypoplasticity:

1. Geological composition and initial state of the ground are specified, using data for w, S_r and c_u and allowing for spatial scattering;
2. Installation of walls, excavation and bracing are simulated for design variants (pre-dimensioned with c_u);

Soils with hard grains, and negligible viscosity therefore, are incorporated by means of hypoplasticity. Numerous validations have been presented (Gudehus 2003b, Karcher et al. 2003), strong earthquakes are included (Cudmani et al. 2003).

Crude estimates with visco-hypoplasticity can show whether a proposed procedure is needed for safety or rewarding for economy. A sound geological investigation is indispensable, defaults of it cannot be bridged by monitoring and back analysis. The permeability can be quite irregular, e.g. due to varves or cracks, this requires simplifications and monitoring. It goes without saying that engineering judgment is of importance, but this requires a sound physical base.

As any theory, hypoplasticity and visco-hypoplasticity have certain ranges of validity. Shear localization has been incorporated for sand with polar quantities (Nübel 2003). This has yet to be done for clays, pore water diffusion has to be allowed for. For low initial OCR the gradient of excess pore pressure cares for a lower bound of shear zone thickness (e.g. Fig. 8). For high initial OCR the gradient of suction enhances shear localization. Negative p_w can lead to cavitation cracks, e.g. during shrinkage or creep rupture. This formation of fractals remains to be modelled, and also the evolution of condensation bridges. Unsaturated soils are implied as yet only for the case of intergranular bubbles. Macropores due to dominant van der Waals or capillary forces are also not implied, thus extremely collapsible soils are excluded. The basic assumption of hypoplasticity, viz. that volume fractions and partial stresses determine the state, cannot cover the whole spectrum of fabrics, but it suffices evidently for a substantial part of it.

ACKNOWLEDGEMENT

Figures 1 to 8 have been calculated by Ana-Bolena Libreros-Bertini, Thomas Meier worked out Fig. 9. More details can be obtained via e-mail: Libreros-Bertini: ana.libreros@ibf.uni-karlsruhe.de, Meier: thomas.meier@ibf.uni-karlsruhe.de

REFERENCES

Bjerrum, L. (1973): Problems of Soil Mechanics and Construction on Soft Clays. State-of-the-Art Report, *Proc. 8th Int. Conf. SMFE*, Moscow, p. 1–53

Cudmani, R., Osinov, V., Bühler, M. und Gudehus, G. (2003): A model for the evaluation of liquefaction susceptibility in layered soils due to earthquakes. *Proc. 12th Panam. Conf. SMGE*, Cambridge/USA

Gudehus, G. (1996): A comprehensive constitutive equation for granular materials. *Soils and Foundations*, Vol. 36, No. 1, p. 1–12

Gudehus, G. (2003a): Prediction of deformations due to various geotechnical actions by means of hypoplasticity. *Proc. XIIIth Eur. Conf. Soil Mech. Geot. Eng.* Prag

Gudehus, G. (2003b): How to control deformations of partly technogeneous ground. *Proc. XIIIth Eur. Conf. Soil Mech. Geot. Eng.* Prag

Gudehus, G. (2003c): A Visco-Hypoplastic Constitutive Relation for Soft Soil. Submitted to *Soils and Foundations*, Febr. 2002

Herle, I. and Gudehus, G. (1999): Determination of parameters of a hypoplastic constitutive relation from properties of grain assemblies. *Mech. Cohes. Frict. Mater.,* Vol 4, p. 461–486

Hvorslev, M.J. (1937): Über die Festigkeitseigenschaften gestörter bindiger Böden. *Ingeniörvid. Skrifter* A, Nr. 45, Danmarks Naturvid. Samfund, Dissertation, Kopenhagen

Kolymbas, D. (1978): Ein nichtlineares viskoplastisches Stoffgesetz für Böden. *Veröff. Inst. Boden- u. Felsmech.* Univ. Karlsruhe, Heft 77

Krieg, St. (2001): Viskoses Bodenverhalten von Mudden, Seeton und Klei. *Veröff. Inst. Boden- u. Felsmech.* Univ. Karlsruhe, Heft 150

Leinenkugel, H.-J. (1976): Deformations- und Festigkeitsverhalten bindiger Erdstoffe. Experimentelle Ergebnisse und ihre physikalische Deutung. *Veröff. Inst. Boden- u. Felsmech.* Univ. Karlsruhe, Heft 66

Niemunis, A. (1996): A visco-plastic model for clay and its FE-implementation. In *Resultats recents on mécanique des soils et des roches*. XIth Colloque France-Polonais, Politechn. Gdanska, p. 151–162

Niemunis, A. (2003): Extended hypoplastic models for soils. Politechnica Gdanska, *Monografia* Nr. 34, Gdansk/Poland

Nübel, K. (2002): Experimental and numerical investigation of shear localization in granular matter. *Veröff. Inst. Boden- u. Felsmech.* Univ. Karlsruhe, Heft 159

Peck, R.B. (1969): Advantages and Limitations of the Observational Method in Applied Soil Mechanics. *Géotechnique* Vol. 19, No. 2, p. 171–187

Schofield, A. (2000): Re-appraisal of Terzaghi's Soil Mechanics, *Proc. XVth ICSMGE*, Vol. 4, p. 2473–2482

Schmidt, B. (1966): Discussion of Earth Pressure at Rest Related to Stress History. *Canad. Geot. Journ.* Vol. 3, p. 239–242

Terzaghi, K. and Peck, R. (1947): *Soil Mechanics in Engineering Practice*. Wiley

Wu, T.-H. and Berman, G. (1953): Earth pressure measurements in open cut: Contract D-8, Chicago subway. *Géotechnique* Vol. III, p. 248–258

Numerical analysis of block movements at the Spis Castle

L. Baskova & J. Vlcko
Faculty of Natural Sciences, Department of Engineering Geology, Comenius University Bratislava, Slovak Republic

ABSTRACT: Spis Castle (Eastern part of Slovakia) is built on a travertine mound overlying the Paleogene soft rocks. The travertine formation represents an erosion remnant of an originally larger complex precipitated during the Miocene/Pliocene epoch. The physical and mechanical properties of travertines are strongly influenced by jointing, weathering and karstification. The travertine body is affected due to block spreading slope failure. The slope deformation and slope failure mechanism using 2D numerical modelling (UDEC – professional code) have been analyzed and investigated.

1 INTRODUCTION

The Spis Castle, Natural cultural Monument, under Patrimony of UNESCO represents the largest medieval fortification system in Central Europe (Figure 1). It was founded in 1120. The historic development of the castle was rather complicated showing traces of many historic epochs up to the Baroque. In 1780 the castle burnt out and since that time it was abandoned and the process of destruction caused both by the natural and man-made factors was going on.

2 GEOLOGICAL SETTING

From a geological point of view the studied area is located in a zone referred to as Hornadska kotlina

Figure 1. Spis Castle.

Basin (Eastern Slovakia). Spis Castle is built on a travertine mound, which is underlain by Paleogene soft rocks formed by claystone and sandstone strata (flysh-like formation). The travertine body reaching more than 52 metres in thickness reflects several features of destruction and is disturbed by a series of faults, cracks and joint systems. Two prevailing joint sets can be found, sub-vertical joints striking approximately NW to SE with a general dip to SW (dip direction/dip 220°–250°/80°–90°) and joints striking approximately N–S dipping to the W (250° to 270°/85°). The origin of the Temna jaskyna Cave is strongly bound to this system as well. The destruction is the result of gravitational slope failures due to block spreading of rigid travertines on relatively plastic claystone strata. The central part of the travertine rock is formed by a block rift (traver-tine cliffs separated by persistent tensional joints and cracks), the marginal parts of the castle rock are formed by a block field consisting of displaced and tilted cliffs reaching the height from 25 to 30 m, sloping at an incli-nation of 70° to 80°, in some places up to 90° with a number of overhangs. The absence of a block field in the SW part of the castle rock is due to the uplift of Palaeogene claystones along the fault line (220°/80°) which inhibited total disintegration of the block field, followed by rock falls, toppling and tilting of huge cliffs of travertines (Vlcko et al. 1993, Vlcko 2002).

3 NUMERICAL MODELLING, PRINCIPAL RESULTS

The model has been analysed using the Universal Distinct Element Code (UDEC, ITASCA 1993). This program simulates the mechanical behaviour of the

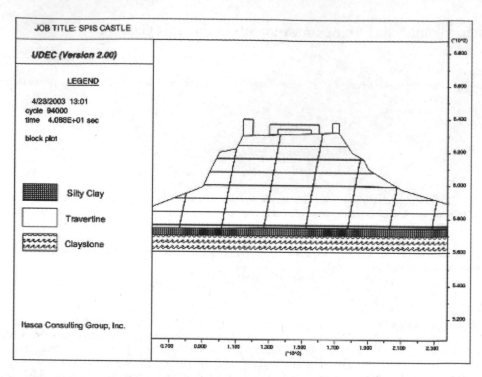

Figure 2. Numerical presentation of a representative geological profile.

discontinuous medium represented as an assemblage of discrete blocks subjected to either static or dynamic loading. The main features of the code can be summarised as follows:

- The discontinuities are treated as boundary conditions between blocks; finite displacements along discontinuities and rotation of block are allowed;
- Blocks may be rigid or deformable; contacts are always deformable;
- The program recognizes new contact as the calculation proceeds;
- Several constitutive behaviour models following linear or non-linear laws are available for both joints and blocks;
- The program can simulate steady or transient fluid flow through the discontinuities.

As introduced in the UDEC code, vertical sides of the model have been assumed to move vertically only and the horizontal ones only horizontally. The rock material and the discontinuities are assumed elastic-perfectly plastic when the Mohr-Coulomb failure envelope or tensile failures are reached. Travertine and bedrock have been assumed as fully deformable blocks and then discretized into finite difference triangular elements. The model ran over a number of iterations until the initial equilibrium conditions were attained.

Table 1. Physical and mechanical parametres.

Lithological types	ρ (kg·m^{-3})	σ_c (MPa)	E (MPa·10^3)	ν
Travertine	2500	63	56,6	0.19
Silty clay	1850	10	17,0	0.35
Claystone	2310	33	20,0	0.25

In geotechnical terms, the stratigraphical sequence can be (schematically) represented as the superimposition of a rigid travertine body over claystone strata, in the upper parts highly weathered. Numerical presentation of a representative geological profile is in Figure 2.

A stepwise modelling procedure we adopted was based upon the back analysis comprised:

a) Simulation of travertine sequences over the Paleogene rocks until the state of equilibrium (initial state of stress) was determined. The travertines were assumed as an ideal homogeneous rock body.

b) Introduction of gradual decrease of bedrock material properties (weathering, softening) as well as gradual decrease of tensile strength along the joints in travertines (mainly joint normal stiffness and shear stiffness) were considered.

The physical and mechanical parameters as input data for modelling are summarized in Table 1.

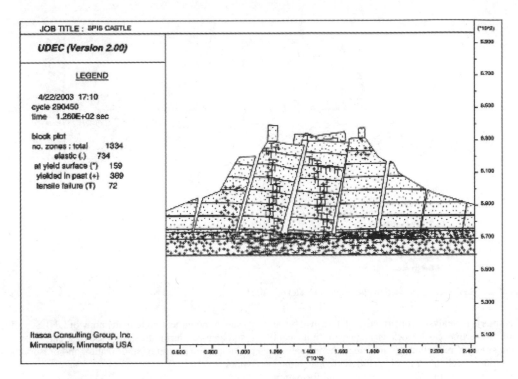

Figure 3.   Developments of tension cracks.

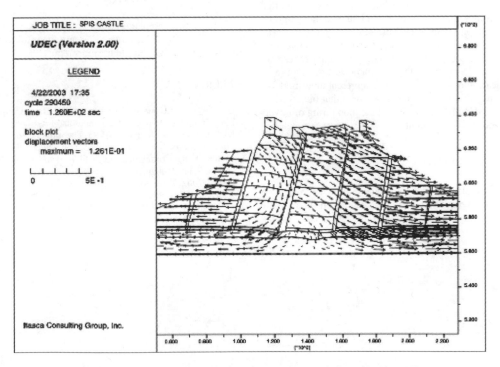

Figure 4.   Character of deformations of travertine's cliff, displacement vectors indicate block spreading.

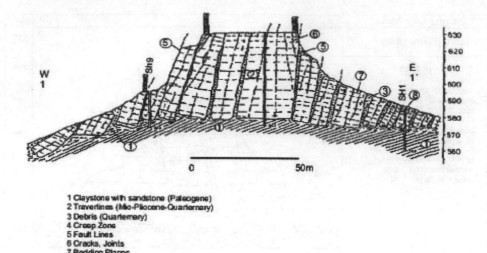

1 Claystone with sandstone (Palaeogene)
2 Travertines (Mio-Pliocene-Quaternary)
3 Debris (Quaternary)
4 Creep Zone
5 Fault Lines
6 Cracks, Joints
7 Bedding Planes
8 Boreholes

Figure 5. Geological cross section after Malgot in Vlcko et al. 1993.

Numerical analysis confirmed that the instability of the travertine castle rock is related to significant shear strength reduction of the subgrade formation. The failure mechanism involves differential subsiding of the travertine cliffs into soft bedrock and formation of persistent tension cracks in the central part of castle rock. The development of tension cracks transmitted the strain to the castle walls and inhibited their rupture, in some places even collapse (Figure 3). On the marginal parts of the castle rock, the cliffs are toppled upslope (western part) while in the eastern part these are toppled down-slope. Displacement vectors (Figure 4) indicate block spreading as a representative mode of failure, which is in coincidence with that observed by the field study (Figure 5) and the monitoring of travertine cliffs displacements.

4 CONCLUSION

To understand better landslide failure mechanism a numerical modelling can significantly contribute to the hazard/risk assessment at various sites. In case of Spis Castle, a monument under patrimony of UNESCO, the results gained during the study are of great value and are helpful in design of stabilization and preservation works.

ACKNOWLEDGEMENT

This paper was prepared as part of the VEGA grant project No 1/9159/02. The authors are thankful to the grant agency VEGA for their kind support.

REFERENCES

Itasca, 1993. UDEC version 3.10–user manual. Itasca Consulting Group Inc., Minneapolis. Minnesota.
Vlcko, J., Baliak, F., Malgot, J. 1993. The influence of slope movements on Spis Castle stability. Landslides – Seventh International Conference and Field Workshop: 305–312. Rotterdam: Balkema.
Vlcko, J. 2002. Monitoring – an Effective Tool in Safeguarding the Historic Structures. International Symposium Landslides Risk Mitigation and Protection of Cultural and Natural Heritage: 267–278. Kyoto: Kyoto University.

Geotechnical Measurements and Modelling, Natau, Fecker & Pimentel (eds)
© 2003 Swets & Zeitlinger, Lisse, ISBN 90 5809 603 3

Weak rocks in engineering practice

T. Durmeková & R. Holzer
Department of Engineering Geology, Comenius University Bratislava, Slovak Republic

P. Wagner
Geological Survey of Slovak Republic, Bratislava, Slovak Republic

ABSTRACT: Weak or soft rocks occur widely in geologic conditions of the Slovak part of the Western Carpathians. It is practically impossible to avoid these rock materials in engineering constructions. The paper deals with the basic definition of weak rocks, characteristics of individual lithological types and it illustrates problems with this rock group in engineering practice.

1 INTRODUCTION

Unfavourable physical and mechanical properties of weak rocks such as the low strength, high porosity, anisotropy of mechanical properties, disintegration in contact with water, swelling, and low durability are significant for this rock group. Field and laboratory research of weak rocks requires specific approach; the quality of obtained geotechnical properties depends essentially upon the sampling methods.

Weak rocks are usually considered suitable as foundation material. Their utilization as a construction material is often problematic.

Several highway tunnels are planned in mountainous areas of Slovakia where mainly weak rocks such as claystone, marlstone, marly limestone, but also weathered sandstone, conglomerate and crystalline rocks or various tectonites will be encountered. The important volume of muck obtained during the tunnel excavation in weak rocks may often be a heterogeneous mixture of gravel-size particles, soils and individual blocks of rock. In this paper, the best possible utilization of the muck from the exploration tunnels is discussed.

The next item discussed is the durability of building or ornamental stone of historic monuments (cathedrals, churches, castles, etc.). Favourable workability of weak rocks has led to very frequent use of such material in the past. Due to the external influences (weathering), rock blocks with weaker structural bonds loose their compactness, they disintegrate and parts of objects often carry signs of visible deterioration and decay. The presented results of investigation of building stone properties aiming to protect historic monuments were achieved from the Bratislava St. Martin Cathedral and the Bratislava Castle.

2 BASIC CHARACTER OF WEAK ROCKS

The term "weak rock" is used only in applied geological disciplines, such as the engineering geology, soil and rock mechanics and geotechnics. The main reasons to distinguish such special rock group are different physical properties and the resulting, often unfavourable, geotechnical behaviour. Weak rocks are typically present in all lithological formations of the Slovak Carpathians as a product of lithogenetical processes, but mostly as result of retrograde changes of originally hard rocks.

There is no appropriate and unified definition of "weak rock" in engineering geology or in the in sciences dealing with rock properties. "Classics" of Czech and Slovak engineering geology Záruba and Mencl (1974) write: "there exist a rock group on the transition between non-cohesive soils and hard rocks which causes many difficulties in constructions. It is a group of rocks with the unconfined compressive strength less than 50 MPa". Rocks such as claystone, marlstone, sandstone, soft limestone, soft slate, shale, with the increasing moisture become progressively softer, they weather and disintegrate very fast, eventually show large changes in volume. According to the definition (Hrašna et al., 1987), used by the engineering geologists in Slovakia, weak rocks are known as rocks with less favourable mechanical properties than hard rocks due to the low degree of lithification or due to the decay of structural bonds. Their properties have deteriorated progressively by external influences. According to this definition rocks such as weak metamorphosed rocks, phyllite, phyllonite, mica schist, would not be included. According to the Slovak Technical Standard (STN) 72 1001 *"Designation and description of rocks and soils in engineering geology"* the rocks are divided into two groups – soils and hard rocks. Weak rocks are mentioned only as a special group in the footnote: "Weak rocks are transitional rock types between hard rocks and soils. Their unconfined compressive strength varies between 1,5 and 50 MPa". STN 73 1001 *"Foundation of structures. Subsoil under shallow foundations"* introduces an additional classification criterion. The rocks are divided according to deformation mode (ratio of modulus of deformation to the unconfined compressive strength). Weak rocks belong to the group of plastic materials. The most of rock classifications are based on values of unconfined compressive strength. Comparing individual classifications (Coates, 1964, Deere & Miller, 1966, Bieniawski, 1973, Broch & Franklin, 1972, IAEG, 1979, ISRM, 1981, Brit. Geol. Soc. BS 5930, 1981) in review compiled by Hawkins (1998), classifications show large differences, although in some of them the classification terms like very weak to weak or groups with very low or low strength (of medium strength) vary between 1,5 to 50 (70) MPa. Selby (1993) classifies the very weak and weak rocks in the range from 1 up to 50 MPa.

3 UNFAVOURABLE ENGINEERING BEHAVIOUR OF WEAK ROCKS

Physical, mechanical and technological properties of weak rocks are considered usually as very unfavourable in the engineering practice. Structural inhomogeneity and anisotropy of mechanical properties changing from place to place (even in the frame of the same outcrop, or an investigation site, etc.) cause the disparity in the strength characteristics (which are usually very low). Weak rocks are typical by their partly high compressibility, high porosity, disintegration in contact with water, swelling, low durability and low resistance against various weathering factors. Weak rock sampling is complicated. Field drilling with water circulation (with portable drilling machine) disintegrates the rock core significantly, specifically in metamorphic and sedimentary rocks with a bedded or foliated structure. Laboratory preparation of regular (cylindrical or cubic) specimens from rock monoliths extracted from outcrops brings about similar problems. The described working procedure provides usually inadequately low numbers of specimens (if any) so that obtained values of physical properties are often controversial. Beside this, there is no reliable method of experimental testing

because according to current technical standards the procedures are more suitable for the group of hard, mechanically or chemically non-decomposed or otherwise weakened rocks.

The currently used investigation and testing methods for weak rocks are considered inadequate and the need to find methods of appropriate sampling, suitable specimen preparation and laboratory testing is very important.

4 WEAK ROCKS IN TUNNELING

When week rocks are encountered in tunnels two basic problems have to be considered:

a) Weak rocks comprise an uncertain environment for the complicated underground construction;
b) Use of weak rocks from the tunnel excavation as the construction material or its deposition as bulk waste may be inappropriate because of unfavourable physical properties.

Based on the experiences from highway tunnels construction in Slovakia, several examples are presented.

The new highway net in mountainous part of Slovakia includes the construction of several tunnels (Fig. 1). After construction of the tunnel Branisko in Eastern Slovakia (expected to be finished in 2003), the tunnel Horelica in the Flysh unit is under construction at present. The pilot gallery for tunnel Ovčiarsko (near the town Žilina) with the length of 2368 m has been finished in 1998. In autumn 2002 the pilot gallery for the longest designed highway tunnel in Slovakia – tunnel Višňové (length of 7410 m) has been finished. Both tunnels are ready for construction.

From the geological point of view the tunnels are located in different geological environment – while the tunnel Ovčiarsko is designed totally in Flysh complexes of the Carpathian Paleogene, the tunnel Višňové crosses the Malá Fatra core mountain. A major part of the tunnel will intersect the Paleozoic rocks. Weak rocks have

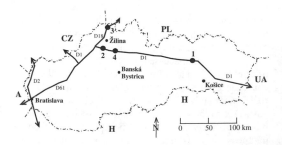

Figure 1. Location map – proposed highway net (D1, D2, D18, D61). Proposed tunnels: 1 – Branisko, 2 – Ovčiarsko, 3 – Horelica, 4 – Višňové.

been confirmed in the western portal part (Paleogene claystone, siltstone and sandstone), in Mesozoic formations and partly as disintegrated rocks in fault zones in granitic core rocks. The character and extent of problems connected with weak rocks is different in both underground structures.

4.1 Complications related to driving of tunnels in weak rocks

Driving of the pilot gallery for tunnel Ovčiarsko encountered following problems:

- Re-activation of ancient landslides near the western portal in claystone and marlstone. The re-activation required implementation of urgent corrective measures;
- Variability of the rock environment required frequent changes of the driving technology – TBM in weak rocks and blasting methods in hard rocks or the Belgian method of underpinning in tectonic faults;
- Swelling of claystone and siltstone containing 20 to 35% (locally up to 50%) of the fine fraction (clay and silt) caused the most serious problems during the driving. The results of mineralogical analysis of the clay fraction indicated the clay mineral association of smectite, illite and chlorite with the dominance of smectite (Ondrášik et al., 2002). Due to swelling, the deformations of the gallery walls and ceiling progressed and opening of joints in the tensile zone in the roof occurred. With the moisture content increasing the volume changes in loosened rock mass also increased and several parts of the lining (shotcrete) had failed (Fig. 2).

Failures in gallery walls appeared several months after the gallery construction. The walls of the gallery had to be reinforced by the additional lining elements. From the total gallery length, several sections (in the whole length of 645 m) needed remedial measures to deal with the rock pressures related to the volume changes.

Similar problems have been encountered during driving of the pilot gallery Višňové. The rocks are lithologically very heterogeneous as shown schematically in Figure 3 and Table 1. Serious stability problems were encountered in the western portal in Paleogene weak rocks. Slip surfaces of ancient landslides have been found in the distance about 60 m from the gallery entry. Most difficulties during the gallery driving have been related to unfavourable groundwater conditions in the crystalline rock mass.

4.2 Weak rocks as a construction material

An excavated material (muck) obtained during driving of the underground openings in weak rocks is often a very heterogeneous mixture of a material major part of which having a particle size of aggregates, soils and individual blocks, the size of which corresponds to quarry stone. Granular heterogeneity is usually complicated by the lithological heterogeneity.

Beside these general facts related to the lithology, the structure of the rock mass and methods of the tunnel driving, other characteristics are important. They include sensitivity to water, volume instability, and a great variability of mechanical properties. These are

Figure 2. Shotcrete failure in the pilot gallery due to the claystone swelling. Photo: A. Matejček.

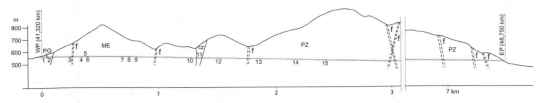

Figure 3. Schematical geological cross-section along the pilot gallery for the tunnel Višňové (simplified according to Ondrášik et al., 2000). PG – Paleogene, ME – Mesozoic, PZ – Paleozoic, f – principal tectonic faults, cz – contact zone, 1 to 15 – sampling points, WP – west portal, EP – east portal.

187

Table 1. Višňové pilot gallery – characteristics of rock environment.

Section [m]	Geological unit	Lithology	Numbers and types of samples	UCS [MPa]	Stability in water
0 – 115	Paleogene	Rhythmical flysh strata: claystone, siltstone, sandstone, sandy claystone (claystone is dominant)	1 – sandstone 1 – sandy claystone 2 – claystone	22–65 20 30	Fairly stable Fairly stable Unstable
115 – 1334	Mesozoic – Jurassic	Heterogeneous complex of strata: limestone, shale, dolomite, carboniferous claystone, tectonic dolomite and clay, sporadically position of evaporite, rocks are in different weathered state and tectonic changed	3 – clayey limestone 4 – dolomite 5 – alterate dolomite 6 – dolomite limestone 7 – dolomite 8 – dolomite breccia 9 – limestone, dolomite 10 – clayey limestone 11 – cataclasite	40–50 71 15–20 45–50 18–21 47 20–53 54 80	Unstable Stable Unstable Stable Unstable Fairly stable Stable Fairly stable Fairly stable
1334 – 1510	Tectonic contact zone	Tectonic dolomite, limestone, rauhwacke, tectonic breccia, cataclasite, mylonitized granite			
1510 – 7410	Paleozoic	Crystalline rocks: granite, granodiorite to tonalite, gneiss (dominance of magmatite above metamorfite)	12 – granite 13 – granodiorite to tonalite 14 – granodiorite 15 – tonalite, granodiorite	85 66 56 98	Stable Stable Stable Stable

Note: UCS – unconfined compressive strength.

very important in deciding of muck utilisation for engineering purposes.

A great variability and heterogeneity of rocks (melange of irregular rock fragments and soils) have been the basic problem of laboratory testing of rocks from the pilot gallery Ovčiarsko. Heterogeneous samples had to be tested separately by methods of rock mechanics and soil mechanics. Primary separation of an important amount of excavated tunnel material before its use in engineering construction was required. Based on results of laboratory testing and the experience from the tunnel driving, it has been confirmed that only an insignificant volume of the material can comply with the standard requirements for its utilisation as a construction material (however, only for core of embankments, protected from direct influence of climate factors). The majority of the muck is not suitable for engineering purposes and it has to be disposed of as waste only. Problematic utilisation of weak rocks is one of the reasons for questionable construction of the highway tunnel. Experience from driving of the Ovčiarsko pilot gallery indicated that the use of weak rock might present a serious ecological problem.

From the viewpoint of utilisation of the excavated weak rocks, less significant problems are expected at the site of the tunnel Višňvé. A simple proposal for using of excavated material according to its physical and mechanical properties is shown in Table 2.

5 WEAK ROCKS AS A BUILDING STONE OF MONUMENTS

Many historic monuments in Slovakia have been built using the easily workable Neogene weak rocks. Leading historic dominants of Bratislava – the St. Martin Cathedral and the Bratislava Castle were originally constructed using the oolithic limestone, calcareous sandstone, and conglomerate. These monuments have been objects of the investigation. Taking into consideration the relatively high structural decay of building and ornamental stone of monuments in Bratislava, the interest how to define the degree of damage and its reasons more quantitatively, is evident. There are many practical solutions from abroad (St. Stephan Cathedral in Vienna, St. Peter Cathedral in Regensburg, etc.) that show more or less microscopic mineralogical and petrological approach concerning the assessment of the building stone decay. Our emphasis was, in addition to the mineralogical study, to assess the rock quality by quantitative technical data, as well.

5.1 Purpose and methodology of the study

The main purpose of the study was to establish an integrated methodological procedure for the building stone quality assessment, determination of physical

Table 2. Suitability of the tested muck for various engineering purposes.

Purpose of the muck utilisation	Sample identification	Length of section in m and %	Note
I. For the concrete and for wearing courses of pavements and railroads	12, 13, 14, 15	5 620 m 76%	
II. For subgrade layers of the pavements and railroads	4, 6	396 m 5%	Suitability to be verified another tests according to technical standards
III. For the embankments generally	8, 9, 11	936 m 13%	Suitability depends on results of additional tests and on a technological design of the embankment
IV. Into the inner part of embankments, or as their less-quality component	1, 10	50 m 1%	
V. Deposition to dumps or utilisation in a simple arrangement field	2, 3, 5, 7	386 m 5%	

and mechanical properties controlling the durability and designation of deteriorative processes influence. Basic methodological steps were devoted to:

– Macroscopic assessment of the actual building stone decay;
– Sampling of cylindric specimens from monuments for physical and mechanical properties assessment in the laboratory (including the petrological and paleontological composition and weathering degree estimation). Problems of how: to obtain satisfactory number of testing specimens from protected monuments;
– Field documentation and rock block sampling from abandoned quarry walls aiming at the specimens preparation for the unified laboratory investigation;
– Laboratory testing to assess individual rock types from quarries and from monuments (necessary comparable studies), as well as the determination of their physical and mechanical properties (real and apparent density, porosity, water absorption, unconfined compressive strength of dry and saturated samples and samples after freezing and thawing cycles, estimation of the softening coefficient and coefficient of freezing and the salt crystallization test in Na_2SO_4 solution simulating severe weathering conditions). Irregular samples were used for the point load test, slake durability test and for the estimation of the resistance to wear in the micro-Deval machine.
– Final analysis of the investigation results, which has been directed towards preservation and monuments restoration possibilities (search of alternative rocks with appropriate color and durability characteristics). The findings of the original building stone quarrying in Austria were important, as no quarries with similar rocks were found in the vicinity of Bratislava at present. But it cannot be excluded that some historic excavations had been on the north-west slopes of Small Carpathians in Slovakia. The

investigation has shown that applicable building stone sites were found in Slovakia like abandoned quarries of sandstone in Sokolovce and conglomerate in Chtelnica in the distance of about 80 km from Bratislava (Greif, unpubl.).

5.2 Lithological characteristics and use of building stone for reconstruction purposes

The upper structure of above-mentioned monuments in Bratislava is in relatively good condition. In historic building periods following types of Neogene rock were mostly used from quarries in Austria:
oolitic limestone from Wolfsthal, *calcareous sandstone and sandy limestone* (Wolfsthal), *carbonate coarse and medium-grained conglomerate* (Hundsheim), *lumachelle limestone* (Wolfsthal), *limestone with sandy admixture* (Mannersdorf).
The most important properties of these rocks determined in the laboratory testing are shown in the Table 3.

Oolitic and siliceous oolitic limestones are greyish or yellowish colored, built up by calcareous ooids with kernels of foraminifera fragments or clasts of minerals with sparite-calcite or quartzitic cement. The building quality is substantially limited because of a potential of rapid weathering (low resistance of the cement) reflected in lower strength and durability (Table 3). The quartzitic oolitic limestone has a better quality. It is suitable as the building stone, also for exteriors.

Calcareous sandstone and sandy limestone are light colored, porous and fine-grained, primarily used because of their easy workability in facades. Most frequently used building stone on monuments in Bratislava and often without expressive structural decay. But due to the selective, gradual weathering, the rock is more suitable for internal reconstruction use.

Lumachelle limestone is mostly white colored with a specific, rough structure. It has been used in the

Table 3. Physical and mechanical properties of building stone (according to Laho, unpubl.).

Lithological type *Locality*	Real density ρ_s [g.cm⁻³]	Apparent density ρ_d [g.cm⁻³]	Porosity n [%]	Water absorption N [%]	UCS σ_c [Mpa]	Coefficient of softening K_1	Coefficient of freezing K_2	Loss of weight after testing in Na₂SO₄ M [%]
1. Oolitic limestone:								
Wolfsthal – quarry	2,70	2,17	19,6	5,1	15,4	0,98	0,97	5,2
Bratislava Castle	2,62	2,12	18,7	1,5	14,7	0,73	0,75	–
Oolitic limestone with quartz:								
Wolfsthal – quarry	2,69	2,52	6,2	1,1	68,2	0,68	0,65	7,2
Bratislava Castle	2,67	2,28	14,6	4,3	43,0	0,48	0,42	8,9
2. Calcareous sandstone and sandy limestone:								
Wolfsthal – quarry	2,70	2,05	23,9	7,8	9,8	0,89	0,68	11,1
Bratislava Castle	2,70	1,77	35,0	13,9	9,1	0,66	0,63	25,0
3. Lumachelle limestone:								
Wolfsthal – quarry	2,69	2,09	22,2	4,1	5,5	0,81	0,85	0,7
Bratislava Castle	–	–	–	–	–	–	–	–
4. Carbonate conglomerate:								
Hundsheim – quarry	2,74	2,40	12,5	2,9	36,3	0,96	0,93	60,6
Bratislava Castle	2,69	2,40	10,8	2,9	59,2	0,62	0,65	37,0
5. Limestone with sand admixture:								
Mannersdorf – quarry	2,70	1,93	28,3	9,2	13,0	0,92	0,72	100,0
Bratislava Castle	2,71	2,07	23,6	5,6	11,1	0,83	0,74	–

Figure 4. Building stone damage on used rock types St. Martin Cathedral Bratislava.

facade of the St. Martin Cathedral in upper ornamental part of the pedestal (Fig. 4). Expressive selective weathering on the facade surface and visibly degraded mechanical properties of the building stone (Table 3) indicate that the rock is only conditionally applicable for the reconstruction.

Carbonate coarse- and medium-grained conglomerate are light-grey organic rock frequently present on Bratislava monuments. Their granular heterogeneity causes differentiated resistance and selective weathering. Medium-grained rocks are aesthetically more suitable for the reconstruction. Both rock types are frequent on monuments and are relatively sound.

Limestone with sandy admixture is typical organogenic rock, yellow-brownish colored. Clasts and tests are weakly cemented by calcite cement. Building stone was for convenient workability used for the construction of St. Stephan Cathedral in Vienna (Müller et al., 1993). Considering the decrease in resistance proved by laboratory testing (mostly irregular, selective weathering), the use of this rock type for reconstruction goals is very limited.

6 CONCLUSION

Weak rocks occur widely in geologic environment on the territory of the Western Carpathians. Their properties assessment is usually complicated and depends on the lithologic type, sampling and rock degradation state, as well as on engineering activity and utilisation.

The paper was worked out in the frame of the Agency VEGA grant project No. 1/0117/03.

REFERENCES

Hawkins, A.B. 1998. Aspects of rock strength. Bull Eng Geol Env, 57, Springer–Verlag: 17–30.

Hrašna, M., Hyánková, A. & Letko, V. 1987. Engineering geological classification and characteristics of weak rocks (in Slovak). *Mineralia Slovaca*, 19: 553–559.

Müller, H.W., Rohatsch, A., Schwaighofer, B., Ottner, F. & Thinschmidt, A. 1993. Gesteinsbestand in der Bausubstanz der Westfasade und des Albertinischen Chores von St. Stephan. *Österr. Zeitschr. Kunst u. Denkmalpflege*, 106–116.

Ondrášik, R., Matejček, A., Holeša, Š. & Vráb l' ová, K. 2000. Gravitational tectonics and slope deformations along the tunnel line Višňové in the Lúčanská Fatra Mts., Slovakia. *Mineralia Slovaca*, 32: 429–438.

Ondrášik, R., Matejček, A. & Šamajová, E. 2002. Mineralogy and swelling of the Inner Carpathian shale in the pilot tunnel Ovčiarsko, Northwest Slovakia (in Slovak). *Mineralia Slovaca*, 34: 329–334.

Selby, M.J. 1993. *Hillslope materials and processes*. Second edition. Oxford: University Press.

Záruba, Q. & Mencl, V. 1974. *Engineering geology* (in Czech). Praha: Academia.

Geotechnical Measurements and Modelling, Natau, Fecker & Pimentel (eds)
© 2003 Swets & Zeitlinger, Lisse, ISBN 90 5809 603 3

Mass movements in the central-southern Marches monoclinal relief (Central Italy): 3D modelling of the Montegiorgio landslide

B. Gentili & G. Pambianchi
Department of Earth Science, Camerino University, Italy

N. Sciarra
Department of Earth Science, Chieti University, Italy

F. Pallotta
Professional geologist, Macerata, Italy

ABSTRACT: The mass movements represent one of the main forces driving the geomorphologic evolution in the central-southern portion of the Marchean region peri-Adriatic belt (Italy). The present work examines the case of Montegiorgio, representative of the main landslide of the area, for which numerous data have been acquired through detailed geological and geomorphologic analyses and in-depth hydrogeological, geotechnical and geophysical studies. A 3D physical model has been reconstructed for use as a numerical analysis based on a finite differences method. The modelling carried out must be understood as a simplified attempt to understand a very complex phenomenon produced over a very long time-span. What is tested is the real evolution, at present time, of the slope only regarding the clayey bedrock and the probable implication of two sandy layers, with high permeability and containing a phreatic surface, on the global system equilibrium.

1 INTRODUCTION

Numerous studies have evidenced the important morphogenetic role played by the action of gravity in the different physiographic contexts of the Marches, seeing this genesis in the wider frame of the morphotectonic evolution in central Italy. In the mountain sector, the action of gravity is essentially manifest in the vast mass movements or, sometimes, in the tectonic-gravitational processes. In fact, in the area more than 500 phenomena with an average extension of ca $2\,km^2$ (maximum up to $20\,km^2$) have been identified and analyzed. Of these, ca. 10% are represented by deep-seated slope gravitational phenomena and the remainder by various landslide typologies. Also for the Adriatic slope piedmont-hilly belt and for some portions of the coastal zone, the slope dynamics prove to be very intense and governed by the recurrence of mass movements of varying typology, extension, and depth. Accident due to landslides mostly affects only the eluvial-colluvial deposits, but the cases of gravitational movement are relatively numerous with their shear zone in the bedrock (Aringoli et al., 2002; Crescenti et al., 2002; Dramis et al., 1995, 2002; Gentili et al., 1995).

The present study analyzes one of these phenomena, the one affecting the summit portion of the Montegiorgio relief in the middle belt of the central-southern Marchean high-hill sector. Since it is distinguished by relative simplicity and regularity in its geological and hydrogeological setting and by specific physiographic and morphodynamic elements, the area can be considered an important one and a representative sample for carrying out in greater depth research into the definition of danger from landslides and the risks connected with it. The analysis and interpretation of the geomorphologic elements have allowed us to describe the evolutionary model for the phenomenon, the basic kinematics of which has been verified by applying a numerical analysis code to the finite differences, also with a view to determining the thrust-deformation correlations inside the deformed and/or moving masses.

2 GEOLOGICAL SETTING

In central Adriatic Italy, only sedimentary rocks belonging to an Upper Triassic – Lower Pleistocene marine succession are to be found (Centamore & Deiana,

1986). The most ancient terrains (Upper Triassic – Priabonian), formed of calcareous and calcareous-marly rocks, crop out in the Apennine ridge; they are folded and overlap with eastward vergence the more recent prevalently marly Oligocene-Tortonian sediments. There then follow the turbiditic Tortonian-Messinian deposits, prevalently with arenaceous-pelitic facies, affected by folds and thrusts, on which rest, mostly in angular unconformity, the marine sediments of the Plio-Pleistocene cycle. On the latter the monocline relief of the study area is sculpted, with its lithostratigraphic successions normally made up of pelitic deposits among which are intercalated, at various stratigraphic heights, coarse-grained clastic deposits represented by polygenic conglomerates or sands having lenticular geometry.

The structuring of the pre-transgressive bedrock was realized essentially in Lower Pliocene and is characterized by ca. N–S trending folds and thrusts, both cropping out and buried. The compressional activity was less intense and continued in Quaternary along the present-day coast, reactivating the above-mentioned structures and producing the further uplifting and emersion of the peri-Adriatic belt. To the phenomena of generalized tectonic uplifting (differentiated in Apennine and anti-Apennine direction) which have affected the study area starting from Lower Pliocene, is connected the (at least at the surface) regular monoclinal setting of the Pliocene sediments (Deiana & Pialli, 1994) with a slope angle varying from 15°–20° in the western sector to 5° in the coastal belt (Fig. 1a). Rare direct faults (Apennine and anti-Apennine) with a weak vertical displacement, and three systems of extensive fracturing dislocate the

monoclinal structure: two of them have a trend similar to that of the faults, while the third trends approximately south (Fig. 1b).

They preferentially affect the stiffer rock bodies and the areas where the Plio–Pleistocene sediments display more reduced thicknesses (the western and eastern belts of the monoclinal relief).

The frequency distribution of the southern fractures is not homogeneous, but rather organized into "belts"; some of these extend continuously (they can be followed for 10 km and more) into the post-transgressive rocky body (seen as a stiff "plate") and cross its entire thickness, sometimes continuing in the underlying Messinian sediments. The spacing, of the order of some tens of meters, is almost constant throughout vast portions of the reliefs, whereas it is even markedly reduced in the vicinity of the edges of the highest scarps, where also vertical fractures are to be found. In no case displacement can be found, while at times a close correspondence can be observed with N–S faults that affect the underlying Messinian "bedrock". At the edges of the "plates", in addition to the fractures of the above-described systems, others can be found trending approximately E–W; these are more recent, discontinuous and less frequent than the former ones, and their origin is associated with essentially gravitational phenomena (Gentili et al., 1995).

3 GEOMORPHOLOGIC SETTING

The first portions of the central Italian Adriatic physical landscape can be correlated with the emersion of the westernmost zone, which probably occurred in

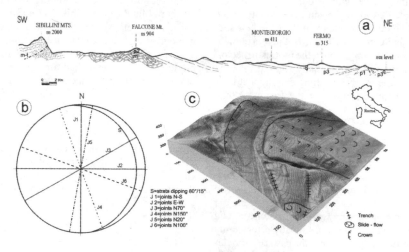

Figure 1. Geological peculiarities. **a**) Anti-Apennine cross section from the Sibillini Mts to the Adriatic sea: m-t = Mesozoic-Tertiary limestone, m = Messinian sandstones and pelites, p1 = Lower Pliocene pelites, p2 = Middle Pliocene calcarenites, p3 = Upper Pliocene, q = Quaternary conglomerates, sandstones and pelites; **b**) Stereonet: main jointing systems related to strata attitude of the monoclinal relief; **c**) Symplified geomorphologic scheme of Montegiorgio.

194

Upper Messinian and continued eastward to be completed in Middle Pleistocene. Until the end of Lower Pleistocene, the "paleolandscape" was modelled by strong areal erosion phenomena connected with arid and subtropical humid climatic conditions, and must have been characterized on the whole by gentle landforms distinguished by weak relief ("paleosurfaces"). With the opening up of the primitive phase of generalized and intense tectonic uplifting, that was to reach its acme in Middle Pleistocene, the first elements of the landscape deriving from it began to take shape: opening up of the intramontane basins and rapid deepening of the hydrographic network, the setting up of which was driven in many cases by tectonic dislocation. The evolution of this dislocation led to a high-relief landscape favoring the activation of intense slope dynamics, which significantly modified the original gentle valley shapes (Coltorti et al., 1996).

In the central-southern Marchean monoclinal relief, various main orographic alignments can be distinguished emerging from the surrounding landscape as a result of the differential erosion that sculpted them into lithic units. Even though variable, the altitudes generally lie between 500 and 700 m a.s.l. and display a generally marked increase going from north to south, where they reach 1100 m a.s.l. with M. dell'Ascensione; another general, but less marked increase in altitude is found when passing from the eastern belt to the western one. The structural control over the morphology in the area is clear and evident, characterized by consequent main valleys that are almost symmetrical and subsequent minor ones distinguished by their marked asymmetry. The deepening of the hydrographic network has shaped the monocline into a succession of triangular planimetrically shaped morphostructures; with their eastward-facing bases and bordered to the west by high structural scarps, they confer on the area its typical "cuesta" morphology (Fig. 1a). The most significant continental covers are represented by fluvial flooding, by rarer glacis deposits and by extensive and, sometimes, powerful landslide slumps (Gentili et al., 1995).

In this context, the occurrence of landslide phenomena of different types is high, translational flows being the most representative kinematics, not from the point of view of frequency, but because they often involve destructively more or less extensive portions of ancient built-up areas. In several cases, they have proved determining for the activation (or reactivation) of movements in recent times, the loss of water pipelines and sewage networks (Dramis et al., 2002).

3.1 The case of Montegiorgio: landslide analysis

The historic center of Montegiorgio, of immense age as all the built-up areas of the Marches placed on the summit of reliefs of the peri-Adriatic belt (Fig. 1c),

represents an example of a town center that has certainly been subject to accident due to water infiltration. It is sited on a hill modelled in the gray-blue Lower Pleistocene marine sediments, the lithostratigraphic succession of which, going from bottom to top, is the following: silty-sandy, gray-blue clays intercalated with thin layers or vela of siltstones; averagely diagenized yellowish sands, in which are intercalated with very variable frequency densely stratified clays; massive gray silty clays; dense alternations of silty clays and fine sands; massive clays with fine sand; yellow-ochre colored medium- to coarse-grained sands intercalated with thin gray-light brown clay layers; and finally polygenic conglomerate layers intercalated with frequent layers of clay and rarer ones of sands. Eluvial-alluvial deposits and landslide slump conceal vast portions of the bedrock.

From the physical–mechanical standpoint, these lithotypes can be grouped into three lithotechnical units: clays, sands, and surface deposits conglobing conglomerates.

The overall bedding of the layers is characterized by a prevalent northeastward dip-direction with a slope angle of from 5° to 12°. Frequent fractures (and subordinately faults), whose orientation is congruent with those of the systems found for the whole monoclinal sector (Fig. 1b) dislocate the geological bedrock.

An extensive portion of the eastern slope, distinguished by bedding dipping out of the slope with considerably lesser angle of inclination than that of the slope, proves to be affected by marked mass movements, to which can be attributed the formation of scarps, counterslopes and trenches, even of considerable size, up to approximately 100 m in width, 40 m in depth, and 1 km in length. In addition, in the southeastern portion, where the bedrock crops out, marked variations in the layer bedding can be observed. These geomorphologic elements allow us to associate the accident in this area with mass movements that are certainly ancient and can be associated with two basic kinematics: *translational slide*, in the portion of the slope including the historic center; *rock block slide* in the southeastern sector (Varnes, 1978).

As in the case of the other historic centers in the area, the first type of gravitational phenomenon displayed significant reactivation above all in the last decades, after the built-up area was supplied by a first aqueduct at the beginning of the 1940s and by a second and bigger one about ten years later, fed by water springs from the calcareous Apennine ridge. This network has meant the direct and continuous supply of water to all the houses.

Local concentrations of surface water, and above all the marked increase in water consumption and the consequent infiltration of considerable amounts of water in the subsoil, deriving from dilapidated sewage works as well as from the aqueduct itself on its way

195

through a tunnel under the hill on which the built-up area rises, have been facilitated by the intense fracturing of the rocky bodies; this has furthermore led to a movement of water between layers and/or levels characterized by different permeability. The result is the marked increase in piezometric levels and in the interstitial pressure values, thereby activating and/or reactivating mass movements.

Geophysical investigations (seismic tomography) carried out inside a medieval tunnel dug 15 m under the built-up area and inside the aqueduct tunnel which is 25 m lower still, have evidenced, below the subsoil and inside the built-up area, some shear zones that have subsequently been reached and identified by mechanical drilling. In the boreholes, a monitoring network has been set up consisting of 7 tiltmeters and 10 piezometers. The tilt data for the months of April and August 2000 evidenced notable shifts (with maximum values of 5 mm) down to a depth of 36 m; the piezometers revealed high phreatic surface values and the presence of aquifers deep down.

3.2 *Modelling*

The dynamic evolution of the area has been investigated utilizing a 3D numerical code based on the difference elements method (FLAC_3D, 2000). This code is an update of the bi-dimensional version and it needs a starting complex work to reconstruct the space coordinates of the model (nodes). So the nodes coordinates have been in advance determined using a commercial software (SURFER_8.0, 2002) by which the topographical surface has been modelled basing on the existing cartography (scale 1:2000). Then, the generated block has been trimmed by a series of planes to construct the physical and geological model (the monoclinalic feature of the formations has simplified the work). In the Figures 2 and 3 it is possible to note the adopted technique. The dimension of the model (433200 zones) and the large number of nodes (450912) have involved a long time of calculus (120 hours for every test) using a bi-coprocessor computer.

Within the sandy layers (Fig. 4), as the geognostic surveys showed, a phreatic surface has been positioned. In particular the upper sandy layer is characterized by a pore pressure about 100.0 kPa, while the lower sandy layer reaches 171.0 kPa.

The physical and mechanical properties of the soils have been investigated by numerous laboratory and in situ tests. In Table 1 the average values utilized in the modelling are reported.

The obtained results allow us to define the successive considerations:

– the superficial cover soils are instable and the velocity vectors are higher on the left part at the top of the slope;

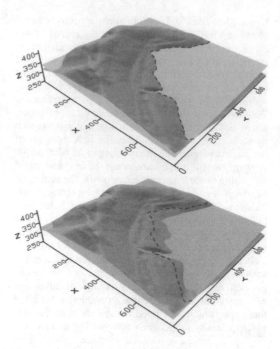

Figure 2. Block generated by SURFER_8 and trimmed with two planes, to construct, in this case, the physical and geological model for the lower sandy layer.

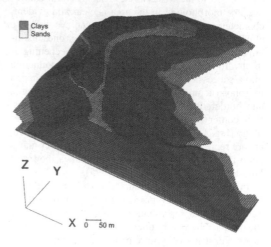

Figure 3. Higher part of the model showing the substratum discretization by FLAC3D code starting from the SURFER procedure, without the cover deposits.

– the sandy formations are basic for the dynamic evolution of the area; the confined pore pressures (Fig. 4) favour the softening of the cover deposits and of the clay formation, primarily in the zones where the sandy layers intercept the topographic surface;

196

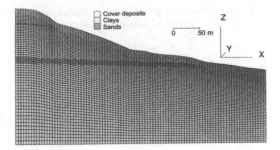

Figure 4. Section with evidenced the sandy formations where a phreatic surface is located.

Table 1. Physical and mechanical soils properties.

Lithotypes	γ (kNm^{-3})	c' (kPa)	ϕ' (°)	E (MPa)	v
Cover soils	18.5	0	23–25	15	.35
Sands	21	0	35	100	.3
Clays	20	15	24	75	.3

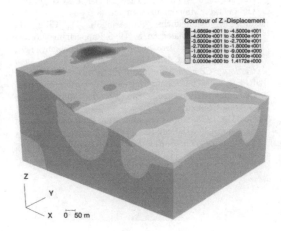

Figure 5. Contour of Z-displacements at rupture. Note the maximum deformation at the top of the slope.

Figure 6. Model perspective with the representation of the displacements vectors.

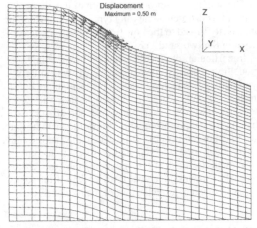

Figure 7. Displacements vectors relative to the cover deposits at the top of the slope.

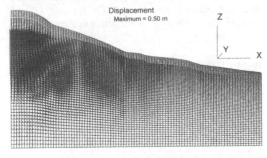

Figure 8. Displacement vectors referred in depth to the clayey formations.

– the top area of the model is subjected to the larger displacements (Fig. 5) considering the deep mechanism of movement (that is evident in Figure 6 that reports the displacements vectors);

– in fact it is possible to distinguish two fundamental and unconnected typologies of movements visible in Figures 7 and 8;

– a first movement characterizes the cover deposits, principally on the top of the slope, where the modelling shows the reach of the rupture conditions, testifying the actual risk of the built-up area (Fig. 7);

– a second movement is revealed in depth (Fig. 8), in the substratum, where the sandy and clayey formations suffer an important roto-translational deformation with the modelling yet in elastic conditions.

4 CONCLUSIONS

The main factors inducing and controlling the large landslide, here studied, are due to the significant geomorphologic peculiarities produced from the Quaternary spreading tectonics and for the hydrogeological conditions.

The tectonic activity produced dislocations (fractures and/or faults) in the bedrock favouring the deepening of the hydrographic net, giving rise the genesis of high values of relief energy.

Then these discontinuities facilitated the water transport in depth, connecting the present formations provided with different permeability. The fractures, also, released the rocky mass predisposing the shear zones at the top of the slope and orienting the geometry of the main and intermediary scarp.

The activation and/or reactivation of the phenomenon and its different states and typology of movement are strictly linked to the pore pressures variations confined in the two sandy layers.

In particular, the results obtained by the numerical modelling confirm the presence of two different typologies of deformation and movement in the investigated system; a first movement regards the cover deposits, uniformly distributed over the all surveyed area, in limit equilibrium and so near the yielding status (the above area of the slope, where the historic centre is sited, presents the higher risk); a secondary movement is localized in depth, within the clayey bedrock, showing an actual involvement of the sandy layers in the general behaviour of the modelled case.

ACKNOWLEDGEMENT

Work printed with MIUR-COFIN 2002 funds, Proff. B. Gentili. The AA thank Dr. D. Aringoli for his help with the editing, and Dr. M. Calista and Dr. M. Mangifesta for their valid assistance in the modelling.

REFERENCES

Aringoli, D., Gentili, B., Pambianchi, G. & Sciarra, N. 2002. Geomorphological analyses and modeling of complex mass movements in the Marches area (Central Italy). In: *"Geomorphology: from Expert Opinion to Modelling"*.

CERG Editions (European Center on Geomorphological Hazards – Council of Europe), Strasbourg (France), 77–84.

Centamore, E. & Deiana, G. 1986. La geologia delle Marche. *Studi Geologici Camerti*, Num. spec., 145 pp.

Coltorti, M., Farabollini, P., Gentili, B. & Pambianchi, G. 1996. Geomorphological evidences for anti-Apennines faults in the Umbro-Marchean Apennines and in the peri-Adriatic basin, Italy. *Geomorphology*, 15, 33–45.

Crescenti, U., Gentili, B., Pambianchi, G. & Sciarra, N. 2002. Modeling of complex deep-seated mass movements in the central-southern Marches (Central Italy). In: Rybar, Stemberk & Wagner (eds) *"Landslides"*. Swets & Zeitlinger, Lisse, The Netherlands. A.A. Balkema Publishers, 149–155.

Deiana, G. & Pialli, G. 1994. The structural provinces of the Umbro-Marchean Apennines. *Mem. Soc. Geol. It.* 48, 473–484.

Dramis, F., Farabollini, P., Gentili, B., & Pambianchi, G. 1995. Neotectonics and large-scale gravitational phenomena in the Umbria-Marche Apennines, Italy. In: Slaymaker O. (Ed.) (1995), *Steepland geomorphology*. J. Wiley & Sons Ltd, 199–217.

Dramis, F., Gentili, B., Pambianchi, G. & Aringoli, D. 2002. La morfogenesi gravitativa nel versante adriatico marchigiano. Studi Geologici Camerti, Nuova Serie 1/2002, EDIMOND, 103–125.

FLAC_3D 2000. Manual_Release 2.1. Itasca Consulting Group, Inc., Minneapolis, Minnesota.

Gentili, B., Pambianchi, G., Aringoli, D., Cilla, G., Farabollini, P. & Materazzi, M. 1995. Rapporti tra deformazioni fragili plio-quaternarie e morfogenesi gravitativa nella fascia alto-collinare delle Marche centro-meridionali. *Studi Geologici Camerti*, Vol. Spec. 1995/1, 421–435.

SURFER_8.0 2002. Contouring and 3D surface mapping for scientists and engineers. *Golden Software, Inc.* Golden, Colorado.

Varnes, D.J. 1978. Slope movement: types and processes. In: Schuster R.L. & Krizek R.S. (1978), *Landslides analyzes and control. Transp. Res. Board. Spec. Rep.*, 176, Nat. Acad. of Sci., 11–33.

Geotechnical Measurements and Modelling, Natau, Fecker & Pimentel (eds)
© 2003 Swets & Zeitlinger, Lisse, ISBN 90 5809 603 3

Experiences about the behavior of stress in a brown coal pit

R. Glötzl
Glötzl GmbH, Rheinstetten, Germany

J. Krywult
OBR, BG Budokop, Myslowice, Poland

ABSTRACT: The guaranty of the stability of slopes in open-cast mining is one of the main tasks of engineering. Suddenly and uncontrolled occurring landslides represent the biggest danger for men and machines. In order to avoid such incidents, EDP-model calculations and geodetic measurements are executed. In the brown coal mine Belchatow seams are mined under difficult circumstances on an extensive mine field and in big depths. Because the geological circumstances were hard enough to explore, a pilot project started an in-situ-measurement program in the area Belchatow in 1999, which collect stress as well as deformations in the underground.

1 GEOLOGY AND TECTONICS

The brown coal deposit in the area Belchatow had been formed in a narrow, 2,5–3 km long tectonic depression, the Klesczower ditch. A row of rift valleys form the main direction NW–SE and WSW–ENE from which result the south and the north border faults. Besides many defects cross the ditch in direction SE–NW (Kasinski, 2000).

The so-called ditch second degree is a narrow, 300–700 m wide tectonic gorge, which is developed between the border faults ub. 1 and ub. 2. In this zone the roof of the layer complex grows by the double. The cap rock with a thickness of 150 m is made of clayey tertiary formations, loose rock and silt.

The mining of the brown coal began in the ditch second degree. In this area signs of neo-tectonic movements were noticed in form of seismic activities and of sliding of tertiary and quaternary layer groups in direction NE. Thirty-four seismic quakes were noticed since 1995 with an energy between $1,0 \times 10^4$ and $3,0 \times 10^8$ J. Earthquakes like that result from the change of the stress ratio during the mining (Gibowicz, 1984).

It is to be supposed that in the underground zones are located, there the lateral pressure is much higher than the pressure which results from the covering. These compressed zones are the sequence of the geological processes which had occurred in this region.

During the mining of the brown coal landslides can come off every time. The structural analysis of slopes stability should be leaned on in-situ measuring results. These results should be recorded during the mining by measuring instruments which are on call at every time.

2 INSTRUMENTATION AND INSTALLATION

The measuring program contains 3 inclinometers and 3 bore-hole-cells or stress monitoring system (SMS) (Fig. 2), which had been arranged to the north of the most massive brown coal seam – the ditch second degree (Fig. 1). The depths of the inclinometers are:

IN-1S = 88,5 m; IN-2S = 93,0 m; IN-3S = 100,0 m (OBR BG Budokop, 2001).

In the borehole 1259B-BIS with a diameter of 173 mm two bore-hole-cells in form of borehole measuring heads with hydraulic pressure cushions and pressure sensors type EBKO K50/3V of the company Gloetzl were installed.

The measuring head 1 is located in the brown coal layer – depth 42–43 m; the head 2 in a coaley clay layer – depth 28,5–29,5 m.

Additional a bore-hole-cell was installed in a marl layer in the borehole 1408-BIS in May 2001 (Fig. 2).

The measuring head could be installed directionally oriented in the borehole with the help of guide rods.

To prevent that injection material – an cement-bentonit mixture – flows out to the surrounding soil the body of the instrument was supplied with a geotextile tube.

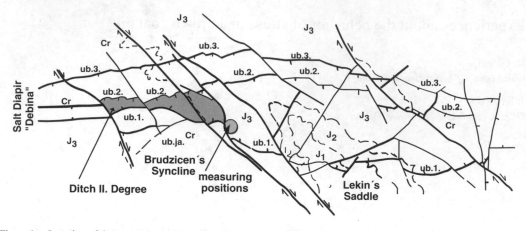

Figure 1. Location of the measuring points.

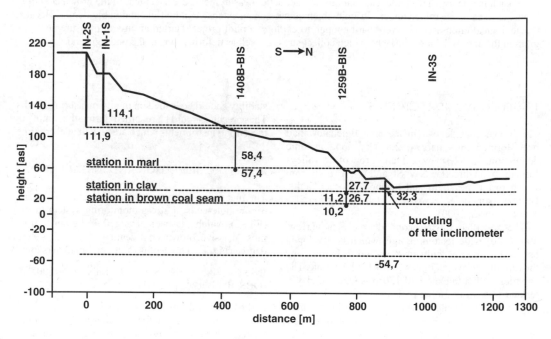

Figure 2. Location of the instruments in the borehole cells.

An uncontrolled refilling of the borehole would lead to an amelioration and by that cause a change of the stress correlation.

The effectiveness of the geotextile tube was guarantied by a control of the quantity of the injected cement-mixture.

The important second injection was executed after 6 weeks of waiting since the installation. For this purpose every cell owns on the border a perforated steel line; during the first injection this line is closed by capsules. The coverage of the perforations of the injection lines will be cracked by the top leading conduction; in this way the injected material can soak into the surrounding soil. By the post-injection, hollows which had been built during the first injection shall be filled and in addition to this it is possible to execute a form of plate load test by adjusting several heights of pressure. During the second injection 14 kg synthetic resin mixture had been injected with a pressure of 0,6 MPa. A complete and controlled injection

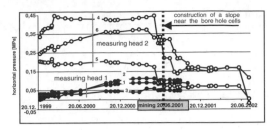

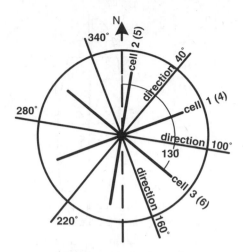

Figure 3. Orientation of the cells.

Figure 4. Stress development (from December 1999 to April 2002) in the clay layer (MH 1) and the brown coal seam (MH 2).

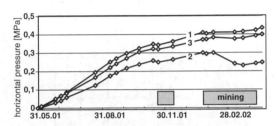

Figure 5. Stress development in the time from June 2001 to April 2002 in the marl layer (MH 3).

is guaranteed if there is a linear correlation between the injection pressure and the cushion pressure – which is measured during the injection.

The measuring heads of the bore-hole-cells contains in each case three vertical directed cells, which are symmetrically arranged – the directions are 0°; 120°; 240°, see Fig. 3. The azimuth of the main measurement direction is 340° and is positioned rectangular to the slope. The cells 1 and 4 collect the horizontal soil pressure in this direction. With the help of the remaining cells it is possible to construct some principal stress ellipses.

The stresses are collected periodically with a readout unit VMG.

3 STRESS DEVELOPMENT

The graphic (Fig. 4) shows the stress development of measuring head 1 in the brown coal seam and of measuring head 2 in the clay layer which lies over the seam.

Fig. 5 shows the stress development of measuring head 3 in the marl layer. The data of every single cell are marked with blue, green and red. The stress development of a single cell corresponds to the predominant stress correlation in the underground.

In the time from December 1999 to the beginning of February 2001 and from October 2001 to April 2002 there were no mining near the measuring positions. The mining happened here from February 2001 to October 2001 in a height between +55 m and +35 m a.s.l.

In the graphics of the measuring heads 1 and 2 (Fig. 4) these proceedings can be recognised very clear.

Noteworthy is that as well the built stress as the stress loosing, which depends on the mining, in the clay and

the marl layer are much higher than that stress in the brown coal seam. At the end the stresses have been equalised after the mining of the brown coal seam.

Unfortunately this effect could not longer be observed, because in October 2001 all cells of measuring head 1 failed.

It is to be supposed that a local landslide had happened and destroyed the connection cable to the connection box.

Additional and at the same time the inclinometer IN-3S, which is installed in 100 m distance was buckled so heavy that measurements below 13 m were no longer possible.

The stress correlation in the soil had been developed also without mining activities. This development is illustrated about months until April/May 2001 by several stress ellipses (Figs. 6 and 7).

It is easy to recognise, that the principal stresses had been changed as well by size as also by direction.

4 CONCLUSION

The mining of the brown coal pit in the area Belchatow, which reaches to large depths, is faced by the unfavourable tectonic circumstances with the problem of slope stability. In view of the complicated formation processes of these brown coal seams different stress fields, which have an unfavourable effect to the

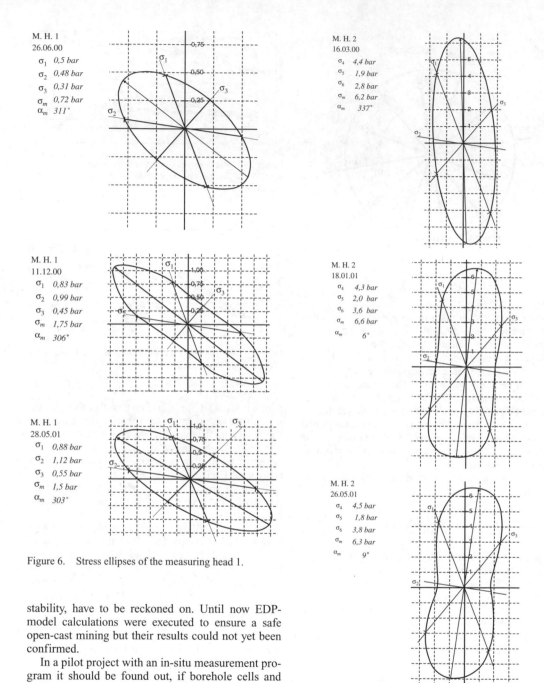

M. H. 1
26.06.00
σ_1 0,5 bar
σ_2 0,48 bar
σ_3 0,31 bar
σ_m 0,72 bar
α_m 311°

M. H. 1
11.12.00
σ_1 0,83 bar
σ_2 0,99 bar
σ_3 0,45 bar
σ_m 1,75 bar
α_m 306°

M. H. 1
28.05.01
σ_1 0,88 bar
σ_2 1,12 bar
σ_3 0,55 bar
σ_m 1,5 bar
α_m 303°

Figure 6.   Stress ellipses of the measuring head 1.

M. H. 2
16.03.00
σ_4 4,4 bar
σ_5 1,9 bar
σ_6 2,8 bar
σ_m 6,2 bar
α_m 337°

M. H. 2
18.01.01
σ_4 4,3 bar
σ_5 2,0 bar
σ_6 3,6 bar
σ_m 6,6 bar
α_m 6°

M. H. 2
26.05.01
σ_4 4,5 bar
σ_5 1,8 bar
σ_6 3,8 bar
σ_m 6,3 bar
α_m 9°

Figure 7.   Stress ellipses of the measuring head 2.

stability, have to be reckoned on. Until now EDP-model calculations were executed to ensure a safe open-cast mining but their results could not yet been confirmed.

In a pilot project with an in-situ measurement program it should be found out, if borehole cells and inclinometer measurements in the mining zone can result in useable values for the stability calculations and therefore make the planning of an improved slope geometry possible.

Based on experiences from over two years it could be recognised, that the stress–deformation-behaviour of a brown coal pit can be determined by well-directed measurements. Especially the precise installed borehole cells bring information about the stress development in every single rock-layer near to the mining. With these recorded values the EDP-model calculations could be confirmed or if necessary be optimised. Therefore the stability during the mining of the coal seams can be improved.

The gained experiences lead in the brown coal deposit "Belchatow" besides the planning of the slope geometry further to an introduction of a control system, which monitors the stability of slopes at risk during the work.

Therefore the planning method proceeds as following:

- The source geometry of stationary slopes will be designed by help of stability calculations by the principle of slope equilibrium. Additional the gained experiences and stability criterions have to be considered.
- For the designed slopes EDP-model calculations will be prepared by acceptance of the known geological circumstances. By their help it is possible to establish allowable deformations and possible dangerous zones of stability loosing; then a special in-situ measuring program will be worked out for these zones.
- For the stability monitoring of skid-endangered slopes it is necessary to execute apart from the geodetic precision measurements in the hole mining area more the following measurements:
  - deformation measurements in the underground (inclinometer measurements with digital measuring heads)
  - soil pressure measurements with bore-hole-cells

The developed EDP-models will be modified continuously by the obtained measuring results.

REFERENCES

Czarnecki L., Dynowska M. & Szymanski J. 2002. Badania stanu naprężeń w górotworze jako element kontroli stateczności skarp w KWB Bełchatów. (Examinations of the stress behaviour in rock as component of the stability proof of slopes in the brown coal pit Belchatow). Magazine: Węgiel Brunatny, 3/2002.

Gibowicz S.J. & Kijko A. 1984. Ocena zagrożenia sejmicznego rejonu kopalni "Bełchatów". Evaluation of the seismic threats in the area of the brown coal pit Bełchatow. Technika Poszukiwań Geologicznych 1984.

Kasinski J., Czarnecki L., Frankowski R. & Piwocki R. 2000. Geology of the Bełchatów lignite deposit and environmental impact of exploitation. 4th European Coal Conference 2000.

OBR BG Budokop. 2001. Badania stanu naprężeń górotworu "in situ" poduszkami Glötzla w przekroju geologicznym SN tj. w strefie zejścia z eksploatacją do rowu II rzędu. (Examination of the stress behaviour in rock by In-situ-measurements by the help of pressure cells system Glötzl in geological cross-section SN, in the area of the mining descent to the ditch II. degree). Myslowice, Oktober 2001.

Investigation and remediation of the leaking areas in the large navigation locks

J. Hulla
Slovak University of Technology, Bratislava, Slovak Republic

P. Slašt'an
CarboTech Slovakia žilina, Slovak Republic

J. Hummel
Watermanagement Construction, Bratislava

ABSTRACT: Underground parts of civil engineering structures, located under the groundwater level, should be waterproof. Permeable positions occur in constructions due to defective water isolations out of different reasons, in working or in dilatation joints, that must be sealed additionally. Such problems have to be solved in the Gabčíkovo navigation locks. This paper describes methods, which were used at the locations of permeable positions, new knowledge obtained by the additional sealing works with polyurethane resins and their efficiency testing.

1 INTRODUCTION

The Gabčíkovo Danube Project navigation locks began to leak shortly after having been put into operation (in 1992). It had become necessary to complete and modernize the monitoring system, so that this would enable a more precise localization of the leaking positions and also monitoring of efficiency of the performed rehabilitation activities.

Additional sealing works were carried out in several stages. Joints were sealed with polyurethane resins and the degraded gravel subsoil was sealed with traditional grouting.

2 GEOLOGY, STRUCTURE, MONITORING

Quaternary gravel soils reach into the depth of 400 m in the locality where the main objects of the Water Power Project Gabčíkovo were constructed, below them lie neogene clays, groundwater level is in the depth of approx. 2 m. Gravel soils permeability varies considerably with respect to the different depths; permeability coefficients cover a considerable span from 10^{-4} m/s to 10^{-1} m/s.

Main structures had to be built in sealed excavations in such geological conditions due to the fact that their depths had to be substantial -33 m for the water power station and 22 m for the navigation locks.

The sealing walls were made of a self hardening suspension, the bottoms were grouted. Wall stability was provided by slopes inside the excavation, the grouted bottom stability against breaking as a consequence of uplift was secured by a sufficient thick soil layer between the excavation bottom and the grouted layer.

Depth dewatering of the excavations was done by drilled wells. The amount of 80 l/s of water have been drawn from the navigation locks' excavation for a long period of time. In the project had been assumed up to 1000 l/s, thus the sealing elements were created in a very high quality; the permeability coefficients of the sealing walls were approximately 10^{-8} m/s, and 3×10^{-7} m/s of the grouted bottom.

Two large navigation locks were built in Gabčíkovo (Fig. 1). Each of them is 34 m wide, 275 m long, and of an entire height of 39 m.

The dilatation joints were sealed by rubber stripes with the width of 500 mm in one position. The dilatation joints width has been 20 to 50 mm, it has been secured by polystyrene and heraclite during the construction.

Very important elements concerning the locks' stability were stripe foundations underneath all dilatation joins.

Each lock consists of six blocks, upper and lower nosing, separated from each other by dilatation joints (Fig. 2).

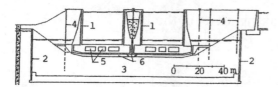

Figure 1. Sectional view through the navigation locks: 1–rubber sealing of dilatation joints, 2 – sealing walls, 3 – grouted bottom, 4 – observation boreholes, 5 – channels, 6 – stripe foundations under the dilatation joints.

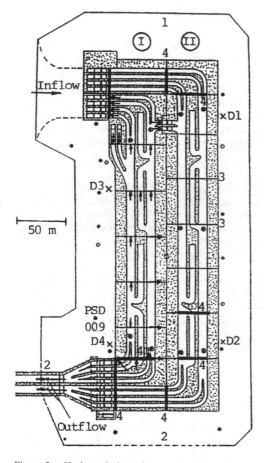

Figure 2. Horizontal view of the navigation locks through the channel system: I – right lock, II – left lock, 1 – sealing walls, 2 – the removed upper parts of the walls, 3 – dilatation joints, 4 – leaking joints positions from the modelling, o – observation boreholes permeable above the foundation of the locks, • – observation boreholes permeable under the foundation of the locks, D1 through D4 – deformation measuring boreholes, × – entrance openings to the channels, ↑ – leaking positions after first stage of sealing in empty right lock.

Water flows into and out of locks by the help of special objects and canals under the bottoms, in their lowest sections.

The monitoring system has been complemented and improved gradually since 1992. Presently observation boreholes are available with permeable parts over or below the navigation locks' foundation. That enables monitoring of development of ground water surface and velocity flow regimes in the vicinity of the boreholes. Each block is equipped by four points for monitoring of horizontal and vertical movements by the help of automatic tachometers. Depth dependencies of vertical deformations of gravel soils in the vicinity of the locks are monitored in special boreholes (D1 and D4).

Development of levels and movements can be observed by the means of automatic sensors and personal computers also in the course of filling and emptying of the locks. Velocity regimes and deformations in the special boreholes are monitored occasionally, due to the need.

The sealing walls and the grouted bottom have, naturally, remained in the earth. The upper parts of the walls were removed only in the surrounding of the outflow object and below the lower channel for ship navigation (dot lines at the Fig. 2) These were substantial boundary conditions for analyses of water flow in the vicinity of the navigation locks.

3   LEAKING AREAS LOCATION

The members of the Techno-Security Supervision Staff discovered shortly after the navigation locks have been put into operation that groundwater levels have increased in the surrounding of the navigation locks only while some of the locks was being filled in with water. This has been the first signal on leakage existence; the groundwater level should not exceed the outlet canal water level under normal conditions.

Figure 3 presents level development in the observation borehole PSD-009 that had been built close to the outflow object. Filling of the left navigation lock has taken 15 minutes in the year 1994, after the lock had been full for the time of 3.5 hours the level in the borehole increased by 3 meters. The right lock has been filled afterwards, and the increased level in the borehole PSD-009 reached the level of 6 m after another 3 hours. Similarly increased the levels in all observation boreholes, in the course of locks' filling.

A comprehensive survey (Hulla et al., 1994) has shown that the level in the direct vicinity of the locks had reached a level of 2 meters above the level in the outlet canal (Fig. 4), and that the average filtration velocity had the value of 1×10^{-5} m/s. This enabled to conclude that the contact between the locks and the input canal had not been quite tight either.

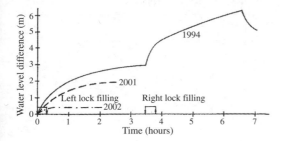

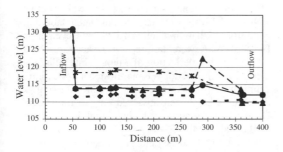

Figure 3. Developments of water levels in the observation borehole PSD-009 in the year 1994 (permeable joints), in the year 2001 (after first stage of additional sealing) and in the year 2002 (after the second stage of additional sealing of the right lock).

Figure 4. Water level development in surrounding of the right wall of the right lock before and during additional joint sealing: –♦– both locks empty, –*– original state 5.2.1994, –▲– situation after the first stage sealing 29.1.2000, –●– situation after the second stage of additional sealing 25.8.2002.

It was considered in 1995 to close the in and out flows of one navigation lock, and to pump out water completely, so that it would have been possible to get in and check the leakage positions also in the canal system below the bottom of the locks visually. But this had not been possible, there would have been an inflow of a great amount of water through the leakage positions (approx 6 m³/s) into the closed lock, and at that time existing pump system would not have managed it.

Mucha and Banský (1998) simulated different positions of the leaking dilatation joints consequently, and compared calculated stream fields with fields, gained on the basis of direct measurements of levels in the observation boreholes. The best match resulted in leaking dilatation joints, as presented in the Fig. 2 with thicker lines and marked by the number 4. These joints became the basis for additional sealing in the first stage.

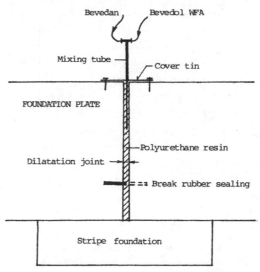

Figure 5. Scheme of additional sealing of the dilatation joints by polyurethane resins.

4 ADDITIONAL SEALING

Evidently leaking positions in the vertical dilatation joints from top to the bottom of the navigation locks have been additionally sealed from 1993 till 1998 using different materials. Yet, efficiency of these works has not been manifested in the level and velocity regimes in the vicinity of the locks in a noticeable way.

First stage of additional sealing in the canals under the bottom of the navigation locks took place by the end of the year 1999. To enable work under water without endangering health and lives of the divers it was needed to drill entrance openings through a ferroconcrete bottom of the locks into a complicated canal system (Faix 1999). These entrances are marked by × in the Figure 2.

A two-component hydrophobic polyurethane resins Bevedan–Bevedol WFA provided by the Carbo-Tech Slovakia company were used for additional sealing of the dilatation joints in the first stage (Fig. 5). A covering tinplate was anchored over the joint before this has been cleaned by pressure water. The Bevedan and Bevedol WFA components were supplied separately, they were mixed directly above the tinplate, and the whole dilatation joint has been filled gradually.

The divers worked 15 meters below the water surface, during winter months, in aquatic environment with low visibility, thus in very demanding conditions. Efficiency of the work was manifested positively by a overall decline of the groundwater levels (Fig. 4, marks–▲ –) but better results have been

207

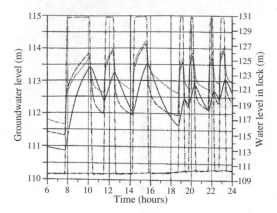

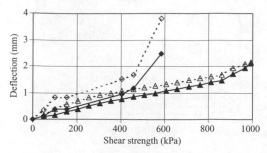

Figure 7. Shear tests of polyurethane resins: ◇ ◆ – Bevedan–Bevedol WFA, △ ▲ – Bevedan–Bevedol WF, ——— – transducer 80, - - - - - transducer 81.

Figure 6. Time dependencies of water levels during normal operation of locks after the first stage of sealing (29.1.2000):——– left lock, ——·— right lock, ——— borehole PR-5, – – – borehole PR-3, ······ borehole PSD-009.

expected a local extreme level increase has even appeared in one of the boreholes in the distance of 290 m. Due to the analyses of the vertical water flow in this borehole, and of the depth development a conclusion was reached that one of the reasons had been silting of the borehole by sand particles but as well as the existence of an unsealed joint.

Development of levels in some observation boreholes while the right navigation lock has been in operation (the left one being empty) has shown after the first stage of additional sealing from January 29th 2001 (Fig. 6) that increases of levels have been dependent from time in course of which has the lock been full, but as well as from cycling of the lock operation modus what also brought about increases of the levels in the vicinity of the lock.

The water level of the PSD-009 borehole increased by approx. 2.5 m around the permeable dilatation joints 2 hours after filling of the right lock in 1994 (Fig. 3). In the year 2000, after the first stage of additional sealing was the increase in this borehole 2.0 m (in time between 8 and 10 hours – Fig. 6), what also indicated and spoke for a lower efficiency of the dilatation works.

Even in spite of an overall discontent with the first stage of additional sealing results important accomplishments have been reached. Technical conditions of both locks equalized, a complete out pumping of the right navigation lock was enabled (the pumped amount was just 0.3 m³/s) and thus it became possible to check-up the dilatation joints in the canals under dry conditions. Leakage positions are marked by ↑ in the Fig. 2; it can be seen that they were placed in joints that have not been successfully sealed but new leakage position were also detected, that have been previously not localized.

Doubts on qualities and correctness of the choice of the polyurethane resins occurred as well. That was why special tests were carried out under laboratory conditions. Dilatation joints filled by Bevedan and Bevedol WFA were failured by a shear tension of 600 kPa with a deflection of 3 mm, joints filled with Bevedan and Bevedol WF manifested a deflection of 2 mm only by a shear tension of 1000 kPa (Fig. 7); deflections have been observed on two transducers. Detailed information is to be found in the report Hulla et al., (2003).

Careful vertical settlement measurements of the individual blocks of the locks in the course of operation have shown differences in the range of 2–3 mm, thus the polyurethane resins Bevedan–Bevedol WFA fully complied with the requirements, and could be used for the second stage of the additional sealing too.

It was possible to gain valuable information on durability of the polyurethane resins. According to Naudts (2003) and Vrignaud et al. (2003) it is possible to take a long term durability in case of hydrophobic polyurethane resins for granted.

The second stage was carried out under dry conditions only in the right navigation lock till now. Additional sealing was done just for positions in the dilatation joints which have been localized precisely, having applied the same process as depicted in Figure 5. Leakage positions were manifested by concentric water inflows with over-pressure of approx. 150 kPa. One joint was not able to be additionally sealed under this conditions (Fig. 8), and it had to be sealed by divers after the water pumping had been finished and after a consequent pressure equalisation.

The additional sealing of dilatation joints of the right navigation lock in the second stage provided with valuable information that will be fully used also in the process of additional sealing of the left lock. The second stage in the left lock will be probably carried out in the year 2003.

Figure 8. One of the joints in the surrounding of water outflows from the lock which was not sealed in the second stage, under dry conditions successfully.

5 ADDITIONAL SEALING EFFICIENCY

Considerable amounts of water leaked out through the leakage into the surroundings, level increase, as well as flow velocities around them during filling of and operation of the navigation locks. Vicinity of the navigation locks is partially sealed by cut off walls and by the grouted bottom, thus are the processes of level increases and flow velocities intensified in the observation boreholes.

The original state of the navigation locks and efficiency of the additional sealing, characterised by the level development, are depicted on the Figures 3 and 4; level differences reduce in a certain time period due to the influence of successful additional sealing.

After the second stage of additional sealing of the right lock has the character of the level development changed during the periods of lock filling and operation considerably (Fig. 9). Compared with the situation before the second stage (Fig. 6) the level differences did not continue to increase, they get stabilize shortly after beginning of the lock filling.

The level differences in the PSD-009 borehole were 2.5 m 2 hours from beginning of filling of the right lock after the first stage (Fig. 6), after the second stage only 0.5 m (Fig. 9) thus the additional sealing efficiency was very good. The biggest level differences remained in the borehole PR-4: approx. 2.5 m after the first stage, 2.0 m after the second one; this borehole had been constructed close to the position that was not managed to be sealed properly in the second stage (Fig. 8).

Figure 10 represents average level increases in the observation boreholes 2 hours after filling of the locks. Before the first stage of additional sealing had the levels increased by 4.5 m while the left lock had been full (the right one had been empty), after finishing of the first stage by 2.5 m; this stituation can be considered to be the starting point before the additional sealing of the left lock in the second stage.

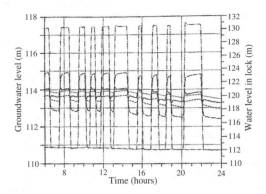

Figure 9. Time dependencies of water levels at the normal operation of locks after the second stage of sealing (25.8.2002): —·— left lock, —··— right lock, ——— borehole PR-3, – – – – borehole PR-4, - - - - - - borehole PSD-009, ········· borehole PSD-007.

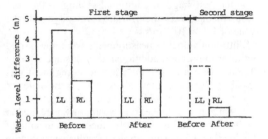

Figure 10. Efficiency of the sealing in the changes of average level differences in the observation boreholes at the 2 hours full left (LL), or right lock (RL).

The levels increased by 2 m in average in the observation boreholes before the first stage of additional sealing, with the right lock full for the period of 2 hours (the left one being empty); technical conditions of the right lock had thus been better than of the left one. The differences between the levels increased during additional sealing and they reached 2.5 m at the end of the first stage.

Pessimistic opinions appeared immediately that the state of the right lock did not improve but worsen. This was not true, increase of the level differences was related to the decrease of the drainage effect of the left chamber after its evident additional sealing, water could not flow in such an amount from the full right lock to the empty left one, it had to flow into the outside surrounding around the locks and increased the levels in the observation boreholes over there.

The average level difference decreased considerably to the value of 0.5 m after the second stage of right lock additional sealing, and further decrease of this value can be expected after the additional sealing of the permeable position by the emptying object.

Groundwater velocity flow has been monitored in the obsevation boreholes using single borehole tracer methods (Halevy et al., 1967). The used tracer was a natrium chlorite solution. Vertical velocities were measured in the boreholes, the continuity formula was the basis for calculating of vertical discharges and from their depth dependencies, after having taken into consideration the borehole drainage influence, resulted filtration velocities in gravel soil in the surrounding of the observation boreholes. Detailed information of the methodical character were published in the paper Hulla et al., 1992.

By two hours full right lock in 1994 with the greatest range of leakage, the filtration velocity median value was 2.4×10^{-5} m/s, the maximal filtration velocity value was 9×10^{-4} m/s.

In the year 1995 while both locks have been empty, median was the filtration velocity value of 1×10^{-5} m/s. It was mentioned already that have been caused by a leakage contact with the inflow canal.

After first stage of additional sealing in the year 2000, median was the filtration velocity with the value of 1.3×10^{-5} m/s.

Filtration velocity measurements were not carried out after the second additional sealing stage. Due to the positive changes in the water level regime it was possible to assume, that values close to the state for both locks empty would have been obtained.

Localization of the leakage positions was related to the highest filtration velocity values. Even in spite of the fact that the interpretation of results was not quite without problems in a strongly heterogeneous gravel environment, a relatively high match was reached with the strongest leakage positions, which were fully localized after pumping out of the right lock before the second stage of additional sealing.

The measurements enabled to determine also the maximal filtration velocity values, and to compare them with critical values when the sand particles begin to move (Hulla, 1996). Such positions were localized in the observation boreholes. It was assumed that processes of washing out sand from the gravel skeleton could develop in a dangerous way in the surroundings of leaking dilatation joints what would threaten the navigation lock stability. A sparing navigation lock operation mode had to be introduced together with gradual additional sealing of the disturbed dilatation joints. Degraded positions in the gravel subsoil of the locks would be additionally sealed as the third stage, if needed.

6 CONCLUSIONS

Dilatation joint sealing of the navigation locks had been probably disturbed during the construction already. Large amounts of water leaked into the vicinity of the full locks through the untight positions, the water levels increased, flow velocities rose and sand

particles have been washed out from the gravel soils. These processes have threatened the stability of the locks and thus it was needed to seal the disturbed dilatation joints additionally.

The extreme size of the navigation locks and heterogeneous gravel medium were a challenge by solving of complicated problems related to localization of the leakage positions. These problems have been solved quite successfully and the first stage of additional sealing was carried out. The sealing agent has been polyurethane resins Bevedan–Bevedol WFA, the work was done by divers in a depth of approx. 15 m under the water level.

The second additional sealing stage was carried out only in the right lock under dry conditions till now. Very good results have been achieved, and so the same procedure will be used also for the left lock in the year 2003.

Degraded positions of the gravel subsoils of the locks would be grouted, if necessary, in the third additional sealing stage.

ACKNOWLEDGEMENT

This paper is a partial result of a grant project of the Ministry of Education of the Slovak Republic No. 1/9066/02.

REFERENCES

Faix, D. 1999. The Gabčíkovo Dam – Rehabilitation of the Foundation Slab of Navigation Locks (in Slovak). *Eurostav* 5 (8): 47–48.

Halevy, E. et al. 1967. Borehole Dilution Techniques: a Critical Review. *Isotopes in Hydrology*: 531–534. Vienna: IAEA.

Hulla, J. et al. 1992. Application of Artificial Tracers in Groundwater Research. In H. Hötzl & A. Werner (eds), *Tracer Hydrology*: 105–108. Rotterdam: Balkema.

Hulla, J. et al. 1994. *Groundwater Flow in Surrounding of Navigation Locks* (in Slovak). Bratislava: Slovak Univ. of Technology.

Hulla, J. 1996. Filtration Stability at Extreme Hydrodynamic Loading. *Slovak Journal of Civil Engineering* 1: 36–43.

Hulla, J., Slašt'an, P. & Janíček, D. 2003. Sealing of Dilatation Joints with Polyurethane Resins. In L.F. Johnsen, D.A. Bruce & M.J. Byle (eds), *Grouting and Ground Treatment*: 1254–1265. New Orleans: ASCE.

Mucha, I. & Bansk", L. 1998. *Modeling of Water Flows in the Foundation Zone of the Gabcíkovo Navigation Locks* (in Slovak). Bratislava: Groundwater co.

Naudts, A. 2003. Irreversible Changes in the Grouting Industry Caused by Polyurethane Grouting. In L.F. Johnsen, D.A. Bruce & M.J. Byle (eds), *Grouting and Ground Treatment*: 1266–1280. New Orleans: ASCE.

Vrignaud, J.P. et al. 2003. Selection Criteria of Polyurethane Resins to Seal Concrete Joints in Underwater Road Tunnels in the Montreal Area. In L.F. Johnsen, D.A. Bruce & M.J. Byle (eds), *Grouting and ground treatment:* 1338–1346. New Orleans: ASCE.

Geotechnical Measurements and Modelling, Natau, Fecker & Pimentel (eds)
© 2003 Swets & Zeitlinger, Lisse, ISBN 90 5809 603 3

Measurements of the deformations of a high retaining structure

N. Meyer & F. Bussert
Institute for Geotechnical Engineering and Mine Surveying, Technical University Clausthal

H. Obermeyer
CERES, Geological Services, Staffort

ABSTRACT: To prevent failure of a rock bar in a quarry a geosynthetic reinforced soil structure (GRSS) was built. Measurement results up to today do not indicate large strains in the geosynthetics as calculated during design. Most of the geosynthetic strains were induced during compaction of the layer directly above the geosynthetic. Construction of the GRSS is still in process, at present 50% of the final height are reached. Until now no excessive deformation occurred. Geosynthetics have been placed successfully by employees not familiar with the material. Measurements of the construction are proceeded continuously.

1 INTRODUCTION

The quarry Werk Karl Majer of the bmk Steinbruchbetriebe GmbH & Co. KG is located next to Gundelsheim, a small town in the southern part of Germany next to the river Neckar.

In the quarry rocks of the Upper Muschelkalk, middle trias, are exploited for the open-cut mining of gravel, mainly used for the local road construction. Keuper layers, upper trias, of low thickness, which are lying over the Upper Muschelkalk have to be removed before exploitation. They have to be redeposited in the quarry. Per year 400,000 to 600,000 tons of limestone are exploited. Dependent on the condition 20–30% of the material is rejected and has to be redeposited as well. Additional overburden from recent and future developments has to deposit in depleted parts of the quarry.

Due to this large amount of material and the limited volume available a special method of redepositing was used. The initial idea was to steepen slopes by deposition behind large rock bars.

The rock bars are also used as roads for quarry workings as well as material transportation.

2 MEASUREMENTS ON THE ROCK BAR

After examination of the rock bar a monitoring program was designed, with which the actual movement rate, possible failure mechanism (tilting, overthrow, sliding), material parameters as well as the hydrological and meteorological influences can be observed.

In the northeastern part of the quarry working panel one rock bar is located next to the western quarry border. This 150 m long rock bar had a height of 33 to 35 m above extraction level. The base width was 10 to 12 m, top width 6 m respectively (Figure 1).

In July 1999 fissures had been detected on top of the deposited material as well as the rock bar. They indicated destrengthening and associated mass movement of the material behind.

Figure 1. Rock bar next to quarry border.

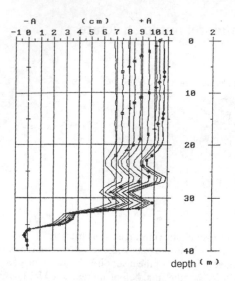

Figure 2.   Result of inclinometer measurement.

Figure 3.   Rock bar after blasting with foresets.

Measurements, risk assessment as well as protection system and reconstruction advise was conducted
by CERES Geological Services, GmbH.

Besides geodetic survey fissure detectors and
inclinometers were installed. To detect movement rate
seven fissure detectors, constructed from angular
shaped steel bands, placed left and right of the ca.
110 m long fissure were installed. Measurements
were made using a calliper gage.

The installation of the three inclinometers should
provided information on the possible turning movement as well as detects possible sliding surfaces.
Inclinometers were placed up to depths of 41.0 and
44.0 m.

As rain fall has a tremendous effect on the moving
rate of the rock bar, weather information (air pressure,
temperature, air humidity, rainfall) were recorded in a
weather station in the quarry.

Measurements indicated a steady creep of 2 to
5 mm/day, measurements from inclinometer and fissure detector indicate good agreement. Inclinometer
measurements indicated that the rock bar was moving
on a slightly tilted bedding plane with reduced shear
parameters, 36 m below the rock bar crest (Figure 2).

This joint plane is located 3 m below the depositing level behind the rock bar and 3 m above the actual
excavation level in the berm. In this height the rock
bar width is around 20 m.

Even no increase in crushed stone activity was
observed, it was obvious that the rock bar was about
to fail. It was expected that even small changes in
weather conditions (rainfall) would increase the moving rate and would lead to failure.

To ensure mining work and to reduce the potential
of seismic risk, it was decided to partially blast the
rock bar and to initiate a controlled slope failure.

After blasting a 20 m translatory movement of the
remaining rock bar into the excavation level occurred,
while the deposited material moved downwards by
19 m (Figure 3). Foresets were raised in front of the
remaining rock bar to increase stability.

3   STABILITY CONCEPT

On base of these observations, risk assessment and
deposition concept was too revised. It became evident
that stability of rock bars depends on thin layers of
claystone. These layers, which have a thickness of not
more than one centimetre, are good sliding planes,
particular when wet. During revision adjacent rock
bar, representing the western quarry border was
assessed as overstrained in future. There was a probability, would not have sufficient stability either and
may fail as well (road on the right side, Figure 3). To
reduce the earth pressure on the rock bar two solutions were possible. A shallow slope (a lot of space for
depositing material would have been lost) or the
deposited material had to stabilise itself. As space is
limited, the second solution was preferred.

Demand by the quarry owner was a safe and economical construction, deformations even within
decimetre range was not of interest. A geosynthetic
reinforced soil structure (GRSS) behind the rock bar
was proposed and designed by CERES Geological
Services GmbH. The structure was build to retain
forces by self weight of the deposited material. Due to
the GRSS the slope above the rock bar can be build
steeper and more material can be deposited.

The rock bar representing the quarry border is
thought not to act as a retaining wall and is due to

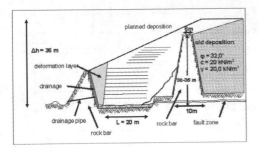

Figure 4.   Stability concept.

recent experience not to absorb limited earth pressure. For the slope stability determination the joint with the lowest shearing parameters (horizontal boulder clay seam) was taken as potential failure plane.

The additional weight of the deposited material has to be retained by the GRSS itself.

Between the rock bar and the GRSS a fill of loose material was placed as a deformation layer. Due to the large pore space in this layer, deformation of the GRSS does not increase earth pressure on the rock bar (Figure 4).

On the backside of the rock bar a 1 m thick drainage layer was placed to ensure that no water pressure increases acting force on the rock bar. Present water can be drained normally through the jointed rock. Additional drainage pipes (d = 100 mm) were drilled through the rock bar to ensure enduring drainage of the construction. As expected, no additional water was observed.

Dimensioning of the 90 m long GRSS was done according to BS 8006 as well as the German suggestions given by EBGEO. All possible failure mechanisms were observed. In the lower part geogrids with a short time strength of 180 kN/m were used, 90 kN/m in the upper part. The length was chosen to 20 m. Distance between the layers was calculated as 1 m, no secondary reinforcement was used. With this layout an economical utilisation of the geosynthetics under working load was expected. According to the codes of practice some of the layers were overstressed. As the design of a knock over was not necessary a further increase in economical design was achieved by simply placing the geogrids on the ground by the local workers.

The front slope of the GRSS is 75°, the final slope of the upper part is planned with 30°, and clearance with the local authorities is still required.

The excavation level where the reinforced slope is build on was due to deposed rejectamenta (dust, clay) from production and rainfall of very low bearing capacity and not stable enough to act as sound working level. To stabilise this layer, a geosynthetic mattress of one layer non-woven geotextile and three layers geogrid with a short time strength of 180 kN/m

were placed with 0.4 m distance. These layers were not considered in the stability analysis.

4   CONSTRUCTION

Construction of the GRSS started in September 2001. Fill material was placed and compacted in layers of 0.3 m thickness. Topsoil, material which doesn't fulfil the requirement as well as all materials that cannot be sold (dust) are used as fill material. Therefore the material is very inhomogeneous, ranging from dust to stones with 30 cm diameter. Fill material is placed with local water content; no quality control according to the German standard (ZTVE-StB 94/97) has been carried out. Compaction is done by sheepsfoot and smooth-wheel roller (12 to) with a minimum of four passes. In February 2003 the GRSS had a height of around 18.0 m, nearly the same height as the rock bar that needs to be prevented from failure. Until now no unplanned deformations occurred.

5   INSTRUMENTATION AND MEASUREMENTS

As the GRSS is higher than normal structures and the geosynthetics are chosen on a very economical basis, a monitoring monitoring program according to German standard DIN 1054 to prove security and stability of the structure was launched.

To observe deformation rates of the entire construction installation of two inclinometers is planned with reaching future surface level.

To measure the stress–strain behaviour within the structure two instrumented sections are installed. In two different heights vibrating wire strain gauges were installed as a redundant system to measure strain distribution over the whole length of the geotextiles and overall deformation of the structure (Figure 5). Vibrating wire strain gauges were chosen as they proved signal stability in other projects over a long time. They are resistant against mechanical and chemical degradation and are very useful by taking continuous measurements over a long time. This is extremely important, as long term behaviour of the geosynthetic (creep) wants to be observed.

Data collection started immediately after strain gauge installation using a stationary data collector. Since installation data is collected every two hours. On base of continuous measurement data can be analysed and compared with progress in deposition. Increase in strain with following construction sequence is observed and a direct nexus is achieved.

Preliminary tests were conducted in the laboratory to prove that the instrumentation of strain gauges doesn't influence stress–strain behaviour of the

Figure 5. Plan view on instrumented sections, 1st level.

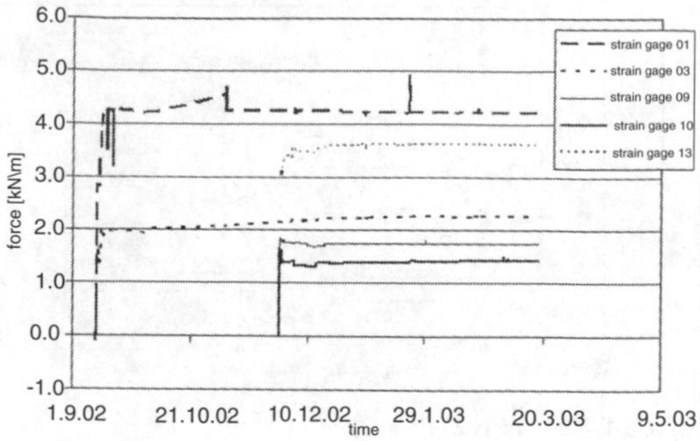

Figure 6. From strain calculated forces in the geosynthetic.

geosynthetic. Strain gauges were connected to the geotextiles and loaded axially. Deformation and force was measured using calibrated load cell and position encoder. As a result of these laboratory tests strain gauges where not glued directly on the geosynthetic as the connection was not stable enough to resist even small strains and strain gauges flaked. Strain gauges were welded on thin steel plates and force-fit connected to the geosynthetic. Laboratory tests showed less influence of the different stiffness of steel plate and geosynthetic as the number of longitudinal ribs increased. In the field strain gauges were connected to two longitudinal and one transverse rib.

Measurement of strain and strain gauge temperature are recorded continuously since construction of the structure started. We observed that soil temperature that exists during compaction is stored for a long time in the ground. For the 1st instrumented layer high temperature due to sunshine is conserved even surrounding temperature decreased. The same is observed for low temperature installation of the 2nd instrumented layer. After some month a mean temperature of around 10–13° Celsius, dependent on the local conditions will be reached, as known from other measurements (HUCH et al., 2001; MURRAY, FARRAR, 1988). Stable temperatures at low levels are good for long-term durability of geosynthetics as creep is decreasing with low temperatures. As temperature in the ground in middle Europe is nearly constant it has to be proven whether creep strain test according to German standard DIN EN ISO 13431 and ISO 554 (progressed at 20°C, relative humidity at 65%) is on an uneconomical side.

All together 15 strain gauges were installed in the two measurement sections. Only a small amount was destroyed directly after fill material was placed. Reasons are thought to be due to the sharp fill material leading to cable break.

Measured strains increase directly after placement and compaction of fill material. Independent of

geosynthetic short time strength strains in an order of less than 0.5% is measured. Similar results have been found by other instrumentations (FLOSS, 2001). Forces calculated from the strains lie within 1.5 to 6 kN/m (Figure 6). A nearly linear increase in strain and therefore force is observed at all strain gauges.

This increase is small compared with the minimum earth pressure ($k_{ah} = 0.2$) calculated from the surcharge. In Figure 7 an example is shown. Filling heights are calculated into surcharge pressure, a very low strain increase is observed. This is less compared with the theoretical consideration by EHRLICH, MITCHELL (1994). On the other side these findings agree very well with observations made by DELMAS et al. (1988). They even stated that in GRSS 70–95% of final geosynthetics strains are induced during construction, especially during compaction of the following layer.

We observed that forces in the geogrid are becoming more unique and are distributed equally when overburden increases. Forces over the whole geogrid length are rearranging when overburden pressure increases. Where due to high compaction energy high forces are introduced in the geogrid, we found that the increase in overburden lead to small increase in force as well as slightly reducing forces in the specific strain gauge. Lower compaction energy and therefore lower strain (force) in the geogrid lead to a more significant increase in strain with increase in overburden. A nearly equally distributed strain behaviour is observed afterwards in the geogrid layer. Further strain distribution in the geogrid will be observed continuously.

Design based on actual design codes seems to lead to an uneconomical utilisation of the geosynthetics used. These observations were proved to take part in a lot of temporary and permanent reinforced soil structures as shown especially by (BRAEU, 2001). Safety factors higher than 40 have been calculated and failure could still not be achieved after the loading

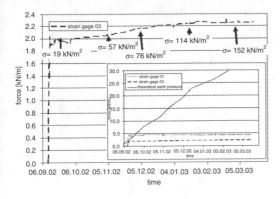

Figure 7. Force development with increasing surcharge.

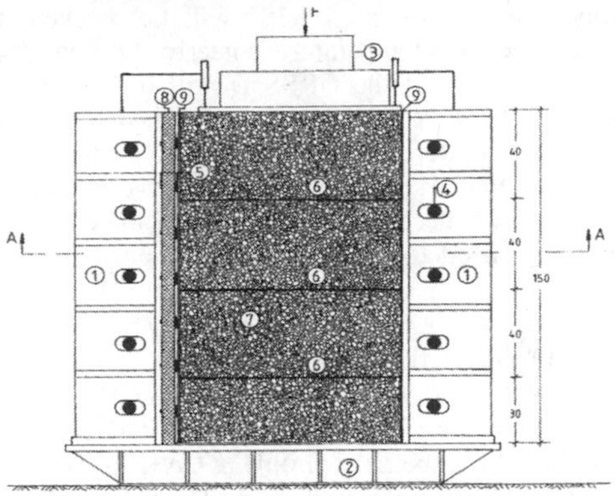

Figure 8. Biaxial apparatus (NIMMESGERN, 1998).

system failed before excessive deformation of the temporary retaining structure was observed.

Comparison with the minimum earth pressure in Figure 7 shows clearly the existing discrepancy of behaviour of the bonded system and actual design methods. It seems to be useful to conduct basic examinations on the behaviour of the bonded system to understand the mechanical behaviour between geosynthetic and soil in the structure. With this it should be possible to describe the shear force development inside the structure and the strains taking place under working stress conditions. This is thought to serve as an optimized economical design method for these kinds of structures. Maybe it is even possible to base the design of geosynthetic soil structures on the deformations compatible with the structure (NIMMESGERN, 1998).

6 PERSPECTIVE

Long-term observation for the construction described here is planned. The results will be analysed to draw conclusions on the deformation type and the stress–strain behaviour of the geosynthetics. Focus will be put on long-term influence of geosynthetic creep on the deformation of the structure. Comparison of measured strains in the geosynthetic and inclinometer deformations will indicate the deformation type of the retaining structure.

In addition we conduct large scale laboratory tests at the Institute for Geotechnical Engineering and Mine Surveying, Technical University Clausthal, to examine the mechanical influence of each material in the composite structure. Soil, geosynthetic, spacing of layers as well as geosynthetic stiffness will be varied. With this a better understanding of the mechanical action of each component will be achieved. It is

assumed that with this information a more realistic and economical design concept can be developed.

The large scale tests are accomplished in a biaxial testing device in which a part of a geosynthetic reinforced soil structure can be simulated under plane strain conditions. In the biaxial apparatus a $1.5\,\text{m}^3$ $(1.0 \times 1.0 \times 1.5$ – width, length, height) large compound structure is buildt with several geosynthetic layers. Due to a movable front side, horizontal deformations can be adjusted in steps of $0.1\,\text{mm}$. Horizontal and vertical deformations as well as inside deformation, strains in the geosynthetics and retaining forces developed in each direction are measured.

7 SUMMARY

In the quarry Gundelsheim rocks of the Upper Muschelkalk, middle trias, are exploited for the mining of gravel. For redepositing rock bars are left in place behind which the fill material is placed. Due to a fault zone one of these rock bars moved significantly so that a failure of the rock bar and the redeposited material was likely to happen.

To prevent failure of another rock bar a GRSS was built. Large deformations and therefore additional forces on the rock bar are prevented by a layer of loose material placed in front of the rock bar.

Measurement results up to today do not indicate large strains in the geosynthetics as calculated during design. Most of the geosynthetic strains were induced during compaction of the layer directly above the geosynthetic.

Temperature measurements taken at each strain gauge position indicate that soil temperatures, present during construction are stored for long time. They are independent of daily changes after a small layer is placed above it. After several months a mean temperature of around 10–13° will be reached. In deeper layers the temperature is nearly constant.

Construction of the GRSS is still in process, at present 50% of the final height are reached. Until now no excessive deformation occurred. Geosynthetics have been placed successfully by employees not familiar with the material. Measurements of the construction are proceeded continuously.

ACKNOWLEDGEMENT

Measurements were taken in the quarry Gundelsheim (bmk Steinbruchbetriebe GmbH & Co KG). The support during instrumentation as well as preparation of all required information and old measurement results and geological report is highly appreciated. Especially Mr. Fierlings personal interest in this project is acknowledged.

LITERATURE

Braeu, G. & Bauer, A., 2001: Versuche mit gering dehnbaren Geogittern, 7. FS-KGEO, München

Delmas, P., Gourc, J.P., Blivet, J.C. & Matichard, Y., 1988: Geotextile-reinforced retaining structures: A few instrumented samples, Int. Geot. Symp. On theory and Practice of Earth Reinforcement, Fukuoka, Japan

Ehrlich, M. & Mitchell, J.K., 1994: Working stress design method for reinforced soil walls, Journal of Geotechnical Engineering, Vol. 120, No. 4

Floss, R., 2001: Geotextil-Stützbauwerk im Zuge der BAB A9 am Hienberg-Aufstieg, Straße + Autobahn 4/2001

Huch, T., Kauter, R. & Schallert, M., 2001: Bau der Doppelsparschleuse Hohenwarte – Messkonzeption, Erfahrungen und bisherige Ergebnisse, 1. Siegener Symposium: Messtechnik im Erd- und Grundbau, Siegen

Murray, R.T. & Farrar, D.M., 1988: Temperature distributions in reinforced soil retaining walls, Int. Journal of Geotextiles and Geomembranes, Vol. 6

Nimmesgern, M., 1998: Untersuchungen über das Spannungs-Verformungs-Verhalten von mehrlagigen Kunststoffbewehrungen in Sand, Schriftenreihe Lehrstuhl und Prüfamt für Grundbau, Bodenmechanik und Felsmechanik der Technischen Universität München, Heft 27

Geotechnical Measurements and Modelling, Natau, Fecker & Pimentel (eds)
© 2003 Swets & Zeitlinger, Lisse, ISBN 90 5809 603 3

Geotechnical measurements for the assessment of the kinematic and long-term behaviour of unstable mountainsides

M. Moser
Chair of Applied Geology, University of Erlangen-Nuremberg, Erlangen

ABSTRACT: Long-term measurements to determine the deformation behaviour of unstable mountainsides either in low mountain ranges or in alpine areas have rarely been taken. The investigations of the kinematics of the reservoir bank instabilities (e.g. Gepatsch reservoir/Tyrol, Durlaßboden reservoir/Salzburg) must be stressed especially in this context. With regard to the long-time behaviour of mostly deep-reaching creeping and sliding processes on infrastructures and settlements the creeping regions in the area of Bündner Schist should be mentioned. Particularly the assessment of the controlling factors concerning deep-reaching and large-scale instable mountainsides (e.g. sagging of mountain slopes) has been proved to be difficult. In very rare cases – it could be clarified, which of the controlling factors (e.g. slope water level, annual precipitation, snow melt processes) are important for the temporal variation of the creeping–sliding process. The complex correlations will be shown and discussed by various alpine large-scale slope movements.

1 INTRODUCTION

The investigation concentrates above all on large-scale and deep-seated mass movements such as sagging of mountain slopes ("Talzuschübe"), rockfalls in combination with mountain splitting areas ("Bergzerrei-ßungsfelder") and earthflows ("Schuttströme"). The key elements for the elucidation of essential aspects of the kinematics and the mechanism of such deep-seated mass movements are the following parameters:

– morphological conditions
– geological–geotechnical and mechanical properties of the rock masses in the different zones of the slope movement
– course of the creeping–sliding process on the surface
– horizontal and vertical displacement vector
– distribution of strain, change in inclination and displacement vector at depth (vector profiles along drillholes)
– spatial geometry of the creeping–sliding mass
– influence of external factors
– time prediction of failure (especially for mountain splitting areas combined with rockfalls)

It is obvious that it is only possible to achieve reliable information on kinematics and mechanisms with expensive, sometimes technically difficult geotechnical measurements over a long period of time.

Thus, it is not surprising that the data taken from the literature are usually considered to be incomplete and seldom satisfactory, since they cannot be interpreted clearly with respect to the objective of a detailed time-dependent and spatial description of the kinematics of the whole unstable mountain slope.

For the detailed assessment of the kinematics of deep-seated mass movements only the combination of monitoring on the surface and measurements at depth are aiming. Surface monitoring should respect spatial conditions (pointwise, linewise, areal; Kovari 1988) and temporal variation (discontinuous and continuous).

It is obvious that the amount of information for example "pointwise and discontinuous" to "areal and continuous" is increased considerably. Due to financial and technical reasons there are constraints especially in cases of large-scale mass movements.

Although measurements at depth can clarify the course of movement, velocity and displacement vector distribution (e.g. "continuous creep, discontinuous creep, translational sliding", Haefeli 1967) results of investigations are rare.

Concerning deep-seated mass movements Noverraz (1996) remarks that the inclinometer casings constitute the only 100% objective criterion to determine the position and the distribution of movements in depth, allowing thus a correct analysis of the mechanism and of the mode of failure. Especially the Trivec probe (Kovari 1988) has become an important tool to measure the distribution of all three displacement vector components along a borehole.

2 ESSENTIAL ASPECTS OF THE KINEMATICS AND THE MECHANISM OF LARGE-SCALE AND DEEP-SEATED MASS MOVEMENTS

Brief description of the investigated mass movements:

Sagging of mountain slopes ("Talzuschübe")
The sagging of mountain slopes is the result of large-scale, imperceptibly slow rock creep and sliding processes (depth sometimes >200 m), which affect dominantly metamorphic rocks (schists and phyllites).

Although the sagging of mountain slopes clearly influences and in many cases even destroys engineering structures and settlements, our research findings could only be improved upon in some aspects. The emphasis on "the lack of information" (Hutchinson 1988) is not surprising, since only quantitative studies carried out over a longer period of time can lead to results.

Mountain splitting areas combined with rock falls ("Bergzerreißungsfelder")
The investigation of the last few years shows that especially reference has been made to the system

"brittle on weak" ("Hart auf Weich", Poisel & Eppensteiner 1988).

Examples of the spreading of mountain ridges show a characteristic behaviour in the development of mass movements. Initiated by loss of stability in the underlying weak sequences (ranging from the compressive yield strength up to sliding fracture), the brittle slab is disintegrated along a set of discontinuities by troughs and cracks into a "rock labyrinth" (e.g. Stepanek 1992). Failure mechanisms at the edge of brittle rock slabs result in toppling and sliding movements of the pinnacles as well as earthflows and debris flows in the weak sequences (Lotter et al. 1998).

Earthflows ("Schuttströme")
Earthflows are mass movements in predominantly deeply weathered claystones and marl formations. According to Hutchinson (1988), these movements are relatively slow moving, commonly lobate or elongate masses of accumulated debris in a softened clayey matrix that advance chiefly by sliding on discrete bounding shear surfaces.

The first two parameters ("morphological ... and geological–geotechnical ..."; see Chapter 1) are not discussed here, although essential aspects can be deduced for the kinematics of such deep-seated slope movements.

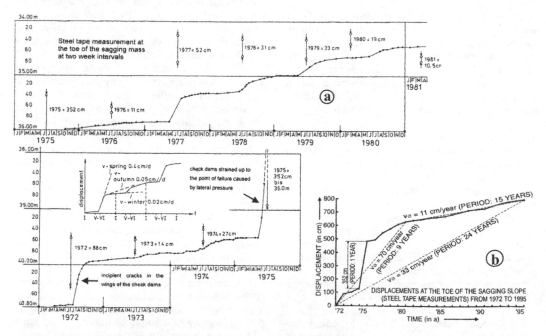

Figure 1. Displacement–time curve of the steel tape measurements between 1972–80, showing the distinct regressive cyclic behaviour of the mass movement (a); (b): velocities between 1972–95. The measurements took place at interval of 2 to 3 weeks.

2.1 Course of the creeping–sliding process in time on the surface

With respect to the time-dependent behaviour, the following questions are of great importance:

– how can the magnitude of rock mass creep be estimated for large-scale rock creep of slope?
– does the rock creep proceed continuously?

If the course of creep does not proceed continuously,

– is it stick slip (occurs in increments)?
– is the course of creep progressive or regressive cyclic?
– are there acceleration periods (in different time scales)?

Are these acceleration periods

– bound to certain time periods?
– is the course of creep motion apparently controlled by external factors and which ones should be considered?

Corresponding the above listed possibilities usually the following concepts are applied:

– pointwise discontinuous (geodetic)
– pointwise continuous (steel tape measurement, extensometer)
– linewise discontinuous (geodetic)
– areal discontinuous (geodetic, only for a section of the large-scale mass movement)

Example: sagging slope Gradenbach/Carinthia

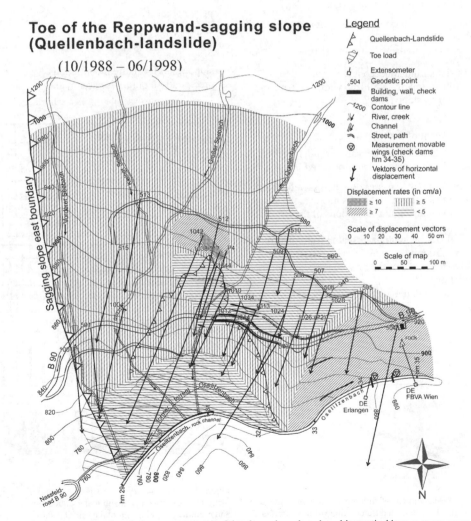

Figure 2. Displacement rates (geodetic measurements) of the channel reaches show kinematical homogenous areas.

Properties:
bedrock: phyllites, schists
height: 1,100 m, length: 2,350 m, width 1,100 m
max depth: ~200 m, slope (°): 27, Vol.: 150 × 10⁶ m³

Aim:
rates of rock mass creep, exact course of motion, relation to external factors.

Figure 1 shows the pattern of movement at the toe of the sagging slope between 1972 and 1995.

The first glance at the curve yields the information that apparently no stationary motion exists even over an annual interval. Rather, the time displacement curves clearly show accelerated deformation, which starts between the 15th of May and July and is completed by the end of October.

In the following winter months only very small amounts of motion can be observed.

Example: Toe of the sagging slope Reppwand/ Carinthia

Properties:
bedrock: slate, sandstone, conglomerate
height: 700 m, length: 1,800 m, width: 800 m, max. depth: >100 m, slope (°): 17

Aim:
areal assessment of the velocity (cm/a).

Figure 2 shows that we can define kinematical homogenous areas which are represented by:

– pattern of the rock masses (e.g. completely disintegrated and shattered rock mass due to slope tectonics or only with slightly alterations of the structural pattern of the original fabric)
– surficial slumps with deep-seated rotational retrogressive failures
– positive influence of the new constructed bedrock channel or
– possibility of undercutting processes

Aim:
movement of the channel embankments and of the check dams

Measuring concept:
– pointwise discontinuous (geodetic), pointwise continuous (extensometer)

2.2 *Influence of external factors on movement behaviour*

The external factors discussed in this paper include precipitation, snow melt, slope water level and undercutting at the toe of the unstable mountain slope.

Example: sagging slope Gradenbach/Carinthia

Aim:
causes of the exact course of motion, influence of slope water level.

Monitoring concept:
– slope movement: pointwise continuous (extensometer)
– slope water level: areal continuous
– precipitation: pointwise continuous

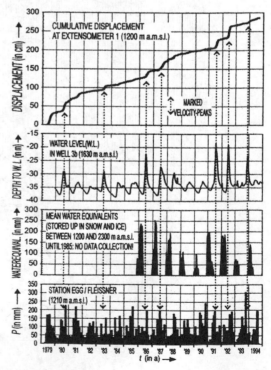

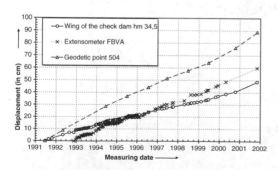

Figure 3. Displacement of the check dams and adjacent channel embankments with different kinds of measurements; toe of the sagging mass Reppwand/Carinthia.

Figure 4. Synoptic representation of cumulative displacement, slope water level, mean water equivalents at hillslope and monthly precipitation totals; sagging slope Gradenbach/ Carinthia.

Analyses carried out lead to the following conclusions up to now (Fig. 4):

– The direct influence of precipitation on the deep-seated movement is weak. Regarding the measuring period between 1980 and 94 only the event of November 1993 can be directly attributed to the influence of rain.
– Time of occurrence and course of time displacement curve of the unstable deep-seated creeping mass are determined primarily by the level of the slope water.
– The maximum movement rates follow the maximum water level after a time-lag of up to 4 weeks.

Especially high annual amounts of rock mass creep are linked to meteorological events, which apparently can provide an extremely quickly available supply of water in addition to the normal supply of water from the snowmelt in the spring months (Weidner & Moser 1998).

The regressive cyclic behaviour, which is apparently caused by snow melt processes and an increase in the slope water level, can also be clearly seen from studies of the slope movements on La Clapière/Maritime Alps (Blanc et al. 1987).

Example: Front-area of the mountain splitting area Treßdorfer Höhe/Carinthia

The active Treßdorfer Höhe landslide within the Carnic Alps (Carinthia, Austria) is a typical example

of a disintegrated inclined brittle rock slab, tens of meters thick, resting upon a mixed-layered sequence of weak and brittle rocks.

A huge potential unstable rock pinnacle about 40 m high (cube 60,000 m³), called Block 1A, stands at the convex slope edge. The most unstable part of this block (Block 1A, cube 7,000 m³) is separated by an increased rate of displacement prior to failure (Fig. 5).

Monitoring concept:
– slope movement: pointwise continuous (extensometer)
– precipitation: pointwise continuous

Apparently linear velocities within a year (0.35 mm/d at site 9/10 in 1997) can be subdivided into a seasonal dependence with acceleration periods reflected by different displacement rates over different periods of time. Characteristic medium-term (monthly) accelerations (0.55 mm/d from April to June 1997 caused by snow melt; 0.53 mm/d in November 1997 caused by rainfall) limit the linear movements to weeks when no water is supplied. Synchronous monitoring of displacement, temperature and precipitation clearly identify a short-term change in velocities within hours to days when directly triggered by water (Fig. 6).

2.3 The spatial displacement vector

Preliminary remarks:
The spatial displacement vector provides us with clues to the geometry of the creeping mass in particular and hence gives information on the kinematic behaviour and the mechanism of movement. Unfortunately, due to financial constraints and technical difficulties, such studies are rather the exception than the law (Kovari 1988).

Example: Toe of the sagging slope Reppwand/Carinthia

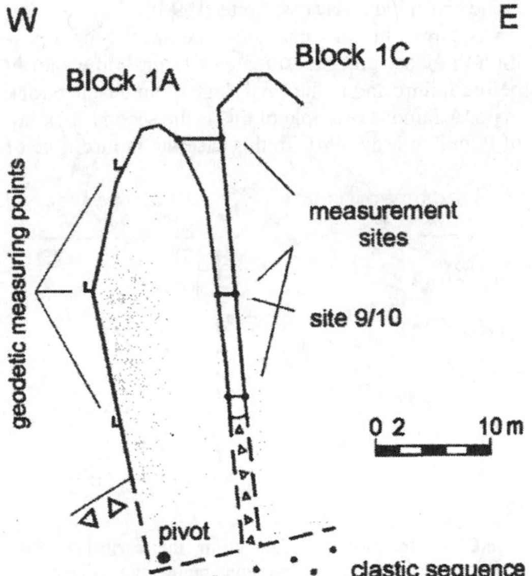

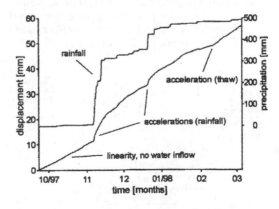

Figure 5. Cross section of Block 1A including the monitoring sites (schematic). Site 9/10 is indicated; Front of the mountain splitting area Treßdorfer Höhe/Carinthia.

Figure 6. Synchronous electronic record of displacement and precipitation at site 9/10; front-tower of the mountain splitting area Treßdorfer Höhe/Carinthia.

Aim:
situation and kind of basal shear surfaces.

Measuring concept:
– linewise discontinuous (geodetic)

Owing to geodetic measurements and geological–geotechnical investigations a basal shear zone inclined at a low angle could be assumed which outcrops at the bottom of the channel.

In this context the deep-reaching process is overlapped by surficial partly rotational failures (up to 20 m thickness).

2.4 The situation and kind of shear surfaces; geometry of the creeping mass and vector profile to depth

The lack of information concerning a detailed understanding of movement processes of deep seated-mass movements in the depth has been reported by Hutchinson 1988, Kovari 1988 and Noverraz 1996. This is especially true for sagging slopes.

As Noverraz (1996) reported recently, the difficulties lie in the fact that "all these models suffer from the same lack of in-depth investigations and a general lack of information related to horizontal and vertical displacement vector". He concludes that a large number of deep-seated mass movements are triggered by the retreat of glaciers as "true giant slides". The authors draw this conclusion from the evidence of 7 large sagging slopes, which have been studied within the context of the project "Versinclim".

More detailed information is provided for vector profiles to depth of less deep-seated mass movements (e.g. earthflows). The following conclusions can be drawn from the results recently conducted from investigations in British Columbia and in Southern Apennines:

– Although many earthflows look like slow viscous deformations, particularly when seen in aerial views, very often a basal sliding zone appears to predominate; the measured velocity profiles exhibit a "plug-like" movement (Bovis 1985).

– a few vertical velocity profiles, however, also illustrate rather strong internal deformation (Fig. 8b).

During investigations of several earthflows in the region of the Southern Apennines, Giusti et al. (1996) came to the conclusion that viscous-like profiles are more likely to be linked with the greatest movement values, whereas "rigid" profiles typically reflect slow displacement rates.

2.5 Failure-time estimation of rockfalls

This paper aims to give a general introduction to this topic. Rock slope failures often result in considerable economic consequences and therefore time predictions of this phenomenon are of major importance to minimize their economic impact. Despite the large number of rock slopes monitored, detailed reports of the kinematic and phenomenological behaviour of the rock mass prior to slope failure are quite rare.

Case histories of detailed analyses and interpretations of the pre-failure kinematics of rock slopes, which have been used successfully to predict a failure, are even less common (e.g. Voight & Kennedy 1979, Zvelebil 1985).

It is difficult to provide precise details on the volume as well as the time window or point in time of such a slope failure. This has been evident in studies beginning with the predicted rock fall in Kilchenstock/Switzerland of ca. 250,000 m³ (Heim 1932) up to the time prediction of rock slope failure in the Treßdorfer Höhe/Carinthia (Glawe & Lotter 1996).

As shown in other examples, we are only in a position to give a precise estimate of failure-time shortly before failure and in the final stage of an unstable rock mass. A famous example of this is the second rock fall of Randa in May 1991. In this case the failure-time of

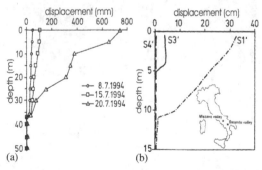

Figure 8. Inclinometer data from the earthflow Falli Hölli/Switzerland in the borehole below the village one month after the initial measurements of 24.07.1994; several slip surfaces were detected at 10, 20 and 37 m (Fig. 8a; after Vulliet & Bonnard 1996). Figure 8b (right): Inclinometer profiles measured in earthflow No. 2 (Basento valley) from 03.12.1992 to 21.01.1993 show great internal deformation of the hole unstable mass (after Giusti et al. 1996).

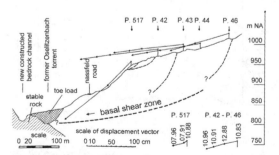

Figure 7. Spatial displacement vectors; toe of the sagging slope Reppwand/Carinthia.

the second event could be precisely predicted owing to the study of the pre-failure behaviour.

3 CONCLUSION

A set of different methods necessarily have to be applied to large-scale mass movements in order to investigate the deformation behaviour over a longer span of time. Morphological, geological and geotechnical qualitative analysis methods can be used to get a first overview. Especially important factors like the course of the creeping–sliding process, the spatial displacement vector, the vector profile to depth, the influence of external factors and time prediction of failures, require quantitative investigation methods, which are often restricted by financial supplies of the research projects. High-resolution temporal measurements can only be achieved by complex investigation programs. This is especially the case for precisely assessing the influence of external factors on the kinematics of deep-seated mass movements.

REFERENCES

Blanc, A., Durville, J.-L., Follacci, J.-P., Gaudin, B. & Pink, B. 1987: Méthodes de surveillance d'un glissement de terrain de très grande ampleur: la Clapière, Alpes Maritime, France. *Bull. Int. Ass. Eng. Geol. 35: 37–44*. Paris.

Bonnard, C., Noverraz, F. & Dupraz, H. 1996: Long-term movements of substabilized versants and climatic changes in the Swiss Alps. *Proc. 7th Int. Symp. on Landslides, Trondheim 3:* 1525–30. Rotterdam: Balkema.

Bonzanigo, L. 1988: Etude des mécanismes d'un grand glissement en terrain cristallin: Campo Vallemaggia. *Proc. 5th Int. Symp Landslides, Lausanne* 2: 1313–16. Rotterdam: Balkema.

Bovis, M. 1985 : Earthflows in the Interior Plateau, southwest British Columbia. *Can. Geotechn. Journ.* 1 (22): 313–334.

Giusti, G., Iaccarino, G., Pellegrino, A., Russo, C., Urciuoli, G. & Picarelli, L. 1996: Kinematic features of earth-flows in Southern Apennines, Italy. *Proc. 7th Int. Symp. Landslides, Trondheim* 2: 457–462. Rotterdam: Balkema.

Glawe, U. & Lotter, M. 1996: Time prediction of rock slope failures based on monitoring results. *Proc. 7th Int. Symp. Landslides, Trondheim* 3: 1551–55. Rotterdam: Balkema.

Hutchinson, J.N. 1988: General report: Morphological and geotechnical parameters of landslides in relation to geology and hydrogeology. *Proc. 5th Int. Symp. Landslides, Lausanne*, 1: 3–35. Rotterdam: Balkema.

Kovari, K. 1988: General report: Methods of monitoring landslides. *Proc. 5th Int. Symp. Landslides, Lausanne* 3: 1421–33. Rotterdam: Balkema.

Leobacher, A. & Liegler, K. 1998: Langzeitkontrolle von Massenbewegungen der Stauraumhänge des Speichers Durlaßboden. *Felsbau* 16 (3): 184–193. Essen: Glückauf.

Lotter, M., Moser, M. & Glawe, U. and Zvelebil, J. 1998: Parameters and kinematic processes of spreading of mountain ridges. *Proc. 8th Int. Congr. IAEG, Vancouver*, 2: 1251–57. Rotterdam: Balkema.

Noverraz, F. 1996: Sagging or Deep-Seated Creep. Fiction or Reality? *Proc. 7th Int. Symp. on Landslides, Trondheim* 2: 821–28. Rotterdam : Balkema

Poisel, R. & Eppensteiner, W. 1988: Gang und Gehwerk einer Massenbewegung Teil I: Geomechanik des Systems "Hart auf Weich". *Felsbau* 6 (4): 189–194. Essen: Glückauf.

Stepanek, M. 1992: Gravitational deformations of mountain ridges in the Rocky Mountain foothills. *Proc. 6th Int. Symp. Landslides, Christchurch* 1: 231–6. Rotterdam: Balkema.

Tentschert, E. 1998: Das Langzeitverhalten der Sackungshänge im Speicher Gepatsch (Tirol, Österreich). *Felsbau* 16 (3): 194–200. Essen: Glückauf.

Voight, B. & Kennedy, B.A. 1979: Slope Failure of 1967–69. Chuquicamata Mine, Chile. *Rockslides and Avalanches* 2: 595–632.

Vulliet, L. & Bonnard, Ch. 1996: The Chlöwena landslide: Prediction with a viscous model, *Proc. 7th Int. Symp. Landslides, Trondheim* 1: 397–402. Rotterdam: Balkema.

Weidner, S., Moser, M. & Lang, E. 1998: Influence of Hydrology on Sagging of Mountain Slopes ("Talzuschübe")– New Results of Time Series Analysis. *Proc. 8th Int. Congr. IAEG Vancouver* 2: 1259–66. Rotterdam: Balkema.

Ziegler, H.J. 1982: Die Hangbewegungen im Lugnez, am Heinzenberg und bei Schuders (Graubünden).–Geologie und Geomechanik. *Unveröff. Diss.* pp. 106, Bern.

A multidisciplinary integrated approach to geological and numerical modelling of the Lodrone landslide (Oriental Alps, Trentino Region, Italy)

M.L. Rainone & N. Sciarra
Dipartimento di Scienze della Terra, Università "G. d'Annunzio",
Chieti Campus Universitario, Chieti scalo (CH), Italia

ABSTRACT: This paper deals with a multidisciplinary integrated study of the Lodrone landslide, a large mass movement located on the oriental slope of Monte Macaone, in Trentino Region (North Italy). This area is affected by processes and forms linked to gravity and probably connected to deep-seated deformation. The obtained modelling and simulations are illustrated, discussed and related to the data observed and recorded by monitoring-net installed on the investigated area.

1 INTRODUCTION

It's well known that reconstruction of underground geometric characters and physical-mechanical properties is one of the main purposes in landslides investigation, especially if these studies are finalized to the numerical modelling of the landslide. Although the "models do not represent reality, but our idea of reality" (Anderson & Crear 1993), a quantitative approach by numerical modelling can considerably contributes to the knowledge of the complex relationship that controls these kind of natural phenomena. The Lodrone landslides is located in Trentino region (north–east Italy), on the oriental slope of Monte Macaone, at the left side of the Idro lake (figs. 1 and 2). multidisciplinary integrated analysis (geological, geomorphological, hydrogeological, geophysical and geothecnical) for defining the geological model, were performed. In particular, geophysical and geognostic surveys were utilised with the aim of reconstructing the underground geometries, verifying the possible presence of a depth potential slip surface and characterizing some physical-mechanical and hydrogeological properties of material interested by landslide phenomena: two high resolution seismic profiles (about 1 km of total length), two refraction seismic profiles (about 300 m of total length), three geognostic holes.

These surveys were carried out in the area interested by mass movements and allowed to define an exhaustive underground geological model.

Figure 1. Investigated area, Lodrone (TN).

2 GEOLOGICAL AND GEOMORPHOLOGICAL OUTLINES

From a geological point of view, in this area of Monte Macaone, the bedrock is represented by siltiti of the Collio Formation and ignibrites (Permian) overlain by detritic deposits. Geomorphological features are characterized by processes and forms caused by gravity action and probably connected to deep-seated deformation. The instability phenomena involve the detritic deposits and the bedrock too. It is possible to recognize several complex mass movements (translational

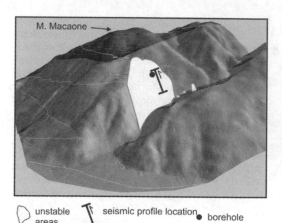

unstable areas   seismic profile location   ● borehole

Figure 2.   Monte Macaone D.T.M.

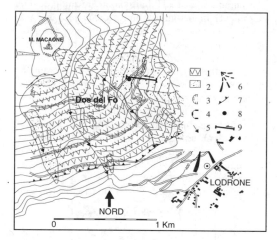

Figure 3.   Geomorphologic sketch of the Lodrone area: (1) landslide scree tongue (rock slide); (2) landslide scree tongue (detrital flow); (3) landslide scarp edge (h < 5 m); (4) landslide scarp edge (h > 5 m); (5) concentrated erosion; (6) alluvial fan; (7) Val Trompia thrust; (8) borehole; (9) seismic profile location.

slides) that interest the detritic deposits (about 15–20 metres of thickness) and the bedrock too (fig. 3).

3   SUBSURFACE INVESTIGATION

A geophysical and geognostic surveys were performed in the area reconstructing the underground geometries and to characterize some physical–mechanical properties of materials interested by landslide phenomena. In particular, for the geophysical prospecting, refraction seismic methodology and high resolution seismic

Table 1.   High resolution acquisition parameters with P waves.

Sampling rate: 0.25 msec
Antialiasing filter: 500 Hz
Record length: 1024 msec
Number of channel: 12
Number of P wave geophones for channel: 3
Natural frequency: 4.5 Hz
Group interval: 5 m
Shot interval: 5 m
Offset: 12 m
Spread: off-end push-increase
Coverage: 600%

Table 2.   High resolution seismic reflection processing parameters (P waves) sequence.

Editing
Sorting
Frequency filtering
Deconvolution
Velocity analysis
Normal move out correction
Stacking
(Migration)

reflection methodology was utilised. This kind of geophysical prospecting, if correctly applied, represents a powerful tool to underground investigation (Mc Cornack et al., 1984; Stümpel et al.,1984; Anibaldi et al., 1994; Rainone et al., 1994). Seismic profiles were carried out utilising P waves due to the presence of a strong difference of seismic impedance between the surface deposits and bedrock, or, generally, in presence of lithoid materials (Rainone et al., 2002). Acquisition and processing parameters of high resolution seismic profiles are shown in tab. 1 and tab. 2. Fig. 4a shows one of the final stack sections obtained. Many reflectors are distinguishable; one of these, is very strong, at about 120 msec (two way time), and is located in the bedrock and it is probably linked to a fractured and potentially slip surface. Fig. 4b represents a reconstructed schematic geological section. A geognostic hole, bored in the upper part of the seismic profile, confirmed the presence of a very rubbled and strongly fractured zone in the bedrock, between 69 and 65 meters of depth (fig. 4c).

4   LANDSLIDE MODELLING

The geological and geognostic surveys permitted to reconstruct the technical framework of the landslide that is schematized in fig. 5.

226

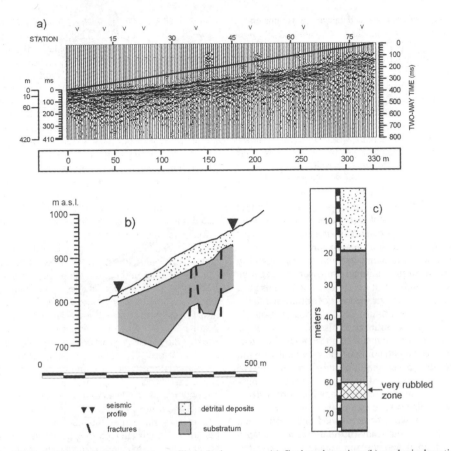

Figure 4. High resolution reflection seismic profile in Ladrone area: (a) final stack section; (b) geological section reconstructed by seismic profile; (c) stratigraphy of borehole.

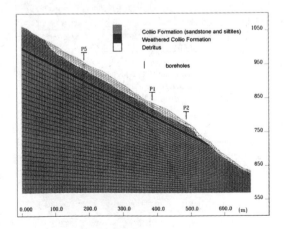

Figure 5. Simplified technical model of the investigated slope.

On this model some numerical modelling were carried out to study the dynamic evolution of the slope. In fact, basing on the inclinometer results, two main typologies of movement were revealed: a superficial roto-translation within the detritus and a slow deformation in depth, primarily translative, within the Collio Formation constituted by sandstones and siltites.

The used code (FLAC 2D, 2000) is a two-dimensional method of numerical analysis of finite differences for calculations of mechanics of continua. FLAC is based on a "*lagrangian*" calculation scheme which is very adaptable in the modelling of large-scale deformations and collapse of materials (Sciarra, 2000, Sciarra & Calista, 2001).

The general analysis carried out consisted in an overall re-equilibrium of the system and therefore in the study of failure conditions. The analysis of the process of global re-equilibrium is divided into three phases. The first phase analyses the filtration within the model to identify, once the hydraulic equilibrium

Table 3. Values of physical and mechanical parameters used in the modelling.

Lithotype	γ (kN/m^3)	c' (KPa)	φ' (°)	ν	E (GPa)	k (m/s)
Collio formation	24	200	45	.35	3.6	10^{-6}
Fractured collio formation	23	100	40	.35	3.6	10^{-4}
Detritus	20	20	40	.35	.25	10^{-4}

γ (weight unit); c' (cohesion); φ' (friction angle);
ν (Poisson coefficient); E (Bulk modulus); k (permeability).

has been reached, the negative pressure distribution irrespective of any mechanical effect. In the second phase the model operates a mechanical adjustment using the values of pore pressure obtained from the first phase and imposing zero the compressibility module of the water (any variations of the negative pressures is avoided). In the third phase there is contemporary analysis of the process of filtration and the relative mechanical adjustment. At the end of the process of global equilibrium it is possible to represent the vectors of movement. In this way it is possible to observe the behaviour of the system. Following the phase of global equilibrium and wishing to analyse the conditions of limit equilibrium, it is possible to move onto the analysis of the stress-strain behaviour of the system.

With this aim we have performed two verifies; the first is relative to the study of the dynamic evolution of the superficial detritus, the second to analyse the movements in depth. In all cases the influence of the water pollution has been investigated due to a broken hydraulic tunnel realised to produce hydro-electric energy. At the top of the slope so we imposed a constant pore pressure of about 200 kPa obtained from in situ measurements. The physical and mechanical parameters of the rocks, utilised in the modelling, are reported in tab. 3. Studying the flow vectors trend we deduced, during the transient, the typology of filtration (fig. 6). We can evidently observe that the weathered and fractured layer, localised in the S4 borehole between 60 and 64 m in depth, becomes a preferential flow zone.

During the stationary state (fig. 7) we verified the behaviour and the stability of the slope. The results show the instability of the superficial detritus (fig. 8). These are totally mobilised (see also fig. 9 in which the maximum shear strains are reported) and numerous failure surfaces are generated.

Subsequently the detritus mobilization, within the fractured zone of Collio Formation further shear strains concentrate determining a new and more deep failure surface (fig. 10). These strains produce in the slope a translational movement less important than the

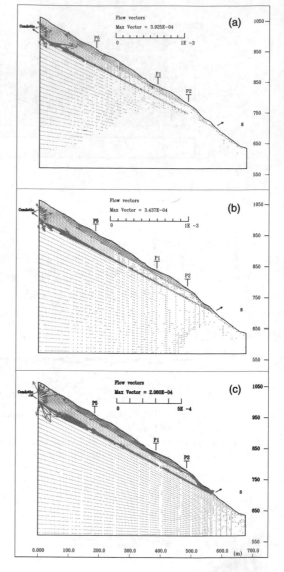

Figure 6. Flow vectors at three steps of transient analysis: initial phase (a), after some days (b) and when the flow reaches the present spring at the bottom of the slope (c).

displacements of the detritus but significant. The presence of a flow is so of particular importance for the stability of the slope.

5 CONCLUSIONS

The numerical analyses on the Lodrone landslide, permitted to confirm the reliability of the physical model

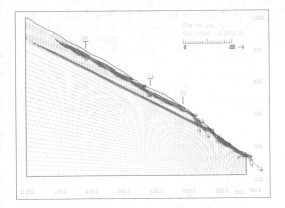

Figure 7. Flow vectors during the stationary state.

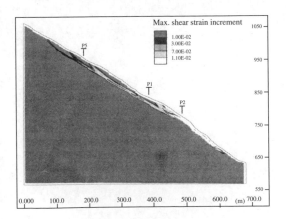

Figure 8. Displacements vectors in the detritus.

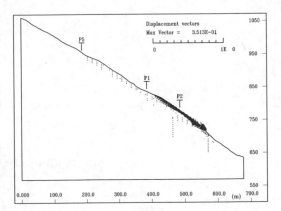

Figure 9. Max shear strains in the detritus during the stationary state.

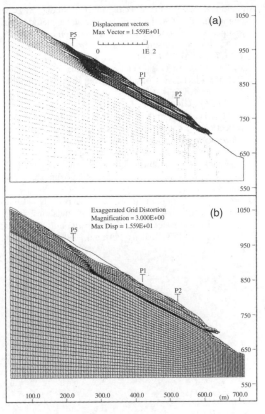

Figure 10. Displacements vectors within the weathered Formation of Collio (a) and meshes deformation (b) clearly showing the rupture zones in depth.

obtained by geophysical prospecting and to define the dynamic evolution of the slope processes. Specifically, the mass movements on the oriental slope of Monte Macaone (both surface and deep movements) were evidenced. These data confirm that the numerical modelling provides good results if the geological model is as accurate and precise as possible.

ACKNOWLEDGMENTS

The AA. thank Dr. M. Calista and Dr. M. Mangifesta for their valid assistance in the modelling, and Dr. P. Sammartino for his help in the editing.

REFERENCES

Anderson, G. M. & Crerar, D.A. 1993. Thermodynamics in geochemistry; the equilibrium model. Oxford University Press, New York.

Anibaldi, A., Massetani, A., Pellegrini, F., Rainone, M.L. & Signanini, P. 1994. Un'applicazione della metodologia sismica a riflessione ad alta risoluzione con onde SH per lo studio dei versanti instabili. *Atti XIII Convegno GNGTS*, Vol II, 779–790.

Flac_2D 2000. Fast Lagrangian Analysis of Continuo. Release 4.0. Itasca Consulting Group Inc. Minneapolis, Minesota.

Gasperini, M., Giorgetti, F., Rainone, M.L. & Signanini, P. 1994. SH wave high resolution and ultra resolution reflection prospecting: some examples. *Atti 1° "European Congress on Regional Geological Cartography and Information System*. Bologna.

Mc Cornack, M.D., Dumbar, J.A. & Sharp, W.W. 1984. A case study of stratigraphic interpretation using shear and compressional data. *Geophysics*, 49, 5, 509–520.

Rainone, M.L., Signanini, P. & Sciarra, N. 1994. Geophysical methods for the study of landslides: some applications. *Proceeding 7th International Congress I.A.E.G.*, 175–183, Lisbona, Portugal.

Rainone, M.L., Signanini, P. & D'intinosante, V. 2002. Application of P and SH waves high resolution seismic reflection prospecting to investigation of unstable areas. *Landslides,* Rybář, Stemberk & Wagner (eds), Lisse (ISBN 90 5809 393 X).

Sciarra, N. 2000. Rigid formation over plastic substratum: modelling of a lateral spread in central Italy. *Proceedings of the 8th ISSMGE & BGS International Symposium on Landslides*, Cardiff, Wales, 3, 1339–1346.(ISBN: 0 7277 2872 5).

Sciarra, N. & Calista, M. 2001. Modellazione delcomportamento di formazioni rigide su di un substrato deformabile: il caso di Caramanico Terme. *Mem. Soc. Geol. It.*, 56, 139–149, Roma (ISSN 0375 9857).

Stümpel, H., Käler, S. & Meissner, R. 1984. The use of seismic shear waves and compressional waves for lithological problems of shallow sediments. *Geophys. Prosp.*, 32, 662–675.

Geotechnical Measurements and Modelling, Natau, Fecker & Pimentel (eds)
© *2003 Swets & Zeitlinger, Lisse, ISBN 90 5809 603 3*

Monitoring and physical model simulation of a complex slope deformation in neovolcanics

J. Rybář & B. Košťák

Institute of Rock Structure and Mechanics, Czech Academy of Science, Prague, Czech Republic

ABSTRACT: Rigid basalt lava flows alternating with tuffs and tuffites have been found deformed on a slope of the Labe River Canyon in the České Středohoří Mts., Bohemia, in an unusual complex configuration. The top section of the deformation with huge basalt blocks was separated from the lower one by a conspicuous platform inclined backwards. This raised a question about the origin and development of such a deformation, as well as about present movements. A series of maps were prepared for the area, and a series of longitudinal and cross sections were surveyed. Besides, monitoring in both the sections was arranged that evidenced present movements. The lower section was found in a slope movement of standard development inhibited by a local brook erosion. The higher section, however, showed very low atypical movements. As for the platform, physical models were prepared simulating the process of structure formation in a stratified system of alternating rigid and ductile beds. The models indicated formation of the platform as of the result of specific deep deformations due to the process of long-term river erosion in the canyon. Apart of that, different deformation conditions in the two separated sections could be observed in the models. This analysis came to a final view explaining the present morphology of the slope in accordance with monitoring results and other observations.

1 INTRODUCTION

Neovolcanites in the hilly country of the České Středohoří Mts. (North Bohemia) provide favourable conditions not only for slope deformations of landslide type but even for some of an unusual character. Regarding slope movement classification of Nemčok et al. (1972) such deformations belong under deep creep movements. They represent very slow long-term slope movements of rock. Usually, they are quite deep covering an extensive area and often represent a preparation phase of faster movements of the type of landsliding or rockfalling.

In the České Středohoří Mts. rocks of volcanic origin are represented by rigid basalt and trachyte lava, pyroclastics, as well as by Tertiary sedimentary intercalations. Often even Upper Cretaceous sediments are present covered or injected by volcanic bodies. These are claystones, siltstones and sandstones. A characteristic form can be often found in the slope relief represented by large platforms that are often slanted backwards.

2 MODEL LOCALITY OF ČEŘENIŠTĚ VILLAGE

The authors concentrated their efforts to clear up conditions leading to the origination and failure mechanism at the model locality of Čeřeniště Village near the town of Litoměřice (Leitmeritz). The locality appears one of the most interesting ones in the České Středohoří Mts. The area is built by superficial volcanic products. They contain intercalations of clayey and silty sediments, including sediments of organic origin. Bazanite lavas are frequently altered. The area deformed by gravitational movements is 1050 m long and 700 m wide (Fig. 1). The upper edge of the separation zone is found at a level of about 600 m a.s.l., frontal section of the accumulation zone at a level of 340 m a.s.l., general slope inclination is 14°. The area belongs to the drainage area of the Rytina Creek. Creek confluence with the Labe River is about 3 km distant from the separation wall at a level of about 140 m a.s.l.

Slope deformations of Čeřeniště Village are of an atypical character. One can differentiate three sections in the relief (A, B, C). The upper portion of the

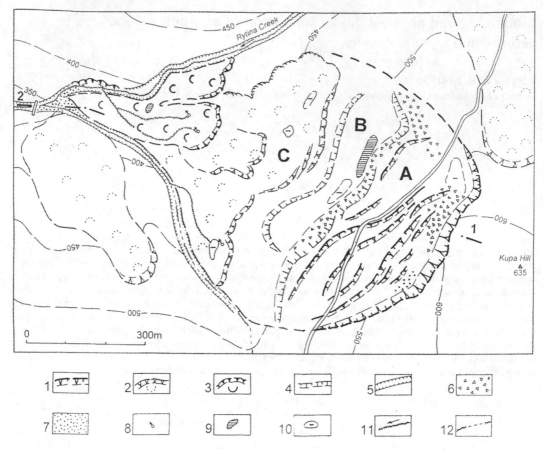

Figure 1. Situation of slope deformations near Čeřeniště Village. A, B, and C – individual sections; 1– scarps and trenches in basalt rocks, 2 – recent potential and old landslides, 3 – recent active landslide, 4 – outstanding edges in the relief, 5 – erosional walls, 6 – debris, 7 – alluvial sediments, 8 – spring, 9 – lake, 10 – undrained depressions, 11 – waterflow, 12 – creek periodically dry.

slope (section A) is faulted by creep movements of lateral character that are at least 100 m deep. It is separated form the central section by a conspicuous sub-horizontal platform (section B) slanted moderately backwards. The platform can be followed without interruption across the sliding area in its full width. To throw light upon the origin of such a platform was one of partial aims of the research. As for the lower portion of the slope (section C) relatively fast land-sliding in rocks of weathered cover takes place.

The upper section A developed a distinctive separation upper scar area. It is limited by a continuous rock wall up to 40 m high, rimmed with a debris mound at its toe. The separation wall shows a skewed course in plan. It is about 600 m long in total. The wall is rimmed with a distinctive depression at its toe. The difference in level between the depression

bottom and the upper wall edge reaches maximum at its central part with a value close to 70 m.

A huge block of basalt about 400 m long and up to 200 m wide is separated from the wall by the depression. The block is broken and a system of parallel transversal trenches separates individual pieces. The pieces between depressions are found as a series of sunk steps. Each intermediate trench is distinctively bordered with a steep outcrop of an antithetical slip plane (Fig. 2). In trench bottoms one finds dry depressions frequently.

Investigations proved that the volcanic complex, which suffered deep slope deformations, is built by at least six lava flows up to 30 m thick. The flows are separated by sedimentary intercalations, mostly of clayey character. As a result of X-ray analysis, the main clayey mineral in all the intercalations was

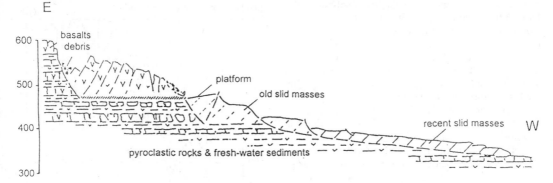

Figure 2.   Geological cross-section through the slope deformation of Čeřeniště Village.

identified as smectite (Cílek et al. 2000), which is extremely unstable in volume and comes to produce weakened planes and zones of decreased shear strength. Such predisposed planes make development of gravitation failure in the volcanic bodies easier.

Early deformations were produced clearly as cooling cracks due to volume reduction in the time when hot volcanic outbursts occurred. It is likely that even unevenness of the basement under the volcanic masses contributed. Besides, gravitation movements of volcanic masses into spaces cleared up by lava hearths could not be avoided.

The development of the Čeřeniště slope deformation into the present form was affected by deepening of the Labe River canyon in Pleistocene, when volcanites were cut through completely up to the Cretaceous basement. Then, unloading of the valley bottom initiated a process of bulging in plastic Cretaceous beds and degradation of stability conditions in valley slopes.

The deepest cut of the Labe River took place in Eem Interglacial period (Riss–Würm); the Labe River reached a level of up to 16 m down the present bottom. An intensive back erosion followed the cut and modelled its tributaries also, i.e. even the valley of the Rytina Creek, where Čeřeniště Village is located. One can expect that this was the top phase period of slope deformations here. However, movements have been active since. They are active even now. The intensity comes to increase in wet periods of Holocene. The periods rich of precipitation result in intensive erosion along the Rytina Creek and slope movements are rejuvenated, which takes place in the lower section mainly.

The occurrence of landslide calamities in the České Středohoří Mts. during the last 300 years was studied in connection with individual climatic factors (Rybář & Suchý 2000). One period extremely wet occurred by the end of the 19th century. Archives revealed that the Rytina Creek and other Labe tributaries developed repeatedly dangerous earthflows, which troubled railroad on the right bank of the Labe River and even the

river transport. To cut down these destruction events by the Rytina Creek, a system of stone baffles was build up at the beginning of the 20th century. As a result the erosion base of the creek was lifted by about 10 m at the frontal part of the landslide area, which, no doubt, depressed landslide movements considerably.

3   METHODS OF INVESTIGATION

A series of maps was produced for the model locality of Čeřeniště Village under the volcanic body of Kupa Hill. It was an engineering-geological map at a scale of 1:2 000, and a map of broader vicinity 1:5 000, then geomorphological and geological maps 1:10 000. A series of longitudinal and transversal profiles were produced.

Monitoring of deformation effects in different parts of the landslide area started in 1998 (Suchý 1999). In the trenches of the field under the main separation wall, i.e. in the slope section where present movements were not anticipated, two highly sensitive mechanico-optical dilatometric gauges TM71 were installed. It was as early as in 1999 when general trends of present movements could have been defined in this section. In the lower section of the landslide area, which appeared much more active a tape extensometry network of observation and reference points were established. Detected movements correlate with geodetic measurements.

The upper and central sections of the landslide area were subjected to applied geophysics surveying in three phases (Hofrichterová et al. 2000). Primarily, it was symmetrical resistance profiling and vertical electrical probing, then the method of spontaneous polarisation SP and radiometry that were found successful.

Geomorphological investigations were oriented to find relation between the evolution of the Labe River valley and Čeřeniště slope deformations. Geological and Quarterly-geology methods (Cajz 2000, Cílek et al. 2000) contributed significantly in solving problems of the evolution of gravitational slope failures in the neighbourhood of the model locality.

Dendrochronology could be applied in co-operation with Italian partners to investigate growth of selected trees found affected by older slope movements. The aim was dating of individual phases of landsliding in the last century (Fantucci et al. 2000).

After having analysed all the interdisciplinary findings from the model locality of Čeřeniště Village a prognosis of the failure mechanism was elaborated in several variants and reliability of individual solutions have been verified using photoplastic physical models.

4 MONITORING RESULTS

The monitoring system includes following methods: tape extensometry, geodesy, and 3D dilatometry using precise extensometric devices – crack gauges TM71.

Tape extensometry and geodetical methods are employed in the lower active part (section C) of the landslide area. Rate of movements reaches here 200 to 250 mm per year being affected by seasonal climatic variations.

Important data came from the two gauges TM71 installed in summer 1998 in the trenches of the block field under the main separation wall, i.e. in the slope section where present movements were not anticipated originally. The crack gauges TM71 used here are 3D mechanical optical devices working on moiré principle, useful for indication of very slow movements on faults and cracks. Indication concerns relative displacements between two bodies of rock including angular deviations between the two rock faces.

The measurement indicated vertical trench sinking at a rate of up to 155 mm per first year (Suchý 2000). Later some variations developed and after five years of measurement it reads about 3 mm per 5 years, which can be considered the average sinking rate in both the trenches (Fig. 3). Such a finding is congruent with the relief evaluation made by Quaternary geology in the main scarp area, where sharp forms indicate Holocene phase of movements.

The trenches investigated by the gauges lie in a wide-open sinking zone under the main separation wall. Sinking deformation can be explained by the reduction of material washed out by underground water seeping toward spring areas found on both the sides of the landslide. The process completely overrides the slope movement. The trenches do not get wider in our days, they are rather stable, under horizontal pressure.

The pressure does not decrease in time, rather the opposite, as witnessed by low angular deviations and almost stable zero horizontal trends registered in the investigated trenches (Fig 4). This finding is in good

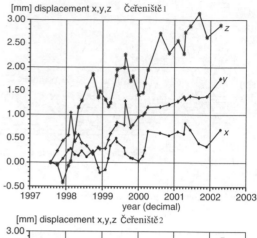

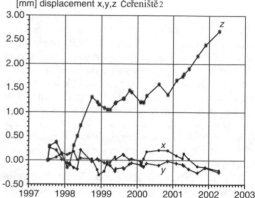

Figure 3. Displacements registered in the two investigated trenches of Čeřeniště Village landslide. Sinking of internal blocks in the trenches represented by z vertical component prevails while trench widths represented by x is rather small, constant at Č2 and tending to compression at Č1.

agreement with a general idea that debris and rock blocks has to become arching due to sinking and deprivation of support in the basement. In that horizontal compression must originate in the whole area of the upper separation scar, which is just opposite to general assumptions of slope movements.

One can make a deduction that the upper section of the slope is presently independent from lower section slope movements inhibited by the creek erosion. Therefore, the upper section should be evaluated independently. This interpretation is supported also by experiments with physical photoplastic models.

5 PHYSICAL MODELS

Engineering-geological mapping, surveying in longitudinal and transversal profiles, geological mapping

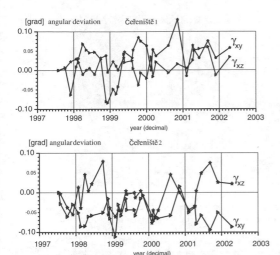

Figure 4. Very low angular deviations $\gamma \in (-0,1; +0,1)$ registered in the trenches indicate compression not allowing much movement but the sinking. Compression due to arching is an opposite finding to general expectations of tensile stresses in the upper part of the landslide body.

and Quarterly geological investigations together with results of geophysical surveying provided basic input data for physical models. These were oriented to throw light to the way how some basic phenomena of the deformation originate, notably the strange phenomenon of the platform, in the process of deep failure of the Čeřeniště Village volcanic complex.

A method of photoplastic modeling was employed (Košt'ák 1977, 1982). Basic materials of the models were prepared using agar-agar gels with special additives. Models in transparent vessels were subjected to the impacts of body weight under the main boundary condition of unilateral pressure release when elastoplastic deformations and failure could develop. The process was observed in polarised light when optically sensitive model materials can show stress increase and relaxation in the model. The method has been developed in Prague photoplastic laboratory of IRSM and it is found very instructive in solving the problems of geological structure formation (Košt'ák & Zeman 1982, 1990, Rybář et al. 1997, Košt'ák 1999).

The data we have obtained in our field investigations represent a description of the present late phase in which the complex occurred. Models cannot reverse the function of evolution in time. Therefore, a series of models were prepared on the basis of some introductory assumptions, the deformation process was then observed and results in individual phases of evolution compared with the present state in nature, and, particularly, with the monitoring results. Models inconsistent with observations were rejected.

The models were produced generally as horizontally stratified bodies of alternating layers with slippery intercalations, which was found as an appropriate simulation of the volcanic massif. The deformation was triggered by a gradual side release of the model body in the vessel, which simulated the process of valley deepening due to water erosion.

To form models four basic variants were assumed and then checked in the model series. The variants differed in material properties and arrangement in the beds representing the original form of the massif that was subjected to deformation. A lot of information was collected from the models. As said before, not all the models tended to create forms indicated in nature, which resulted in rejection of the particular models, and appropriate structural features with deformation processes produced as well. The landslide is too complex to allow for chances to produce one model, which would successfully simulate all the detailed features that were indicated in nature. However, two basic phenomena were successfully simulated in more detail using two separate models. The first more detailed, featured the formation of the main upper scar, the second more general, featured the process in which the strange platform had been produced.

The variant representing best the findings of the interdisciplinary field investigations can be then represented well with two model pictures. The first shows formation of a deep upper scar with rigid volcanic blocks sinking slowly back into the scar while block steps and trenches develop in the marginal zone (Fig. 5). This coincides with the morphology, and the process is well confirmed by monitoring as the main deformation process in the zone above the platform that is going on at present inhibited with outwash in the root area.

The second model comes to represent the process of the platform formation. It shows a special model representing deep failure in the complex stratified body of the volcanites due to slow deepening of the Labe River valley. A general process of shear does not produce only regular slip planes with downward slope orientation but also the antithetic ones. There was a phase observed when several alternating shear plane couples dominated and persistently formatted the model body. The process led to formation of a wedge-like structure in which wedges based on wide bottom tend to general disintegration while wedges up-side-down disintegrate in their narrowing lower noses only leaving the flat upper part almost stress free and uninterrupted, floating freely in the mass (Fig. 6). This is going to explain the presence of platforms in the landslides like that of the Čeřeniště Village. The up-side-down wedges separated due to antithetic shear planes in the slow process of sliding provide favourable conditions to find the platforms as relatively peaceful islands in the slope. The phenomenon is strengthened

Figure 5. Upper scar area of the Čeřeniště Village landslide produced in the first representative model. Morphology shows separation wall, deep scar, inclined sinking blocks, trenches in the marginal zone (right) and voids under the hill (left).

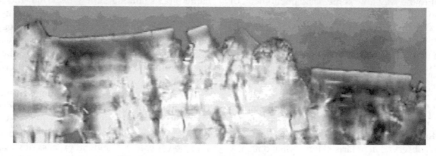

Figure 6. The second model of the Čeřeniště Village landslide type showing general development of the deformation in two subsequent advanced stages after platform formation. Two platforms originated as wedges separated from other mass by a couple of antithetical shear planes. The platform is almost stress free (dark colour) and may slip down without any disturbance (cf. two subsequent stages).

with the presence of stratified and largely heterogeneous rock body of the volcanites.

Summarizing, the authors have compiled field observation data with the models and put forth following characteristics of the process leading to landslide deformation of Čeřeniště Village (Suchý & Košt'ák 2000).

The deformed area can be described as a complex of slope deformations that developed in fractured lava flows alternating with volcanic sedimentary bodies.

Synvolcanic tectonics of N–S orientation resulted in breaking down the massif into quasihomogeneous units. As a whole, basalt flows are found rigid while volcanic tuffs are more plastic so that the whole structure is very sensitive to sliding.

Water erosion created valley slopes. Fast deepening of the erosion basis accommodating to the level of the Labe River resulted in internal pressure release in the massif. Destabilized rock units on the slopes started to move. Deformations reached deep zones of the massif. A deep upper scar separated the landslide body from the main separation wall. Individual lava flows broken tectonically into blocks were moving slowly down the slope. The basement zone of the blocks was fissured with many pseudocarstic voids and disintegrated rock materials. Deep-seated shear zones predisposed different development in the upper and lower sections of the slope with a specific structure in between. This structure is separated from the other body of the massif by shear zones where an antithetic shear zone provides conditions that the structure is preserved here from destruction, creating a relatively stable platform.

In the lower section close to the toe of the landslide all slide bodies become more and more disintegrated and transformed to soil materials. This lower section is then deformed by shallow slides along rotation-planar surfaces. In the upper section above the platform the main scar continues its development by underground erosion – suffusion, individual blocks being caught by the development in the upper scar. Steps and transversal trenches develop. Thus, the upper section takes a different character than the lower section under the platform.

As demonstrated by the experiments with photoplastic models, the deformation process leads to a strange formation of landslides with platforms. In such a way the unusual slope deformation found at Čeřeniště Village can be explained.

6 CONCLUSIONS

An interdisciplinary research oriented to the complex deformation of Čeřeniště Village landslide and to several other regional localities allowed to throw light upon some theoretical aspects connected with deep slope failure in the engineering-geological region of Bohemian Massif neovolcanites. Results can be partially applied even to the region of volcanites of Western Carpathians. Methodically, the approach integrated and verified several new field investigatory as well as laboratory procedures.

ACKNOWLEDGEMENTS

Authors acknowledge support granted by Czech Grant Agency, project Nos *205/97/0526, 205/00/0665* and also by the Ministry for Education, Youth and Physical Culture, the Czech Republic, project *COST OC 625.10.*

REFERENCES

Cajz, V. 2000. Geological setting of the Čeřeniště slope movement locality (in Czech). *Zpr. geol. Výzk. v r. 1999*: 177–179. Praha.

Cílek, V., Sýkorová, J., Melichárková, E. & Melka, K. 2000. The sedimentary intercalations the superficial volcanics of the Středohoří Complex and their influence on slope stability (in Czech). *Zpr. geol. Výzk. v r. 1999*: 183–186. Praha.

Fantucci, R., Rybář, J. & Vilímek, V. 2000. Dendrochronological analysis of Čeřeniště landslide (Czech Republic). – In Lollino, G. ed. *"Geological and geotechnical influences in the preservation of historical and cultural heritage"*: 493–500. Torino.

Hofrichterová, L., Müllerová, J. & Suchý, J. 2000. Geophysical survey at Čeřeniště Village locality (in Czech). *Sborník vědeckých prací VŠB-TU 2000 Ostrava* 46 (3): 73–81. Ostrava.

Košťák, B. 1977. Photoplastic slope deformation models. *Bull. IAEG* 16: 221–223, Krefeld.

Košťák, B. 1982. Evaluation of long-term geological mechanical processes with the use of photoplastic models (in Czech). *MS IRSM, Ac. Sci. Czech R.* 62 p. Prague.

Košťák, B. 1999. Model and field studies into the dynamics of block slope structure formation.– *Acta Montana IRSM AS CR (1999)*, A 14 (114): 125–140. Praha.

Košťák, B., Zeman, J. 1982. Tectonophysical model of deformations in a structure with an elastic competent layer (in Czech). *Čas. Min. Geol.* 27(1): 51–59. Prague.

Košťák, B., Zeman, J. 1990. Experimental structure induced by rheological deformation in folding. In *Sychenthavong, S.P.H.* ed. *"Crustall evolution and orogeny"*: 303–312. Oxford a. IBH Publ. Co., New Delhi, Bombay, Calcutta.

Nemčok, A., Pašek, J. & Rybář, J. 1972. Classification of landslides and other mass movements. *Rock. Mech.*, 4 (2): 71–78, Wien, New York.

Rybář, J., Košťák, B., Málek, J. (1997): Geomechanische Modelle zur Erforschung von Rutschungen. *Geotechnik* 20 (4): 266–272. Essen.

Rybář, J., Vilímek, V. & Cílek, V. 2000. Process analysis of deep slope failures in České Středohoří neovolcanites. *Acta Montana IRSM AS CR*, AB 8 (115): 39–46. Prague.

Rybář, J. & Suchý, J. 2000. The influence of climate on the České Středohoří Mts. slope deformations – data analysis since the 18th Century. *GeoLines* 11: 69–72. Prague.

Suchý, J. 2000. Slope deformations of the Labe river in the České Středohoří Mts. *Acta Montana IRSM AS CR*, AB 9 (122): 9–34. Prague.

Suchý, J. & Košťák, B. 2000. Physical modelling of the slope deformation Čeřeniště Village (in Czech). *MS Internal Report IRSM AS CR*, Prague (unpubl.).

Geotechnical Measurements and Modelling, Natau, Fecker & Pimentel (eds)
© 2003 Swets & Zeitlinger, Lisse, ISBN 90 5809 603 3

Time settlement behaviour of a ship lock with a flat foundation

T. Schanz
Laboratory of Soil Mechanics, Bauhaus-Universität Weimar, Germany

D. Alberts
Bundesanstalt für Wasserbau, Hamburg, Germany

ABSTRACT: "Traditional" (that means according to the regulations of the national German codes) and numerical methods for calculating settlements are compared for the case of the *Rothensee Lock* (located at the Magdeburg waterway intersection). Extensometer measurements were used to calibrate the numerical model. Construction began in the summer of 1997, and the lock was opened for shipping four years later.

1 INTRODUCTION

The new Rothensee Lock is located at the Magdeburg waterway intersection and provides an elevation descent of approximately 16 m from the "Mittellandkanal" to the Elbe River and the harbor of Magdeburg. The lock has a flat foundation which is situated on moderately overconsolidated series of changing cohesive soils. A reliable settlement prognosis was necessary for calculating the static's and finally designing the lock.

Therefore the following aspects (boundary conditions) were taken into account:

- Load and deformation conditions at the chamber walls and at the stiff foundation plate depend on the water level in the lock. The momentum acting on the foundation plate depends upon the sub base conditions which in turn depend on the relative rigidity between plate and subsoil and the degree of overconsolidation of the subsoil.
- Depending on the rigidity/strength of the construction-foundation soil system, the chambers and lock heads experience settlement differences and thus double or triple axial tilting which must be compensated for by the system of joints (Figure 2). Otherwise, it must be feared that future system breakdowns und reduced levels of utilization might occur. All calculations were based on characteristic (normalized with respect to the stress state stiffness determined from odometer tests.

Three six fold extensometers were installed beneath the lock since construction began. In order to interpret their results, it was necessary to ascertain the consolidation and stress conditions in the layers causing settlement as a function of specific loading situations. In a first approach the calculations were based on one dimensional consolidation theory. Subsequently, the system's behavior was considered to be modelled two dimensional by applying the FEM. On the basis of the interpreted measurements, the significant relevant parameters (stiffness and permeability) could be inversely determined. Sensitivity analyses were made with respect to the strengths and permeability's (anisotropy) of the various layers.

Figure 1. Lock Rothensee, situation, aerial view.

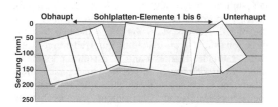

Figure 2. Relative deformation between blocks.

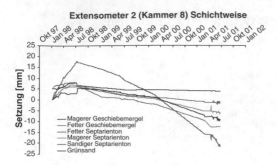

Figure 3. Time series of extensiometer measurements in the section analyzed (per layer and total settlement).

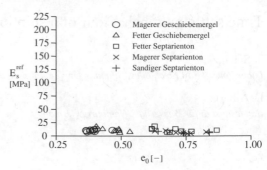

Figure 4. Normalized coefficients of compressibility for initial loading (reference stress 100 kPa).

2 PROCEDURE

Extensometer measurements were available at various locations of the construction site since the early beginning of the construction. In the following, only the measurements for the middle section of the construction will be evaluated (see Figure 3). They include all settlement relevant foundation soil layers. In general engineering practice the question of how good (realistic?) the stiffness parameters are that were determined in the laboratory is of paramount importance for the following settlement calculations. In general, it is common practice to compliment laboratory tests with field experiments. In the present case, the lock construction itself serves as a large-scale loading experiment/test, due to its size and geometry and the relevant loads. In addition of having knowledge of the foundation soil characteristics, it is extremely important to record the various construction stages, so that the time-settlement may be calculated. This includes not only the construction stages of the sections where the extensometers are installed, but also taking into account the loads spreading to surrounding areas, to adjacent parts of the construction, to fill areas, etc.

In the design and construction structural analyses, the soil stiffness were determined as safe averages. The settlement calculations according to DIN 4019, i.e., with respect to an elastic half space, used the average stress dependent un-/reloading stiffness. In this paper, we used the concept of "normalized" or reference stiffness (original concept by OHDE in the thirties, unfortunately often referred to JANBU's famous paper from 1963). Here the rigidity is determined from the effective stress state and the respective loading direction (initial or un-/reloading), and the overconsolidation ratio (see Figures 4 and 5). It is thus possible to qualitatively determine, in addition to the excavation dependent unloading and reloading, the influence of the overconsolidation. Characteristic values of the reference stiffness were ascertained. Parameter intervals resulted from statistical analysis which are directly

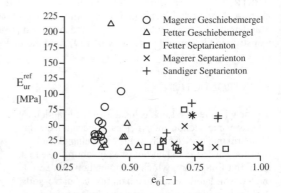

Figure 5. Normalized coefficients of compressibility for un-/reloading (reference stress 100 kPa).

related to the size of the assumed interval of confidence. The at first glance seemingly strongly spreading values do indeed conform to a normal distribution. The established confidence interval allows for varying the soil coefficients in a realistic stochastic/statistical framework without the influence of randomly increased or decreased parameters.

In addition, approximately 40 CPT soundings were evaluated (see Figure 6). Their results make it possible, using an empirical correlation of the two measurement (toe resistance – oedometer stiffness) methods, to provide an extensive description of the foundation soil properties which goes beyond the basis points of the oedometer tests.

It is possible to determine final settlements with all of the calculation models used. The 1D calculations provide the most flexible possibilities for time-settlements. The influences of the construction stages can only be approximated with the 2D-FEM due to the plane strain type of the approach where all stages of construction become active at the same time.

3 COMPARISON OF THE DETERMINED RIGIDITIES

The 1D recomputation of the extensometer measurements (inverse analysis) showed that the initial loading stiffness lies in the range of the maxima of the characteristic values. The corresponding unloading and reloading stiffness are more of the order of the minimum values.

4 SETTLEMENT CALCULATIONS

The construction of the complete lock was carried out in a large number of different phases ("stages of construction"). In the first phase, a 0,60 m thick diaphragm wall was constructed subsequent to the necessary excavation work, which enclosed the lock's entire construction pit. Construction of the diaphragm wall had no direct influence on the settlement of the structure. Lowering of the groundwater was started two weeks after the diaphragm wall had been completed.

After some delays, the construction pit was excavated. Installation of the three six fold extensometers was completed in 1997, when two thirds of the construction pit had been excavated. Thus the measurements only cover the last third of the excavation but the complete following construction of the concrete structure.

Subsequently the pouring of concrete for the lock's foundation and chamber walls commenced. The individual chamber blocks were constructed using the so called step by step method ("Pilgerschrittverfahren") (for the influence of adjacent chambers, see Figure 8, use of characteristic points for settlement calculations applying stiff loads, see Figure 7).

The concrete was poured in 13 steps. Shortly after completing the section foundation up to a level of 35.85 m NN, the construction pit area was sealed and refilled to base level of the chambers. After all lock chambers were completed, the sides of the pit were filled and sloped to the chambers' upper edge. Summarizing the performed calculations pertain to three distinct phases: unloading due to the excavation of the building pit, loading due to the erection of the lock structure, and loading due to the back-filling of the chamber walls.

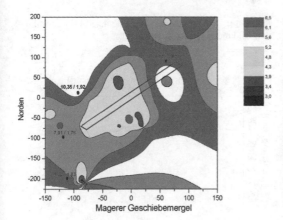

Figure 6. Isolines of resistance (sandy boulder marl).

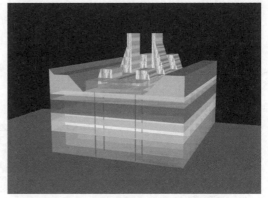

Figure 7. Concept of characteristic points to determine settlements of "stiff" sections (1D consolidation).

Table 1. Soil stiffness from inverse 1D consolidation analysis (reference values for 100 kPa) to capture field measurements, all units in MPa (first interval: initial loading, second interval unloading/reloading).

	1. Design ("safe mean values")	2. Design charact values (EC 7)	Inverse analysis (1D consolidation)
Magerer Geschiebemergel	40,0	9,5–11,2 34,5–60,5	11,2 54,2
Fetter Geschiebemergel	18,0	8,4–12,4 15,6–34,9	12,4 15,6
Fetter Septarienton	14,0	9,8–14,7 9,3–31,3	9,8 10,9
Magerer Septarienton	29,0	5,9–8,5 15,8–40,3	8,5 15,8
Sandiger Septarienton	80,0	5,5–8,0 49,5–77,6	7,9 49,5
Grünsand	120,0	–/–	60,0/120,0

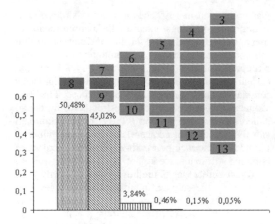

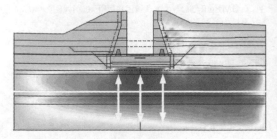

Figure 10   Excess pore water pressures after completion of the lock from 2D FEM consolidation analysis.

Figure 8. Influence of adjacent chambers section on settlements at specific location of the extensiometer (1D consolidation).

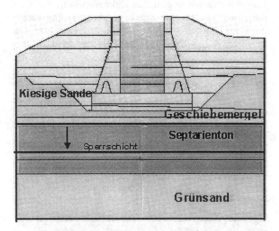

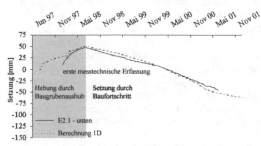

Figure 11.   Comparison of the calculated (1D consolidation) and measured total settlement.

Figure 9   Drainage boundary conditions for different intermediate layers from 2D FEM calculations.

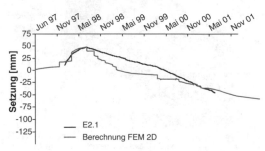

Figure 12.   Comparison of measured and calculated time settlements from 2D FEM consolidation analysis.

4.1   1D consolidation

A modified consolidation theory was used in this paper where a single loading step is divided into a number of sub-increments. Within a sub-increment, the layer thickness, the porosity as well as the soil stiffness are updated according to the former changes of the state parameters from the last increment. A significant limitation of this procedure is the absence of pore water pressure measurements. An improvement of this situation is obtained by performing a running adjustment to the latest measurements. In the framework of the outlined procedure, additional assumptions must be made with respect to drainage boundary conditions. This question could be resolved with the 2D-FEM calculations (see Figures 9 and 10).

There is very good agreement between the calculations and the measurements for the total settlement of the structure as well as for the settlement of the individual layers (see Figure 11).

4.2   2D-FEM calculations

In total 27 construction phases were modelled using the PLAXIS FE-code. All phases were effective for the total length of the structure at the same moment.

There is very good agreement for the excavation period and the settlement to when the lock became operational (see Figure 12).

As a result of assuming the deformation conditions to be of plane strain type, the time-settlements, however, appear too early in the calculations.

5 3D-FEM CALCULATIONS

Supplementary to the described investigations up to now, 3D-FEM calculations were performed (see Figure 13). Doing so it was our goal to determine the order of magnitude of the tilting of individual chambers. The degree of tilting is the pertinent criterion for the design of the joints.

For these calculations, the lock and its back-fill were modelled using 3D volume elements (lock: ideally elastic, soil: isotropic hardening plasticity model). The soil beneath the foundation was modelled as a non-linear

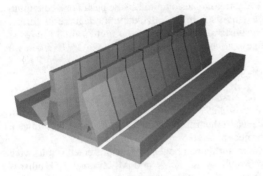

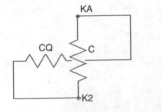

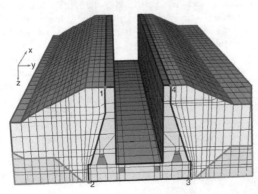

Figure 13. 3D-FEM calculations – above discretization, middle: contact modelling, below: joint layout of the construction.

elastic half space. Its constitutive characteristics resulted from the spatial distribution of the CPT sounding based soil coefficients. Contact elements were placed between the lock and the back-fill and the foundations subsoil. An additional earth pressure resulting from soil compaction behind the walls was not taken into consideration. Two types of joints can be found between the individual chambers: continuous joints separating the chamber walls as well as the foundation plate, three additional joints are situated within each chamber which only separate the chamber walls. Both types of joints were described using contact elements applying a ideal elastic – ideal plastic constitutive formulation. Due to the calculation code used only final settlements could be ascertained. A verification of the numerical model using the current measurements is thus not possible. The prognosticated final settlements of 184 mm lie in the same order of magnitude as the results of the 1D and 2D calculations. Deformations for the maximum loading of the joints were on the order of 15 mm (extensions of the gap) and 6 mm (reduction of the gap). The chosen joint design had no problem tolerating these deformations.

6 SUMMARY AND OUTLOOK

The following orders of magnitude for time and final settlements result from the various calculation methods in connection with the corresponding groups of material parameters (see Table 2 with *experiments*: characteristic values based on the design planning, *measurements*: recalculations using extensometer measurements for inverse analysis).

The following conclusions can be drawn from the above results:

- All methods show that the final settlement will be attained in approximately 30 to 60 years.

Table 2. Comparison of the determined final settlements and duration of primary consolidation.

Source of parameters	Final settlement/ duration	Settlement at start of operation 47 mm
BAW 1st design	150 mm	–
Final design	161 mm 30–60 years	–
1D consolidation (Experiments/ measurements)	147 mm/153 mm min.73–max.176 mm 25 years	13[1] mm/49[2] mm
FEM 2D (Experiments/ measurements)	173 mm/181 mm 30 years	38[1] mm/45[1] mm

[1] Original permeabilities, [2] optimised permeabilities.

- When using the 1D analysis, the permeability must be modified with respect to the intervals of the characteristic values (due to the in situ 2D/3D nature of the consolidation process, i.e., resulting in shorter consolidation times). They do, however, lie within the bandwidth of the characteristic values.
- The final settlement calculated during the design period of the construction are in good agreement with the values of these investigations.
- One must take into account, however, that in the design process, we are dealing with "probable" settlement. The calculated settlement was reduced by a factor based on experience of 0.75.
- Our calculations showed results of a comparable order of magnitude without such a reduction from experience and thus preclude the use of a empirical reduction factor.

The following conclusions and recommendations can be stated for further work in this and future projects:

- It was shown that both laboratory and field experiments are necessary for a successful analysis applying numerical methods.
- The proposed approach of combining results of oedometer tests, CPT soundings, and extensometer measurements is highly recommended.
- Additional measurements of the (excess) pore water pressure before, during and after construction of the lock are indispensable for an unambiguous interpretation of the above mentioned measurements.

- During the conception of the investigation program, it must be verified whether or not the soil permeability is expected to show an anisotropic nature. Accordingly tests have to be planned.
- It is a helpful and necessary idea to explore the complete region lying reasonable outside of where the future structure will be located. Having a number of reference points (where several independent methods of soil investigation are applied, both field and laboratory tests) for the various investigations is a necessary prerequisite for the extrapolation of the various parameters.
- The position of cross-sections where in situ measurement are performed must be geometrically simple as possible.
- Relevant construction phases must be documented.
- The installation of field sensors prior to construction begin is highly advisable.
- Taking into account the cyclic behavior of the foundation soil in the calculation models merits further substantial consideration.

ACKNOWLEDGEMENTS

The first author wants to thank "Bundesanstalt für Wasserbau" (Hamburg). The continuous support with data and the fruitful discussions during the referenced project phase is very much appreciated.

Important components of the presented results were provided by K. Kratz and M. Zimmerer (Bauhaus-Universität Weimar).

Increasing safety of unstable slopes by unconventional pore pressure release technique

R. Schulze & H.J. Köhler
Federal Waterways Engineering and Research Institute (BAW), Karlsruhe, Germany

ABSTRACT: Transient pore pressure distributions may be estimated by regarding submerged soil as a three-phase medium (gas, water and solids). Thus new approaches to release pore water pressure have been developed in order to stabilise endangered slopes in low permeable soil. The applied concept of using reversely inclined bore holes to release pore pressure in the vicinity of potential shear zones is described. The proposed way of pore pressure release in order to increase safety of unstable slopes is widely applicable. Examples of applications to increase safety include cut slopes as well as river dikes. Additionally results of calculations and field measurements are provided and discussed.

1 INTRODUCTION

In this paper pore water pressure release techniques are described, which may be used to increase slope stability in low permeable soils. Transient pore pressure distributions are initiated by applying such concepts. This paper will focus on the practical estimation of the increase of slope stability.

First a cut slope will be discussed. Pore pressure measurements have been undertaken to observe the effectiveness of bore holes allowing dissipation of pore pressure. Results of the measurements are presented providing input for safety assessment calculations.

The second example describes pore pressure reduction in order to increase slope stability of existing river dikes, used for flood protection. Improvements of existing structures are often requested due to increasing probability of higher flood waves than originally anticipated.

2 PORE PRESSURE RELEASE TECHNIQUE

It is widely assumed that drainage pipes need to withdraw water visibly in order to be effective. Thus apparently dry drainage pipes are often viewed falsely as being ineffective. Especially in low permeable soil even the removal of small quantities of water (e.g. evaporation into the pipe) will reduce pore water pressure surrounding the drainage pipe.

Pore pressure reduction has been applied in the past at many occasions usually considering steady state conditions. To increase slope stability the following aspects are applied in this paper.

Pore pressure reduction may occur regardless of the inclination of the bore hole or the water level inside the bore hole. Furthermore a new approach and main target is the estimation of the transient development of pore pressures covering the gap between the steady state conditions of the initial and final state. The method which is based on the application of the three-phase-model is described in Schulze et al. (2003). Even reversely inclined bore holes or bore holes entirely filled with water may be able to decrease pore pressures.

Removing water from the bore holes increases effectiveness, because barometric pressure is allowed to be transferred directly into the soil. A successful application of this concept depends on the original magnitude of the pore pressures to be reduced.

Looking at a bore hole filled entirely with water the piezometric level all over the bore hole is constant and solely determined by the geodetic level of the bore hole mouth (flow velocity in the bore hole assumed to be negligible). In a bore hole which is filled with air (water removed), the local piezometric level (hydraulic boundary condition along the bore-hole) will be the local geodetic level.

In many practical cases these facts allow a much more effective placement of the drainage pipes into the shear zone. In accordance with Terzaghi's principle of effective stress the stability of the slope will be increased directly as pore water pressure is reduced.

3 STABILISING AN ENDANGERED CUT

3.1 General information and soil properties

The instable cut is about 20 m deep and located at a navigable canal nearby Lühnde (Germany). Indications of slope movements led to the current geotechnical investigations, which were initiated in 1995, resulting out of the intention to deepen the canal to allow larger ship vessels to pass. Pore water pressures, slope movements and rainfall are observed by automated data acquisition systems.

Falling barometric pressure has been identified to be a decisive factor being able to trigger slope movements (Schulze et al. 1999). Using this factor, forecasts as to when movements of the slope will be accelerating have repeatedly been successful in the field. Further details referring to that specific aspect, as well as soil properties, instrumentation and data, have been published (Köhler et al. 1999, 2000, 2002a).

The cut is located in stiff, fissured Lower Lias clay, $w_L = 0.58$, $w_P = 0.22$, clay 40%, silt 60%. Narrow limestone bands are embedded occasionally. The permeability is considered to be about 10^{-10} to 10^{-11} m/s, although fissures and limestone bands increase the large scale permeability of the soil.

3.2 Field test and pore pressure measurements

A field test has been designed and performed to verify the anticipated effectiveness of the proposed method. The test field has been located at a section that had already been used for several years to monitor slope movements.

In a test field located within the endangered slope a pattern of three differently inclined pore pressure reduction borings was selected. Figure 1 shows sections and ground plan. The geometry of three inclined bore holes depicted in the general section is repeated at a distance of about 6 m and has been chosen in order to influence significant parts of the shear zone.

To enhance an existing monitoring scheme additional pore pressure sensors have been positioned in between the bore holes. The location of seven sensors (including sensor W30) is indicated in Figure 1. It was intended to place the sensors at the furthest distance from the bore holes in order to measure at locations least influenced by the bore holes.

Sensors measuring absolute pressures were installed. As described in previous papers a packer system was used to measure pressures with an extremely small time lag. The pressure sensors have been installed months ahead of the installation of the bore holes to allow general equilibration of the pressures.

The bore holes (diameter 178 mm) were drilled and equipped with drainage pipes according to DIN 4262-1 (2001): DN 100, circular type R2, in general totally perforated pipes were installed. Prior to installation the drainage pipes were wrapped with a geosynthetic filter material.

During drilling most drains appeared to be dry. After drilling about 40 m of bore hole D1, water was reported discharging from the bore hole at a rate of 20–30 l/h. When D4 was being drilled at a depth of about 20 m, water appeared here as well at about the same rate and discharge from D1 diminished. This observation may be explained with water bearing strata (limestone bands) which are occasionally embedded

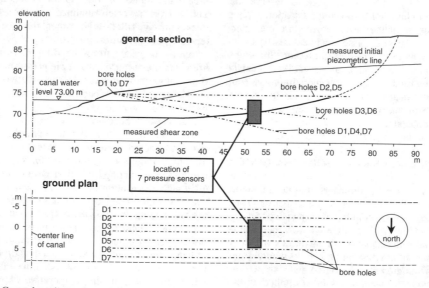

Figure 1. General section and ground plan of test field.

in the clay formation. Another observation which supports this explanation is a highly variable pore pressure measured at pressure sensor W30. Presumably this sensor is located in or in close vicinity of a water bearing limestone band.

Occasionally the drains have been visited to check for water discharge. According to those observations water flow varies with the seasons. At D2 and D6 water discharge of about 10 l/h (each) has been observed in mid-December 2002. Most other bore holes rarely discharge water visibly, but all have been found filled with water to high levels.

In Figure 2 the development of the measured pore water pressures with time is presented. An average of the piezometric level of all 7 pore pressure sensors is shown. Because of extreme fluctuations of the sensor W30, a second line is presented showing average values omitting W30. Measurements were taken every 30 minutes by an automated data acquisition system. Those original readings have been modified by subtracting the barometric pressure which was acting at the time of the measurement, assuming 100% of the barometric pressure having actually reached the sensor. As described in a previous paper (Köhler et al. 1999) fluctuating barometric pressure may not reach beyond certain depth levels. This is due to delayed pressure spreading which depends on the velocity of barometric pressure changes and the soil permeability. Due to this effect the piezometric level depicted in Figure 2 is accurate within about ±0.20 m, which is

perfectly acceptable regarding the subject of this paper. Small data scatter (e.g. in February 2002) may be explained by this effect.

Originally piezometric levels of about 79.5 m NN have been measured. The pore pressures had been fairly steady before drilling started and relate to the initial piezometric line (see Fig. 1). Drilling of the bore holes and installation of the drainage pipes took place between December 17th, 2001 and January 9th, 2002. Work had been interrupted between Christmas and New Years Day. During drilling fluctuating pore pressures have been observed, which may be mainly a result of volume changes due to the drilling process or the interconnection of water bearing limestone bands. While drilling a considerable drop in pore pressure was observed.

Following the installation period, Figure 2 shows a noteworthy phase of relatively continuous decline of pressures until mid-October. Increases have been recorded in April 2002, mid-June, mid-July and in August which corresponds to extremely wet weather causing flooding elsewhere in Germany. Evidently sensor W30, which may be located in or in the vicinity of a water bearing limestone band recorded major pressure differences due to extensive rain, while the remaining sensors are influenced much less by such events. Starting in late October and continuing into spring 2003 the pressure sensors show an increase averaging about 5 kPa which corresponds to wet winter conditions.

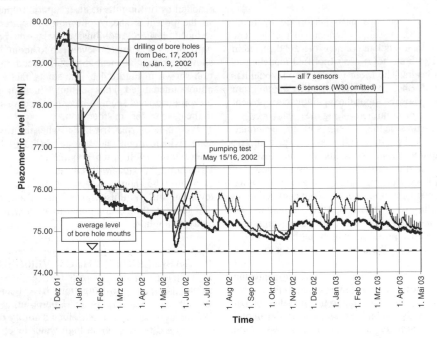

Figure 2. Results of pore pressure measurements.

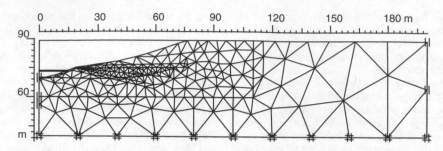

Figure 3.   Finite element model of the endangered cut slope.

In mid-May 2002 a pumping test was performed. Water was removed from the bore holes repeatedly and the velocity of the rising water levels in the bore holes was measured. These measurements allow conclusions regarding hydraulic properties of the water bearing limestone bands and possible hydraulic connections to the canal water.

During all times the piezometric line above the drains has not yet changed. This has been measured with additional sensors which have not been covered in this paper. The piezometric line is expected to be extremely lowered approaching final steady state conditions. The measurements continue to observe the size of seasonal deviations of the pore pressures which have to be taken into account concerning slope stability.

3.3   Calculations

Safety assessments have been performed by using PLAXIS (1998), a finite element (FE) code. The geometry of the model and dimensions are depicted in Figure 3. The bore holes are simplified modelled as continuous layers. The Mohr-Coulomb soil model has been selected. Shear parameters used by the program are constant. Thus the input shear parameters represent a value averaging conditions along the shear zone.

Pore pressures acting in the soil are generated depending on the position of the assigned piezometric line. Compared to pressure distributions resulting from flow calculations this procedure remains on the safe side concerning safety assessments. Above the piezometric line pore pressures are assumed to be zero (equalling barometric pressure). Remaining on the safe side, negative pore pressures (pressures below barometric pressure) which might exist in the field (Köhler et al. 2002a) are not considered in the calculations.

In a first step the model was validated by analysing a situation before the bore holes existed. Pore pressures were assigned according to the measured piezometric line. In Figure 4 a contour lines of incremental displacements are shown, giving the geometry of the potential shear zone. Comparing the geometry of the calculated potential shear zone with the position of the measured shear zone (dotted line) shows good agreement. The safety factor $f_s \sim 1.0$ represents the unstable initial situation. Attempts to deepen the canal would result in global failure.

To calculate the safety for a transient state, pore pressures acting in the soil have been modified. In a region encircled by a dotted line pore pressures have been assigned corresponding to a piezometric level of 75.0 m NN (see Fig. 4 b). This specific level was reached in late August 2002 (see Fig. 2). The region is estimated to reach 2 m beyond the limits of the outer bore holes (see Fig. 4 b). This estimation is based on observations of pore pressures sensors not covered by this paper. It remains on the safe side since pore pressure reduction will spread with time.

At transient state an increased safety factor will be reached by drilling the drains deeper into the ground. In the potential shear zone pore pressures will be reduced earlier. Additional calculations performed with higher piezometric levels (up to about 77 m NN) show the potential shear zone still forced into a position shown in Figure 4 b. The safety factor remains almost unchanged. Thus seasonal fluctuations of the piezometric level seem to have no decisive influence in the case under consideration.

When the piezometric line reaches the final steady state (which might take a long period of time) the safety factor will reach about 1.18 (see Fig. 4 c). Deepening the canal will reduce this factor of safety by about 0.05. Thus safety will remain at a very low level but will minimise movements of the slopes. To add security the observational method will continuously be applied.

4   STABILISING EXISTING RIVER DIKES

Safety assessment investigations have been carried out for case study reasons (Köhler et al. 2002a, 2002b) in order to compare slope stability of a river dike caused by seepage at high water level and in a worst case situation of a high crest water level.

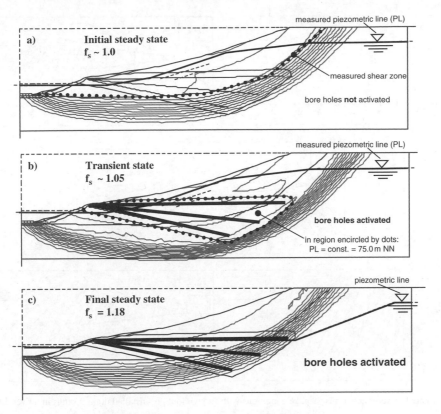

Figure 4.   Contour lines of incremental displacements resulting from FE calculations.

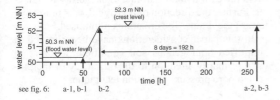

see fig. 6:   a-1, b-1   b-2   a-2, b-3

Figure 5.   Water level history of a flood event acting on a river dike, modelling a worst case scenario (with reference to Fig. 6).

Figure 5 shows the water level history, which has been assumed to act on the river dike, rising from high water level at 50.3 m NN to high crest water level at 52.3 m NN within 20 h, staying at this level for 8 days.

In order to model a worst case scenario, in which the highest possible water level may reach the elevation of the dam crest level of 52.3 m NN, the crest water level is assumed to remain unchanged at this level within the following next eight days. Such a loading situation may usually not be expected at a natural flooding event.

Even under such extreme loading conditions a dam break ought to be prevented.

Figure 6 describes different seepage situations with and without the installation of reversly inclined drain borings, showing the acting pore pressure distribution at initial (Fig. 6 a-1 and b-1) and final (Fig. 6 a-2 and b-3) steady state seepage conditions.

Figure 6 b-2 describes a pore pressure condition at transient state at the time step t = 70 h, after rising the water level from 50.3 m NN (see Fig. 6 a-1) to 52.3 m NN (see Fig. 6 b-2) taking place within 20 h. During the phase of rising water level, reversely inclined drain borings have been activated.

As depicted in Figure 6 a-1 the factor of safety f_s has remarkably dropped down to $f_s \geqslant 1.1$ already at flood situation with a high water level at 50.3 m NN. With a further increase in water level up to high crest situation at 52.3 m NN bank stability decreases rapidly (see Fig. 6 a-2). The calculated safety factor f_s against slope failure reaches values below 1.0 as soon as water infiltration may saturate the whole dam body up to a certain level above the seepage condition shown in Figure 6 a-1. This situation is unsafe leading to failure. A lot of emergency work needs to be

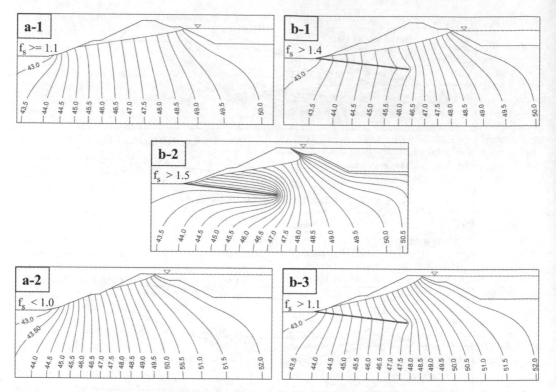

Figure 6. Pore pressures at various seepage conditions with and without pre-installed drain borings in a river dike.

undertaken in such flood situations to ensure dam stability under such conditions. Usually the top load of the dam embankment surface is increased by e.g. placing sand bags or additional filter layers on to the downstream side of the saturating dam body. In order to avoid catastrophic events during high crest situations, pre-installed drain borings may be an adequate improvement of existing dikes. Bank stability at the required level of service can be ensured.

As shown in Figures 6 b-1 and 6 b-3, the applied pore pressure release technique by using pre-installed drain borings would provide a significant increase of stability even at steady state condition of a long lasting high flood situation. The drain borings may be entirely filled by sand or gravel fulfilling the requested filtration criteria to ensure long term stability and service. It is recommended to select drain material in the borings with a permeability k which should be about 100 times larger than the permeability of the material to be drained.

In comparison to the occurring discharge at flood situation without the additional drain borings, estimations of the water discharged by the drains, result in an increase of no more than 50%. Usually such an increase of discharge may easily be handled, even remaining unnoticed.

If the drain borings are installed ahead of a flood event the dam will not saturate to dangerous high levels. Further the direction of seepage flow is redirected toward the drains. Toe erosion is minimised.

In this paper considerations have been limited to simplified homogenous soil conditions in order to draw primarily attention to the efficiency of the presented pore pressure reduction technique.

5 CONCLUSIONS

Possibilities of selecting reversely inclined drain borings are indicated to reduce pore pressures in the region of the potential shear zone. Application of this technique leads to increased slope stability. Submerged soil under water is regarded as a three-phase medium (gas, water and solids). Based on this concept and on the results of field measurements numerical investigations have been performed for two practical examples regarding transient pore pressure conditions.

Application of drains in low permeable soil are encouraged. Further aspects of technological development need to be considered, e.g.:

- The selection of the type of drain pipes needs to be evaluated. If the bore holes are filled with appropriate filter material (sand, gravel) no pipes are needed at all.

- Continuous removal of water collected in the drains improves effectiveness. With the application of vacuum safety gains are even larger.
- Water discharged from the drains needs to be led away properly, winter conditions should be taken into consideration (freezing etc.).
- Advanced drilling methods may enable to apply curved bore holes, allowing to place drains even more effectively along the potential shear zone.

REFERENCES

Köhler, H.-J., R. Schulze & I. Feddersen 1999. Influence of barometric pressure changes on slope stability – measurements and geotechnical interpretations. *Proc. 5th International Symposium on Field Measurements in Geomechanics*, Singapore, 1–3 December 1999, 381–386. Rotterdam: Balkema.

Köhler, H.-J. & R. Schulze 2000. Landslides Triggered in Clayey Soils – Geotechnical Measurements and Calculations. In Bromhead et al. (eds.*), Proc. of 8th International Symposium on Landslides*, Cardiff (Wales), UK, 26–30 June 2000, 837–842. London: Th. Telford.

Köhler, H.-J., R. Schulze & K. Asami 2002a. Protection measures in order to increase safety of unstable clay slopes by unconventional pore pressure release techniques. In Rybář et al. (eds.), *Landslides – Proc. 1st European Conference on Landslides*, Prague, 24–26 June 2002, 597–601. Lisse: Balkema.

Köhler, H.-J., R. Schulze & K. Asami 2002b. Flood crest situation and its influence on bank stability and protection measures. *MITCH Mitigation of floods, droughts and landslides risks, WORKSHOP 3*, Potsdam, 24–26 November 2002, http://www.mitch-ec.net/workshop3/ Presentations/ powerpoint_kohler.pdf.

PLAXIS (1998), Professional Version 7.2.

Schulze, R. & H.-J. Köhler 1999. Landslides in overconsolidated clay – geotechnical measurements and calculations. In F. B. J. Barends et al. (eds.), *Geotechnical Engineering for Transportation Infrastructure – Proc. 12th European Conference on Soil Mechanics*, Amsterdam, 7–10 June 1999, 601–608. Rotterdam: Balkema.

Schulze, R. & H.-J. Köhler 2003. Stabilisation of endangered clay slopes by unconventional pore pressure release technique. *Proc. 5th International Symposium on Field Measurements in Geomechanics,* Oslo, 15–18 September 2003, Lisse: Balkema.

Risk assessment and management

Geotechnical Measurements and Modelling, Natau, Fecker & Pimentel (eds)
© 2003 Swets & Zeitlinger, Lisse, ISBN 90 5809 603 3

Numerical simulation of sinkhole formation using the UDEC computer code: influence of horizontal stress

A. Abbass Fayad & M. Al Heib
INERIS, Ecole des Mines de Nancy, Parc de Saurupt, Nancy-Cedex, France

Ch. Didier
INERIS, Verneuil en Halatte, France

Th. Verdel
LAEGO, Ecole des Mines de Nancy, Parc de Saurupt, Nancy-Cedex, France

ABSTRACT: Sinkhole can be responsible for varying degrees of damage depending on its dimensions, the type of land in which it occurs and ground occupation. Sinkhole is often predicted on the basis of current practice and experience feedback. We have sought to simulate the rupture mechanism and predict the form of the shape-bell (arch), which results from its propagation to the surface. This simulation, carried out using the distinct elements method (UDEC), is supported by observations. It has been applied to a real case, using data from the Paris region and is based on a methodology of progressive generation of the fractures in the cap rock strata. To help understand this complex phenomenon, we have presented the results obtained from study of the sensitivity of sinkhole formation to variations in horizontal stress. This study confirms that the methodology proposed is a valid method for predicting the formation of sinkhole.

1 INTRODUCTION

Sinkhole is a progressive phenomenon that starts when the roof of an underground cavity gradually collapses to form a cone of fallen rocks that invades the initial void. When the bell-shape (arch) which progressively forms on the roof of the cavity comes to the surface, it gives rise to what is called sinkhole as illustrated in figure 1; in the final stage, the bell-shape generally has the following characteristics: height (h), angle at the base (α) and a base diameter (D), while those of sinkhole are: diameter at the surface (ϕ) and depth (P) (Vachat, 1982; Tincelin and Sinou, 1978; Tritsch, 1995).

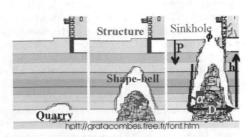

Figure 1. Development and characteristics of sinkhole.

The formation of sinkhole is generally predicted on the basis of empirical or analytical formulae (Vachat, 1982; Mandel, 1959; Tincelin and Sinou, 1962; Tritsch, 1987; Didier, 1997). Although they are pragmatic, these approaches are not accurate enough since they do not take sufficient account of the geotechnical environment of the land.

We set ourselves the goal of improving the prediction of sinkhole formation and its consequences on the surface by using modern methods to determine the probable shape of sinkhole and specify the influence of horizontal stress, which remains a very difficult parameter to assess, on the behavior of the ground. The tool chosen was distinct element numerical simulation (UDEC (ITASCA, 1997)), which was used to simulate the formation of sinkhole (A. Fayad et al., 2002; Thoraval, 2000).

2 CHOICE OF THE COMPUTER CODE AND METHODOLOGY

The formation of sinkhole is often associated with a stratified, fractured roof. These conditions led us to choose UDEC. UDEC treats a discontinuous area as an assemblage of blocks separated by joints. The resilient behavior of the joints is determined by the normal and shear stiffness and their entry into plasticity is defined by a given rupture criterion. The blocks can likewise be assimilated to deformable elastoplastic blocks. Thus, during the simulation of the formation of a cavity, so-called plasticity points can appear in the strata. Since UDEC cannot generate new fractures in the rupture zones, we have developed a methodology designed to change plasticity zones into induced fractures. With this method, we can simulate the fall of the roof of a cavity and, if applicable the formation of sinkhole.

When plastic points develop all along the cross-section of a stratum, we can consider that a fracture has been created. (A. Fayad et al., 2002). The opposite case, where the plastic points do not completely cross the section, is referred to as partial rupture (partial fracturing). In this case, we consider that the stratum in question keeps its integrity (intact block). The proposed approach is iterative. The first stratum of the roof is dealt with initially. If there is a fracture, the simulation is repeated with the presence of these fractures. Then, if there has been a fall of blocks, the 2nd stratum of the roof is analyzed in the same way, and so on. Depending on the geotechnical conditions, the procedure leads to the formation of sinkhole or a state of equilibrium.

The behavior chosen for the different strata of the massif is elastoplastic behavior with negative hardening (Hoëg and Prévost law (1975)). Behavior of the joints obeys the elastoplastic law established by Barton et al., (1985) with variable spring depending on the normal stress at the joint [this type of behavior is better suited to rocky joints (see Barton et al., 1985)].

3 CASE ANALYSIS

3.1 Description of the site and properties

The case chosen is that of sinkhole occurring on the surface of an old coarse limestone quarry in the Paris Basin (Vachat, 1982). The procedure used is specific. The example chosen corresponds to a quarry 8 m wide (l_0), 2 m high (w) and 8 m cap rock (H), giving a ratio H/w = 4. This is therefore, according to the geological cross-section (fig. 2), cap rock formed of nine stratified non-fissured layers. These stratified layers, of varying thicknesses, are formed of limestone, marl and clay sand. The geometrical and geo-mechanical properties of these materials and the interface joints are taken from a bibliography on the Paris Basin i.e. (Filliat, 1984; Fine, 1993) and given by A. Fayad et al., (2002).

Figure 2. Geometrical properties of the real case studied.

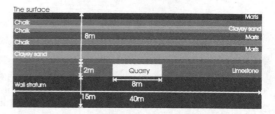

Figure 3. Cross-section and materials in UDEC.

3.2 Simulation of the rise of the bell-shape using UDEC

Figure 3 shows the geometry chosen with UDEC. The thickness of the roof strata is the result of a sampling log (Vachat, 1982).

In this basic case, studied by A. Fayad et al. (2002), the ratio K (= σ_{v0}/σ_{h0}) is equal to 0.3, and the authors showed the ability of the method to reproduce sinkhole. The procedure used and the main results obtained are given briefly below.

Excavation causes a change in the state of equilibrium. The simulation of the rupture and fall of the roof strata is divided into the following phases:

– the first phase consists of analyzing the rupture zones caused by the digging of the quarry, given by UDEC as plastic points. The completely plastic zones of the first roof stratum are then located. For example, figure 4a (1st stage) shows the existence of a completely plastic central zone. Two fractures are then introduced on either side of this zone in the initial simulation and the calculation is started again from the beginning (digging of the quarry). At the end of this new calculation that leads to the fall of the central block, the two parts of the 1st stratum that remain in place are subjected to a ≪cantilever≫ (fig. 4b, 2nd stage). This entails a new distribution of plastic points, in the 1st stratum, in the walls (extremities). If these lateral zones are completely submitted to plasticity, we introduce new fractures in

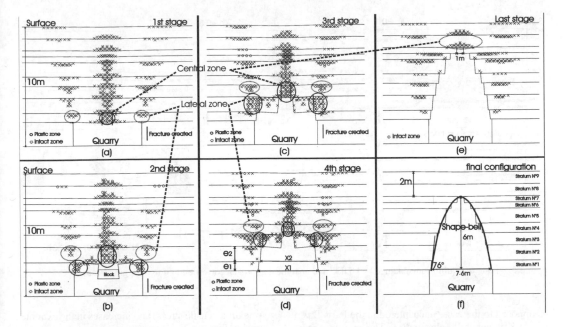

Figure 4. Procedure of the creation of induced fractures up to final equilibrium (final configuration).

them and start the calculation again from the beginning. The calculation leads to the fall of a number of blocks of the 1st stratum and the redistribution of the plastic points in the simulation as a whole, leading us to repeat the same fracture creation operations in the 2nd stratum of the roof;

– the second phase therefore consists of applying the previous method to the 2nd stratum then iteratively to the other strata of the roof. Figures 4c, 4d (3rd stage and 4th stage) show the following stages. And, in this case, the calculations result in final equilibrium shown in figure 4e.

In the case shown in figure 4, it can be seen that the development of the void stopped with a cap rock of 2 m (fig. 4f, final configuration).

This stop is the result of meeting a resistant stratum and a decrease in the span of the stratum exposed to the void (stratum 7: $X = 1m$) that results, according to the assumption used, in the absence of zones completely submitted to plasticity in the stratum. The shape-bell determined at the end of the calculation has a height (h) of 6 m, an angle at the base (α) close to 76° and a diameter at the base (D) of 7.6 m.

This data partly agrees with that given by Vachat (1982), in particular the angle (α):

$$\alpha = \text{arctg}[(e_1 + e_2)/(X_1 - X_2)] = \text{arctg} [2/.5] \approx 76°.$$

Where (fig. 4d): X_1 and X_2 = broken span lengths of strata numbers 1 and 2; and, e_1 and e_2 = thickness of strata numbers 1 and 2.

In fact, in the site studied, sinkhole 2.5 m in diameter occurred (Vachat, 1982). The difference obtained by calculation can result from a number of factors[*]. Consequently, to better understand this complex phenomenon, we propose to study the influence of horizontal stress (σ_h) on the occurrence of sinkhole.

4 INFLUENCE OF HORIZONTAL STRESS ON THE FORMATION OF SINKHOLE

According to Mestat (1998), the ratio $K = \sigma_{h0}/\sigma_{v0}$ (pressure ratio of land at rest) is a parameter connected with the site, however, it is not intrinsic to a material. Experiments have shown that it varies according to depth and depends on the type of land and the history of horizontal and vertical stress. The ratio K is thus affected at each stage of the formation of ground or a rocky massif.

According to several authors (Herget, 1988; Amedei and Savage, 1985), for thin cap rock, horizontal stress generally tends to be greater than vertical stress (massif already exposed to tectonic movements for example),

---

[*]It is noted that a decrease of 30% in the resistance of the two strata remaining in place (due to the deterioration of their resistance in the long term, the development of microfissures, water seepage, etc.) causes them to ruptue and allows sinkhole close to 2m in diameter to occur (consistent with observations).

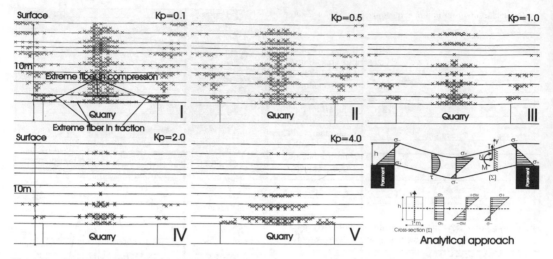

Figure 5. State of equilibrium obtained by UDEC and analytical approach corresponding to a sagging beam.

as opposed to the cases encountered in the Paris Basin where K is generally close to 0.5. The calculations of horizontal stress resulting from the weight of land ($\sigma_{h0} = (v/v-1) \cdot \sigma_{v0}$) give values of horizontal stress which are always low (especially when Poisson's ratio is low). Direct measurement of this stress is therefore often the only way to assess it. Since it seems to play an important role in the possibility of sinkhole occurring, we propose a general study of its influence. For this, the case presented above was re-worked with different values of K**: 0.1, 0.5, 1, 2 and 4.

4.1 Analysis of the state of equilibrium

After simulation of excavation, plasticity points corresponding to those zones subjected to plastic strain appear in the strata of the cap rock and the procedure presented above is applied to each of the simulations until final equilibrium is obtained (figure 7).

It can be observed that the density and location of plasticity points in the cap rock depend on the value of K. They decrease in number as K increases (between 0.1 and 2) then increase for K = 4.

Analysis of the first stratum of the roof reveals the following facts:

– in the central part, the number of plastic points diminishes as K increases in the extreme lower fiber

(maximal tensile stress) and increases in the extreme upper fiber (maximal compressive stress);
– at the extremities, the location of maximal tensile and compressive stress is inversed and respectively the variation, depending on K, of the number of corresponding plastic points.

This variation of the density of plastic points and their location, depending on K, confirms the theory of Peng et al. (1981) which assimilates the strata of the roof to girders subjected, through vertical stress σ_v and horizontal stress σ_h, to compound bending and shear (τ, fig. 5: analytical approach). They generally rupture because of compound bending. Analytically, a section is said to be subject to compound bending, when it bears both a bending moment (M) and normal strain ($N = \sigma_h \times h$). Longitudinal stress (σ) which acts upon these cross-sections (Σ) is determined, in resilience [linear distribution of σ, fig. 5 (analytical approach) and fig. 6], by the following equation:

$$\sigma = \sigma_h \pm \sigma_M \tag{1}$$

where σ_M = stress due to the deflection moment; $\sigma_h + \sigma_M = \sigma+ \leqslant \sigma_c$ (compression); $\sigma_h - \sigma_M = \sigma- \leqslant \sigma_t$ (tension); σ_c and σ_t being compressive and tensile strength (data seized for each stratum).

On the other hand, in plasticity, the distribution of σ which depends on σ_c and σ_t is uniform. Consequently, the height of the plastic zone (plastic points) depends on the resistance values (fig. 6).

According to equation (1), it can be noticed that σ decreases in tension as K increases as well, therefore, as the height of plasticity (h_p, fig. 6). This was in fact confirmed by digital simulation.

** It should be pointed out that the real variation margin of K for shallow quarries in the Paris region is probably between 0.4 and 2. We have increased this margin, from 0.1 to 4, to carry out a more general study on the sensitivity of sinkhole ooccurrence according to K.

4.2 *Analysis of the final configuration*

According to the methodology of induced fracture generation presented above, the existence of zones completely submitted to plasticity in the 1st stratum and in the case of simulations I, II, III and V (tab. 2) leads to the appearance of fractures in this stratum (fig. 5). On the other hand, an absence of continuity of plasticity points is obtained in the 1st stratum in the case of simulation IV.

Simulation IV, having K equal to 2, can therefore be considered as stable. On the other hand, the other simulations may show a change in the development of the void and shape-bell formation. Figure 7 shows the final form of the shape-bell obtained for the different values of K, while digital results accompany them in table 1.

Figure 7 shows that we can distinguish two categories of behavior depending on horizontal stress and the distribution of plastic points:

– category A: corresponds to a value of K below 2 where the plastic points are more present in a zone of traction than in a zone of compression (fig. 5). The appearance of the first plastic points in the cross-sections of the stratum is therefore linked to its tensile strength σ_t (fig. 7, simulations I, II and III);

– category B: corresponds to a value of K greater than 2 where plastic points are concentrated in a compression zone (fig. 5). The rupture of the stratum is then linked to longitudinal stress exceeding compressive strength σ_c (fig. 7, simulation V).

However, simulation IV represents the boundary between these two categories where plastic points are distributed virtually equally in zones of traction and compression, hence the most stable configuration (no shape-bell formation).

Furthermore, it can be observed that no surface sinkhole occurs in any of the simulated cases (stopping of

Table 1. Variation of the shape-bell according to K.

| Model | K (σ_{h0}/σ_{v0}) | Shape-bell | |
		Height, h (m)	Angle, α (°)
I	0.1	7.0	76
Basic case	0.3	6.0	76
II	0.5	5.0	70
III	1.0	3.8	66
IV	2.0	0.0	00
V	4.0	2.0	80

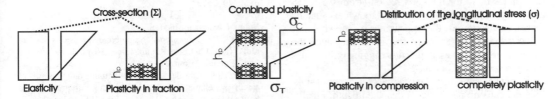

Figure 6. Plasticity in cross-sections depending on the distribution of longitudinal stress (σ).

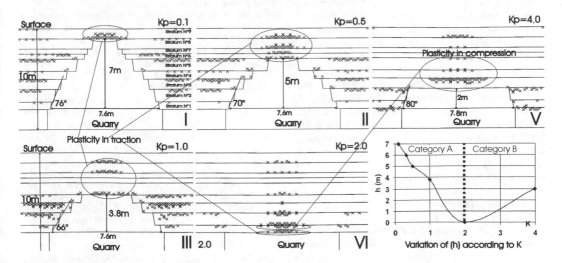

Figure 7. Final configuration of the simulations (I, II, III, IV and V).

the propagation of the rupture of the roof strata). The height of the shape-bell varies significantly depending on initial horizontal stress. It goes from 7 to 2 m, in a virtually parabolic relation between K and the height (h) of the shape-bell developed with simulation IV as origin (K = 2 and h = 0).

5 CONCLUSION

Shallow underground voids are responsible for a great deal of sinkhole. Its frequency and far-reaching consequences have led us to develop a simulation methodology based on distinct element simulations. This method simulates the phenomenon of shape-bell rising by the manual creation of induced fractures in order to predict the shape and development of shape-bell rise to the surface.

To gain a better understanding of this phenomenon, we have emphasized a geotechnical parameter which is difficult to measure, namely horizontal stress. Thanks to digital simulation, we have been able to estimate its influence on sinkhole formation.

The digital simulations were carried out using the distinct element method, well suited to the simulation of stratified, discontinuous areas. Taking account of the real values of K for the Paris region (from 0.5 to 2), the study of horizontal stress has shown us that its increase encourages the appearance of sinkhole. It reduces the development of the shape-bell since the strata of the roof are assimilated to sagging girders subjected to compound bending.

This conclusion shows how far the occurrence of sinkhole depends on parameters other than depth. Finally, the approach implemented for the calculations is an innovative method of predicting the occurrence of sinkhole, allowing its induced surface consequences to be better evaluated.

ACKNOWLEDGEMENT

Part of work presented falls under the research program of the department of ground and underground hazards of INERIS (ETAT-DRS-02) financed by the Ministry for Ecology and the Durable Development MEDD, "Analysis, prevention and control of ground movement risks arising from the presence of underground voids".

BIBLIOGRAPHY

Abbass Fayad, A., Al Heib, M. & Didier, C. 2002. Modelling propagation of sinkhole, in both slow and dynamic modes, using the UDEC computer code, Methodology and Application. Conference of the 5th North American Rock Mechanics Symposium (NARMS) and the 17th Tunnelling Association of Canada (TAC): Mining and Tunnelling Innovation and Opportunity, Toronto, Canada, 7–10 juillet 2002: 695–704.

Amedei, B. & Savage, W.Z. 1985. Gravitational stresses in regulary jointed rock masses. A Keynote lecture, in Proc. Int. Symp. On Fundamentals of Rock Joints, Bjorkliden, Centek Publ, Luleä, 1985: 463–473.

Barton, N., Bandis, S. & Bakhtar, K. 1985. Strength, deformation and conductivity coupling of rock joints. International Journal of Rock Mechanics and Mining Science & Geomechanics Abstracts, vol. 22, issue 3, June 1985: 121–140.

Didier, C. 1997. Principes de fermeture des anciens ouvrages miniers débouchant en surface. INERIS intermediate report for France coal mining. INERIS-SSE-CDi/CS-97-25EP35/R04.

Filliat, G. 1981. La pratique des sols and fondations: Ch. 29, géologie et géotechnique de la région Parisienne. Monitor edition: 319–355; 1315–1351 and 1190–1284.

Fine, J. 1993. Le soutènement des galeries minières: 27–42. ARMINES edition: Centre de géotechnique et d'exploitation du sous-sol. 35 rue saint Honoré, 77305 Fontainebleau.

Herget, G. 1988. Stresses in rock. A.A.Balkema, Rotterdam ISBN 90 6191 685 2.

Hoëg, K. & Prévost, J.H. 1975. Soil mechanics and plasticity analysis of strain-softening. Géotechnique, vol. 25, n°2: 279–297.

ITASCA 1997. Itasca Consulting Group. UDEC 3.1 Manual, Minneapolis, USA.

Mandel, J. 1959. Les calculs en matière de pressions des terrains. Imp. La Loire Républicaine, 16 pl. Jean-Jaurès – Saint Etienne – France.

Mestat, P. 1998. Etat des contraintes intiales dans les sols et calcul par éléments finis. Bulletin des Laboratoires des Ponts et Chaussées: N°215, May–June 1998, 4188: 15–32.

Peng, S.S. & Cheng, S.L. 1981. Predicting surface subsidence for damage prevention. Coal Min Process, V18, N5, May 1981: 84–95.

Tincelin, E. & Sinou, P. 1962. Effondrements brutaux et généralisés (coups de toit). Revue de l'Industrie Minérale, April 1962.

Tincelin, E. & Sinou, P. 1978. Mode d'action et règles du boulonnage. R.I.M, October 1978.

Thoraval, A. 2000. Etude de sensibilité des facteurs conditionnant la stabilité du toit d'anciennes cavités salines. Unité de Modélisation et Evaluation des Risques géotechniques – Direction des Risques du Sol et Sous-sol, INERIS-DRS-00-23008RN01.

Tritsch, J.J. 1987. Carrière souterraine de Belle Roche: Examen des conditions actuelles de stabilité. Cerchar Industrie, GAI-JTr/JS 87 (1)-98 71-1833/01.

Tritsch, J.J. 1995. Evaluation des méthodes et du coût de la mise en sécurité des populations menacées par les risques d'effondrement de carrières. INERIS, SSE-JTr/CS 24EA-03/R02.

Vachat, J.C. 1982. Les désordres survenant dans les carrières de la région parisienne. Mémoire Diplôme d'Ingénieur CNAM, Paris, 1982.

Geotechnical Measurements and Modelling, Natau, Fecker & Pimentel (eds)
© *2003 Swets & Zeitlinger, Lisse, ISBN 90 5809 603 3*

Development and use of knowledge-based systems for risk assessment in geotechnical engineering

J.U. Döbbelin

Dr.-Ing. Orth GmbH, Karlsruhe, Germany

ABSTRACT: As the development of construction shows an increasing complexity, a changed mode of operation during project handling was conditioned in the last decade, which characterises itself by additional planning activities. As a result a multiplicity of interfaces exists, which have to be linked by suitable information and communication methods. In the last two decades a significant number of knowledge based systems and neural network approaches has been developed for the use in geotechnical engineering. Some of them for the classification of soils, some concerning foundations, earth retaining structures, tunnelling etc. This short report describes some contributions to the risk management within tasks of geotechnics, as well as to the reduction of building damages and their subsequent costs. It reviews two knowledge based computer programmes, which have been developed for risk assessment by combining the possibilities of expert system technology with the advance of the theory of fuzzy-logic.

1 INTRODUCTION

1.1 *General*

Extensive investigations of construction damages in geotechnic – especially in foundation engineering – showed that a substantial proportion of damages (technical and economical) is not only based on the unawareness of constructional aspects and on errors in the execution of a construction, but also based on an unsuitable planning of the building and lack of co-ordination of the project-partners due to insufficient or missing information and communication. Specifications e.g. by quality control or certifying according to DIN ISO 9000ff are often not sufficient – in particular for small and middle companies and enterprises – in order to meet the necessary information and communication requirements during project handling, and to make an effective contribution for the minimisation of construction damages.

Commonly geotechnical engineering is known as an "imprecise" area of engineering due to the fact, that we are dealing with a material produced by nature (the ground). In many circumstances, our understanding of soil and rock behaviour still falls short of being able to predict how the ground will behave. Under these circumstances, expert judgement plays an important role, and empirical approaches are widely used.

Since the techniques of knowledge based computation can make use of heuristic knowledge (empirical knowledge/experience) for the solution of problems, they should be ideally suited for applications in the field of geotechnical engineering.

By analyses of the planning and execution processes of buildings new practice-oriented aids have been developed, regarding a minimisation of construction damages. The knowledge-based computer programmes PLURIS and FLORAN were developed as uncomplicated information and communication tools for a small selection of construction methods in geotechnics – especially in foundation engineering. Additionally, these programmes enable a quantitative estimation of the risk, which must be taken into consideration during the planning of a construction, as well as the early detection of construction damage relevant interfaces.

For the first time, concerning this range of topics, methods of theory of the "Fuzzy-Logic" were used, which enable the transfer of inaccurate linguistic (verbal) information and indistinct values into mathematically accurate relations.

The knowledge-based computer programmes PLURIS and FLORAN contain work and planning aids for planning and executing staff equally. These aids are especially co-ordinated with the requirements of small and middle enterprises and construction companies. Different subtasks are considered e.g. the co-ordination, the pre-investigation, the investigation of soils, the planning, the preparation of execution, the planning of the execution and the check of the execution process. The programmes were developed independently from a project or a company, so that it is possible to transfer these programmes into a larger target.

1.2 *Construction damages research in foundation engineering*

This still comparatively young area of research has its origin within the insurance economy. Building damages have to be settled by the insurers, when they become visible, this means during the construction or later on at the finished building (mostly in the first two years). For this reason already in the 1980th several billions of DM were spent annually by German insurance companies. The economical damage was considerably high. At the beginning of the 21st century it was estimated to be 3% to 5% of the entire building costs – about 10 billion EUR annually.

This led to a number of questions:

- Where do building damages really come from?
- Which errors have to be made, in which places, to induce the occurrence of a damage?
- How can these errors be avoided?

The research began with high expectations, but the early results were quite disappointing, as none of the involved could – probably for reasons of competition – admit own errors. Therefore the inducing influences of construction damages could not be determined directly, but only by a systematic analysis of seized data, this means by the evaluation of construction damage events. Thus specialised insurance companies opened their archives for a neutral research work. Thousands of documents of building damages in foundation engineering were seized and processed in a data base structure, which was developed for this particular reason.

The data of the individual cases consists of partly verbal, partly graphical and partly numerical damage-specific information of most different sources. In order to be able to link these diverse criteria with one another, and to permit an access also on partial data, for this data base a special grid was compiled, after which the individual data of each case could be categorised. Based upon the individual characteristics of single damages, flow charts were developed which were useful to filter the damage-relevant errors.

In particular the direct confrontation of the number of the examined damages to the determined damage costs showed that the small number of planning faults caused about 60% of the entire damage costs.

This was one of the most significant results, and it became clear, that additional work had to be done for the minimisation of these costs. In this coherence it was to be guaranteed that any tools, which would have been developed should be integrated as early as possible in the operational sequence of a construction measure.

Therefore the investigations of damages were extended. Out of this deepened evaluation flow charts were developed, which point out fundamental damage-relevant connections, by a systematic backtracking from damages to their source. For the first time it was possible to uncover sources of errors and to derive single construction-damage-relevant factors. Furthermore their reciprocal effects during the different working steps (pre-investigation, planning, execution) were seized. These results opened the way for systematic risk analyses in foundation engineering.

2 RISK AND RISK ANALYSES

2.1 *General*

During the process of building risks arise not only from technical-constructional reasons, but also from technological-economical or ecological reasons. Some times situations exist, in which the possibility of a damage event or the costs resulting from it appear to be small. In these situations the risk the occurrence of a construction damage is accepted. With the exceeding of a certain mostly individual tolerance-limit, however, measures for the prevention of damages and the associated costs become necessary. During an evaluation of the situation, and with knowledge of the size of the risk, which has to be expected, effective preventing measures can be introduced for the reduction of the endangerment. Therefore a risk assessment is no self-purpose, but an aid for decision-making on a rational basis.

Often risk analysis, if performed at all, is undertaken in a casual manner, and risk is not communicated effectively. But planning and control of risks (risk management) can open important potentials concerning construction period and construction costs. For example, if activities on the critical path cannot be finished in time, since staff and material resources are not available to planned or required extent.

A way for the determination of the individual tolerance-limit is a risk analysis, and the most difficult part of it the individual risk evaluation, which must be done by the use of knowledge, experience, methodology and intuition.

2.2 *Estimation of risks*

Based upon the knowledge – acquired by the earlier construction damage analyses – as a further aid for the minimisation of construction damages a methodology for the fast analysis of the "residual risk" of construction measures was compiled. In this coherence the "residual risk" means the risk, which remains at a certain moment during planning/execution of a construction after having done (or not) the time-specific activities.

The estimation of the residual risk of a construction measure allows the referring to the possible occurrence of a construction damage (risk-potential). From the knowledge of the residual risk and its sources conclusions on necessary changes can be derived.

For this purpose the complete planning and execution of a construction was divided into different activities and processes (defined as parts), which can be faulted. They were evaluated independently in each case regarding their damage-inducing influence. With these partial influences the residual risks of a construction measures can be estimated.

Thus the new aims of the research work were defined:

- Determination and description of specific risk-potentials including their mutual influence/dependence
- Development of methods for collection and handling of information about the possibility of occurring risk situations including their period in the schedule of a project
- Development of concepts for storage and relocation of such information.

3 DEVELOPMENT AND USE OF SOFTWARE

3.1 General

The newly developed methods had to be integrated in concepts of the project management and in applications of EDP for a quantitative estimation of the residual risk potential, as well as for an early recognising of damage-relevant weak points. All this with the emphasis to reduce risks and to cause only little additional work and costs.

3.2 PLURIS

Most of the results of the earlier construction damages research were implemented in applications of EDP. But the existing systematic was adapted to the new requirements, and the data acquisition was extended so that new ranges of topics could be described by adding further categories.

The developed program line was named PLURIS (PLanning sUpport and estimation of RISks). It contained models for the estimation of risk-potentials concerning constructions in foundation engineering, as well as the range of invitations to tender and allocation of construction measures in general. This application for the risk management gives an overview to the user, within which ranges risks have to be expected. Thus a simple tool was developed for the quality

management as well as for the early adjustment and control of construction measures.

One of the most important elements for the implementation of this program was the knowledge acquisition. Concerning the different steps during planning and execution of a construction a risk evaluation had to be done. This was realised by a questioning of experts according to the DELPHI method, by a sequence of several written questionings, each of them with the opportunity of discussion and correction of the results of the earlier questioning.

As an interface with the user the earlier developed and well introduced check-list form in the sense of an actual comparison was maintained for the PLURIS program. By filling out these question catalogues the residual risk-potential of the examined project can be estimated. For this reason to each given answer a weighted risk value was assigned, which was extracted out of results of the expert questioning. The higher this value, the more largely is the influence of the answer to the resulting risk-potential.

For the estimation of the risk-potential for a completely filled out question sequence those values are summed up. From this sum the allocation to the resulting risk potential is made.

The user can access basic question-sequences for:

- Securing of excavation with Berlin-type walls
- Securing of excavations with sheet pile walls
- Grout anchoring
- Dewatering of excavations
- Invitation to tender and allocation concerning sheet pile constructions
- Invitation to tender and allocation concerning engineering structures

The newest advancements of the PLURIS program allow the user to modify check-lists according to his own purposes, based upon the basic versions, which are contained in the program, or even to edit completely new check-lists and modify these again. This makes the program more flexible, since it can be adapted also to special cases. Also a World Wide Web application to deliver key components of PLURIS has been developed and tested.

A disadvantage of the check-lists, however, is that project-specific linkages, as well as boundary conditions can not be considered (closed check-lists), and during more complex projects, it is necessary to enter more specific and more physically measured variables into the estimation of the risk potential, than it is possible with a simple check-list inquiry.

These considerations influenced the extension of the program PLURIS by components, which are able to consider such linkages during an estimation of the risk-potential. The search for suitable methods for this purpose led to fuzzy-control.

3.3 *FLORAN*

The so-called fuzzy-logic – based upon the "Fuzzy Set Theory" – is an extension of the conventional bivalent logical formulations by the introduction of decision nuances, which allow a sliding transition of statements. Fuzzy-control again is an already successfully used method for the control of technical processes using the methods of the fuzzy-logic.

For the extension of the program family of PLURIS a system of fuzzy-controllers was developed and implemented in the computer program FLORAN (Fuzzy LOgical Risk ANalyses). The development of this new model again required the support of experts, who were willing to contribute their time and knowledge in another sequence of written questionings according to the DELPHI-method.

Due to this the opportunity was created to model inexact formulations and dimly focussed (fuzzy) verbal information, as well as measured and other project-specific parameters. With the help of the fuzzy control this tool unites the different parameters in a global instrument of decision support concerning the above mentioned topics.

For this reason the existing check-lists were modified (open Check-lists) and integrated into the program. Their input values become so-called linguistic variables.

The handling of FLORAN is – although a complex topic is treated – very simple. The program surface leads the user in thematically arranged order from question to question. Owing to regulation bars typing errors are almost impossible. Supported by colourbars (red to green) an additional signal was created to make the user attentive on risks.

FLORAN provides the user the opportunity to investigate possibilities, whilst getting instant feedback on his decisions. By offering the user alternatives and highlighting factors that have to be considered it is thought to give the user a better understanding of the processes behind the black box. Whilst it is backed up by reference material in the form of help files, a certain degree of understanding of geotechnics and foundation engineering is expected. The program is therefore seen as a supplement to the planner, not a replacement.

It provides a risk estimation and enables the user to become aware of the effects of the endangerment in order to be able to seize suitable counter measures. The weak points of the examined project can be recognised relatively fast by a descriptive diagram of the results.

4 CONCLUSION

The development of building is influenced in the last years by an increasing complexity. Thereby a changed mode was conditioned in planning and execution, which characterises itself by additional planning activities on the part of all project-partners. As a result of these circumstances a multiplicity of interfaces exists, which have to be defined and to be linked by suitable information and communication media.

Since the early 1990th check-lists were developed for construction measures in foundation engineering. These were constantly updated and developed. They serve as supporting instruments during planning and execution, and are thought to make a contribution to the minimisation of damages.

In consequence of these research activities simple tools were developed for the estimation of risks in foundation engineering.

The PLURIS program allows the user, based on the available basic versions, to edit the check lists for own purposes. Also a World Wide Web application of PLURIS has been developed.

As the development requires a lot of time (for knowledge acquisition), the resuming FLORAN program is limited to a few topics of foundation engineering at present. For the future it can be expected, that results and experiences of new research can be used in order to strengthen the FLORAN program. Maybe the different equations, available and used for risk estimation, will have to be re-examined. New data and information should be obtained and used to evaluate which equations are best suited for use within FLORAN, and improved sets of equations and parameters should be implemented.

PLURIS and FLORAN are knowledge-based computer programmes, which were developed for the estimation of risks concerning foundation structures. They are designed for use as simple tools and offer a critical assessment of the decision after it has been made. Damage-relevant factors can be eliminated by suitable measures regarding the feasibility of a construction measure and the associated risk.

After all it is recognised, that these knowledge based computer programmes are valid for some small aspects of solving engineering problems. For the most of our applications in geotechnics other typical (conventional) approaches are best suited. Therefore it is strongly suggested that knowledge based systems should be developed as support tools, rather than attempting to replace human expertise.

REFERENCES

Chur, H. 1998. Selektionsverfahren für Projektierungs- und Bauaufträge, *Schweizer Ingenieur,* Heft 19, S. 7–10, Zürich
Döbbelin, J. 2000. Zur geotechnischen Anwendung wissensbasierter Systeme mit Elementen der Fuzzy-Logik, Dissertation, *Mitteilungen des Instituts für Grundbau, Bodenmechanik und Energiewasserbau, Universität Hannover,* Heft 53, Hannover

Goebel J. 1995. Kooperation der Beteiligten am Bauprojekt auf der Grundlage eines gemeinsamen Informationssystems, Dissertation; *Universität der Bundeswehr München; Fakultät für Bauingenieur- und Vermessungswesen,* München

Hall, N. C. G. 1986. Using fuzzy expert system for decision support of the strategic planning process, PhD thesis, *Ann Arbor UMI,* Atlanta

Katzenbach, R., Boley, C. 1999. Baugrundrisiko, Planungs- soder Ausführungsfehler?, *Der Sachverständige,* Heft 6, S. 13–21, Bonn

Rizkallah, V. et al. 1998–2001. *Informationsreihe Institut für Bauschadensforschung e. V.,* Hannover, Hefte 1 bis 17, Hannover

Rizkallah, V. & Döbbelin, J.U. 2002 Empfehlungen zur Vermeidung von Planungs- und Ausschreibungsfehlern bei Ingenieurbauwerken – Spundwandbauwerke, *Mitteilungen des Instituts für Grundbau, Bodenmechanik und Energiewasserbau, Universität Hannover,* Heft 58, Hannover

Schneider, J. 1996. Sicherheit und Zuverlässigkeit im Bauwesen: Grundwissen für Ingenieure, *Hochschulverlag AG an der ETH Zürich und B.G. Teubner, Stuttgart,* Stuttgart

Sommer, B. 1999. Risikofelder bei Auftraggeber und Auftragnehmer, *Bauwirtschaft,* Heft 8, S. 22–23, Wiesbaden

Steiger, A. 1999. Bauherrenrisiken und Schadensmanagement, *Schweizer Ingenieur und Architekt,* Heft 35, S. 720–725, Zürich

Zadeh, L. A. 1977. Fuzzy sets as a basis for a theory of possibility, *Electronics research laboratory, University of California,* Berkeley

Risk management during construction of the Gotthard Base Tunnel

H. Ehrbar
AlpTransit Gotthard Ltd., Lucerne, Switzerland

J. Kellenberger
Ernst Basler + Partners Ltd., Zurich, Switzerland

ABSTRACT: A new railway tunnel through the Swiss Alps is currently being constructed which will reach 57 km length and represents the most significant component of the new Alp-crossing railway connection (NEAT). An integral management system was implemented to cover the issues quality management, environmental management and work safety. Risk management is comprehensively being addressed on a strategic and an operative level involving all project partners. It covers the assessment of threats and opportunities including their causes, which have a potential long-term influence on planning and execution of the new connections as a total. As crucial project requirements in the context of operational risk management were identified: functionality, costs, construction scheduling, environmental impact, work safety, and project organization. Finally, the terminology "risk" was extended to include threats *and* opportunities contained in the project.

1 GOTTHARD BASE TUNNEL

Since 1996 the longest tunnel of the world is being constructed: The new railway tunnel through the Swiss Alps (Figure 1). The length of the tunnel will reach 57 km and represents the most significant component of the new Alp-crossing railway connection (NEAT) through the Swiss Alps. On one hand, NEAT will serve the transportation of people, on the other hand, it is a crucial measure in the context of Swiss traffic policy to

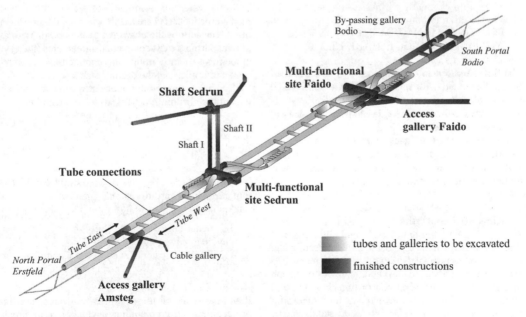

Figure 1. Gotthard base tunnel: length 57 km, status of work progress in spring 2003.

switch from road to railway for the transportation of goods.

With the new Gotthard Base Tunnel the Alps will be crossed at the lowest possible altitude. The vertex lies at an altitude of 550 m above sea level. This is 600 m lower than the original railway line that was constructed in 1882 with a tunnel length of 15 km.

The slopes of the ramps leading to the old tunnel reach 26‰ while the base tunnel will only reach 4 or 7‰ respectively. Hence, a flat railway line is being realized through the Alps.

Due to safety reasons the base tunnel consists of 2 single-track tubes. At the location of so called "multifunctional sites (MFS)" which are installed at the intermediary construction sites Sedrun and Faido, the trains can change the tube in case of maintenance works. The MFS are located after one third and two thirds of the tunnel length. In addition, the MFS provide the infrastructure for emergency stops with high-performance smoke escape ventilators and the feed of fresh air from outside. Connections between the two tubes are constructed all 325 m.

The tunnel has been designed for passenger trains with a speed of 200–250 km/h and freight trains with 100–160 km/h. The capacity for such mixed traffic amounts to about 300 trains per day in both directions.

During the past seven years the intermediary construction sites Amsteg, Sedrun and Faido, as well as the by-passing gallery close to the portal in Bodio were completed. In this year the actual drilling and excavation of the single-track tubes was started. Based on the current time schedule the tunnel will be put into operation in 2013.

2 INTEGRAL MANAGEMENT SYSTEM AT ALPTRANSIT GOTTHARD LTD.

The Swiss Confederation (government) has mandated the public limited company AlpTransit Gotthard (ATG) with the planning and execution of the new railway sections on the Gotthard axis. ATG is a 100% subsidiary of the Swiss Federal Railway.

In the context of its function as project-owner ATG has implemented an integral management system. The system covers the issues quality management, environmental management and work safety.

3 RISK MANAGEMENT

The risk management discussed in the present paper forms a component of the quality management. It is established on condition that each involved contractor, planning or engineering consortium has its own company-related quality management (CQM), for instance based on the ISO 9001:2000 standard. The

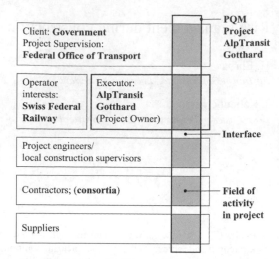

Figure 2. Company-related and project-related quality management in the overall project AlpTransit Gotthard.

CQM is completed by a project-related quality management (PQM) (Figure 2).

By the enforcement of a systematic PQM it is ensured that:

- interfaces between project partners are defined and secured,
- the risk situation is early recognized, evaluated and appraised,
- determining and crucial project requirements are defined as main quality emphasis.

The quality management and thus also the risk management at ATG are highly valued and therefore in the direct duty of the chairman of the board. With all project partners exist contractual agreements (the QM-agreements) which oblige the contractors to comply with and implement the defined standards.

The risk management is oriented after the objectives of the corporation ATG, which is given below:

"All constructions which are part of the Gotthard axis are to be realized according to the quality agreed (with the government), as fast as possible and at minimum cost."

Aiming at this objective two questions can be formulated with regard to risk management:

- Which factors impair or even prevent the accomplishment of the objective? (Recognize threats, characterize threats, control threats)
- Which factors foster or support the accomplishment of the objective? (Recognizing and making use of opportunities)

Risk management is strictly implemented over all project phases from planning to realization (Figure 3).

3.1 Strategic risk management for the Gotthard Axis

Apart from the Gotthard axis, the NEAT consists also of the Zimmerberg Base Tunnel (total length 20 km) and the Ceneri Base Tunnel (15 km) as well as open railway tracks. The total investment amounts to 10 Billion Swiss Francs (~6.7 Billion Euro). The total time period for planning, design and execution covers about 20 years.

On the level of the management board ATG a strategic risk management is pursued. This risk management covers the assessment of threats and opportunities including their causes, which have a potential long-term influence on planning and execution of the new connections as a total.

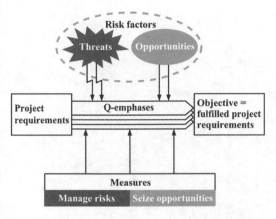

Figure 3. Basic principles of the risk management at ATG.

In the latest project progress report of the management, which is published twice a year and addresses the board of directors of ATG, the supervising authorities and political organizations, the threats and opportunities are listed as shown in Figure 4.

Particularly, the report identifies two "mega"-threats and a "mega"-opportunity, each of them showing a high probability of occurrence and a great extent of impact:

- Threat T1: The financial reserves have been used by the government for so-called purchase order changes, such as to enhance operational safety (e.g. establishment of two single-track tubes for the Ceneri Base Tunnel) or to enable technical quality changes. By the allocation of the reserve funds a financing gap could be created which may have an adverse impact on the realization of the two base tunnels at Zimmerberg and Ceneri, originally planned for 2006.
- Threat T5: If the construction permit for the northernmost section of the Gotthard Base Tunnel including the portal in Erstfeld is not obtained as scheduled by the end of 2003, the deadline for putting into operation the whole base tunnel cannot be met.
- Opportunity O5: The railway lines adjacent to the Gotthard- and Ceneri Base Tunnels have not yet been legally defined, which creates strong political opposition in the regions concerned and influences the construction process of the base tunnels. ATG has recognized the "mega"-opportunity to settle this conflict in a legally binding form under the label NEAT II.

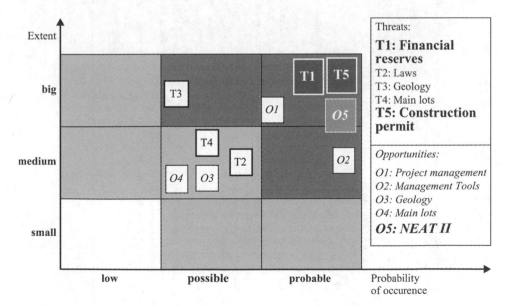

Figure 4. Risk situation in the overall project as of December 31, 2002.

Table 1. Excerpt from the risk analysis and measurement planning lot 360 Sedrun.

No.	Threats	Risk (neg.)			Planned measures (according contract/project/CQM) – Additional measures	Residual risk			Resp.	Date	Monitoring
		P	E	R		P	E	R			
100	**Ground Conditions**										
110	**Geology**										
112	Reuse of excavated material worse than predicted										
112a	TZM South	2	2	4	Early initiation of gravel mining in Val da Claus	2	2	4	öBL/OBL	On-going	ArSi Ko Team MBK
112b	GM	2	3	6	Gravel mining in Val da Claus – Possibly: Proposal of quality reduction, AVOR: Adaptation of construction schedule, lot boundaries	2	2	4	öBL/OBL	On-going case-by-case	ArSi Ko Team MBK
113	Mica content in excavated material higher than predicted										
113a	TZM South	3	1	3	Adaptation of treatment technology (mica flotation) – Possibly: Proposal of quality reduction	3	1	3	öBL/PI	On-going case-by-case	Monthly report lot 374
113b	GM	2	2	4	Adaptation of treatment technology (mica flotation) – Possibly: Proposal of quality reduction	2	2	4	öBL/PI	On-going case-by-case	Monthly report lot 374
115	Mine temperature higher than predicted	1	3	3	– Increase cooling performance, possibly reduced working hours	1	2	2	UN öBL	Case-by-case	Daily report
120	**Disruption zones**										
122	Distinctive disruption zone in the Nalps area	2	3	6	Preliminary investigation, injections, long protected explorations – Additional sondage, additional injections	2	2	4	öBL/PI *ATG öBL*	2006 *2004*	*Additional study Instruction öBL*

270

In order to prevent the occurrence of the identified threats and to seize the recognized opportunities, comprehensive measures have been proposed, mainly on a superior management level.

3.2 Operative risk management for the construction of the gotthard base tunnel: risk analysis and mitigation measures

Based on the risk analysis that was carried out by the project-owner at the beginning of construction planning for the Gotthard Base Tunnel, the following crucial project requirements were identified in the context of the operational risk management:

Functionality (safety of load-bearing structures, practical capability), costs (cost minimization, supplementary charges by contractors, etc.), construction scheduling, environmental impact, work safety, and project organization of all involved parties (process management, implementation of contractual agreements, CQM).

The threats and opportunities including their influence on project requirements from the perspective of the project-owner were disclosed and communicated to the contracted project engineers, construction supervisors and construction firms by means of the quality master plan.

Subsequently ATG discussed the risk analyses in detail and elaborated action plans for mitigation measures with the contracted project engineers and construction supervisors. With this procedure ATG aims at defining the major threats and opportunities and pursuing the risk issues in a coordinated, systematic way jointly by all involved parties during project realization.

In the applied methodology for the risk assessment, the risks are quantified as product of occurrence probability and extent of the damage/benefit. Probabilities and damage extents are estimated according to the scale given in Table 1.

Measures with regard to responsibilities, deadlines, reference documents and verification protocols in the context of risk management and realizing opportunities are defined in the master plan.

ATG has developed a three-stage valuation matrix shown in Table 2 which can easily be applied.

The following definitions are being used:

Risk $R = P \times E$: Value after implementation of planned measures from the contracts, project and CQM, i.e. remaining risk potential.
Residual risk $= P \times E$: Value after implementation of additional measures.
Threats: reduced damage potential (residual damage)
Opportunities: seized opportunities (supplementary benefit).

3.3 Strategy of action of ATG

The strategy of action of ATG is based on the following principles, which are visualized in Figure 5:

1. If opportunities or threats reach a risk value of >6, measures are imperatively taken.

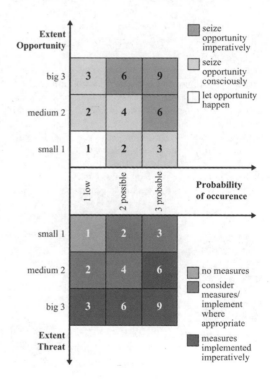

Figure 5. Strategy of action.

Table 2. Applied valuation matrix.

	1	2	3
Probability of occurrence (P)	Low (not expected)	Possible (cannot be excluded)	Probable (occurrence assumed)
Extent of damage/benefit (E)			
Costs	Low (below CHF 1 Mio.)	Medium (CHF 1–10 Mio.)	High (over CHF 10 Mio.)
Schedule	Low (below 2 months)	Medium (2–6 months)	High (over 6 months)

2. For all threats with a low probability of occurrence, but a high extent of damage ($P = 1$; $E = 3$; $R = 3$) measures are imperatively taken.
3. Opportunities and threats with a risk value of 1 don't require any planning of measures, because the optimum target has already been reached.
4. In all other cases measures are considered and implemented where appropriate.

4 CONCLUSIONS

Risk management for the Gotthard Base Tunnel forms a central element in the integral management system of the project-owner AlpTransit Gotthard Ltd. Risk analyses and the planning of mitigation measures are performed for all project phases and construction execution, tailored to the specific level. The more comprehensive these steps are carried out, the smaller is the surprise to be unprepared if an undesired event is encountered. On one hand, risk management as an instrument develops its force by systematic application, on the other hand its significance is enhanced by the implementation not only for the project-owner, but for all involved partners.

Through the comprehensive harmonization of the applied methodology and a common terminology of all subcontractors a sound control of all interfaces was established which also facilitates communication. During application of the comprehensive risk management system it was revealed that the project not only contains threats, but also opportunities which should not be lost. Thus, the terminology "risk" was extended to include threats *and* opportunities.

Geotechnical Measurements and Modelling, Natau, Fecker & Pimentel (eds)
© 2003 Swets & Zeitlinger, Lisse, ISBN 90 5809 603 3

The usage of the multi-task learning concept in landslide recognition with artificial neural nets

T.M. Fernández-Steeger, J. Rohn & K. Czurda
Department for Applied Geology, Karlsruhe University (TH), Germany

ABSTRACT: Within the framework of landslide hazard analysis the knowledge of the existence and location of hazard zones is an essential prerequisite. Besides the well-known deterministic and statistical methods, the use of artificial neural nets (ANN) is a very promising approach for susceptibility mapping. The advantage provided by neural nets is their flexibility and their ability to handle non-linearities. The problem resulting from this is the control of the model complexity. An interesting approach to control model complexity as well as improving the overall performance of the ANN is the application of the multi-task learning (MTL) concept. The tests, carried out to identify landslides in test areas in the Eastern Alps, reveal that the best nets have classified up to 89% of the areas correctly. It is of particular interest that ANN are able to identify different types of mass movements with one and the same net.

1 INTRODUCTION

Landslides represent a considerable danger for human beings and infrastructure. Therefore it is necessary to develop protection concepts for affected areas.

The most efficient and economic possibility to prevent danger caused by landslides is to avoid the endangered areas and to use them only extensively. Therefore it is necessary to obtain detailed information about present and potential landslide areas.

For this reason strong efforts were made in the last 10 years to simplify and objectify the detection of landslide areas, applying quantitative analysis together with mechanical data processing (Aleotti & Chowdhury, 1999).

However these techniques often cause difficulties regarding the generalisation abilities of the corresponding models. Therefore a separate model has to be developed for each landslide type. (Carrara et. al., 1995). Another difficulty of deterministic modelling is the absence of models for various landslide types. Furthermore it is not possible to determine all necessary parameters required for deterministic or statistic analysis at acceptable costs.

The application of methods derived from mechanical learning is at least an approach to cope with this problem in a better way. As the statistic landslide recognition corresponds with the classical pattern recognition, the use of artificial neural nets is of particular interest. Neural nets have already been used successfully in pattern recognition e.g. in science for a long time (Zell, 1994, Bishop, 1995).

The following presents, how different types of shallow landslides in debris or soil may be recognized using artificial neural nets. Moreover a method will be described to handle the generalisation problems related to this task. These models were tested in a complex alpine environment in the Eastern Alps (Fig.3).

2 NEURAL NETS

The theory of artificial neural nets (ANN) was intended by the idea to simulate biological information processing, especially biological learning, with mathematical methods.

Simplified, ANN (Fig. 1) are information processing systems, consisting of a number of highly connected simple processing units, so-called neurons (s. Fig. 2). They "learn" to solve classification tasks by optimising the weights between the units in an iterative process. During the learning process by adjusting the weights w_{ij} of the input S_j (Fig. 2) the output S_i is being optimised. Here the RPROP learning algorithm (Riedmiller, 1994) was used to train the nets.

Very abstract ANN may be compared to a very complex regression with n-dimensional interactions of the input parameters. The advantage is that the neural models are able to realize this regression of n degree without any prior knowledge about the functional context of these parameters. This means the neural nets are able to handle extremely complex functions as well as non-linearities. This also implies that the ANN are able to cope with noisy data as errors in data sets or weak parameters.

Neural learning does not signify the exact reproduction of the learned only, but furthermore the integrated generalisation ability of the ANN is of particular interest. This means that the optimization of the weights has to be done in a way, that the nets are able to act even in an unknown situation as correctly as possible.

In the present case only complete feed forward nets, also called multi layer perceptron (MLP), have been used (Fig. 1). They process information from the input to the output and each neuron of one layer is connected to all neurons of the next layer.

The nets are trained with one set of data and "learn" to separate this set into e.g. stable or unstable areas. The

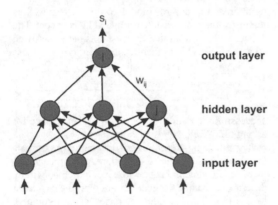

Figure 1. Complete feed forward net with a 4-3-1-topology, where the processing units, called neurons are arranged in layers. The four-dimensional input vector is handled in the 3 neurons of the hidden layer and provides a one-dimensional output.

learned abilities are tested with another set of data to check the generalisation performance of the trained net.

A more detailed introduction to the theory of neural nets may be found in Bishop (1995) and Zell (1997).

3 DATA

Modelling landslide recognition using ANN takes advantage of the fact that the detection of landslides is essentially a classification problem. This means, that due to its characteristics each site in the terrain can be classified as stable or unstable. This proceeding is essentially comparable to the proceeding of the geological fieldwork and can be simulated with neural nets.

For the analysis of the area a simplified virtual image, a digital terrain model (DTM) is created in a geographical informationsystem (GIS). For the development of the model, two test areas in the Eastern Alps (see Fig. 3) close to Bad Ischl/Austria have been chosen, which show a similar geological environment and for which detailed geotechnical investigations and maps were available (Rohn, 1991, Resch, 1997, Xiang, 1997). The landslides in these study areas are mainly earth flows and shallow landslides of different types in soil and debris.

The data in this DTM are the basis for further analysis. Basically the DTM consists of 6 main homogeneous information layers, namely 1. the digital elevation model (DEM), 2. the lithology, 3. the stream net, 4. the vegetation, 5. the main scarps and 6. the pre-classified landslide areas (Fig. 4). Additionally data as for example friction angle or calculated safety factors are added. The DTM permits to obtain further parameters from these input data, for example slope angle or distance to brooks. For the analysis in the neural models the data are subsequently gridded

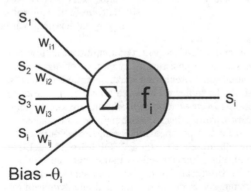

Figure 2. Schema of a processing unit, a so-called neuron. During learning by adjusting the weights w_{ij} of the input S_j, the output S_i will be optimised.

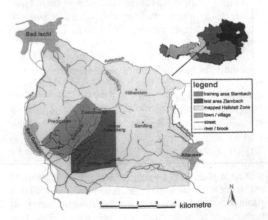

Figure 3. Location of the study areas in the Eastern Alps close to Bad Ischl/Austria.

(20 m resolution), normalized and the relevant features (s. Fig 5) exported to the SNNS format (s. Zell, 1997).

4 INSTABILITY FACTORS

The reasons for mass movements or landslides are changes in the equilibrium of the downslope forces and the resisting forces. The causes leading to changes in the equilibrium can be divided in 1. geological, 2. morphological, 3. antrophical and 4. physical causes. Usually not only the change of one factor but furthermore the interaction of different factors lead to an increase of shear stress. For this reason Popescu (1994) suggests to separate this stability influencing factors into preparatory factors and triggering factors.

This leads to a systematic approach to identify landslide affected or potentially affected areas. Today driven by the technical development it is a lot easier to map the permanent preparatory factors than the variable triggering factors. This is acceptable as the triggering factors are mainly important for the analysis of the temporal occurrence of mass movement. Their potential influence is still included e.g. as locations where potentially high pore pressures are realistic.

Based on these facts the relevant parameters for landslide recognition may be searched and selected from the DTM either by expert knowledge, statistical methods or by trial and error. In this case a preliminary selection was made on the basis of expert knowledge which later on was modified by statistical methods as for example correlation analysis as well as trial and error during the neural net development. The features used for this model (s Fig. 5.) take into consideration that normally not all features required for stability analysis are usually available. This means that as few parameters as possible, corresponding to their public availability were used. For example the digital

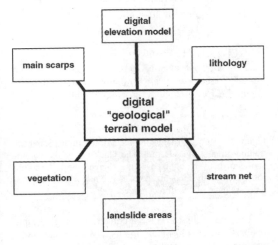

Figure 4. Elements of the digital "geological" terrain model (DTM).

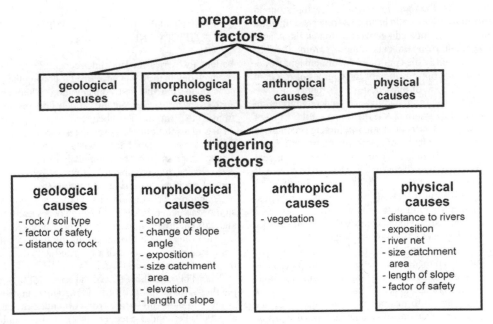

Figure 5. Causes that may influence slope stability and their subdivision in preparatory and triggering factors. The features used in the neural models are displayed in their context to the preparatory factors.

275

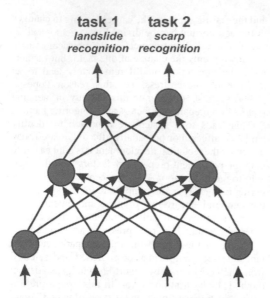

task 1
*landslide
recognition*

task 2
*scarp
recognition*

Figure 6.   Schematic sketch of an MTL net. The net learns to solve two related task in one model.

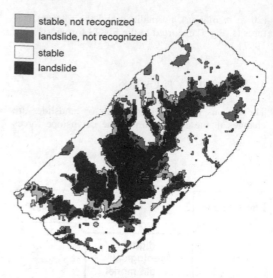

- stable, not recognized
- landslide, not recognized
- stable
- landslide

Figure 7.   Result of landslide recognition in the Stambach training area with MTL net. The black and white colours show correctly recognized cells, whereas the light colour show incorrectly recognized cells. The major landslide systems are generally well recognized.

elevation model (DEM) applied in the test area was the one of the national survey service of Austria.

5   NEURAL MODELS

As mentioned the main focus while developing good performing ANN should be to develop nets with good overall performance and generalisation skills. A specific problem of neural nets is their extreme flexibility. Consequently the control of the dimensions of the model and the optimization of the degrees of freedom in the models are of outstanding importance.

Besides structural stabilization and regularization the usage of the multi task learning concept (MTL) of Coruana (1993, 1994) is an interesting method to control the complexity of neural models.

This concept implies that in a common neural model different but similar or related tasks have to be solved. It takes advantage of the fact that the same information often is important for the solution of related problems. Here the net has to identify cells with main scarps and cells with landslides simultaneously (Fig. 6). Both tasks are related to each other because main scarps are the starting point of landslides and zones of instability. This means that there might be information or combinations of information which are important for both tasks. These connections will be strengthened. Apart from this other informations will only be important for one task or even only for a special type of scarp or landslide. This information or these connections will be weakened.

The nets had an 17-11-2 architecture using 13 input features coded as 17 binary inputs. The complete model development and validation was done in the Stambach area, the Zlambach area was selected only for testing.

6   RESULTS OF THE MTL-TRAINED ARTIFICIAL NEURAL NETS

Various test series for landslide recognition have been conducted with the ANN developed following the MTL concept to find the best configuration. Although the analyses were carried out in various grid resolutions from 20 to 50 m, the described results always refer to a 20 m grid resolution. Moreover all tests were realized by varying the number of hidden neurons in order to obtain a better impression of the complexity of the tasks. The performance of the trained nets is normally indicated in percent of correctly recognised patterns (hit-rate).

To avoid overfitting and to obtain a better impression of the generalisation skills, the nets were trained and validated with the data from the Stambach training area only. The performance of the trained and so optimised nets was additionally tested with data from the Zlambach test area.

The MTL nets show that in the course of the training they have developed skills to recognize landslides as other models did before (ref. Fernandez-Steeger et al., 2002, Fernandez-Steeger 2002). In the Stambach training area up to 85.1% of the cells are classified correctly (Tab. 1). Figure 7 shows that the ANN has

Table 1. Results of the MTL net in the training area Stambach. Overall the net recognized 17,779 or 85.1% of the cells correctly (printed in bold).

Odds ratio		Training area Stambach	
		Stable	Unstable
Model	Stable	**12,482 cells** **59.7%**	1711 cells 8.2%
	Unstable	1409 cells 6.7%	**5297 cells** **25.4%**

Table 2. Results of the MTL net in the test area Zlambach. Overall the net recognized 20,294 or 86.7% of the cells correctly (printed in bold).

Odds ratio		Test area Zlambach	
		Stable	Unstable
Model	Stable	**12,028 cells** **51.4%**	1390 cells 5.9%
	Unstable	1731 cells 7.4%	**8266 cells** **35.3%**

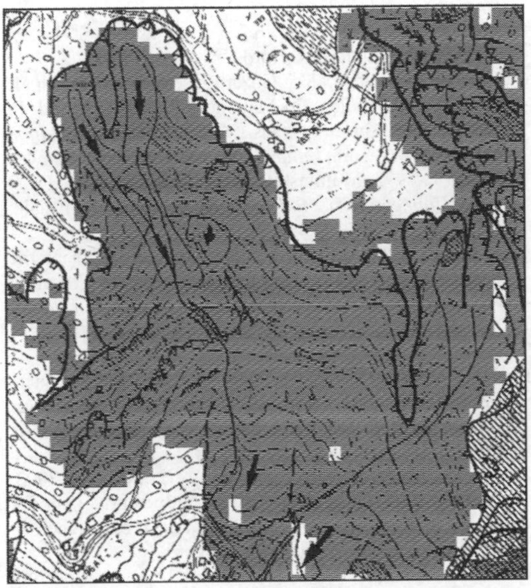

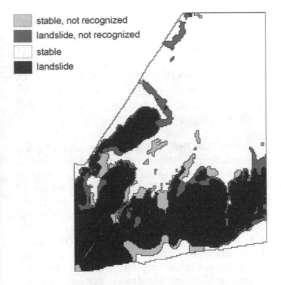

- stable, not recognized
- landslide, not recognized
- stable
- landslide

Figure 9. Result of landslide recognition in the Zlambach test area with MTL net. The black and white colours show correctly recognized cells.

Figure 8. The shaded areas are recognized as instable areas in a big landslide structure.

identified all main landslide structures and classified them correctly.

In particular the model recognized landslide systems related to mount torrent systems as well as landslides in glacial oversteep slopes precisely and coherently. It is also interesting that the nets are capable to identify rotational, translational slides and moreover earthflows as unstable areas. Only in the border areas (s. Fig 8.), in very young mount torrent systems and singular slumps the net has difficulties with the correct classification of the cells.

Testing the performance of nets, previously trained at the Stambach area, with data from the Zlambach area, the rate of recognition rises slightly up to 86.7% (Tab. 2). This indicates that the neural nets based on their learned skills in the Stambach area have classified the unknown data from the Zlambach test area correctly.

The more homogeneous geotechnical environment of the Zlambach test area might explain these better results. While all mount torrent systems in the Zlambach area have a similar formation and age, except one, in the Stambach area the nets also have to identify singular landslides, young mount torrent systems and landslides related to post glacial relief compensation.

Some of the misinterpretations made by the ANN may be corrected by neighbourhood analysis and rule based algorithms in the post processing. As the nets are only trained to recognize shallow landslides in debris or soil, rock areas can not be affected by this type of landslides. These mistakes may be rectified easily. Moreover statistically it is not very probable that simple stable cells are completely surrounded by instable cells. These "errors" may be corrected in a simple way, if wanted. Nevertheless we have to take into consideration that these "corrections" might cause additional or

277

Table 3. By post processing optimised results of the MTL net in the test area Zlambach. Overall the net recognized 20,864 or 89.1% of the cells correctly (printed in bold).

Odds ratio		Test area Zlambach	
		Stable	Unstable
Model	Stable	**12,569 cells** **53.7%**	1361 cells 5.8%
	Unstable	1190 cells 5.1%	**8295 cells** **35.4%**

new errors. Table 3 shows however, that the overall classification performance in the Zlambach area has improved up to 89.1%. In the Stambach area the improvement was not that high and the classification performance ameliorated only up to 85.8%.

7 DISCUSSION

The Multi Task Learning approach in neural models represents in general an interesting and very promising method of landslide analysis. It can help to improve the overall classification performance of neural nets as described in Coruana (1993, 1994). The comparison of MTL nets with neural nets switched in line and normal MLP (Fernandez-Steeger, 2002) show that MTL nets are able to improve the overall classification performance and show best generalisation abilities. But not only the calculated classification rate is the best, but also the plausibility of the results in an geological context is very high.

Moreover, neural nets are capable to identify different landslide types and mechanisms with one and the same net. Since the nets in the frame of the training do not provide "absolute knowledge" but mainly associative skills, they have a quite good generalisation performance. So they are able to transfer their learned skills to the second area.

Generally the landslide recognition with ANN as data driven method has one main disadvantage. The ANN can only learn to recognize landslide types that are represented in the training data sets. Landslide types caused by completely different triggers or mechanisms, which are not presented adequately in the data sets (training areas) will not be recognized satisfyingly. This is for example the case with the huge single slumps area or insufficiently developed young

mount torrent systems in the Stambach area. Here the nets show a unsatisfactory performance. It has to be mentioned, that in the course of net development for landslide recognition (s. Fernandez-Steeger et. al., 2002) the highest increase in performance was achieved by a better coding and the use of new or different input features.

REFERENCES

Aleotti, P. & Chowdhury, R. 1999. Landslide hazard assessment: summary review and new perspectives. *Bull. Eng. Geol. Env.* 58 (1): 21–44.

Bishop, C.M. 1995. *Neural Networks for Pattern Recognition.* Oxford. Oxford University Press.

Carrara, A., Cardinali, M., Guzetti, F. & Reichenbach, P. 1995. Gis Technology in Mapping Landslide Hazard. In Carrara, A. & Guzetti, F. (eds.), *Geographical Information Systems in Assessing Natural Hazards.* Dordrecht: Kluwer.

Caruana, R. 1993. Multitask Connectionist Learning. *Proc. 1993 Connectionist Models Summer School.* Hillsdale: Erlbaum.

Caruana, R. 1994. Learning Many Related Tasks at the Same Time with Backpropagation. In Tesauro, G., Touretzky, D.S. & Leen, T. (eds.), *Advances in Neural Information Processing Systems (7).* Cambridge: MIT Press.

Fernandez-Steeger, T.M. 2002. *Erkennung von Hangrutschungssystemen mit Neuronalen Netzen als Grundlage für Georisikoanalysen.* Doctoral Thesis Karlsruhe University. Karlsruhe.

Fernandez-Steeger, T.M., Rohn, J. & Czurda, K. 2002. Identification of landslide areas with neural nets for hazard analysis. In Rybar, J. Stemberg, J. & Wagner, P. *Landslides.* Lisse: Swets & Zeitlinger.

Poppescu, M.E. 1994. A Suggested Method for Reporting Landslide Causes. *Bull. of the IAEG* 50: 71–74.

Resch, M. 1997. *Geologische und Ingenieurgeologische Untersuchungen der Talzuschub- Systeme am Hohen Raschberg bei Bad Goisern (Oberösterreich).* Diploma Thesis Karlsruhe University. Karlsruhe.

Riedmiller, M. 1994. Advanced supervised learning in multilayer perceptrons – from backpropagation to adaptive learning algorithms. *Int. Jour. of Comp. Standards and Interfaces* 16: 265–278.

Rohn, J. 1991. Geotechnische Untersuchungen an einer Großhangbewegung in Bad Goisern (Oberösterreich). *Schr. Angew. Geol. Karlsruhe* 14. Karlsruhe.

Xiang, W. 1997. Der Einfluß der Kationenbelegung auf die bodenmechanischen Eigenschaften von Tonen am Beispiel einer ostalpinen Großhangbewegung. *Schr. Angew. Geol. Karlsruhe* 48. Karlsruhe.

Zell, A. 1994. *Simulation neuronaler Netze.* Bonn: Addison-Wesley.

Geotechnical Measurements and Modelling, Natau, Fecker & Pimentel (eds)
© 2003 Swets & Zeitlinger, Lisse, ISBN 90 5809 603 3

Rock cleaning along Bavaria's road network

P. Jirovec
Autobahndirektion Nordbayern, Nürnberg, Germany

ABSTRACT: Active and passive safety measures are set out in this paper, to ensure the safety of traffic along a stretch of motorway of approximately 3 km. Nowadays one also has to take account of the environmental impact of stabilization works against rock fall faces.

1 INTRODUCTION

Stabilization works against rock fall along motorways must be planned and accomplished in order to eliminate any danger possiblity due to such an event. A falling rock could have serious consequences for human life in today's traffic density.

2 AREAS CLOSE TO THE ROAD

On the A 9 motorway from Berlin to Munich prominent rock formations along a 3 km stretch can be found in the area of Kindiger Berg. There are also many single boulders further above the carriageway. In geological terms these are limestone stratified in thick beds and dolomite fractured in large blocks.

The rocks were so weathered and loose that shotcrete and rock anchors had to be used to secure them.

Cement that can treat the chalk with a similar colour to the dolomite should be preferred in order to produce shotcrete that blends in with nature. Hikers are not aware of such artificial measures (Fig. 1).

Figure 2. Wedge of rock.

Figure 1. Rock formation.

In this instance, cement from the Passau area was used, which did not meet the specifications of the authorities.

Only cement from Harburg, Swabia, ensured that the shotcrete matched the colour of the dolomite. The concrete was applied carefully and sparingly; every area, fissure and disturbance was chosen by the engineer. The natural beauty of the Altmuehl valley was thus not affected.

An observer would not notice the artificial measures needed to preserve the prominent rock landscape.

Further measures to secure the safety of the traffic in this area were passive in nature, such as fences, net, embankments and gabions.

A large wedge of rock was observed on climbing the face and thoroughly examined (Fig. 2). This rock was completely detached from the rock face, safety factor against sliding about 1. A gabion was erected, and shortly afterwards the wedge of rock did indeed fall, and was caught by the gabion. This measure had prevented a potential catastrophe, as the motorway is extremely busy day and night.

Gabions have the advantage of absorbing the kinetic energy into a deformation of the wall, as can be seen in Figure 3.

The rock face forms an uneven arch next to the rock formation above the motorway. The wall is between 4–20 m high. Tests were performed to establish whether the carriageway could be affected by a natural weathering process.

Despite the soft soil of the forest, stones that fall from above still pass over to reach the carriageway.

There were two reasons for not using shotcrete – the significant cost and concern for the environmental impact. The topography made an optimal solution possible. A 2 m high embankment was created and planted with greenery and bushes.

Behind the embankment was created an extensive buffer zone. A 10 m high wall borders the embankment for 350 m. This wall consists mainly of thin limestone layers, with thicknees between 10–25 cm. At the lower half part of the area this layers are intercalated with marl layers, which tends to from hollows. The gap between these large ravines is between 5–10 m. The wall was covered with a double-galvanised netting, made of 3 mm wire, with a hexagonal mesh of about 5 × 5 cm. For aesthetic and mechanical reasons the wire netting should be flush against the rock face. This

Figure 3.   Gabions.

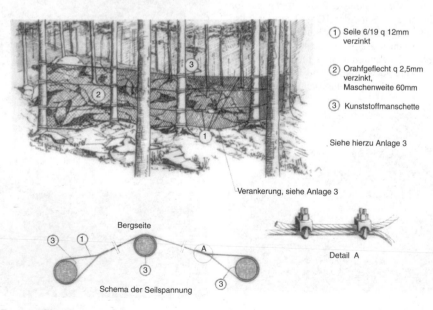

① Seile 6/19 q 12mm
verzinkt

② Orahfgeflecht q 2,5mm
verzinkt,
Maschenweite 60mm

③ Kunststoffmanschette

Siehe hierzu Anlage 3

Verankerung, siehe Anlage 3

Bergseite

③ ①

A

Detail A

③

③   ③

Schema der Seilspannung

Figure 4.   Fence profile.

was achieved by cementing short anchors 60 cm deep into the rock. These nets are contact pressure across the surface of the rock, sufficient to prevent stones becoming detached and falling.

Personally I do not approve of the way the Deutsche Bahn covers rock walls with nets, leaving a gap between the rock and the net of approximately 40 cm.

It is not advisable to cover walls with plastic-coated wire netting for securing the rock face. These nets are made of significantly thinner wire, and damage to the plastic coating allows corrosion of the wire. They do not acquire apatina to allow them to blend in with their surroundings.

The passive approach to supporting works against rock falls has two advantage – it is in harmony with nature and repairs are easy, nets can easily be replaced at any time.

3 AREAS FURTHER FROM THE CARRIAGEWAY

In areas further away from the carriageway one finds unstable rock in the woods. These were secured using fences (Fig. 4). The fences were assembled on the spot using cables and thicker wire netting. Existing trees were used as supports for the nets. Only the healthiest beech trees were used as supports – these trees are extremely elastic and supple, reaching an average age of 120–160 years, and so can be used longer than steel

Figure 5.   Fence in natural surroundings.

posts. To prevent damaging the bark with the cables, a soft, plastic protective layer was used. Thus protective rings were employed between the trunk and the cable. Figure 5, taken from the road, shows how well these fences blend in with their environment. They do not stand out.

4 CONCLUSION

In summary, we can say that supporting works against rock fall faces must nowadays conform with environmental requirements. This need not imply higher costs for the authorities, if engineers thoroughly examine.

Field intensity measurements of erosion instability in flysch deposits in Istria, Croatia

V. Jurak & D. Aljinović
Faculty of Mining, Geology and Petroleum Engineering, University of Zagreb, Zagreb, Croatia

J. Petraš
Faculty of Civil Engineering, University of Zagreb, Zagreb, Croatia

D. Gajski
Faculty of Geodesy, University of Zagreb, Zagreb, Croatia

ABSTRACT: The phenomenon of excessive erosion represents a geohazard. According to its effects the erosion may be considered either as a hydrotechnical problem due to the rapid reservoir loading with sediment or as a geotechnical problem due to the erosion instability of slope cut (e.g. in road construction). In both cases it is a question of earth material and field degradation that requires geotechnical remedial measures. It is to be particularly pointed out that the goal of research has been to quantify the erosion instability of flysch type clastic deposits. Erodibility of the deposits under consideration is found to be extremely high and it represents an excessive erosion that determine the geotechnical condition of certain terrain.

1 GENERAL

Erosion is to be observed in three extents:

- the extent of a sample – intact rock – where the erosion is manifested through the erodibility/durability of rock material (Jurak & Belošević 1999);
- the extent of an outcrop or an engineering structure, for instance in road cut i.e. in the rock mass extent where the erosion is manifested by a deficit of rock mass due to surface wash, slaking and rockfall which characterize the begining of load production (Jurak et al. 2002);
- in the extent of a catchment area where erosion is manifested by a change of configuration, the sediment discharge as well as sediment delivery ratio and where the erosion is closely connected with hydrological phenomenon – overland flow.

When erosion represents a geotechnical problem its effects are usually considered in the first two mentioned scales. As an example of the degradation of road cut the works of Romana & Roman (1997) and Miščević (1994) can be cited. In the latter the mentioned phenomenon in the Eocene flysch deposits is described being almost identical with the described deposits in this paper.

When erosion represents a hydrotechnical problem then the entire catchment area is considered and erosion is treated as a geohazard event on a larger scale (Jurak 2000; Jurak & Fabić 2000).

Regardless the extent of the object under consideration the erosion susceptibility of rocks and respectively the deficit of rock mass that takes place within a certain time interval (time base) is measured. In this way the results of a complex physical–geological process – erosion, are quantified.

It is well known that the main factors of erosion by surface wash and slaking are rainfalls of certain intensity and insolation which represent the characteristic climate elements of a region.

2 INTRODUCTION

The region of central Istria has all the physiogenetic predispositions necessary for the development of water erosion which is excessive in many places. The water erosion is manifested there in various forms of pluvial erosion (surface wash), particularly in denuded areas so-called badlands, by torrential outflow regime and accordingly by discharge and delivery ratio of the sediment which is mostly deposited in watercourses.

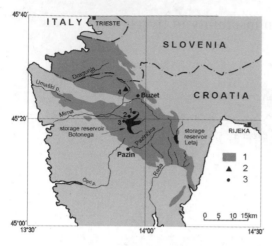

Figure 2. Position of denuded area in the surrounding terrain representing a significant topographic element – source of parent rock excessive erosion.

Figure 1. Location of erosion photogrammetric monitoring in Istria. Legend: 1 – Flysch deposits; 2 – Abrami test area and meteorological station; 3 – Photogrammetric location (badlands).

The research project covering the investigation of ground erosion in the catchment area of the Botonega multipurpose reservoir in the central Istria has been launched in order to determine preventive measures for the reservoir protection against the excessive sediment loading. Within the project in 1995 started the investigation of excessive erosion on the denuded flysch slopes. Three typical denuded flysch slopes with exposition enabling monitoring by terrestrial photogrammetry were selected for research (Fig. 1).

The investigation methodology by using terrestrial photogrammetry and computerized technology of terrain digitalization on the basis of stereopairs is described. Computation of the mass imbalance (i.e. sediment delivery ratio by slaking and slope wash) was performed for one of the locations, the one situated near St. Donat (location 1 in Fig. 1), during the period of 1995–2001. The results are compared with the ones obtained in earlier investigations on the test plane in Abrami near Buzet.

Monitoring comprised also the steep denuded areas – test parcel 1 – at the gauge station measuring soil erosion in Abrami near Buzet. In the period from 1969 to 1978 measuring of pluvial erosion was performed on the erosive test parcels within an investigation project launched by FAO at UN (Rula & Stefanović 1977). In the paper the investigation goals of the excessive erosion of flysch complex in the denuded area near St. Donat are described (Fig. 2).

The investigation methodology by using terrestrial photogrammetry and computerized technology of relief digitalization on the stereopairs basis is described as well as the computation of the mass debalance

(sediment delivery ratio). The applied methodology enables monitoring of the intensity of continuous physical–geological process – parent rock erosion. The results of the investigations performed so far on the denuded area near St. Donat are compared with the ones performed in the 70s on the location of the test area in Abrami as well as with the later investigation results obtained on the same location (Petraš et al. 1999).

3 DESCRIPTION OF DENUDED SURFACE

The object of photogrammetric data acquisition and analysis represents a completely denuded right bank of the ravine which forms the northern tributary of the Botonega reservoir. The ravine cuts the hill in a length of 800 m approx. at an elevation between 200 and 300 m a.s.l. The result was a completely denuded steep gullied surface exposed towards north-east. Thus deposits section with clearly distinguished stratification with south-west inclined layers at an angle of 10° approx. was revealed. The persistence of layers has been observed along the entire length of the denuded area (Fig. 2).

For the repeated photogrammetric data acquisition the right (north-west) part of the denuded area from Figure 2 was selected. In this way a section 60 m high and 60 m visible in its depth was comprised (Figs 3 and 4).

On the denuded surface flysch deposits crop out predominantly consisting of fine marl intervals with presence of sandstone intercalations. The marl intervals are rather thick while thickness of the sandstone layers is lesser ranging from 2 to 5 cm approx. (seldom exceeding 10 cm). By some later investigations the deposits of the mentioned petrographic properties have been defined within the flysch deposits as a globigerinid marls unit (Bergant, pers. comm.). The mentioned unit represents the sediments of the turbidite fan of distal part of the basin or more precisely Tc–Te intervals sensu Bouma.

Figure 3. Photogrammetric image of the denuded area in flysch – object of interpretation. Distribution of lithotypes and positions of topographic–lithologic sections.

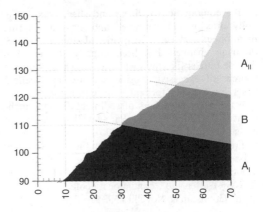

Figure 5. Topographic–lithologic section – PRO1.

Figure 4. Lateral view of topographic elements of denuded surfaces – ridges.

In the visible part of the sedimentological column three lithotypes (deposits sections) may be distinguished (Figs 3 and 4):

A_{II} – top part ... rhythmic alternation of sandstones and marls, however, sandstone intercalations are thicker and occur more frequently than at lithotype A_I; according to the position in the turbidite fan it would pertain to the central part of the fan.

B – central part ... domination of pelitic intervals. Sandstone layers continue to be tabular, however, they are thinner in relation to the lithotype A_I. Homogeneous or horizontally laminated sandstones prevail (Td interval). Marl intervals (Te) are thicker than at lithotype A_I (average thickness is 19 cm). The content of $CaCO_3$ is higher and amounts 64–73%. The main difference in relation to the lithotype A_I represents

the occurence of very thin sandstone layers (usually only 1–2,5 cm thick) and sand and silt laminae within the marl intervals. Poorly lithificated laminae contribute to the nonhomogeneity of marls and to the weak lithification of fine intervals in general.

A_I – basal part ... sediments of this part of the analyzed column correspond to the lithotype A_I and are composed of the intercalations of ripple-cross laminated very fine sandstones and thicker marl intervals. Only seldom the siltstone intercalations may be also found. Sand and siltstone intercalations are tabular and sometimes discontinuous, lensoid. The lower bedding plane is predominantly sharp, planar and the upper plane is wave-shaped. In the layer intersection a ripple-cross lamination (Tc interval) is established and it sometimes changes into plane, parallel lamination (Td interval). Also a gradual change from very fine sandstones to siltstones is present. The sandstone intercalation thickness is ranging from 1 to 10 cm, but usually it is 2 to 3 cm approx. The sandstones correspond to the types of lithic arenites (in the inferior quantity of graywackes) sometimes with substantial content of carbonate lithoclast. The marl layers are homogeneous and have a very high content of $CaCO_3$ (from 54 to 62%). The average thickness of marl intervals is 7 cm.

The described deposits are dominant parent rock of the major part of the central Istria and the observed object is a typical topographic element of the so-called "grey Istria". A selective erosion in the mentioned deposits clearly differentiates the more resistant lithologic members – sandstones/siltstones from the less durable silty marls and marls.

Consequently, the flysch areas are extremely liable to all kinds of erosion. The denuded surface occured by the combination of linear and sheet erosion i.e. gully erosion, surface wash and slaking. On the mentioned surface a network of the first-order drainage channels has been already formed. The denuded areas like the mentioned one represent the sources of excessive erosion and are an excellent basis for torrents formation. Accordingly they are also the main producers of sediments in the catchment of the Botonega reservoir as well as in the other parts of central Istria (Jurak 2000).

From the engineering standpoint the observed rock mass represents a lithologic complex in which most lithotypes pertain to the group of hard soil–soft rock (HS–SR) after the "continuous geotechnical spectrum" (Johnston & Novello 1993).

4 DESCRIPTION OF INVESTIGATION METHODOLOGY

For the quantitative monitoring of erosion on the selected test areas the method of analytical comparison of two states of the same exposed area surface in two distinct time intervals was chosen. Namely, it is known that erosion is a continuous dynamic physical–geological process and its uninterrupted qualitative monitoring is technically almost impossible. Therefore the data on the area subjected to erosion in discrete time intervals are adopted. Great number of measurements necessary to define geometrically the exposed surfaces and the possibility of precise determination of the capturing time require the application of photogrammetric method. This method enables capturing of all data visible from the exposure station in a very short time interval (exposition time). Measuring of the great number of areal points is performed later on in the cabinet supported by the automatic or semi-automatic methods. By the interpolation of the photogrammetric data the digital terrain models (DTM) are developed. They describe the real exposed area in one particular time interval (capturing moment). The analytical intersection of two DTM results in a digital difference model (DDM) which is the basis for the determination of the disposition and quantity of the eroded rock masses.

5 PHOTOGRAMMETRIC DATA ACQUISITION

Photograms of the exposed area were taken from the ground with the photogrammetric camera UMK 10/1318 (ZEISS production, Jena). Since the time base between the repeated photogrammetric recording was relatively short (30 months only), a high accuracy level of measurement was required (± 5 mm) and so a photo scale of 1:700 was chosen for the remotest part of the area under monitoring. The images were taken

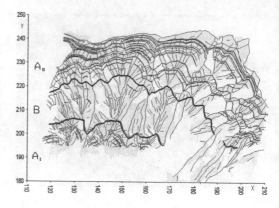

Figure 6. Structure of the measured data with lithological interpretation, ground plan – situation 2001.

on low-speed photo-plates "ORWO Topoplatte TO1" for the highest resolution and lowest deformation.

At the null point of photogrammetric data acquisition (June 12, 1995) the points of reference network were stabilized outside the field of erosion intensive acting in order to define reference coordinate system necessary for the observation of erosion on the selected location at later measurements (December 11, 1997 and February 12, 2001). Prior to every photogrammetric data acquisition geodetic surveying was done because of the deformation analysis of the reference network and the evaluation of its reliability for further monitoring of erosion.

The data acquisition for the DTM was done by analytical stereoplotter LEICA BC3 in the Institute of Photogrammetry, Faculty of Geodesy, University of Zagreb.

Two methods were used:

- stereoscopic measurement of structure lines which define discontinuity of the terrain surface;
- direct measuring of DTM to define the form of the surface between structure lines. A regular grid (dimensions 0.40×0.50 m) was measured and a total of 17,498 points was collected for one DTM (Fig. 6).

Among structure lines the ones being the result of selective erosion are predominant – in the vertical succession the competent rocks – sandstones and siltstones – stacked out.

Density of the collected points in the regular grid defines the degree of generalization of the real terrain surface elevation in relation to the interpolated surface. There are several methods to evaluate the necessary point density (Kraus 2000). In this work the density has been determined by measuring the minimal curvature radii along the characteristic profiles for each profiling direction separately.

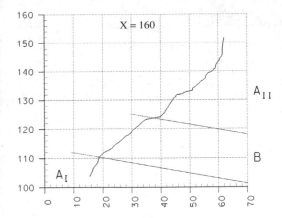

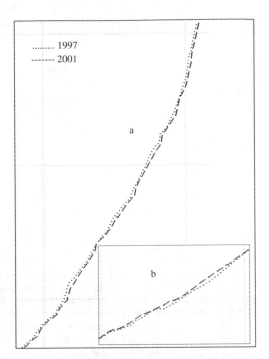

Figure 7. Interpolated profile with lithological interpretation.

6 INTERPOLATION OF DTM

On the basis of the data collected for the DTM the surface representing the denuded area was interpolated and one model for every time interval was developed. The extent of the test area was roughly $X = 90, Y = 60$ and $Z = 60$ m. Since the purpose of DTM is the computation of volume, the method of linear interpolation between neighbouring points was used for the interpolation of DTM (Delaunay triangulation).

Although the relatively large areas have been excluded from the interpolation because they were hidden by the details from the front plan or by vegetation, a great number of precisely defined points (17,498) on the denuded area offers the reliable analysis of erosion on the observed test area. Much better covering could be achieved by additional photogrammetric recording from different photogrammetric stations. The interpolation of DTM for two time intervals (1995–1997 and 1997–2001) can be used to quantify geometrical changes of the observed object within the mentioned period. Beside the quantity analyses the interpolated profiles also provide a good basis for more accurate lithologic interpretations (Fig. 7). On the profile three segments of the profile line are distinguished which is entirely in accordance with the lithologic structure of the denuded area.

The change of topography of the observed area as a result of the permanent physical–geological process – here excessive erosion – is represented in Figure 8.

7 RESULTS AND DISCUSSION

The intensity of physical–geological processes – eroding of the denuded area in the period from 1997 to

Figure 8. Details of profile X = 150.

2001 – is represented by the deficit of rock mass. The mass moved in the period between the first and the second recording, from 1997 to 2001 (time base of 38 months) is:

– eroded: 361 m³ on a planar area of 2292 m²;
– accumulated: 47 m³ on a planar area of 478 m²;
– transported outside the area of erosion monitoring: 314 m³.

The total planar area where the analysis has been carried out is 2770 m² which corresponds to the exposed surface area of 4791 m².

The accuracy evaluation of the determination of the rock-mass deficit by photogrammetric method:

– indirect determination by means of functional difference model of two DTM in different time intervals in which the mass deficit is observed:

$$\Delta V = V_{2001} - V_{1997} = \sum_{i=1}^{n} B_i (h_{2001} - h_{1997})_i \qquad (1)$$

where ΔV = total difference of volumes during the period of analysis; B = planar area of the base of ith elementary prism; h = height of ith elementary prism; n = total number of elementary prisms on the entire location.

Table 1. Comparison of the measurement and monitoring results of excessive erosion.

Location	Method	Time base (year)	Eroded quantity (m^3)	Exposed surface area (m^2)	Planar area (m^2)	Average annual deficit of rock mass per exposed surface area ($m^3/m^2/year$)
Abrami – erosion parcel I	Capturing of eroded mass	(1970–1976) (76 months) 6.333	~1.024*	~30	15.08	0.005
Abrami – erosion parcel I	Terrestrial photogrammetry	(1995–1997) (30 months) 2.5	0.445	8.9**	3.70	0.020
St. Donat	Terrestrial photogrammetry	(1995–1997) (30 months) 2.5	215	3856	2191	0.022
St. Donat	Terrestrial photogrammetry	(1997–2001) (38 months) 3.167	361	4239	2292	0.027

* Based on 158 rainfalls which are representatives in the computation of sediment delivery ratio for every parcel (in vegetation period); ** the upper one third of the original surface.
The data for the first row from Rula & Stefanović (1977); for the second and the third row from the report "Photogrammetric Monitoring of Erosion in Istria on the Locations Abrami and St. Donat" (1999) (in Croatian).

Assuming the presence of the random errors of the measured values h only and applying the error propagation law it follows:

– standard deviation of the height difference

$$m_{h(2001-1997)} = \pm\sqrt{m^2_{h(2001)} + m^2_{h(1997)}} \quad (2)$$

– standard deviation of the mass deficit determination

$$m_{\Delta V} = P \cdot m_{h(2001-1997)} \quad (3)$$

where P = planar area of observed surface.

In addition to the standard deviations of photogrammetric determination of heights which reads: $m_{h(1997)} = \pm2.5\,mm$ i $m_{h(2001)} = \pm3.4\,mm$ the standard deviation of the determination of the rock-mass deficit between the mentioned time intervals is obtained in an absolute amount of $m_{\Delta V} = \pm11.6\,m^3$ which in a relative amount is $m_{\Delta V} = \pm3.2\%$ (for the location St. Donat in the period from 1997 to 2001).

On the observed area a part of the eroded mass has been accumulated in the places where, at least temporary, the sediment can be retained (on the concave segments of the profile line). Differentiation of the topographic profile according to the basis of prevailing process is assumed. In the upper, steeper part of the denuded area (Fig. 8a), crumbling and slaking prevail as a result of thermodynamic process while in the lower part (Fig. 8b) washing – pluvial erosion – prevails. This fact could offer perhaps the explanation for the discrepancy of the results obtained by the two methods applied at test parcel 1 in Abrami (Table 1).

Selective erodibility of the intact rock – sandstones and siltstones on one side and silty marls and marls on the other side – at the end results in erosional instability of the entire exposed rock face. Thus the outcrop/denuded area becomes the source of excessive erosion in the surrounding terrain.

The measurement result represents the deficit of rock mass that integrates all the erosion processes of all lithologic members of a rock complex, flysch in this case. In this way the measurements by terrestrial photogrammetry may be considered as in situ geotechnical measurements.

REFERENCES

Jurak, V. 2000. Torrential catchment raindrop erosion in the Central Istria (Croatia). In G.B. Carulli & G. Longo Salvador (eds), Riassunti delle comunicazioni orali e dei poster. 80a Riunione estiva, Trieste, 6–8 September 2000: 296–298. Trieste: Edizioni Università di Trieste.

Jurak, V. & Belošević, V. 1999. Durability of the "Macelj Sandstones" (In Croatian). In I. Jašarević et al. (eds), Rock Mechanics and Tunnelling; Proc. Sci. Symp., Zagreb, 30 September–2 October 1999, 1: 47–52. Zagreb: Faculty of Civil Engineering and Faculty of Mining, Geology and Petroleum Engineering.

Jurak, V. & Fabić, Z. 2000. Torrential Catchment Raindrop Erosion in the Central Istria (In Croatian). In I. Vlahović & R. Biondić (eds), Proc. 2nd Croat. Geol. Congr., Cavtat – Dubrovnik, 17–20 May 2000: 603–612. Zagreb: Institute of Geology.

Jurak, V., Petraš, J. & Gajski, D. 2002. Research into excessive erosion of bare flysch slopes in Istria by use of terrestrial photogrammetry (In Croatian). *Hrvatske vode* 10(38): 49–58. Zagreb: Hrvatske vode.

Johnston, I.W. & Novello, E.A. 1993. Soft rocks in the geotechnical spectrum. In Anagnostopoulos et al. (eds), *Geotechnical Engineering of Hard Soils–Soft Rocks. Proc. Intern. Symp.*, 1: 177–183. Rotterdam: Balkema.

Kraus, K. 2000. *Photogrammetrie Band 3, Topographische Informationsysteme.* Bonn: Dümmler.

Miščević, P. 1994. The Erosion of Surface Made on Flysch Layer (In Croatian). In R. Mavar (ed), *Geotechnical Engineering in Transportation Projects; Proc. symp., Novigrad, Croatia, 5–8 October 1994*, 1: 339–346. Zagreb: Civil Engineering Institute of Croatia.

Petraš, J., Kunštek, D. & Gajski, D. 1999. Application of Terrestric Photogrammetry in Research of Excessive Erosion Processes (In Croatian). In D. Gereš (ed), *Croatian Waters from the Adriatic to the Danube; Proc. 2nd Croat. Conf. on Waters, Dubrovnik, 19–22 May 1999*: 1029–1036. Dubrovnik: Hrvatske vode.

Romana, M. & Roman, F. 1997. Erosion behavior of a great cut in limestones and claystones at Sagunto (Spain). *Intern. Symp. The geotechnics of structurally complex formation, Capri '77*, Vol. 1: 415–421. Associazione geotechnica Italiana.

Rula, B. & Stefanović, P. 1977. Prikaz istraživačkog rada iz oblasti erozije i razvoj delatnosti u Odseku za zaštitu slivova. *Saopštenja Instituta za vodoprivredu "Jaroslav Černi"*, 60–63: 225–230. Beograd.

Geotechnical Measurements and Modelling, Natau, Fecker & Pimentel (eds)
© 2003 Swets & Zeitlinger, Lisse, ISBN 90 5809 603 3

Better understanding of field measurements by means of Data Mining

A. Prokhorova & M. Ziegler
RWTH Aachen, Geotechnical Engineering, Aachen, Germany

ABSTRACT: Data Mining is a process to recognize patterns and structures in a large amount of data and to derive rules for their correlation. These rules provide the prediction of particular parameters. The basic principles of Data Mining and possible applications in the field of geotechnics are illustrated and exemplified by the interpretation of the results of cone penetration tests.

1 INTRODUCTION

Ground investigation takes place with the help of selected explorations only. For densification, the exploration points are expanded with indirect exploratory methods. The advantages lie in an area-wide effect and lower cost intensity. Indirect ground-exploration, however, does not provide explicit values for geotechnical parameters. Rather one tries to transfer the measured parameters from indirect exploration into soil-parameters as base variables for characteristic values, defined for geotechnical calculations. For the comprehension of the coherence between measured and derived values an extensive knowledge is inevitable. Quantitative coherence between the results of an indirect analysis and a geotechnical parameter can be determined deterministically. A high physical complexity of the coherence between some soil-parameters and the measured variables, however, makes a deterministic assignation of the parameters difficult or even impossible. In these cases one tries to acquire geotechnical parameters by means of statistics. The methods based on regression analysis using classical statistics are of limited use, because these methods are often complicated in application and often deliver no transparent results.

A solution to this dilemma can be the application of Data Mining methods. Data Mining defines itself as a process to recognize patterns and structures in a large amount of data and to generate rules. By means of these rules groups can be defined. Through allocation of an element to a group one can predict its behaviour according to the "behaviour model" of the group. Unlike classical statistical analyses Data Mining uses not only single characteristic values of individual, but also information on combinations and correlation of variables features in the search of logical rules. Despite this difference statistical means are applied also within many Data Mining algorithms.

2 DATA MINING ALGORITHMS

The following algorithms are applied in the context of Data Mining:

– Association rules
Association rules tell us that if any event X occurs, event Y is likely to occur.

– Classification methods
The aim of the classification is to find a batch of rules. This batch permits to classify many objects. Classification rules then are required, if the absence of information demands the derivation of the attributes of objects. Such rules are learned by using an assemblage of preclassified training examples.
A possible classification method is a classification by means of decision trees. Herein the criteria or rules are displayed in a tree structure.

– Bayesian classification
This method is based upon the conditional probability rule from Bayes. Given a hypothesis H and an event E compatible to it, the probability of H can be approximated by:

$$P(H \mid E) = P(E \mid H) \cdot P(H)/P(E).$$

– Clustering
In contrast to classification methods the classes are not prior known but are found by arrangement of

objects in groups. Data points in one cluster are to be more similar to one another and data points in separate clusters are to be less similar to one another. This similarity enables to ascertain the unknown attributes of objects from any cluster.

3 EXAMPLES OF DATA MINING – METHODS IN GEOTECHNICS

3.1 *Application of Bayes Theorem for the determination of a soil type on the basis of the cone penetration tests*

With the aid of Bayes rule we can correct the accuracy of soil type prediction by using the knowledge about the typic values of CPT sounding data for any area. It is known that the friction ratio $R_f = f_s/q_t$ of sand and clay has different values. Let us assume that the results for several cone penetration tests for any area exist. Due to these results the probability distribution of friction ratio values can be defined for each soil type.

We can define three soil types as: sandy soil (class S), clayey soil (class T) and mixed soil type (class M). Zhang and Mehmet (1999) take the normal distribution of parameter U (soil engineering classification index) for each soil type in their work. The parameter U has a statistical correlation with soil types defined by a soil type index SI (soil engineering classification index). This type of distribution can also be assumed for the friction ratio R_f. It is known that the R_f value for sand is smaller than for clay and the value of mixed soil lies in-between.

If all distributions are charted, one can notice, that the graphics are overlapping (see Figure 1). That means, for different soil types the same values of R_f are possible but with different probabilities. We divide the range between the minimum R_f, which can be found for sand, and the maximum R_f, being found for clay, in three different sub-ranges, they are R_{fI}, R_{fII} and R_{fIII}, and which boundary values are defined as R_{f1}, R_{f2}, R_{f3} and R_{f4}.

The probabilities of finding the R_f value for any soil classes in any sub-range among CPT sounding results can be derived as follows:

$$P(R_{fI}|S) = \int_{R_{f1}}^{R_{f2}} f_S(R_f)\,dR_f = 0,8$$

$$P(R_{fI}|M) = \int_{R_{f1}}^{R_{f2}} f_M(R_f)\,dR_f = 0,25$$

$$P(R_{fI}|T) = 0$$

$$P(R_{fII}|S) = \int_{R_{f2}}^{R_{f3}} f_S(R_f)\,dR_f = 0,20$$

$$P(R_{fII}|M) = \int_{R_{f2}}^{R_{f3}} f_M(R_f)\,dR_f = 0,5$$

$$P(R_{fII}|T) = \int_{R_{f2}}^{R_{f3}} f_T(R_f)\,dR_f = 0,1$$

$$P(R_{fIII}|S) = 0$$

$$P(R_{fIII}|M) = \int_{R_{f3}}^{R_{f4}} f_M(R_f)\,dR_f = 0,25$$

$$P(R_{fIII}|T) = \int_{R_{f3}}^{R_{f4}} f_T(R_f)\,dR_f = 0,9$$

From this useful information, one can enhance the accuracy of soil type determination for identical basic conditions. If for example the predisposed probability for which the soil type is sandy is 20%, clayey type – 50% and mixed type – 30%, this probability can be corrected based on Bayes Theorem. If for example the R_f value from CPT sounding at a certain place lies in sub-range between the values R_{f2} and R_{f3}, then also the probability of a clayey soil can be calculated as follows:

$$P(T|R_{fII}) = P(R_{fII}|T) \cdot P(T) = 0,1 \cdot 0,5 = 0,05$$

The probability of encountering a sand or mixed soil can be calculated in an analogical way:

$$P(S|R_{fII}) = P(R_{fII}|S) \cdot P(S) = 0,2 \cdot 0,2 = 0,04$$

$$P(M|R_{fII}) = P(R_{fII}|M) \cdot P(M) = 0,5 \cdot 0,3 = 0,15$$

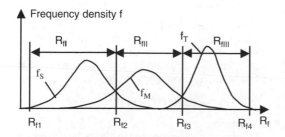

Figure 1. R_f-distributions for all soil types.

Through a standardisation of computed values one expects to get a percentage of probability for each assumption:

$$P(T|R_{fII}) = \frac{0,05}{0,05 + 0,04 + 0,15} = 21\%$$

$$P(S|R_{fII}) = \frac{0,04}{0,05 + 0,04 + 0,15} = 17\%$$

$$P(M|R_{fII}) = \frac{0,15}{0,05 + 0,04 + 0,15} = 62\%$$

The corrected probability for the type "mixed soil" is higher than for the clayey soil type. In that case the significance of predisposed probability would be examined. If for example they are based on expert opinion, the experts are asked to explain their forecast.

As alternative to this method the a-posteriori knowledge about the probable R_f values can be used not in form of a statistical distribution function, but be applied as Data Mining rules.

A difficulty of using this classical statistical procedure, which is often being used by Data Mining algorithms, consists in the necessity of an assumption for the distribution of the parameter. A less traditional method is the classification on the basis of decision trees.

3.2 Soil classification via decision trees

With the help from an extremely simplified virtual example the approach to classification with Data Mining methods will be explained. Six CPT data sets will be used. Each data set contains the CPT sounding data from any point with the values of the following measurements: sleeve friction, cone tip resistance and excess pore pressure. With the help of direct explorations we can assign the first 3 cases to soil type I (cohesive soil) and the other 3 cases to soil type II (non-cohesive soil). For a clear-cut description of each situation, we use 3 variables x_i. The variable x_i has a value 1 if the corresponding parameter has a value within a

Table 1.

Case	$f_s \in$ $[f_s^A; f_s^B]$ x_1	$q_t \in$ $[q_t^C; q_t^D]$ x_2	$\Delta u \in$ $[\Delta u^E; \Delta u^F]$ x_3	Soil type
1	0	1	1	I (noncohesive)
2	1	1	0	I (noncohesive)
3	0	1	0	I (noncohesive)
4	1	0	1	II (cohesive)
5	1	1	0	II (cohesive)
6	0	0	1	II (cohesive)

Table 2.

	x_1	x_2	x_3
S (I)	1	3	1
S (II)	2	1	2
D	1	2	1

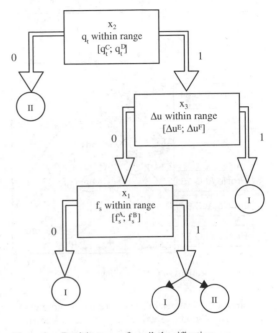

Figure 2. Decision tree of a soil classification.

predefined range of values. In all other cases the variable x_i is of the value 0.

The soil type allocation rules for these cases can also be found. These rules must allow classifying similar cases.

The variable with a maximum importance for categorisation has to be found.

A maximum difference between the sums of values for the soil type I (S_I) and for the soil type II (S_{II}) $D_{max} = \max(|S_I - S_{II}|) = |3 - 1| = 2$ leads to the variable x_2. The classification would be commenced with this attribute. All cases with $x_2 = 0$ belong to the second soil type.

The search for the classification variable would be continued within all cases with $x_2 = 1$. This procedure would be repeated until no further subdivision is possible.

The resulting decision tree is displayed in Figure 2.

From Figure 2, the value of sleeve friction for a soil classification is not always determining. As a computer-aided method we can optimise this method

293

Table 3.

Identifier	Soil type
1	Sand to silty sand
2	Silty sand to sandy silt
3	Sandy silt to clayey slit
4	Clayey silt to silty clay
5	Clay

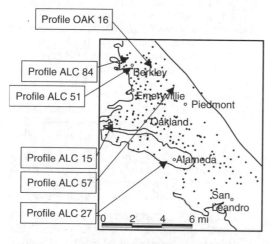

Figure 3.   Locations of CPT soundings.

by variation of ranges of values for the parameters f_s, q_t and Δu. The above written technique is known as ID3 (Interactive Dichotomizer). The principle of these methods consists in a cyclical classification of training data in classes according to the value of the leading differentiating variable. There are more decision tree methods existing: CHAID (chi square automatic interaction detection), that uses the Chi-Quadrat-criterion, CART (classification and regression trees) and others.

The above shown classification method would be applied to CPT sounding data. These data were collected by the USGS Western Earthquake Hazards Team. The data consist of processed CPT soundings from Alameda Country as displayed on the map in Figure 3 (http://quake.wr.usgs.gov).

From several CPT sounding data we have chosen 6 CPT soundings. Locations of soundings are marked by an arrow in Figure 3. Graphic logs of soundings are shown in Figure 4 that display friction ratio, tip resistance and sleeve friction. The stratigraphy of these logs was known. From these data one can also obtain the classification rule for the specific soil classes shown in Table 3.

The decision tree, represented in Figure 5, is evaluated with the help of Data Mining program. See Figure 5.

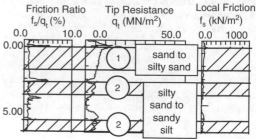

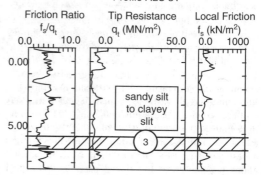

Figure 4.   Graphic logs of CPT soundings.

If one applies the rules for another CPT sounding data from this area, the error of soil classification is about 5%. Such a high accuracy of prediction is caused by the similar geological genesis of the classified soils.

Figure 6 shows the classification diagram for the specific soil type depending on cone tip resistance and friction ratio. The dotted lines and the textual reports are taken from Lunne, Robertson and Powell (1997). The shaded areas corresponding with decision tree from Figure 5 are additionally plotted in Figure 6. The identifiers of soil types correspond to Table 3. Obviously, this representation demonstrates a good accordance of the regional specific rules with the universally valid classification rules.

Profile ALC 57

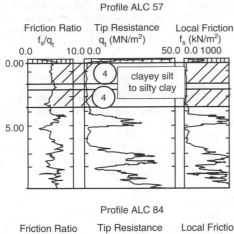

Profile ALC 84

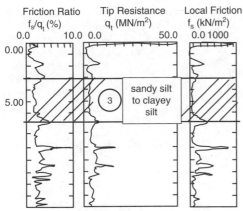

Profile OAK 16

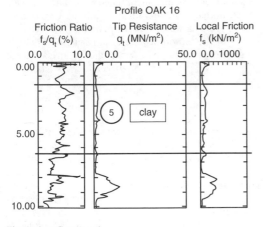

Figure 4. *Continued.*

This example shows that, with the help of Data Mining, one can find fast and less complex classification rules for the regional data. The accuracy of rules depend on the data quality which is warranted due to a lot of sounding logs with known stratigraphy.

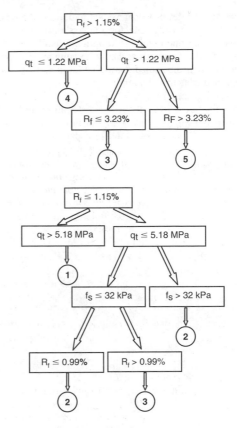

Figure 5. Decision tree of a CPT classification.

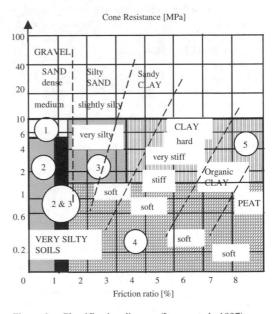

Figure 6. Classification diagram (Lunne et al., 1997).

295

The advantages of the presented method is the possibility to select the data for analysis in a reasonable way on the one hand and to quickly handle a lot of data on the other hand. These advantages minimise the error in the interpretation of large data sets and ensure a user-friendly application.

REFERENCES

Baldi, G., Bellotti, R., Ghionna, V.N., Jamiolkovski, M., Pasqualini, E. 1986. Interpretation of CPTs and CPTUs, Part 2: Drained penetration of sands. *4th International Geotechnical Seminar on Field Instrumentation and In situ Measurements*: 143–156. Nanyang Technological Institute, Singapore.

Chen, B.S.Y., Mayne, P.W. 1996. Statistical relationships between piezocone measurements and stress history of clays. *Canadian Geotechnical Journal* 33: 488–498.

Duk, W., Samojlenko, A. 2001. *Data Mining*. St. Petersburg: Piter.

Lunne, T., Robertson, P.K., Powell, J.J.M. 1997. *Cone Penetration Testing in Geotechnical Practice*. London: Blackie Academic & Professional.

Robertson, P.K. 1990. Soil classification using the cone penetration test. *Canadian Geotechnical Journal* 27: 151–158.

Wroth, C.P. 1984. Interpretation of in situ soil test. *Geotechnique* 34: 126–135.

Zhang, Z., Mehmet, T. 1999. Statistical to fuzzy approach toward CPT soil classification. *Journal of Geotechnical and Geoenvironmental Engineering* 3: 244–265.

Adaptive ground modelling in geotechnical engineering

M. Schönhardt & K.J. Witt
Professorship of foundation engineering, Bauhaus-University Weimar, Germany

ABSTRACT: Geotechnical design depends on parameters got from field investigations. Therefore the development of subsoil models for assessing the behaviour of the subsoil has a strong influence on the reliability of geotechnical structures. The contribution deals with an adaptive improvement of soil models using geostatistical methods, demonstrated for a data set of a simplified practical example. Starting with subjective interpolations based on two data sets from a preliminary a detail investigation, the influence of additional integrated geotechnical model assumption is shown. In a next step further regionally information are integrated into the model. This given soil information are necessary to calibrate the results on any location in the field using the smoothing splines method. For a virtual machine footing the influence of the generated model precision on the dimension of an assumed square size isolated machine foundation is analysed. The uncertainties and spread of all characteristic quantities in the limit state function, an equation for determine the skew position of the footing, is taken into account. Using the Monte Carlo Simulation for uncorrelated values, the reliability of the isolated foundation is determined. Finally, the different predictions of the adaptive modelling are compared with the real soil conditions at the site.

1 INTRODUCTION

One of the most essential tasks of geotechnical engineering is to develop a geological model of the subsoil. With different techniques we make a limited number of observations and measurements and we draw scientifically defensible conclusions from them. Based on different field monitoring data we have to build up a geological model of the site conditions, describing the layers, kind of soil, mechanical properties as well as the spatial variability of those parameters. The reliability of such a model is fundamental for all decisions during design procedure. Thus, the interpretation of field data requires insight into the geology, its properties, details and anomalies but also experience with foundation engineering and construction practice. In the stage of site investigation we usually deal with two typical problems (i) very few observations and measurements due to limited budged and (ii) only rough information about the structure of the building, the quantity and location of loads. Therefore we have to build up the geological model in a more or less adaptive manner, using successive steps of investigation as well as geostatistical methods for controlling, supporting and improving the interpretation. These steps should be demonstrated with a simplified practical example.

For planning a new large industrial plant, the underground was investigated at a stage with less information about the structure, the number and size of the machines, their loads and their foundations. Figure 1 shows a map of the machine hall with some possible areas for the machine placement. From a preliminary course of investigation, using hand driven auger drillings, dynamic probing and some core drillings, we got general data about the subsoil of the overall area. This step delivers level 1 information. In a detailed second level focussed on the machine hall we carried out some more and deeper core drillings and a couple of static cone penetration tests in a systematic screen. All points of investigation are marked in figure 1.

Evaluating and assessing the test results the subsoil surround the machine hall can be subdivided in 3 relatively homogeneous layers with a more or less random stratum: A thin upper silty clay with a soft consistency, a second layer out of a stiff clay reaching down to 8–14 m and under that until a depth of 70 m a sandy clay with a very stiff and hard consistency.

The data of the different steps of exploration are used for an adaptive modelling of homogeneous profiles with help of geostatistical methods. Thus we can demonstrate the influence of limited investigation to the reliability of the prediction. At least we can use this certain levels of prediction to design a virtual isolated

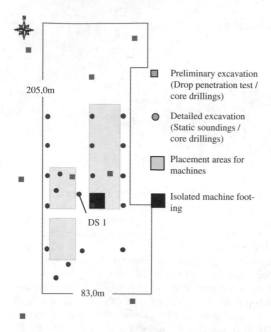

205,0m

Preliminary excavation
(Drop penetration test /
core drillings)

Detailed excavation
(Static soundings /
core drillings)

Placement areas for
machines

Isolated machine foot-
ing

DS 1

83,0m

Figure 1. Arrangement of preliminary and detailed excavation/placement of the foundations.

footing in the centre of the machine hall in a probabilistic manner. To compare the prediction under certain levels of information with the reality, the virtual footing was placed at a point of investigation without using this real data in the process of prediction.

2 HOMOGENEOUS PROFILES

Starting with a subjective estimation of a general homogeneous profile on the base of the preliminary and the detailed investigation the homogeneous profile is specified in the single adaptive steps supported by geostatistical methods under consideration of scattering values. The scattering values and its uncertainties are integrated through a wide range of repetition in the numerical computation. Every parameter set is different from each other. Every quantity is determined by there characteristic value, the standard deviation and the statistical distribution (Schönhardt 2002).

At first the homogeneous profile was estimated subjective using the results of the preliminary investigation. After that the data set has been extended through the data of the cone penetration. This leads to a new homogeneous profile. The reliability of both data sets shall be equal, so that no weighting factors a necessary for bringing together the different field investigations.

The next adaptive step will take the results of the level 1 investigation. By using geostatistical methods

on the base of variographie and universal kriging the homogeneous profile for the foundation was determined more precisely.

In the next step this soil model was extended through the data set of the detailed investigation.

In a further step of the adaptive modelling a consistency proving method was integrated. Here additional regional information in form of a given homogeneous stratification is necessary.

Finally the different created homogeneous predictions are compared with the real soil profile on the foundation place. This information is derived by a separately static sounding.

3 PROCEDURE OF MODELLING

The results of the preliminary investigation are used for a first soil model. The subjective arithmetically estimation of the thickness for each homogeneous layer is mostly too safe. The single estimations of the layers thickness are built into a 3d-soil model. The strata of the homogeneous surfaces are presented in table 1.

Specifically characteristic properties will be only conditional considerate. In the neighbourhood of the foundation we have only a low density of information. The second level of investigation improved this weak point.

From the new greater data set we derived in a second estimation a new set of the absolute heights of the homogeneous layers (table 2).

The evaluation of the results presented in table 1 and table 2 shows that the surfaces form layer 1 and layer 2 are similar. The stratum of the layer 3 is different. The difference is expressed in the standard deviation of the arithmetical determination. Both estimations contain uncertainties, because internal trend and the anisotropic behaviour within the different data are not considered.

A third estimation using geostatistical method takes the anisotropy and trends into account, Christakos 1992. The result is an improved 3d-soil model. Internal computations are necessary to distinguish between local drift and global trend. The drift function is indicated by a linear combination, which extended the normal universal Kriging matrix.

The analysis of the data shows that there is a global trend and different drifts. The trend can be imposed by fluctuated values. In addition there is an anisotropy direction in the area. Every research area has an own anisotropy ratio $a_1:a_2:a_3$. In opposite of the trend is expected value is equally, but the spatial correlation is different. The spatial range in the variogramm computation differs. The anisotropic soil can be seen as an ellipsoid soil model, figure 3.

The prerequisite for geostatistical computations is the isotropy, Clemens 1995. The main isotropy direction here is determined to 350° NNW. Therefore the actual soil properties must be transformed from the ellipsoid to the unit sphere.

After during that on the base of the whole date population from all levels using the self developed software application GEOSTAT a fourth and fifth homogeneous soil profile is generated. The homogeneous profile for the foundation place calculated in this manner, is given in table 3 and table 4 columns 1 and 2.

The sum of thickness of the modelled layers differs. To guarantee the consistency is it necessary to involve further information. An additional given homogeneous profile will be integrated in the soil model using the method of smoothing splines. The thickness of the single layers will be corrected before they are put together in the model, figure 4.

The uncertainties and spread of the characteristic quantities consider by a wide range of single

Table 1. Strata of the layers, using the preliminary investigation [m].

	Stratum	Stand. deviation
Layer 1	129,94	1,648
Layer 2	127,83	1,361
Layer 3	119,76	2,036

Table 2. Strata of the layers, using the preliminary and the detail investigation [m].

	Stratum	Stand. deviation
Layer 1	130,22	1,274
Layer 2	127,94	1,329
Layer 3	117,89	3,193

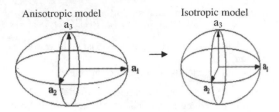

Anisotropic model   Isotropic model

Figure 3. Anisotropic and isotropic soil model simplification.

Table 3. Strata of the layers, using the preliminary investigation combined with geostatistical methods [m].

	Stratum 1	Standard deviation 2	Stratum [consistent] 3	Standard deviation [consistent] 4
Layer 1	129,60	1,412	129,16	1,507
Layer 2	127,44	1,100	126,99	1,411
Layer 3	119,38	1,460	118,22	1,220

Table 4. Strata of the layers, using the preliminary and detail investigation combined with geostatistical methods [m].

	Stratum 1	Standard deviation 2	Stratum [consistent] 3	Standard deviation [consistent] 4
Layer 1	129,73	1,251	129,21	1,148
Layer 2	127,48	1,040	127,06	0,929
Layer 3	118,22	1,810	118,18	0,950

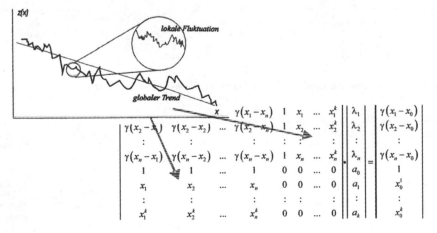

Figure 2. Drift function implemented in Kriging matrix Schafmeister 1999.

299

realisations. The inserted smoothing splines method has to satisfy two conditions

– each homogeneous layer have to exist in the research area
– the homogeneous layer may not intersect.

The sixth and seventh generated soil model based on this mentioned algorithm. The results at the modelled location are presented in table 5 columns 3 and 4. For the results in column 3 is used the reference profile from table 4 and for column 4 the reference profile from DS 1.

Table 6 shows a summary of the strata of the different homogeneous layers (L1–L3). Columns 1–4 present the adaptive soil modelling. To prove the quality of the

single simulation results on the foundation location column 5 shows the result of the measurement.

The geostatistical determination of the homogeneous profile, column 2, has a good validity in layer 1 and layer 2 to the reference profile, but the result is not consistent. The difference in the thickness from layer 1 to 3 can be expressed $|15,72\,\text{m} - 18,00\,\text{m}| = 2,28\,\text{m}$. This difference is not a measure for the quality of the geostatistical soil model.

The additional homogeneous profiles, like mentioned before, improved the estimation. The results get better with the increasing quality of the additional information, the inputted homogeneous profile. So the estimated stratum of layer 3 in column 3 is not better than in the steps before. Is it possible to integrate a

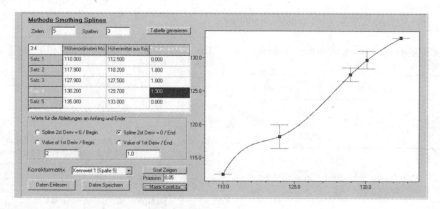

Figure 4.   Exemplary use of the smoothing splines method as matrix adaptation scheme in Geostat.

Table 5.   Summary of the strata of the different soil models at the location [m].

	Preliminary excavation 1	Detail excavation 2	Detail excavation consistent 3	Detail excavation consistent 4	Reality 5
L 1	129,62	130,74	130,18	130,78	130,70
L 2	127,46	127,75	127,28	127,47	127,45
L 3	119,42	115,02	116,45	112,95	112,70

Table 6.   Characteristic quantities and statistical parameter for shear modulus G_o and the stiffness modulus E_s.

	Layer 1	Layer 2	Layer 3
Shear modulus G_o (NV) [MN/m^2] E[x]/E[x]^2	25.017	32.300	115.540
	7.532	4.870	31.401
Shear modulus G_o (LNV) [MN/m^2] E[x]/E[x]^2	23.253	32.600	121.225
	5.317	4.840	34.456
Stiffness modulus $E_{so} = 4,33 G_o$ (NV) E[x]/E[x]^2	108.323	139.859	500.288
	32.613	21.087	135.966
Stiffness modulus $E_{so}^{-1} = 0,23 G_o^{-1}$ (NV) E[x]/E[x]^2	$1,02 \cdot 10^{-5}$	$7,15 \cdot 10^{-6}$	$2,00 \cdot 10^{-6}$
	$2,92 \cdot 10^{-6}$	$1,16 \cdot 10^{-6}$	$6,54 \cdot 10^{-7}$
Characteristic quantity E_{so} [MN/m^2]	98.407	136.063	486.590

reference profile in the near of the planed foundation location, reproduce the soil model a good agreement with the in-situ situation, column 4 and 5.

A geostatistical 3D-Model on the base of a given data set can be used only for limited expressions for chosen single location. In the model important information to the trend and isotropy behaviour in the research area are integrated. These are not enough for expressions to small locations in the field. The additional integrated reference profile can improve the results regionally.

The results of this adaptive improvement used to design of an isolated machine footing at the reference point. The uncertainties and spread of integrated parameters will be involved by using a high number of realisations. As numerical model we use the Monte-Carlo Simulation (MCS).

4 DETERMINATION OF CHARACTERISTIC QUANTITIES

4.1 General

From the results of the cone penetration tests and the static sounding tests the resistance parameter q_{c1} is derived. The sounding results have been evaluated every 25 cm and they have been summarized in one data set to one parent population.

Intermediate values are linear interpolated. As a result the in-situ soil is subdivided in three homogeneous layers, the homogeneous profile. For settlement analysis it is necessary to distinguish between the shear modulus for small strains and for normal strains, because the dependency between the resistance parameter q_{c1} and the shear modulus G_o exist only for small strains (figure 5 and figure 6). The Young's modulus E, the stiffness modulus E_s and the shear modulus G are linear dependent.

The shear modulus can be transformed into the Young's modulus and in the stiffness modulus using a predefined Poisson's ratio ν. For all calculations we used an uniform ratio $\nu = 0.35$.

$$G_o = \frac{E_o}{2 \cdot (1 + \nu)} \tag{1}$$

$$E_{so} = \frac{1 - \nu}{1 - \nu - 2 \cdot \nu^2} \cdot E_o$$

$$= \frac{1 - \nu}{1 - \nu - 2 \cdot \nu^2} \cdot 2 \cdot (1 + \nu) \cdot G_o$$

$$E_{so} = \frac{2 - 2\nu^2}{1 - \nu - 2 \cdot \nu^2} \cdot G_o \tag{2}$$

$$E_{so} = 4,33 \cdot G_o \quad \text{oder} \quad \frac{1}{E_{so}} = 0,231 \cdot \frac{1}{G_o}$$

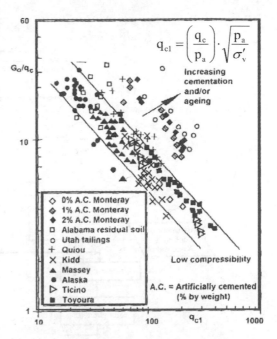

Figure 5. Ratio between tip pressures qc of CPT and the shear modulus G_o for small strains, Robertson 1997.

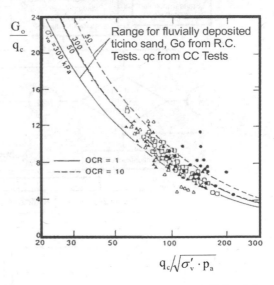

Figure 6. Ratio between tip pressures qc of CPT and the shear modulus G_o for small strains IMAI/TONOUCHU 1982.

4.2 Characteristic quantities

The data included in the different adaptive model steps are combined to a parent population. The characteristic quantities are determined by the scheme of Eurocode 7 (EC 7) and Schuppener 1999. Under the assumption of normal distributed data the scheme is shown in figure 7.

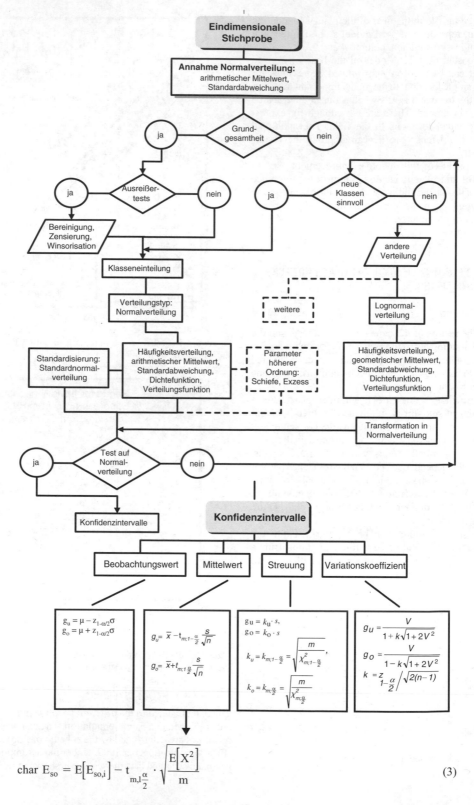

$$\text{char } E_{so} = E[E_{so,i}] - t_{m,1\frac{\alpha}{2}} \cdot \sqrt{\frac{E[X^2]}{m}}$$

(3)

Figure 7. Determination of characteristic quantities.

Without outliers and eliminated trends in the data set the characteristic quantities, the corresponding mean values and standard deviation are shown in table 6.

On the base of statistical tests the data sets are normal distributed. The density functions in the numerical simulation haven't a left side or a right side boundary. The randomly used parameter set for each calculation step is determined from the whole range. This assumption is possible, because the relation between the mean value $E[X]$ and the accompanying standard deviation $E[x]^2$ is

$$E[x] - 3 \cdot E[x]^2 \geq 0$$

The probability, that negative values will be simulated is less than 0,01%.

The following calculations of the skew position of the isolated foundation will use the characteristic quantities. Inside the Monte-Carlo simulation of the limit state function the random values will be determined using the mean value and the standard deviation of each involved data set.

5 FAILURE PROBABILITY OF FOUNDATION

5.1 General

For the design of the isolated machine foundation the bearing capacity and the deformation capability is to investigate. In this example we look for the ultimate serviceability limit state and define the skew position $\tan \alpha$ as limit state.

5.2 Limit state function

The technical guidelines demand a small skew position. As limit state function we use the following settlement equation.

$$
\begin{aligned}
s_m + s_x = \sigma_o \cdot b \cdot & \left[\frac{f_1}{E_{m,1}} + \sum_{i=2}^{n} \frac{(f_i - f_{i-1})}{E_{m,i}} \right] \\
& + \frac{M_y}{b^3} \cdot \left[\frac{f_{x,1}}{E_{m,1}} + \sum_{i=2}^{n} \frac{(f_{x,i} - f_{x,i-1})}{E_{m,i}} \right]
\end{aligned}
\quad (4)
$$

The constant immediate settlements s_m are not of interest. The limited state is defined to $\tan \alpha \leq 0,008$. The function can be simplify for the determination of the skew position to

$$
s_x = \frac{b}{2} \cdot \tan \alpha = \frac{b}{2} \cdot \frac{M_y}{b^3} \cdot \left[\frac{f_{x,1}}{E_{m,1}} + \sum_{i=2}^{n} \frac{(f_{x,i} - f_{x,i-1})}{E_{m,i}} \right]
\quad (5)
$$

or

$$
g\left(\frac{1}{E_{mi}} \right) = \frac{M_y}{b^3} \cdot \left[\frac{f_{x,1}}{E_{m,1}} + \sum_{i=2}^{n} \frac{(f_{x,i} - f_{x,i-1})}{E_{m,i}} \right] - 0,008
\quad (6)
$$

For $g(1/E_{m,i}) \leq 0$ the construction fail and for $g(1/E_{m,i}) \geq 0$ the construction is reliable.

5.3 Stiffness modulus

The stiffness modulus and the shear modulus are elastic parameters and depend on the present strain state, Yamashita 2000. It is to decide between the stiffness modulus for small strains E_{so} and the stiffness modulus for normal strain E_s, Simons 2000.

With the relationship in figure 5 and figure 6 the stiffness modulus for small strains can be determined. As a function we use a bilinear relationship, figure 8. The reference stiffness modulus, determined in laboratory tests, is labelled as $E_{m,min}$.

The skew position of the foundation depends on the present stiffness modulus. The maximum of elastic strains is directly under the isolated foundation, so that the bilinear function reduced the stiffness modulus higher than in deeper layers. Therefore the homogeneous layer 1 and 2 are subdivided in 20 sections. For universal calculations we use a software application, developed on the professorship of foundation engineering. This program considers the characteristic quantities and the statistical parameter set of each data set.

5.4 Failure probability

The permitted failure probability p_f is subjective. Therefore we used the recommendation of Deutsches Institut für Normung 1981. On the base of safety classes the safety index β is classified.

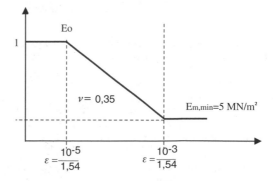

Figure 8. Empirical relationship between the stiffness modulus and elastic strains, Schönhardt 2002.

The failure probability is determined with the MCS-method, because the implemented variables are not correlated.

The value depends on the formulation of the limit state function g(x), the statistical parameter of the integrated variables, especially of the standard deviation $\sigma_{g(x)}$, and on the analytical or numerical calculation method, Schönhadt 2002.

$$P_f = \frac{1}{2 \cdot \pi} \cdot \int_{-\infty}^{0} e^{-\frac{1}{2} \cdot \left(\frac{g(X_i)}{\sigma_{g(X_i)}} - \beta \right)^2} d\left(\frac{g(X_i)}{\sigma_{g(X_i)}} \right) = \phi(-\beta) \tag{7}$$

Table 7. Safety index β and the corresponding failure probability for predefined safety classes.

Limit state	Safety classes 1	2	3
Deformation	2,5 $6{,}21 \cdot 10^{-3}$	3.0 $1{,}35 \cdot 10^{-3}$	3,5 $1{,}35 \cdot 10^{-4}$
Bearing capacity	4,2 $1{,}34 \cdot 10^{-5}$	4,7 $1{,}30 \cdot 10^{-6}$	5,2 $9{,}98 \cdot 10^{-8}$

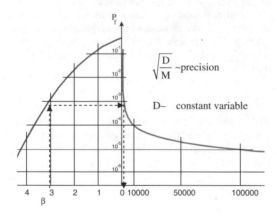

Figure 9. Dependency between failure probability p_f and the number of realisations M, Müller 1979.

For a given failure probability of $p_f = 1{,}35 \cdot 10^{-3} 10^5$ realisations M are necessary. To improve the precision on the factor 10, 100 times more simulations are to do.

5.5 Example

For example we consider a square sized spread machine footing loaded with

– one axial moment m = 500 kNm
– centrically vertical load n = 1.200 kN

The surface has a stratum of 127,50 m. We have assumed that no ground improvement will be done. In table 8 the necessary foundation dimension for a quadratic geometry are summarized.

These foundation dimensions are determined by 10^6 realisations using the different prediction of soil modelling. In this example the incomplete knowledge of the real strata and stiffness of the layers lead to an underestimation of the necessary dimensions to guarantee the allowable relative settlements. The error is expressed in terms of linear difference of the footing size.

6 CONCLUSION

Because of the genesis the description and modelling of the in-situ soil is complicate. Therefore the characteristic quantities of the soil are containing uncertainties and spatial variations. Due limited budgets there are often only few data from soil investigation. Ground models, based on small data sets, will produce inadequate estimations.

With the use of geostatistical methods we can produce better estimations between the given data set than with interpolation. A homogeneous profile at certain location can be determined.

An important aspect for using geostatistical models is the determination of anisotropy and the trend behaviour. Without this knowledge all estimations could not be better than a subjective estimation of expert.

The presented 3d-homogeneous model contains information to correlations between the integrated quantities, the regionally distribution, the anisotropy and the trend behaviour. Nevertheless this model is inconsistent to the reference results. Therefore the

Table 8. Dimensions b of a quadratic shallow foundation in dependency of the adaptive ground modelling and the deviation of the single results to the reference location in Table 6.

	Preliminary investigation 1	Detail investigation 2	Detail investigation consistent 3	Detail investigation consistent 4	Reality 5
b	2,630	2,605	3,205 m	3,530 m	3,450 m
Δ	23,8%	24,5%	7,2%	2,3%	0,0 %

method of smoothing splines is additional integrated. The comparison of model results with the realistic profile is able to improve the estimation of the homogeneous model. Alternative there exist other polynomial methods. The use of them depends on the actual boundary conditions.

REFERENCES

Atkinson, J.H. 2000. Non-linear soil stiffness routine design. *Geotechnique* 50, No. 5, 487–508.

Clemens, T. 1995. Ausgewählte Schätz- und Testprobleme bei räumlich korrelierten Daten, Dissertation, Bielefeld.

Christakos, G. 1992. Random field models in earth sciences. Academic Press, California.

Deutsches Institut für Normung e.V., Normenausschuss Bauwesen 1981. Grundlagen zur Festlegung von Sicherheitsanforderungen von baulichen Anlagen. Deutsches Institut für Normung, 1. Auflage.

Müller, P.H., Neumann, P., Storm, R. 1979. Tafeln der mathematischen Statistik. VEB Fachbuchverlag, Leipzig.

Schafmeister, Marie-Theresia 1999. Geostatistik für die hydrogeologische Praxis.

Schönhardt, M., Wuttke, F., Witt, K.J. 2002. Setzungsprognose mit streuenden und unsicheren Kenngrößen auf der Grundlage geophysikalischer Messungen. in. Ziegler, M. (Hrsg.) Stochastische Prozesse in der Geotechnik. Schriftenreihe, Heft 1, Institut für Grundbau, Bodenmechanik, Felsmechanik und Verkehrswasserbau, RWTH-Aachen.

Schuppener, B. (1999). Die Festlegung charakteristischer Bodenkennwerte – Empfehlung des Eurocodes 7 Teil 1 und die Ergebnisse einer Umfrage. *Geotechnik, Sonderheft: Deutsche Beiträge zur Europäischen Normung*.

Simons, Noel & Menzies, Bruce (2000). A short course *foundation engineering*. 2nd edition, Thomas Telford.

Yamashita, S. and Jamiolkowiski, M. Lo Presti, D. 2000. Stiffness non linearity of three sands. *Journal of Geotechnical and Geoenviromental Engineering*, 929–938.

Geotechnical Measurements and Modelling, Natau, Fecker & Pimentel (eds)
© 2003 Swets & Zeitlinger, Lisse, ISBN 90 5809 603 3

Safety analysis in geotechnical engineering by means of risk simulations

M. Ziegler

Institute for Geotechnical Engineering, RWTH Aachen, Germany

ABSTRACT: It is a characteristic of most geotechnical calculations that due to the inhomogeneity of the naturally layered soil the initial input parameters are usually subjected to a spatial and time scatter. In common calculations for the determination of the safety factor of geotechnical constructions, fixed design values for the actions and the resistance parameters are set as input parameters for the limit state equation. As the safety factor is calculated only for one combination of initial quantities, there is no information about the total probability of the calculated safety factor being exceeded. Information about the overall safety can be obtained by risk simulation. After a brief introduction to the method the application of the method will be illustrated for the calculation of the safety factor for a steep slope reinforced by several layers of geogrids. The method can also be used to investigate the influence of the several input parameters in detail by performing sensitivity studies which allow to identify the most decisive input parameters in order to optimize the design or to take remedial countermeasures in advance. The risk simulation method can be extended to all kinds of computations where the initial quantities are scattered. Besides the calculation of safety factors the determination of schedules and costs in geotechnical projects is a preferred application.

1 INTRODUCTION

Completely homogeneous ground conditions rarely exist. As soil investigations provide only information at certain points, the shear values of the soil, particularly cohesion, are random variables.

For the calculation of a geotechnical safety factor the initial random variables are transformed to certain characteristic values by means of statistical methods provided there is a sufficient database. In most cases, however, the characteristic values of the actions S_k and the resistances R_k in the soil are established by a geotechnical expert using cautious average values which are expected not to be exceeded within the lifetime of the object considered (Eurocode 7 1994). The assessment of the characteristic values is thus somewhat arbitrary as shown exemplarily for the angle of friction in Figure 1.

It has to be realized that with the determination of the characteristic values the further procedure of calculating the factor of safety is fixed. For the calculation of a global safety factor the determined characteristic initial values are directly applied in the problem specific limit state equation. The global safety factor will then be obtained by comparing the existing characteristic resistance forces with the existing characteristic actions. To satisfy the requirement for safety the factor of safety has to exceed 1,0.

$$\eta = \frac{\sum R_k}{\sum S_k} > 1,0 \tag{1}$$

To take into account the individual scattering of the different input parameters it is useful to use the concept of partial safety factors. In this case the limit state equation is evaluated with design values which will be obtained from the characteristic values by multiplying the actions and dividing the resistance forces with partial safety factors γ_S and γ_R, respectively (both larger than one). Sufficient safety is given if the sum of design

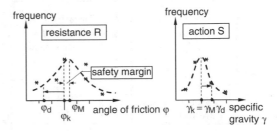

Figure 1. Determination of characteristic values (index k) and design quantities (index d) with possible consideration of a safety margin.

resistance forces exceeds the sum of design actions. In order to know how far the resistances are exhausted the term with the resistance forces can be multiplied by a factor $1/f < 1,0$ so that the limit equilibrium is just reached.

$$\frac{1}{f}\sum R_d - \sum S_d = \frac{1}{f}\sum (R_k/\gamma_R) - \sum (S_k \cdot \gamma_S) = 0$$

(2)

However, both calculation methods have the major disadvantage that the safety factor can merely be determined for only one combination of initial values. There is no information about the total probability of failure and its sensitivity against variation of the input parameters.

2 THE PRINCIPLE OF RISK SIMULATION

The disadvantages mentioned above can be overcome by applying risk simulation computations. For certain probability densities the total probability of failure can be calculated analytically (Mostyn 1993, Pöttler 2001). But in most practical cases the probability density is arbitrary and cannot be given by an analytical expression. Therefore the determination of the total probability requires numerical methods. In the following the principle of the method is shown for a simple virtual example. The application of the method is very simple and its major advantage is the fact that arbitrary probability densities can be treated with the same algorithm.

We consider the two risks A and B in Figure 2. On the horizontal-axis we have the estimation of the possible damage S (e.g. in Mio €) and on the vertical-axis we have the corresponding probability of occurrence p that this risk will take place. The sum of all probabilities of occurrence for the different damage values for one risk is always 100% which means that any of the considered possible damages will really occur. Of course the value of 0 can be assigned to a possible damage which means that in this case there is no real risk.

The risk R itself can be evaluated by the product of S and p. For reasons of simplicity we assume that the total risk can be obtained by adding the two individual risks. Of course, in real application there can be much more complicated connections.

The idea of risk simulation is now just to evaluate all possible combinations of the different risks with respect to their probability of occurrence. If we combine, for example, the third estimation of the first risk with the second estimation of the second risk we get a total damage of $3 + 3 = 6$ Mio € with a probability of occurence of $15\% \times 60\% = 9\%$.

After evaluating all possible combinations the obtained risks are put each in classes with the same damage-size and the probabilities of all values belonging to a certain class are added as done in the table of Figure 3. The corresponding plot of the histogram is also shown in Figure 3. The cumulative representation of the histogram yields the so-called risk profile curve. Each point on the risk profile indicates the probability with which a certain damage will occur.

It is obvious that the number of combinations to be evaluated increases rapidly with the number of risks and the acuteness of the discretisation. If k(i) denotes the number of discretisations for the risk i and n the number of risks, the number of combinations to be evaluated is

Total risk [Mio €]	3	4	5	6	7
Probability of occurrence (%)	8	36	42	11	3
Probability of not exceeding (<) (%)	0	8	44	86	97
Probability of exceeding (≥) (%)	100	92	56	14	3

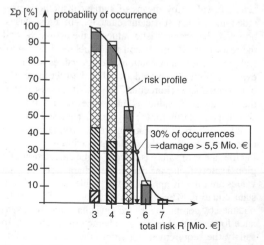

Figure 3.   Histogram and creation of the total risk profile.

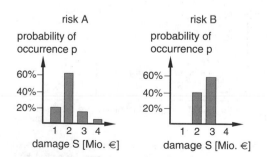

Figure 2.   Assessments of two individual risks A and B.

obtained by Equation 3:

$$N = \prod_{i=1}^{n} k(i) \tag{3}$$

For large problems with a big number of input values the computation time can exceed acceptable limits. In such cases it is sufficient to use a random generator which provides a lower number of random combinations for the evaluation of the limit state equation without loss of accuracy. This is the so-called Monte-Carlo-method (Alen 1998). For the example presented in chapter 3 it was not necessary to use a random generator as all results were obtained within minutes on a usual personal computer despite the extended variation of the geometry of the slip circles.

3 GEOGRID REINFORCED SLOPE

The risk simulation method was applied for the determination of the stability of a steep slope reinforced with several layers of geogrids. The situation and the parameters used in the risk simulation are shown in detail in Figure 4.

Besides the angle of friction φ and the cohesion c of the soil the properties of the geosynthetic material can vary enormously.

The coefficients A_1 to A_4 are reduction factors by which the short time strength $F_{B,k0}$ of the geosynthetic material has to be divided. The coefficient A_1 takes into account that the material has a tendency to creep. If no specific examinations have been done a minimum of $A_1 = 2,5$ should be used according to the recommendations of the German Society for Geotechnics (EBGEO 1997). A_2 stands for a possible partial damage of the geosynthetic reinforcement during transport or installation. For a non-cohesive material A_2 should be at least 1,5. A_3 has only to be taken into consideration if there is an overlapping of the geosynthetic layers in the direction of the acting forces which is not the case here. If there is an enviromental impact a reduction factor of $A_4 = 2,0$ has to be used. The recommended minimum values of the factors A_1 to A_4 can be reduced if special examinations for a specific product have been conducted which have actually shown lower numbers.

Following the recommendations of EBGEO the stability of the slope has besides other failure mechanisms to be examined with a slip circle mechanism in which the resistance force of the reinforcement can be treated as a non pre-stressed anchor as marked in Figure 4.

The anchor contributes to the resisting moments R_M on the one hand by the moment of the anchor-component acting parallel to the slip curve and on the other hand by its normal component which causes

additional friction forces in the slip surface. In Equation 4, which follows the method of Bishop, applied in the German code of practice E DIN 4084 (1990), F_{Ai} is the resistance of the anchor and ε_{Ai} its inclination to a horizontal line. 1/f stands for the degree of how far the resistances are exploited. R_M indicates the resisting moments of all segments of the slip circle.

$$R_M = r\sum_i \frac{(G_i + (1/f)F_{Ai}\sin\varepsilon_{Ai})\tan\varphi_i + c_ib_i}{\cos\upsilon_i + (1/f)\sin\upsilon_i\tan\varphi_i} + r\sum_i F_{Ai}\cos(\vartheta_i + \varepsilon_{Ai}) \tag{4}$$

In the reinforced slope of Figure 4 the reinforcement elements are horizontal, i.e. $\varepsilon_{Ai} = 0$, so that the

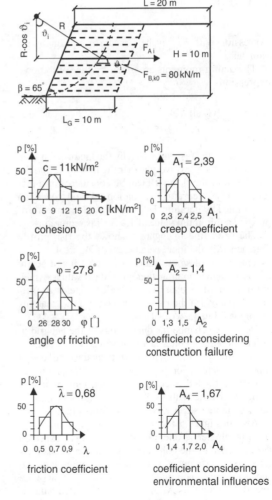

Figure 4. Geogrid reinforced slope with scattered input parameters.

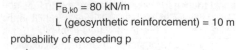

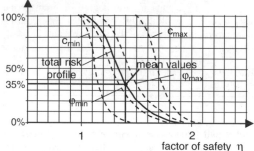

F_{B,k0} = 80 kN/m

L (geosynthetic reinforcement) = 10 m

probability of exceeding p

Figure 5.   Risk profile and sensitivity studies for the angle of friction and the cohesion.

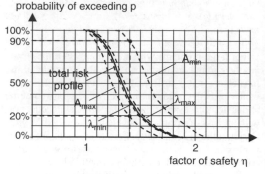

$F_{B,k0} = 80$ kN/m

L (geosynthetic reinforcement) = 10 m

probability of exceeding p

Figure 6.   Risk profile and sensitivity studies for the friction coefficient λ and the reduction coefficients A_i.

expression with F_{Ai} in the first term of Equation 4 vanishes.

The acting moments are given by the well known expression

$$S_M = r \sum_i (G_i \sin \upsilon_i) \qquad (5)$$

If we reduce the resistances R_M by the factor $1/f$ in such a way that $R_M = S_M$ we consider the limit state and the reciprocal of the factor $1/f$ can be considered as a factor of safety. According to Equations (4) and (5) we can determine f by iteration. Following the German code of practice a minimum value of 1,4 for f is required.

The solid line in Figure 5 shows the risk profile obtained with the input parameters of Figure 4.

The factor of safety varies between 1,0 and nearly 2,0. Although a computation with mean values, which are often used as characteristic values, yields a sufficient value of 1,4 the risk profile curve shows that this is only valid in 35% of all cases.

Before we change the dimensions of the geogrid in order to obtain a better factor of safety we investigate the sensitivity of the factor of safety against the several input parameters. For this purpose we fix the parameter to be investigated once with its minimum value and some other time with its maximum value whereas all other parameters can vary in their whole range.

Also in Figure 5 the results of the variation of the shear parameters φ and c are shown. As to be expected the influence of the cohesion c is much more distinctive than the influence of the angle of friction. Figure 6 shows the corresponding results for the variation of the material parameters A_1 to A_4 and the friction value λ activated in the area of the geogrid. It is surprising that the results hardly depend on the value of λ. This

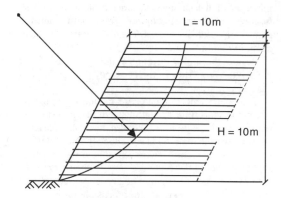

$F_{B,k0} = 80$ kN/m

L (geosynthetic reinforcement) = 10 m

Figure 7.   Decisive slip circle.

indicates that the friction between the area covered by the geogrids and the soil cannot be activated to its full size as the geogrid stripes themselves reach their tensile strength.

On the other hand the results make evident that it is worth to make some effort in the investigation of the parameters A_1 to A_4. Without further investigation the parameters have to be chosen with their maximum value. For this parameter combination the required factor of safety is only obtained in 20% of all cases. If we consider the curve for the minimum values of A_i the probability to get a factor of safety of 1,4 reaches nearly 90%.

In Figure 7 the slip circle corresponding to the minimum value for the factor of safety is plotted. Obviously there cannot be taken an advantage of the length of the geogrid as the slip circle cuts the geogrid

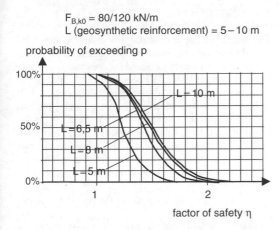

$F_{B,k0} = 80/120 \text{ kN/m}$
L (geosynthetic reinforcement) = 5 – 10 m

Figure 8.   Risk profiles for different lengths of geogrids.

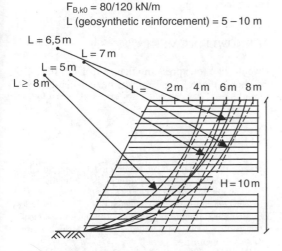

$F_{B,k0} = 80/120 \text{ kN/m}$
L (geosynthetic reinforcement) = 5 – 10 m

Figure 9.   Most unfavourable slip circles for different lengths of geogrids.

approximately in the middle. This suggests to shorten the length of the geogrids and to use two different types of geogrids: one with a tensile strength of 80 kN/m in the upper half and one with a tensile strength of 120 kN/m in the lower part.

In Figure 8 the corresponding risk profiles for different lengths of the geogrids are plotted. Evidently there is hardly a change in the risk profile for geogrids longer than 6,5 m. If we shorten the geogrids, however, e.g. to 5 m, there is a significant loss of safety.

For a better understanding of this effect we consider the corresponding slip circles in Figure 9. As one can see the most unfavourable slip circles coincide for all length greater than 8 m and cut the geogrids nearby the middle.

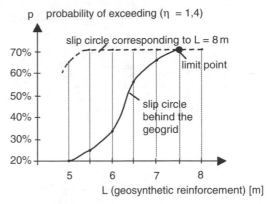

Figure 10.   Influence of geogrid length on safety of slope.

If we shorten the length of the geogrids the slip circles for the most unfavourable cases run behind the end of the geogrids. Obviously the increase in safety due to a more shallow circle is exceeded by the missing forces of the geogrids.

From Figure 10 in which the probability of exceeding corresponding to the required factor of safety ($\eta = 1,4$) is plotted against the length of the geogrids we can see that the point where the most unfavourable slip circle jumps back to the middle is approximately given for a length of 7,5 m.

If we fix the slip circle which correponds to a length of 8 m (or more) and shorten the geogrids in mind we get the upper dashed line in Figure 10. Up to a length of 5,5 m the probability of exceeding remains unchanged, which means that the friction resistance of the geogrid versus the soil cannot be exploited, as the tensile strength of the geogrids is exceeded before. Only if the geogrid has a lower length the pull out resistance of the geogrid is lower than the material strength.

Figure 10 points out precisely that the factor of safety increases with increasing grid length up to the limit point where the slip circle jumps back and the tensile strength of the geogrid becomes dominant. In other words the risk simulation calculation can also be used to find the optimum design for the length and strength of the geogrid.

4   OTHER APPLICATIONS

The method of risk simulation can be used whenever a problem with scattered input parameters has to be solved. Predestinated are all geotechnical calculations as the input parameters are subjected to a more or less marked scatter.

But also management problems like the establishing of time schedules or the determination of costs can be

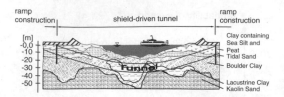

Figure 11.   Geological section of an underwater tunnel.

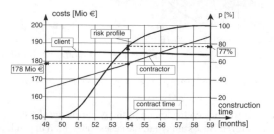

Figure 12.   Cost–risk interaction diagram.

treated in the same way. Instead of the soil parameters as there are the angle of friction and the cohesion we have the estimated costs and the time needed to perform a certain kind of work as scattered input parameters.

Consider the underwater tunnel in Figure 11. Due to the geological situation such a tunnel has to be excavated by a tunnel boring machine. The parameters defining the progress of works like the velocity of the tunneling machine in the different types of soil, the availability of the machine, the time needed for the assembling of the lining segments and the distance at the beginning of tunneling necessary to come up to full performance, and the soil parameters like the structure, the frequency of the occurrence of large stones and their grain size have an enormous effect on the average excavation time. Based on interview with experts the densities of probability of the several input parameters can be fixed and put into a risk simulation computation. As a result the total risk profile for the average excavation time is obtained. From it other important parameters for the project like the total construction time or the total costs including the interest for the construction time can be detected. Furthermore, milestones in the schedule of the construction site can be better assessed.

By means of sensitivity analysis the most decisive parameters could be identified. This enables us to take effective counteractive measures in advance in order to reduce the expected risks. For the example considered, a more powerful rock crusher has been installed in order to reduce the cost and time intensive removal of stones by hand in front of the tunneling machine. For more details see Ziegler (2002).

5   CONCLUSION

The simple example of a stability analysis of a reinforced slope clearly shows that the determination of a single factor of safety obtained by conventional deterministic computations is not sufficient to assess the influence of the scattered input parameters.

Risk simulations offer a way out of this dilemma. The application does not need much more effort, as the algorithm to establish the total risk profile is always the same. It is only necessary to define the limit state equation which is specific for each new problem and to integrate it in the general algorithm as mentioned before. In many cases, particularly if a random generator is used, there are no restrictions on the number of input parameters.

The method of risk simulation can also be easily applied to management problems like establishing of schedules or determination of costs where the soil parameters are only one part of the parameters influencing the problem.

ACKNOWLEDGEMENT

I would like to express my thanks to Dipl.-Ing. Alla Prokhorova who has performed the calculations presented here.

REFERENCES

Alen, C. 1998. Random Calculation models exemplified on slope stability analysis and ground–superstructure interaction. *On Probability in Geotechnics* (1), Department of Geotechnical Engineering, Göteborg.

EBGEO, 1997. Empfehlungen für Bewehrungen aus Geokunststoffen. Deutsche Gesellschaft für Geotechnik. Berlin: Ernst & Sohn.

E DIN 4084, 1990. Gelände- und Böschungsbruchberechnungen. Berlin: Beuth Verlag GmbH.

Eurocode 7, 1994. Geotechnical design – Part 1: General rules. German version ENV 1997 – 1. Berlin: Beuth Verlag GmbH.

Mostyn, G. R. & Li, K. S. 1993. Probabilistic slope analysis – State-of-play. *Proc. of the Conf. on probabilistic methods in geotechnical engineering.* Canberra, 10–12 February 1993. 89–109. Rotterdam: Balkema.

Pöttler, R. et al. 2001. Probability analysis in tunnelling – theoretical background and application. *Bauingenieur* (76): 101–110.

Ziegler, M. 2002. Project calculation by means of risk simulations. In Rudolf Pöttler (ed.), *International Conference on Probabilistics in Geotechnics, Technical and Economic Risk Estimation.* 15–19 September 2002. Graz, Austria. Verlag Glückauf Essen.

New technologies for laboratory and field tests

Geotechnical Measurements and Modelling, Natau, Fecker & Pimentel (eds)
© 2003 Swets & Zeitlinger, Lisse, ISBN 90 5809 603 3

The use of a relative humidity sensor for suction measurement of compacted bentonite-sand mixtures

S.S. Agus & T. Schanz
Laboratory of Soil Mechanics, Bauhaus-University Weimar, Germany

ABSTRACT: Compacted bentonite-sand mixtures have been widely used as barrier material in many applications, such as in waste disposal systems. As they are commonly unsaturated in the field, suction can be used as a measure to characterise the material condition in situ in addition to water content, dry density and coefficient of permeability that are commonly adopted in general practise. Measurement of suction, in this case, total suction, poses great challenges for geotechnical engineers due to incapability and inaccuracy of available methods in measuring the whole range of suction that may occur in the field. This paper presents an alternative means for suction measurement by which almost the whole range of suction encountered in the field can be quantified indirectly. A commercially available relative humidity sensor is used to determine relative humidity of the material such that the corresponding suction can be calculated. The sensor was used to measure total suctions of two bentonite-sand mixtures compacted to standard and enhanced proctor densities. It is shown that satisfactory results can be obtained and a range within which the sensor is likely to be most applicable is proposed.

1 INTRODUCTION

In waste disposal engineering, compacted bentonite-sand mixtures have been widely used due to their low permeability. Intensive researches that have been performed in this field, for instance Alonso et al. (2001), Herbert and Moog (2002), Tang et al. (2002) and many others, mostly deal with their engineering characterisation, some by which parameters required for further analyses, e.g., for numerical analyses, are determined.

In most applications, for instance, as clay liner in landfills and as sealing element in nuclear waste disposal systems, compacted bentonite-sand mixture is commonly found in unsaturated condition. Hence, suction as one of the stress-state parameters in unsaturated soil mechanics, plays an important role in controlling its behaviour especially the coefficient of permeability as it varies several orders of magnitude with suction. It has also been a fact that shrinkage and swelling can take place as the liner or the sealing element undergoes drying and wetting or increase and decrease in suction. Thus, suction can also be used as a measure for characterisation of the material for this types of application as a compliment to dry density, water content and coefficient of permeability that are commonly adopted. Dineen et al. (1999) have measured suctions and volume changes of a bentonite-enriched sand. There seems to be a correlation between suction

and volume change of compacted material upon wetting. Tang et al. (2002) have shown the importance of suction in characterising the behaviour of the compacted bentonite-sand mixture as sealing element for nuclear waste disposal.

Methods of suction measurement, total suction in this case, are challenging as all available methods have their own restrictions. Psychrometers are restricted to their incapability to quantify the whole range of suction that may occur in situ. Only upto a maximum total suction of about 8000 kPa can be measured using thermocouple psychrometer (Fredlund and Rahardjo, 1993). Filter paper method, although it can be used for the whole range, a intolerably long equilibration time makes it impossible to be used when many measurements are to be performed in a short duration of time. Two to three weeks are required to complete one suction measurement (Leong et al., 2002). In addition, for low suction measurement, those above methods suffer from imprecision due to influence of condensation problem.

This paper presents a part of an ongoing research in which an alternative for suction measurement that can be used for the whole range of suction with a very short response time is suggested. Suction is basically determined by first obtaining relative humidity of the compacted bentonite-sand mixtures. A commercially available relative humidity sensor is utilised to determine relative humidity of two bentonite-sand mixtures compacted to standard and enhanced proctor densities.

2 RELATIVE HUMIDITY SENSOR USED AND ITS VERIFICATION

The relative humidity sensor used in this study is a polymer capacitance sensor that is commonly used in meteorological field. The authors have found that it may offer a good chance for geotechnical engineering field applications as it is small, easy to handle and most importantly it gives a rapid response to changes in relative humidity and thus to changes in suction.

As it is basically a warmed-head sensor, high relative humidity or low suction, can be measured without having condensation, which commonly occurs as a result of a change in temperature of water vapour, especially at high relative humidity (i.e., above 95%) causing erroneous readings as the reading always shows 100% relative humidity (RH). The warming function generates heat to keep the head dry and hence, prevents the condensation to occur. However, the readings obtained are not correct if the sensor overheats, thus, an additional temperature sensor is required to give a reference such that the head is only heated to the dew point corresponding to the RH and temperature of the system upon which the RH is to be determined. Therefore, it is also often called a dew point sensor. The main sensor (0.8 cm in diameter and 6 cm in length) is used for RH measurement and is further called RH sensor, while the additional sensor (0.5 cm in diameter and 6 cm in length) is for temperature measurement and is further called temperature sensor. According to the manufacturer's specification, the RH sensor is capable of measuring relative humidity from zero to 100% with a sensitivity of 0.1% RH and an accuracy of (0.5 + 2.5% of the reading) % RH, whilst the temperature sensor measures temperature in the range of $-48°C$ to $180°C$ with a resolution of $0.1°C$ and an accuracy of $0.1°C$. The head of the RH sensor is covered with a stainless steel sintered filter. Typically rapid response of 15 seconds is obtained for both RH and temperature measurements according to the factory's specifications. Readings are shown on a digital display that can be connected to a personal computer to permit automatic data collection.

In order to check its response, the RH sensor was verified using different known RH values generated by several salt solutions each prepared in a glass container measuring 400 ml in volume, each containing 200 ml solution. The corresponding RH values of the salt solutions used can be obtained from Lang (1967), Greenspan (1977) and Lide and Frederikse (1994). The RH sensor was suspended over the salt solution in the container while the temperature sensor was

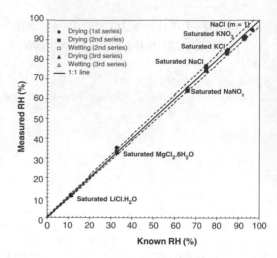

Figure 1.   Measured versus known RH values.

placed in a different container to avoid the influence of heat generated by the heating function. Both containers were placed in a thermal-insulated box with a capability to maintain constant temperature (21.5°C ± 0.2°C). Cyclic drying and wetting series were performed in this study.

The plot of the measured RH versus the known RH from the salt solutions is shown in Figure 1. The error band as specified by the manufacturer is also drawn in the figure. It is shown that the sensor largely gives accurate response despite small deviation from the known RH values. This fallacy seems to increase with increasing RH. Insignificant hysteresis is depicted in Figure 1. The negligible hysteresis is of particular importance for the field measurements as rapid drying and wetting processes that may occur often cause difficulties in obtaining good results.

3 MATERIAL USED AND TESTING PROCEDURES

The sensor was used to determine suctions of two compacted bentonite-sand mixtures. A commercially available calcium bentonite (Calcigel) from southern part of Germany was mixed with 2 mm sieved quartz sand at ratios of 30/70 and 70/30 dry mass. Properties of the materials used are listed in Table 1.

Water was added to achieve the targeted water contents. Mixtures with different water contents were then cured in plastic bags for one week such that uniform water content distribution could be obtained and bentonite hydration was completed. The specimens were compacted to achieve standard proctor densities

Table 1. Basic properties of the bentonite and the sand used in the study.

Properties	Bentonite	Sand
Specific gravity	2.65	2.65
Liquid limit	130	n.a.
Plasticity index	97	n.a.
Clay fraction[*]	83	–
D_{10} (mm)	–	0.25
D_{30} (mm)	–	0.40
D_{60} (mm)	–	0.70
Specific surface area (m²/gr)[**]	69	–
Cation exchange capacity (meq/100 gr)[*]	49	n.a.
USCS classification	CH	SP

[*] Data from Herbert and Moog (2002).
[**] Determined using BET method.

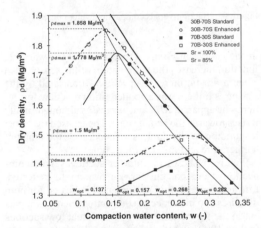

Figure 2. Compaction curves of the mixtures used.

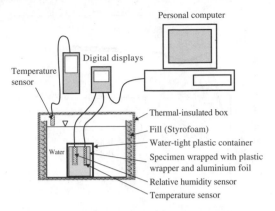

Figure 3. Schematic drawing of testing arrangement.

(compaction energy: 600 kJ/m³) according to ASTM Standard (1997) D 698-91 and enhanced proctor densities (compation energy: 1000 kJ/m³). The compaction curves of the materials used are shown in Figure 2.

After compaction, two holes at a separating distance of 4 cm were drilled in each specimen to embed the RH and the additional temperature sensors. The hole prepared for the RH sensor was 2 cm in diameter and 8 cm in depth while for the temperature sensor a 1 cm diameter hole with the same depth was drilled. A perforated pipe was inserted into the hole drilled for the RH sensor to avoid direct contact between the sensor and the material, which would cause a liquid-phase water transfer from the mixture to the RH sensor during measurements. The compacted mixture with the holes was wrapped with plastic and subsequently with aluminum foil to prevent evaporation.

The RH measurements were performed immediately after compaction. The RH and the temperature sensors were inserted into their respective holes and the openings were immediately sealed. The specimen with the embedded sensors were contained in a plastic container. The container with the specimen and the sensors was placed in the thermal-insulated box where the measurement was subsequently conducted. Another temperature sensor was placed in the box to monitor temperature equilibrium between the specimen and the chamber. The test arrangement is schematically illustrated in Figure 3. The measurement was terminated after equal temperature between the vapour space in the box and that in the specimen had been reached and RH readings were stabilised for sufficient time.

A maximum temperature fluctuation of 0.1°C was found in the specimen holes, slightly lower as compared to the temperature fluctuations observed in the thermal-insulated box. Hence, at worst, a maximum temperature gradient of 0.3°C between the temperature of the water vapour in the specimen and that in the hole, could be expected.

It has to be noted that allowing the switched-off sensor embedded in the specimen for overnight is not advisable since water vapour may condense on the measuring element of the sensor, which is difficult to be dried out by the heating function. In the case of the compacted mixtures used in this study, the condensed water vapour could not be easily dissipated to the specimen causing a higher RH being measured, especially at high RH values. This may not be the case for more porous and permeable materials as although condensation may occur, the excess water vapour on the measuring element can be dissipated more easily and rapidly. Suction, ψ was calculated from the measured RH and temperature using Kelvin's law:

$$\psi = \frac{-RT}{M_w\,(1/\rho_w)}\,\ln(RH) \tag{1}$$

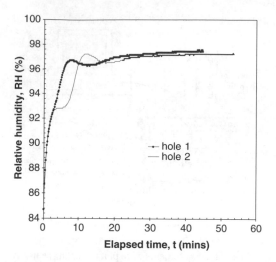

Figure 4. Typical response of the RH sensor.

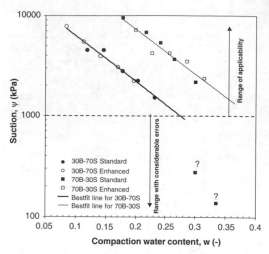

Figure 5. Suction versus compaction water content.

where R is the universal gas constant ($= 8.31432$ J/mol K), T is the absolute measured temperature in K, M_w is the molecular weight of water ($= 18.016$ kg/kmol), ρ_w is the unit weight of water in kg/m^3 as a function of temperature and RH is measured relative humidity defined as u_v/u_{vo} with u_v is partial pressure of pore-water vapour in the specimen and u_{vo} is saturation pressure of water vapour over a flat surface of water at the same temperature.

4 RESULTS AND DISCUSSION

Figure 4 shows results of the RH measurement of the specimen prepared from 70:30 bentonite-sand mixture compacted using proctor standard at a compaction water content of 0.27 with corresponding dry density of 1.42 Mg/m^3. For this particular specimen, two holes were drilled for the RH sensor. Two RH measurements were obtained to verify the repeatability of the readings. Repeatable results are indicated in the figure. Approximately 35 minutes are required by the sensor to give stable RH reading. This response time is considerably longer than that of in the specification, although it can be considered significantly rapid as compared to the filter paper method. The reason for this deference is that the insertion of the RH sensor induces a disturbance to the water vapour in the hole. It takes some times for the water vapour in the hole to re-equilibrate after the placement of the sensor.

An additional investigation was performed by allowing the measurement after reaching stable readings for overnight (with the sensor remained on). It was found that the RH value was essentially similar to the value at first the stable reading was observed. Kink

in each curve in Figure 4 indicates that the warming function increases the temperature of the sensor head when a drastic increase in RH was translated by the sensor as if there was a condensation or a dew formation. The apparent RH value decreases although finally it rises towards the actual value. The elapsed time corresponding to this kink is a function of the ambient temperature and the size of the hole.

Figure 5 shows the plot of suction versus compaction water content for the two mixtures tested. Linear relationship in semilog plot indicated for both mixtures. No influence of compaction energy is apparent in the figure. It is also shown that at low suctions (below 1000 kPa), two data points that belong to the mixture of 70% bentonite and 30% sand deviate from the lines. The authors believe that this could be due to the temperature gradient between the specimen and the vapour space in the hole existing during measurement (i.e., 0.3°C). The temperature gradient causes a change in saturated vapour pressure, which in turn, changes the suction as observed by the RH sensor.

The change in suction resulted from the change in saturated vapour pressure due to temperature gradient can be computed by taking derivative of Equation (1), yielding the following equation (Croney et al., 1952):

$$\frac{d(\ln \psi)}{dT} = \frac{1}{T} - \frac{1}{u_{vo} \ln(RH)} \frac{d(u_{vo})}{dT} \qquad (2)$$

Figure 6 illustrates the change in suction at temperature of 20°C as computed using Equation (2). The 30% suction-error line is also drawn in the figure. If 30% error in suction measurement is assumed to be satisfactory, for temperature gradient of 0.3°C

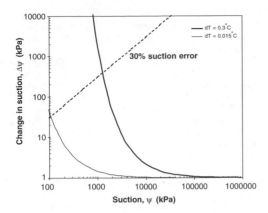

Figure 6. Change in suction due temperature gradient.

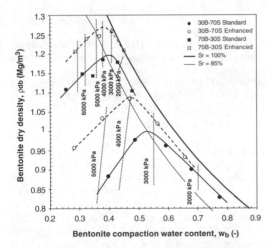

Figure 7. Suction isolines determined for the mixtures.

as occurred in the test, a lower limit of 1000 kPa above which the measurement is likely to give reliable results is defined. Therefore, in the case where the temperature gradient can be maintained at the minimum value of 0.3°C, the sensor can give reliable readings when used for measuring suction as low as 1000 kPa. One can also easily define that measurement of suctions below 100 kPa using this method requires control of temperature within ±0.015°C, which is difficult to achieve even in the laboratory.

Figure 7 shows suction isolines of the mixtures as drawn based on the measurement results plotted together with the relationship between bentonite dry density versus bentonite compaction water content. It is assumed that water is only adsorbed by the bentonite while the sand acts as an inert material. Although

only two different compaction energies were used in this study, which may require further confirmation, the figure also seems to indicate that for the bentonite used, suction is only a function of bentonite compaction water content. This is mostly true for mixture of 70% bentonite–30% sand, while for the other mixture, at low bentonite compaction water content, the suction isolines slightly incline, which may give little contradiction to the previous thought that suction is a only function of compaction water content.

At very high bentonite compaction water contents or very low suctions, the isolines may curve. However, this cannot be observed in this study due to the capability of the equipment used as described above.

5 CONCLUSION AND SUGGESTIONS

The use of a RH sensor in measuring total suction of two bentonite-sand mixtures is described. The sensor has been found to give reasonable results with good accuracy for measuring total suction as low as 1000 kPa. Measurement of suction below that value requires a means to maintain specimen temperature with high accuracy, which is difficult to be implemented in the field.

Typical response of 35 minutes was observed for the sensor. This response is however a function of the measured suction and the size of the hole where the sensor is inserted.

Comparison of the sensor's readings with those obtained from other measurements shall be warranted to give better confidence for the use of the sensor. Part of this work is still underway, therefore, the results cannot be presented herein. It is important to note that to date and to the authors' knowledge, better accuracy for measuring suction via relative humidity measurement can only be achieved by chilled-mirror hygrometer as described in Gee et al. (1992) and Leong et al. (2003), although the method offers no chance for field applications.

The same arrangement cannot be adopted when the sensor is to be used in the field. Rapid flow of soil water or water vapour as a result of a temperature gradient and other factors may cause difficulty in obtaining good readings. The use of the stainless steel sintered filter sensor may not be appropriate in this case. It may be necessary to replace the filter with a ceramic filter as that used at the tip of thermocouple psychometer. It is recommended that the RH and the additional temperature sensors be separated at least 4 cm when employed for measuring suction of the same materials as used in this study. However, a greater separating distance is required for higher permeable materials with additional consideration regarding spatial variability of *RH* and temperature in the field where the sensor is to be used.

ACKNOWLEDGEMENT

An acknowledgement is given to Mr. Michael Kaemmerer who assisted the authors in the experimental works.

REFERENCES

Alonso, E. E., Romero, E., Hoffman, C. and Garcia-Escudero, E. (2001) Expansive bentonite/sand mixtures in cyclic controlled-suction drying and wetting. *Proceeding of the 6th International Workshop on Key Issues in Waste Isolation Research (KIWIR 2001)*, Paris, France, ENPC: 543–597.

ASTM Standard (1997) D 698-91 Test method for laboratory compaction characteristics of soil using standard effort (12,400 ft-lbf/ft^3 (600 kN-m/m^3)). *Annual Book of ASTM Standard*, 04.08, Soil and Rock (II), ASTM International West Conshohocken, PA: 77–84.

Croney, D., Coleman, J. D. and Bridge, P. (1952) *The suction of moisture held in soil and other porous materials*. Road Research Technical Paper, 24.

Dineen, K., Colmenares, J. E., Ridley, A. M. and Burland, J. B. (1999) Suction and volume changes of a bentonite-enriched sand. *Proceeding of Institution of Civil Engineers Geotechnical Engineering*, 137: 197–201.

Fredlund, D. G. and Rahardjo, H. (1993) *Soil Mechanics for Unsaturated Soils*. John Wiley & Sons, Canada.

Gee, G., Campbell, M., Campbell, G. and Campbell, J. (1992) Rapid measurement of low soil potentials using a water activity meter. *Soil Science Society of America Journal*, 56: 1068–1070.

Greenspan, L. (1977) Humidity fixed points of binary saturated aqueous solutions. *Journal of Research of the National Bureau of Standards-A. Physics and Chemistry*, 81A (1): 89–95.

Herbert, H. J. and Moog, H. C. (2002) *Untersuchungen zur Quellung von Bentoniten in hochsalinaren Lösungen*. Abschlussbericht GRS-179, Förderkennzeichen 02 E 8986 5 (BMBF), Gesellschaft für Anlagen und Reaktorsicherheit (GRS)mbH, Germany.

Lang, A. R. G. (1967) Osmotic coefficient and water potentials of sodium chloride solutions from 0 to 40°C. *Australian Journal of Chemistry*, 20: 2017–2023.

Leong, E. C., He, L. and Rahardjo, H. (2002) Factors affecting the filter paper method for total and matric suction measurements. *Geotechnical Testing Journal*, 25 (3): 321–332.

Leong, E. C., Tripathy, S. and Rahardjo, H. (2003) Total suction measurement of unsaturated soils with a device using the chilled-mirror dew-point technique. *Géotechnique*, 53 (2): 173–182.

Lide, D. R. and Frederikse, H. P. R. (1994) *CRC Handbook of Chemistry and Physics*. CRC Press Inc., Boca Raton, FL, USA.

Tang, G. X., Graham, J., Blatz, J., Gray, M. and Rajapakse, R. K. N. D. (2002) Suctions, stresses and strengths in unsaturated sand-bentonite. *Engineering Geology*, 64: 147–156.

Geotechnical Measurements and Modelling, Natau, Fecker & Pimentel (eds)
© 2003 Swets & Zeitlinger, Lisse, ISBN 90 5809 603 3

A solution to concurrent measurement of the normal and tangential earth pressure in model tests

M. Arnold & D. Franke
Institute of Geotechnical Engineering, TU Dresden, Germany

U. Bartl
BAUGRUND DRESDEN Ingenieurgesellschaft mbH, Dresden, Germany

ABSTRACT: In numerous projects in the field of earth-pressure research, the distribution of earth pressure is of interest, as well as the force described by value, location, and direction. Often the tests remain limited to a punctiform measurement of earth pressure, and in many cases only the earth-pressure component acting normally to the surface is measured. Thus, assumptions have to be made for the direction of action. A new type of load cell has been developed at the TU Dresden's Institute of Geotechnical Engineering. This type is capable of concurrently measuring normal and tangential earth pressure and includes larger parts of a surface. It is based on a simple, statically determinate system. The principle construction of the load cells will be presented, including the necessary force transducers. The transducers work on the basis of strain-measurement technology. They have been developed at the Institute for different force levels. For this reason, the load cells may be adapted to the estimated stress level in the model test. With the side-by-side arrangement of these load cells, it becomes possible to determine the distribution of earth pressure and its inclination over the height of the wall. Therefore, the entire surface of the wall or parts of it may be covered. The measurement principle proved the functionality in several completed research projects at the Institute. Some of the applications, together with the adaption to the special task, will be also presented. Typical features, application possibilities, and limits will be discussed.

1 INTRODUCTION

Model tests have been executed in the earth-pressure research for many years. While at first, the interest was on the magnitude of the earth-pressure force, the point of action soon also came into focus. To measure the resulting force on the entire wall in this sense, test walls are supported in a statically determinate way. If the tangential supports are instrumented, too, the additional determination of the direction of the resulting force becomes possible.

With the advances in engineering retaining structures, especially the development of bending stressed ones, the knowledge of the point of action was insufficient and the distribution of the earth pressure itself had to be known. Similar questions to contact-stress distribution also occur in other cases of soil structure interaction, e.g. at shallow foundations. Therefore, contact stresses themselves had to be measured.

This task has been solved in model testing as well as in field measurement by the integration of measuring cells in the wall surfaces. These cells are relatively small in relation to the researched wall. Thus, the result is a punctiform determination of the earth pressure. Assumptions have to be made for the earthpressure distribution between the measuring points. Furthermore the cells are only capable of measuring the pressure acting normally to the wall surface.

A new development to measure contact-stress distributions over the entire surface of an object are thin-film pressure sensors, which are already used in model testing. But like the conventional pressure cells they are not capable of acquiring tangential contact stresses.

2 MEASURING TECHNIQUE

2.1 *Concept*

A new concept of contact-stress measurement has been developed for the earth-pressure model tests of the Institute. The whole test wall is divided into segments. In each one the resulting force as integral sum of the earth pressure acting on the wall in normal as well as tangential direction to the surface is measured.

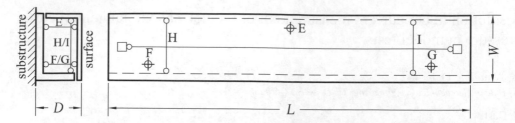

Figure 1.   Basic principle of cell construction: cross-section and front view.

For interpreting the data a constant, average pressure distribution over the segment is assumed. This results in a pressure distribution over the wall height.

2.2   Pressure cell construction

Each segment is formed by a measuring cell. A principal sketch is shown in Figure 1. The cell consists of a surface plate and a base plate; the link between both is statically determinate. An angle profile reinforced by additional fixed profiles is used as a stiff surface plate. The base plate is formed by a simple flat profile, which is used to fix the pressure cell on the wall construction. Both plates are made of aluminium.

The statically determinate support of the surface plate is realized by six simply supported members. The simple support is provided by spikes standing in calottes. The three members placed in normal direction to the surface (E, F, G) and the two ones in tangential direction (H, I) are used as force transducers (see Section 2.3). The sum of the forces in E, F, G and, accordingly, in H and I form the normal and tangential resulting forces.

There is a distance between the lines of action of the tangential pressure and the tangential supports. This leads to a pair of forces, which changes the support forces that act normally to the surface. However, the sum of normal forces remains constant.

Because of the spikes in calottes solution the pressure cell only operates in the compression case. To avoid the cell to fall apart, it is prestressed by springs in normal as well as in tangential direction. Additionally to compression forces small tensile forces can be measured due to the prestress.

While the normal contact stress between cohesionless soil and structure is always compression stress, the direction of the tangential stress may differ. Therefore, the expected direction of the tangential stress has to be considered at installing the pressure cells

2.3   Transducer construction

For economic reasons, self-developed and self-built force transducers were adopted. The electric strain measurement was chosen as the fundamental measuring principle. The advantages of this technology lie in

Figure 2.   Transducer Type 1 without strain gauges.

its small size and in the possibility of simple data logging as well as real-time data interpretation.

However, there is also a disadvantage. When strains are generated in a body, the body must deform. And deformations of the transducer cause deformations of the measuring cell, which may result in changes of the measured contact stresses. So, the deformations have to be minimised. To improve the sensibility of the transducer, the strains should be maximised. By optimising the design of the transducer both demands may be balanced.

Depending on the load range, different types of transducers have been designed. For higher nominal loads a simple compression member has been chosen. Full (Type 1, fig. 2) as well as hollow (Type 2, fig. 3) sections were used.

For the case of lower loads, the transduceres as shown in Figure 4 have been developed, which operate on basis of generating bending stresses. By modifying the section geometry, the fine tuning to the load level is done.

Two strain gauges are applied on every transducer. The strain gauges consist of two grids which are located perpendicularly to each other. One grid determines the longitudinal strains, the other one the transverse strains. The four grids are linked in a complete Wheatstone's bridge circuit to eliminate temperature influences (Hoffmann 1987).

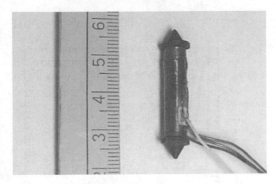

Figure 3.   Transducer Type 2.

Figure 4.   Transducers of Type 3.

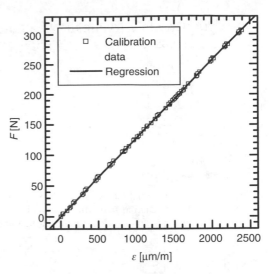

Figure 5.   Calibration curve of a Type 3 transducer.

Table 1.   Pressure cells in executed research projects.

Research project	$L/W/D*$ [mm]	Number of cells	Transducer
Section 3.1	400/60/50	8	Type 1 + 2
Section 3.2	400/75/45	18	Type 3

* See Figure 1.

The strain gauges are located on the place where the highest strains occur. For Types 1 and 2 this is in the middle of the member. On Type 3 transducers, the strain gauges are placed on the outsides (left and right in fig. 4), where the bending stiffness is minimal.

To assign the force values F to the measured strains ε, a calibration of each transducer has been executed, so that variations in the geometry, in the material parameters (e.g. Young's modulus) and in the gauge parameters are included in the calibration coefficient. An example of a calibration curve is shown in Figure 5. The transducers exhibit very good results in linearity and hysteresis as well as only a minimal dependency on temperature. Depending on the load, a little creep occurs. The effect may be neglected for tests lasting less then one day.

3   APPLICATION EXAMPLES

The pressure cells have proved their functionality, reliability and durability already in three research projects at the Institute. Two of them will be presented. Everytime, the basic principle of construction was retained. The geometry and the transducer model were adapted to the special task. Table 1 shows the exterior dimensions of the pressure cells, the number of utilised cells and the transducer types used in the presented projects.

3.1   Mobilization of passive earth pressure

The technology has been first applied in investigating the mobilization behaviour of passive earth pressure (Bartl 1995, Bartl & Franke 1998). A wall (fig. 6) with a height of 564 mm has been used in a model box of 1800 mm length, 990 mm width, and 900 mm height. All parts of the wall surface are made of aluminium. The wall is fixed in tangential direction.

The wall is divided into three parts. The 400 mm wide central one is instrumented. Three supports (A, B, C) in normal direction to the wall, two tangential ones in vertical direction (D, E), and one tangential support in horizontal direction fix the part in a statically determinate way. By measuring the forces in the supports A–E the magnitude, direction and point of action of the resulting earth-pressure force on the wall will be obtained.

Eight pressure cells are included to measure the earth-pressure distribution. By subtracting the sum of the forces on the pressure cells from the resulting earth-pressure force, the earth-pressure force acting on the bottom of the wall can be calculated.

All test have been executed in dry sand. Three different wall movements have been carried out: rotation around top of the wall and bottom of the wall as well as

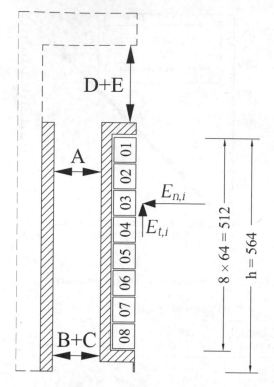

Figure 6. Wall section.

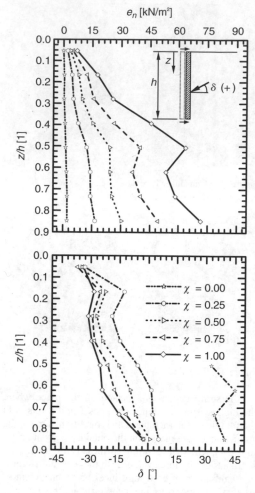

Figure 7. Magnitude of the normal component e_n and the inclination δ of earth pressure in parallel translation.

parallel translation. A test with parallel translation is presented as an example of the acquired results.

Figure 7 shows the passive earth pressure and its direction at different states of mobilization χ. With the normal components of the resulting earth-pressure forces at the initial state $E_{initial,n}$, at passive state $E_{p,n}$, and during mobilization E_n the parameter χ is obtained by

$$\chi = \frac{E_n - E_{initial,n}}{E_{p,n} - E_{initial,n}}. \tag{1}$$

As the results show, the earth-pressure e_n is distributed in a nearly triangular shape as commonly assumed. Of particular interest is the development of earth-pressure inclination δ.

At the initial state there was a positive inclination because of the vertical soil movement, which was caused by the compression of the material. The material was built in with the sand-rain-method. Yet with mobilization of passive earth pressure, the direction of action changes. The very interesting point is the non-constant inclination over the wall height ranging from small positive values at the bottom to negative ones at the top of the wall.

3.2 Cantilever retaining walls

Because of fewer field measurements and experimental investigations to the earth and load pressure acting on these kind of walls, model tests have been executed (Arnold 2000). Dry sand has been used as subsoil and backfilling. The wall surfaces are made of polished aluminium.

The wall model as shown in Figure 8 had an height of 940 mm and a base length of 639 mm. From the 990 mm wide wall, the central 400 mm are used for measurement. The substructure providing the wall its stiffness is located in the side areas (Arnold 2001).

To measure the normal and the tangential component of earth pressure on the wall stem, eleven pressure cells are placed. Additionally, seven cells on top of the wall heel acquire both components of the load

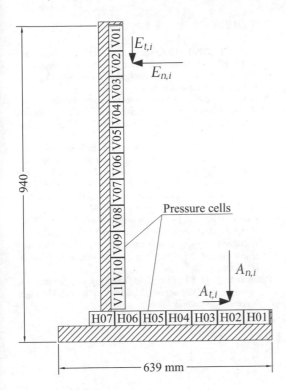

Figure 8. Wall section.

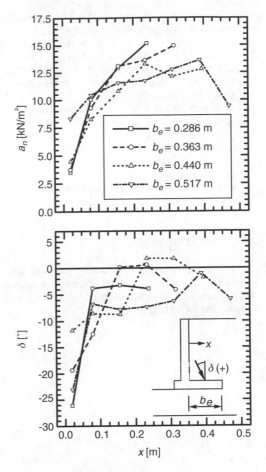

Figure 9. Magnitude of the normal component a_n and inclination δ of load pressure on the heel of a cantilever retaining wall in model tests.

pressure acting on the heel. The wall base has not been instrumented.

Several tests have been executed, e.g. varying wall geometries generated by different positions of the wall stem on the wall base. Some interesting results regarding the load pressure on the heel are shown in Figure 9.

The load pressure does not remain constant over the heel width like it is commonly assumed in wall analysis. There is a typical minimum near $x = 0$ caused by the tangential earth pressure on the wall stem. A further assumption is the vertical action of the load pressure or a little positive direction. But the results show a negative direction of action, especially near the wall stem.

Because of the unexpected direction of action, the orientation of the pressure cells had to be changed after the first test. But no further problems occured with the cells in the test programme. So, the cells have demonstrated their functionality also in cases of low stress levels like in the backfilling soil, which is in a condition between the active and the one at rest.

4 SUMMARY

A new solution of contact-pressure measurement has been presented. It has been applied in a couple of

research projects. It has got several advantages in comparison to other measuring technologies:

- The normal and tangential components of the contact pressure are concurrently measured.
- By larger dimensions of the area in which the pressure is acquired, the effects caused by heterogenities in the soil are reduced.
- The possibility of a side-by-side arrangement of the pressure cells minimises or eliminates the arching effect between the stiff surface of the construction and the pressure cells.
- The entire surface of the test object or large parts of it may be covered by the cells.

Of course there are limits to the application. The transducers showed a small creep, so that the time, in which the measurements are carried out, especially when the transducers are loaded, should not exceed one day.

325

This fact, the size of the cells, and the missing water protection limit the application to model tests.

Transducer deformability reduces the stiffness of the pressure cell. If the cell is used beside a nondeformable surface, some pressure transfer may occur. The magnitude of this effect depends on the problem and should be evaluated.

Some very interesting investigations have been executed with the new measuring technology. Passive earth pressure model tests show noteworthy results to the distribution of the earth pressure and its inclination at different mobilization states. Also, the investigation of other kinds of contact stresses, like the load pressure on the heel of cantilever retaining walls, was made possible by the newly developed pressure cells.

ACKNOWLEDGEMENT

The research projects "Mobilization of passive earth pressure" and "Earth pressure and load pressure on cantilever retaining walls" were funded by the Deutsche Forschungsgesellschaft, DFG.

REFERENCES

Arnold, M. 2000. Untersuchungen zum Erddruck auf Winkelstützmauern. In *26. Baugrundtagung, Hannover: Forum für junge Geotechnik-Ingenieure*: 44. Deutsche Gesellschaft für Geotechnik.

Arnold, M. 2001. Modellversuche zum Erddruck auf Winkelstützwände. In D. Franke (ed.), *OHDEKolloquium 2001*: 23–34. Institut für Geotechnik der TU Dresden, Heft 9.

Bartl, U. 1995. Untersuchungen zum Erdwiderstand auf ebene Wände am Beispiel von 1g-Modellversuchen mit Kopfpunktdrehung. In D. Franke (ed.), *Festschrift zum 60. Geburtstag von Prof. Dr.-Ing. habil. Dietrich Franke*: 201–216. Institut für Geotechnik der TU Dresden, Heft 3.

Bartl, U. & Franke, D. 1998. Zur Mobilisierung des passiven Erddrucks in trockenem Sand. In B. Maric, Z. Lisac & A. Szavits-Nossan (eds.), *Proceedings 11. Danube-European Conference on Soil Mechanics and Geotechnical Engineering*: 189–196. Rotterdam: Balkema.

Hoffmann, K. 1987. *Eine Einführung in die Technik des Messens mit Dehnungsmeßstreifen*. Darmstadt: Hottinger Baldwin Meßtechnik GmbH.

Direct shear testing system

M. Blümel & M. Pötsch
Institute for Rock Mechanics and Tunnelling, Graz University of Technology

ABSTRACT: Direct shear test procedures on jointed and intact rock are being developed and performed on an automated, digitally controlled servo-hydraulic shear test system. New capabilities allowed by the system are greatly simplified performance of calculated control shear tests that were previously difficult to perform, including normal stiffness control during multiple and continuous failure tests in direct shear. Dilation and rotation are determined by measuring the shear box displacement in multiple locations. The accurate measurements of displacements and forces and the ability to create user defined control modes allow performing new types of test procedures, adjusted to specific problems. As an example of application of the direct shear system a shear test procedure for a research on removable blocks and its performance is presented.

1 INTRODUCTION

Servo controlled direct shear tests using variable feedback channels for test control provides the opportunity to implement different boundary conditions to the sample during testing (Blümel & Bezat 1998). Shear tests are typically performed to evaluate the strength of preexisting discontinuities such as foliation planes or joints, while compression tests are used for intact samples (Blümel 2000). For highly anisotropic and weak rocks, the acquisition and preparation of samples for compression tests often results in a highly biased selection of stronger samples due to difficulties in specimen preparation. The preparation effort is much less for direct shear testing, thus allowing the testing of the weaker material, as well as the highly competent material using the same procedures. To evaluate the anisotropic behavior, a sample can be placed at any orientation within the shear box to evaluate the strength and failure processes associated with a shear direction that is not directly parallel to the preexisting discontinuity structures (Button & Blümel 2002). Constant normal load (CNL) shear tests do not really test the rock strength but the resistance to shear at a certain normal load. The use of constant normal stiffness (CNS) testing procedures can be used to define a sample's "Ultimate Shear Strength" which is the sample's natural response to simple shearing. The strength is influenced by a complex interaction between cohesion, dilation and sliding friction. The breakdown of these factors to a residual strength is not easily defined. This interaction continues through the entire straining process to different degrees. The use of sophisticated testing procedures can begin to quantify the processes (Blümel, Button & Pötsch 2002). One parameter that is extremely difficult to quantify is the shear stiffness. This parameter is highly nonlinear and controlled by the interaction of the above-mentioned factors. As this is a typical input parameter in many numerical models the development and degradation of this parameter is currently being more thoroughly investigated.

2 SHEAR SYSTEM DEVICE

There are currently no guidelines for determining the stiffness to be used for a given CNS investigation program. Many CNS testing apparatuses use springs to simulate the stiffness and are thus rather limited. While state-of-the-art apparatuses using hydraulic servo controlled test systems allow any stiffness to be applied, and even provide the opportunity to vary the stiffness with calculated control modes, for example a portion of the stiffness can be controlled by the horizontal displacements. Choosing the appropriate stiffness is important, as the measured peak strength is a function of both the initial load and normal stiffness. The normal stiffness should approximate the expected in situ rock mass stiffness. If an infinite stiffness is used (simple shear) then the "Ultimate Shear Strength" of the sample can be determined. This value is not the maximum shear resistance, but the maximum strength that the specimen possesses in shear.

As an example of application of the direct shear system in scientific work the derivation of a shear test procedure and its performance is presented. The procedure is focused on the characteristics of the stress path that a joint of a removable block undergoes during the displacement when it detaches from the rock mass. There are several factors that influence the stress path such as the stress magnitude, the friction angle of the rock material and the roughness of the joint. This paper addresses the influence of the rock mass stiffness and the detachment angle.

3 GENERAL CONCEPT

Before the excavation of a tunnel takes place the primary stresses are acting on the joints. The stress condition on a joint can be described in terms of normal and shear stress. After the excavation of the tunnel the stress redistributes and the joints may form removable blocks (Goodman & Shi 1985). Unstable blocks begin to displace and tend to detach from the surrounding rock mass. Figure 1 shows the situation of a potentially unstable block in the tunnel roof. Normal and shear stresses are acting on the joints. As the block displaces it tends to detach from the rock mass due to the angle between the joint and the direction of the block displacement. In the succeeding paragraphs this angle is called the detachment angle β. The normal stress is expected to decrease due to the detachment. Since the joint is assumed to be rough, dilation takes places and therefore counteracts the displacement. The normal stress tends to increase. Basically, it

depends on the magnitude of dilation in relation to the detachment angle if the normal stress increases or decreases. Therefore, the magnitude of the stress variation is controlled by the effective normal displacement of the joint (dilation or contraction) and additionally, by the rock mass stiffness. To adequately simulate this kind of shear behaviour the shear tests have to account for the stiffness boundary condition.

4 DERIVATION OF THE SHEAR TEST PROCEDURE

A computer-controlled shear test procedure was developed in order to simulate this kind of behaviour. These procedures have the capability to control the load cylinders in a way that the desired boundary condition is fulfilled. The external condition is expressed in terms of a basis function, which includes and connects the parameters of interest. In this procedure the input parameters of the basis function are the shear displacement u, the normal displacement v, the initial normal stress $\sigma_{n,i}$, the detachment angle β and the confining normal stiffness K which accounts for the rock mass stiffness. Figure 1 shows the steps for the formulation of the basis function. It is assumed that dilation is related to a negative and contraction to a positive displacement.

Generally, the procedure is displacement-controlled. This means that the load cylinders for the shear force are controlled in a way to maintain a constant shear displacement rate. The shear displacement is measured and serves as an independent input parameter

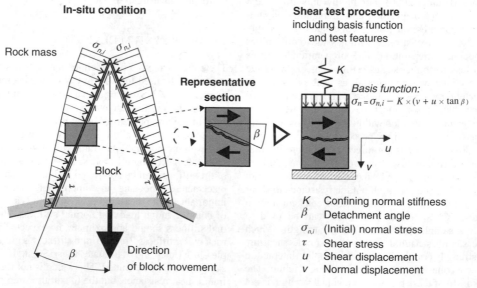

Figure 1. Development of the shear test procedure.

for the basis function. The normal displacement is also measured to provide another independent input parameter. The measured normal displacement results mainly from dilation due to joint roughness. To account for the effect of detachment an inclined joint within the sample would be necessary. It is possible to simulate this behaviour with a joint parallel to the shearing direction using the basis function. The parameters β and K are required to input manually. Since the detachment is proportional to the shear displacement u with the factor of $\tan \beta$, it is only necessary to subtract the value $u \times \tan \beta$ from the measured normal displacement v due to dilation to obtain the stress-effective normal displacement. Therefore, the basis function is:

$$\sigma_n = \sigma_{n,i} - K \times (v + u \times \tan \beta)$$

Hence, the normal load cylinders now are controlled during shearing in a way that the initial normal load $\sigma_{n,i}$ is modified to fulfil the basis function taking into account the current stress-effective normal displacement and the magnitude of the normal stiffness K.

5 PERFORMANCE OF THE SHEAR TESTS

A series of these shear tests on artificial saw-tooth joints were performed (Pötsch 2002). The samples consist of anchor grout. They have a shearing area of about 12×12 cm. The slope of the asperities is about $22°$. The unconfined compressive strength (UCS) of the grout was tested and determined to be 65 MPa with a standard deviation of 4 MPa using unconfined compression tests and Schmidt Hammer Index tests. The residual friction angle of the material was determined to be $37°$ with a standard deviation of $4°$ using multi-stage shear tests under constant normal load.

The shear tests were performed under an initial shear displacement rate of 0,07 mm/min. At an advanced test state long duration tests were smoothly accelerated to 0,2 mm/min. The initial normal stress was at 5 MPa. Variations of the normal stiffness K and the detachment angle β have been carried out. The normal stiffness was chosen with 100, 1.000 and 10.000 MPa/m and the detachment angle with 15, 25, 30, 35 and $40°$. It can be seen that one detachment angle is smaller than the slope of the asperities whereas the other four angles are greater.

6 DISCUSSION OF THE RESULTS

In the following paragraphs the results of the test series are discussed. Figure 2 shows the plots of tests with different detachment angles under a normal stiffness of 10.000 MPa/m. Apart from the stress path the plots of the shear stress, the normal stress and the measured normal displacement along the shear displacement are presented.

These plots show two different failure mechanisms. The first failure mechanism occurs in the test with $\beta = 15°$ where dilation compensates the detachment in the pre-failure region. After the displacement for matching of the asperities the normal and shear stress increase until shearing of the asperities initiates. Afterwards, within the strain hardening, the shear stress approaches the peak value. The normal stress remains at an almost constant level. This is caused by a rotation of the sample that compensates dilation and detachment. After completely shearing through the asperities rotation terminates and the sample continues sliding along the failure crack. Therefore, the normal stress and subsequently the shear stress drop because detachment now dominates the behaviour. The stress path follows a residual envelope toward the origin. This envelope is influenced by the residual friction angle of the material.

The tests with a detachment angle greater than the slope of the asperities show the second failure mechanism. Detachment dominates the behaviour from the initiation of shearing. Hence, the normal stress drops continuously, the greater the detachment angle the greater is its gradient. In the pre-failure region friction mobilizes with the displacement and causes the shear stress to increase. It approaches the peak within a displacement, which is considerably smaller than that of the first failure mechanism. Due to the reduction of the normal stress, affected by the higher detachment angle sliding on the asperities occurs. Hence, the measured normal displacement (dilation) is greater than in the previous case but not stress effective. A peak failure envelope as depicted in Figure 2 limits the peak shear stress. The geometry of the asperities and the resistance of the material influence this envelope. After the peak the stress path approaches the origin within both envelopes. At the beginning of the asperity degradation the stress path follows the peak envelope. On the other hand, when the asperities are predominantly sheared through it follows along the residual envelope. At low normal stress levels (below approximately 0.5–1 MPa) the normal stress gradient abruptly decreases although a drop to zero is expected. This is caused by the test assembly which permits the sample to rotate and therefore to again dilate to compensate the detachment (see normal displacement plot). In reality the joint would completely detach.

Figure 3 shows the plots of tests with different normal stiffness under a detachment angle of $40°$. Apart from the stress path the plots of the shear stress, the normal stress and the measured normal displacement along the shear displacement are presented again. The behaviour is dominated by detachment. This means

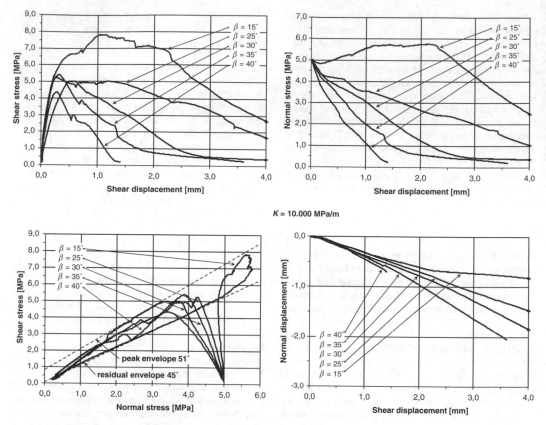

Figure 2. Plots of the test results of the variation of the detachment angle.

that the normal stress continuously decreases. The influence of the normal stiffness is clearly visible. At a normal stiffness of 100 MPa/m the normal stress is not very sensitive to variations in the normal displacement. It remains at an almost constant level. On the other hand, at a normal stiffness of 10.000 MPa/m the normal stress is very sensitive to variations of the normal displacement. It rapidly drops to zero due to detachment. The stress path approaches the peak envelope and then drops to the residual envelope approaching the origin. The mechanisms are similar to the second failure mechanism as described in the former paragraphs.

The shear behaviour of saw-tooth joints under constant normal load is described by Patton 1966. At low normal stress the shear resistance is dominated by the friction angle of the joint φ_j and the slope of the asperities i. Sliding on the asperities takes place. The equivalent in this test series is the peak envelope. At higher normal stress the shear resistance is dominated by the cohesion and friction angle of the material. Shearing through the asperities takes place. The equivalent in this test series is the residual envelope.

Therefore, applying Patton's law the peak envelope has an angle of 59° whereas the residual envelope has an angle of 37°. For comparison, in the variation of the detachment angle the peak envelope has an angle of 51° and the residual envelope an angle of 45°. It was verified that as the normal stiffness decreases, the angles of both envelopes approach the values of Patton's law. Additionally, in the variation of the normal stiffness the peak envelope has an angle of 56° and the residual envelope an angle of 40°.

7   CONCLUSION

A shear test procedure was developed to adequately simulate the behaviour of a detaching dilatant joint under constant normal stiffness. It is the dominant behaviour of removable blocks around a tunnel. The procedure accounts for detachment, dilation, and stiffness of the rock mass and initial normal stress level. With the test series it was shown that a complex interaction between normal displacement, normal stress, normal stiffness and detachment angle influences the

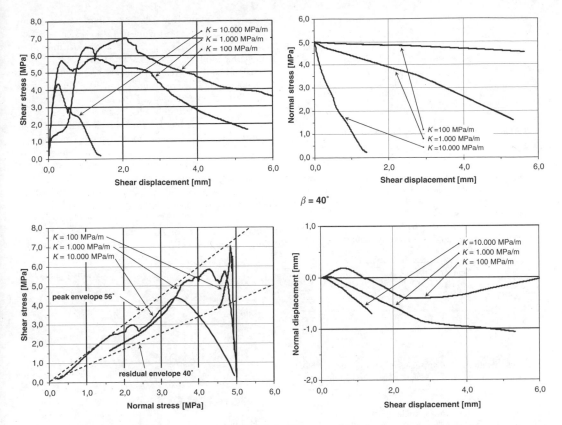

$\beta = 40°$

Figure 3. Plots of test results of the variation of the normal stiffness.

shear resistance of the joint. A higher initial normal stress level increases the shear resistance and the peak displacement required to reach the peak strength. Two failure mechanisms have to be distinguished. When the slope of the asperities is greater than the detachment angle, this leads to a significantly increase of the shear resistance, particularly under high rock mass stiffness. On the other hand, if the detachment angle is greater than the slope of the asperities, detachment dominates the behaviour. High rock mass stiffness leads to a lower shear resistance because the normal stress abruptly decreases. For a practical engineer it is obvious, that such kind of block failure often occurs in tunnel excavations with low overburden or low confining stress and stiff rock mass conditions. Nevertheless a reliable design and prediction needs relevant parameters, which should be quantified under specific boundary conditions. The surrounding rock mass stiffness is one of the major influencing parameter and not just the determined friction and cohesion in classical point of view. The presented test method is appropriate to evaluate these various boundary conditions.

REFERENCES

Blümel, M., Button, E.A. & Pötsch, M. 2002. Steifigkeitsabhängiges Scherverhalten von Fels. *Felsbau Rock and Soil Engineering 20/3*, 22–32.

Blümel, M. 2000. Improved Procedures for Laboratory Rock Testing. *EUROCK 2000* ISRM Symposium Aachen 2000, 573–578.

Blümel, M. & Bezat, F.A. 1998. Advanced Control Techniques for Direct Shear Testing of Jointed Rock Specimens. *Non-destructive and Automated Testing for Soil and Rock Properties*, ASTM STP 1350, W.A. Marr and C.E. Fairhurst, Eds., American Society for Testing and Materials.

Button, E.A. & Blümel, M. 2002. Servo-Controlled Direct Shear Tests on Phyllites. *Proceedings of the 5th North American Rock Mechanics Symposium*, Toronto.

Goodman, R.E. & Shi, G. 1985. *Block Theory and Its Application to Rock Engineering*. Prentice-Hall, New Jersey.

Patton, F.D. 1966. Multiple modes of shear failure in rock. *Proc. 1st Congress Int. Soc. Rock Mech.*, Lisbon, 509–513.

Pötsch, M. 2002. Influence of the three-dimensional stress condition at the tunnel face on the stability of removable blocks. *Diploma Thesis*, Institute for Rock Mechanics and Tunnelling, University of Technology, Graz.

An experimental method to avoid temperature bias on instrumentation measurements

M. Ceccucci, M. Ferrari & A. Roasio
Citiemme s.r.l., Torino, Italy

ABSTRACT: When the magnitude of the variation of the measured parameters is very small, it becomes difficult to detect the trends that characterise the measured quantities because multiple factors can modify the electrical readings of the sensors employed to perform the continuous measurements. The reduction of the effects of such factors is achieved improving the installation procedure and the management of the measurements. In addition, in the field of data reduction, the authors propose the following analysis method, derived from a large data set acquired over more than four years of continuous measurements. The aim is to detect the correlation existing between the measured deformations and the thermal variations to which the sensors are subject, defining a method to reduce the "shadow effect" masking the true variations of the monitored parameters.

1   INTRODUCTION

The continuous development of the electronics instrumentation in the geotechnical field has allowed, in addition to the availability of more and more precise instrumentation at a steadily decreasing cost, to automatize the data taking and to collect large data samples.

The concept of automatic measurement becomes fundamental especially in those cases where a continuous monitoring is required for security reasons: this aspect has increasing importance especially in the field of territorial monitoring (landslides) aiming to swiftly detect significant variations of those parameters deemed to signal disruption. As important is, in the geotechnical field, the management of the alarms when the control done by the instrumentation serves to verify the stability of buildings or of strategically important structures (dams, bridges, etc.). In many cases the interpretation of the data is complex because a large number of variables connected both to the installation conditions and to the site environmental variations come into play.

In addition to the recommendations to be followed for a correct installation of the instrumentation, (Duncliff 1988, Hanna 1985) that represent the base for any type of work aiming to the detection of information via electronics sensors, one has to carefully scrutinize the collected data in order to avoid wrong interpretations of the recognized trends.

In this study, starting from the observations of Youn et al. (2002), particular care has been placed on the role that the thermal excursions have with respect to the readings of the instruments. In particular, starting from the electrical readings of some of the most common sensors employed in the geotechnical field and from the temperatures measured by the thermometers associated to them, data have been compared in order to obtain on site the same precision and accuracy obtained by the instrumentation in the lab during the calibration phase.

2   METHODOLOGY

To clean up the electrical signals of the geotechnical instrumentation from the effects of the thermal excursions of the environment in which it operates is one of the overriding and precautionary activity for anybody managing instrumentation aiming to detect "real" changes of the parameter under control. The thermal excursions are responsible for both the thermal dilatation of the metallic parts of the instrument and for those of the materials on which the instruments are positioned (Košt'ák 2002). This two-fold effect masks the real deformations that can affect the measured quantities.

Monma et al. (2000) suggest to employ the following equation to take into account the effects of the thermal dilatation of the instrumentation.

$$L_{true} = L_{ms} - kT$$

where, L_{true} = true rock displacement; L_{ms} = measured displacement; k = displacement due to temperature change by $1°C$; and T = temperature.

Youn et al. (2002) compute the value of k by means of three different methods:

1. The method of daily temperature correction factor (DTCF);
2. Regression after dividing the data at obvious intermittent slope change;
3. Application of least square method.

The authors propose DTCF as the most suitable method to define the temperature correction factor taking into account the daily thermal excursion and the related displacement.

Starting from these observations in the present paper the hypothesis of a method is put forward to remove the "shadow effect" generated by the noise on the sensor signal for the same operating conditions at the site for the parameter to be measured, and for the same quality and sensitivity to the environmental effects of the employed sensor. Such "shadow effect" seems to be mainly related to the thermal excursions both of the materials forming the instruments and of those on which they are positioned. The aim of a method allowing to clean up the signal from such effects is also to remove some of the anomalies that randomly affect the behaviour of the installations on the field. The proposed approach is based on the procedure followed on the field which considers a few points as fondamental:

- quality of the employed sensor;
- ability to conform to the monitored site;
- positioning of a reference instrument (Check Signals);
- closely spaced sampling of the measured parameters;
- presence of sensors for the temperature recording.

The proposed method tries to correct the DCTF, evaluating it for each series of successive acquisitions without taking into account the values recorded over the entire day; in this case for the calculation of the k coefficient, valid for T in the samplying measure, the following equation has been used:

$$k_i = (L_i - L_{(i-1)})/(T_i - T_{(i-1)})$$

The acquired know how passes then through the realisation of an "Experimental On Site Laboratory" where the performances of the instrumentation is verified evaluating the difference between what emerged during the calibration phase of each component of the installation in the laboratory and the "real" functioning on the field. In this way, one can remove some of the totally unpredictable anomalies that mask the real values detected by the sensors and lead to trends of the parameters under measurement of difficult interpretation.

3 CONSIDERED INSTRUMENTATION AND ANALYSED DATA

The considered instrumentation is represented by some of the most common sensors employed for the evaluation of the displacements both in the geotechnical field for the structural control of buildings and for the monitoring of the territory when it is necessary to keep under survey fractures in rock faces subject to slides or to verify differential movement of portions of terrain subject to complex landslide phenomena.

In particular the presented instrumentation belongs to three installations with different monitoring purposes built and maintained over time:

- Mazze' Dams on Dora Baltea River;
- La Fenice Theater in Venice;
- Ceppo Morelli rockslide.

In detail, the evaluations that follow take into account the data detected from 37 instruments classified in the following typologies:

- 18 Vibrating wire displacement transducers (GEOKON, USA);
- 12 Potenziometric Jointmeters (TER, Italy);
- 7 Potenziometric Wire crackmeters (SISGEO, Italy).

In the first case the measurement of the temperature was done by means of thermistors housed in the sensors, while in the others Platinum RTD of the same type have been installed at various locations within installation in order to allow a systematic evaluation of the temperature on site.

The employed Data Logger belong to the series CR-10 of Campbell Scientific. In the table below the technical characteristics are listed, as given by the manufacturer, for each considered typology.

As far as the data are concerned, the considered sequences have been collected between 1998 and 2002 and they appear in varying number; the time between

Table 1. Technical characteristics of the considered instrumentation.

Instruments	Type of sensor	Accuracy (mm)	Resolution (mm)
GEOKON Displacement transducer	Vibrating wire	0.1	0.01
TER Jointmeter	Linear potenziometer	0.1	0.01
SISGEO Wire crackmeter	Rotatory potenziometer	1	0.03

two samplings varies between 30 and 60 minutes while the whole data taking period for a single instrument varies between 1 and 4 years. Therefore, the number of data points varies from a minimum of 3000 to a maximum of over 30,000.

4 ANALYSED DATA

In figure 1 is shown the trend of the absolute temperature (in grey) detected by a probe in the proximity of the instrument and of the displacement (in black), measured in mm by a jointmeter, computed as:

$$L_r = L_{es} - L_0$$

with L_r = relative measurement; L_{es} = running measurement; and L_0 = reference measurement.

The graph clearly shows a significant displacement signalling the opening of the fracture monitored by the instrument. It is possible to identify two distinct portions labelled A: pre-motion and B: post-motion.

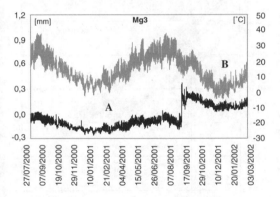

Figure 1. Mg3 – trend of the displacement and of the temperature.

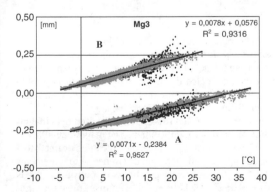

Figure 2. Mg3 – trend of the displacement as a function of the temperature.

In figure 2 the same data are presented but displaying the displacement as a function of the temperature measured at the sampling time.

The marks are placed along two main alignments (in grey): below one identifies the values related to the first section (A) and above those related to the last section (B). The black marks show the month when the displacement occurred.

The trending straight lines (in black) computed on the two dispersions show similar angular coefficient k (0.0071 for section A, 0.0078 for section B) evaluated by minimal square method. The difference between the intercept of the two trending lines allows to measure precisely the value of the displacement [0.0576 − (−0.2384)] = 0.296 mm.

In figure 3 a case is shown where, differently from before, there is not an obvious displacement. Also in this example the data have also been processed displaying the displacement values as a function of the temperature.

The result is shown in figure 4. In both representations the trend is subdivided in several sections, each of different duration and defined by capital letters.

The comparison of the two diagrams allows to correctly define the effective trend of the variation of the spacing that affected the fracture continuously monitored by the instrument.

The entire time span can be subdivided in periods characterised by phases of standstill interspersed by phases of real deformation. Also in this case, in fact, one recognises periods during which the marks (in black) are perfectly placed along the lines characterised by similar slope (in white) connected by "clouds" of scattered points (in grey) showing preferential alignments.

The distribution of the marks during the two external periods show that the instrument has not changed its own characteristics because during the two phases the marks are aligned according to the straight lines with angular coefficient 0.0107 (section AB) and 0.0116

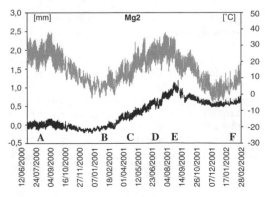

Figure 3. Mg2 – trend of the displacement and of temperature.

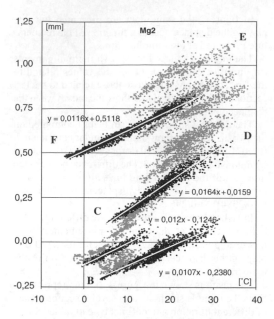

Figure 4. Mg2 – trend of the displacement as a function of the temperature. AB: No evidence of movement, BC: phase characterised by two movements separated by a standstill, CD: very slow movement, DE: movement characterised by an opening phase (prevailing) followed by a closing phase, EF: no evidence of movement.

(section EF). The difference of the intercepts show the "real" displacement that has affected the instrument $(0.5118 - (-0.238)) = 0.7498$ mm. In addition, a series of pulsations that cannot be detected in figure 3 where the relative displacements are reported as a function of time is highlighted.

5 REDUCTION OF THE THERMAL INFLUENCE

In figure 5 is shown what has been detected during approximately four years by a vibrating wire sensor named F7.

The trend of the relative displacements (in black) and that of the temperature (in grey) show the correspondence between the signal given by the displacement transducer and the thermal oscillations. Showing the displacement as a function of the temperature, it becomes clear that most of the marks are placed preferentially along two alignments. To better emphasise, in figure 6 only the data from the period 1998–2000 are shown; the marks related to data collected during 1998 are in grey nearby the origin of the axes (trending straight line $y = -0.0103x + 0.1716$) while those related to 2000 are distributed around the straight line described by the equation $y = -0.0116x + 0.2888$.

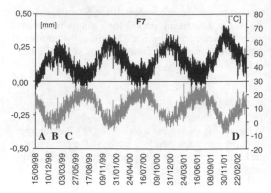

Figure 5. F7 – trend of the displacement and of the temperature.

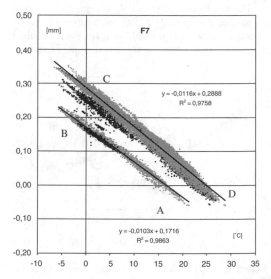

Figure 6. F7 – trend of the displacements as a function of the temperature.

The black marks are the 1999 measurements and show that during that year a displacement has taken place. In the detail shown in figure 7, the dispersion of the data collected during January and February 1999 (in black) clearly show that during those two months four different movements took place.

Of these, the first two are characterised by a magnitude of 0.04 mm and the others by a magnitude of about 0.02 mm. In total the displacement can be defined, to a good approximation, by the difference of the intercepts of the straight lines of the two outward periods $(S = 0.2893 - 0.1716 = 0.1177$ mm). This case highlights the impossibility to detect such displacement from the graph of the normal data analysis in figure 5.

336

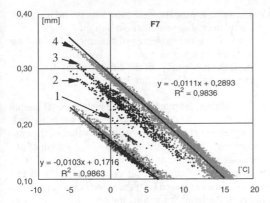

Figure 7.   F7 – detail of the graph presented in figure 6.

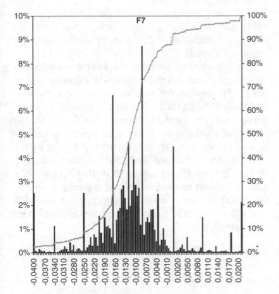

Figure 8.   F7 – histogram of the frequency of the coefficient k_i.

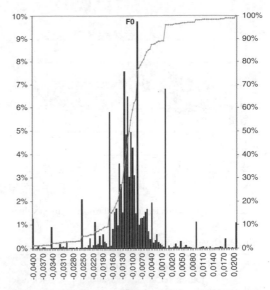

Figure 9.   F0 – histogram of the frequencies of the coefficient k_i.

value is included between 0.001 and −0.03 with a probability of 95%.

Given the current knowledge of the authors, it does not seem possible to define the reasons leading to the oscillation of the k_i value, nevertheless it seems plausible to formulate some hypotheses on this aspect. Firstly, assuming that the measured temperature always coincides with that at the place of installation of the displacement transducer, it appears plausible that among the possible motivations one should include the characteristics of resolution of the instruments and of the thermometer. The ageing of the instrumentation and the relative shift of temperature and precision is another sure factor responsible of the characteristics of the processed curve.

The histogram presented in figure 9 contains the values of k_i computed analysing in the same way the data detected by the instrument F0. This (connected to the same Data-logger) is positioned on a block of material with geo-technical characteristics similar to that on which the other sensors were positioned and has the scope of detecting the noise that affects the instrumentation owing to the thermal excursions and to the anomalies of operation (Check-Signal).

In this case the mean value is −0.0105; the most probable value is still in the bin −0.0085, −0.0090 mm/°C. The k_i value computed taking into account the data taken from the instrument F7 results larger than −0.04 with a probability of 2.5% while is just above 1% in the case of the reference instrument F0. Similarly the bin containing the values of $k_i > 0.02$ has a probability of about 1% for F0 and more than 2%

Thus, the method of thermal influence reduction has been applied starting from the analysis of the data detected by the instrument and defining the relative k coefficient for the examined period and then analysing the period with the statistic method.

In the histogram presented in figure 8 is shown the trend of the k_i value (computed for the instrument F7) subdivided in bins of 0.0005 mm/°C. The mean value is 0.0118 but the bin containing the mean value is detected with a frequency of 2.64% while the most probable value is that included between −0.0085 and −0.0090 mm/°C with a frequency of 8.74%. In addition, the dispersion of the k_i value shows that this

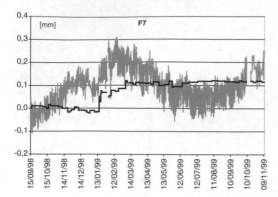

Figure 11. F7 – corrected trend of the displacement values.

Table 2. k values (mm/°C).

	Potenziom. jointmeter	VW displ. transducer	Wire crackmeter
R^2	0.93	0.94	0.78
k_{mean}	0.0102	−0.0110	0.1055
k_{min}	0.0069	−0.0096	0.0924
k_{max}	0.0121	−0.0125	0.1457

for F7. The reason for this is that instrument F7 has detected the crack of the monitored fracture as shown in figure 7.

Based on the considerations reported above, in the following diagram is shown the proposed method to remove the "shadow effect" mainly due to the thermal response of both the material on which the instrument is positioned and of the mechanical and electrical components forming the instrumentation.

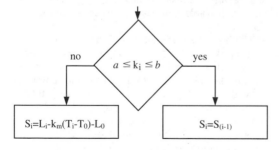

Figure 10. Scheme for the computation of the corrected relative value.

where:

- S_i is the ith displacement;
- L_i is the value of the ith displacement measured at the temperature T_i;

- L_0 and T_0 are the displacement values and temperature detected during the installation;
- a and b are the extremes of the interval including 95% of the k_i values;
- k_m is the mean value of the correction coefficient in the considered histogram.

In addition, the majority of the instrumental anomalies related to data acquisition "spikes" are mitigated.

In practice at each acquisition the value of the correction coefficient k_i is computed as a function of the previous measurement. This value is compared with the extremes of the interval containing 95% of the values of the computed k_i. If the value is included in the interval, the S value remains the same to the one previously computed ($S_i = S_{(i-1)}$); if the value of k_i is outside of the interval, the value of S_i will be computed using the k_m mean value.

In figure 11 is shown the trend of the displacement value detected by the instrument before (grey) and after (black) the correction of the values performed using the proposed method.

The effectiveness of the method to clean up the displacement curve and the consequent improvement to detect also minimal deformations are evident. The correctness of the method is supported also by the comparison between the data obtained from the measuring instrumentation and from the Check-signals, as demonstrated by the same frequency bins of k_i and by the mean values k_m that overlap to the third decimal digit.

Finally the authors propose the following table of k values that can be used when dispersion data for the described instrumentation typology are not available (the mean value of R^2 measured in the considered cases is also reported).

6 CONCLUSIONS

The experience of the authors accumulated in the field of instrumental monitoring has shown some fundamental aspects that have to be considered for a correct data handling:

1. For the comprehension of the trends of the parameters under study is essential to clean up the electrical signals obtained from the sensors from the anomalies connected to the thermal excursions affecting the site under investigation;
2. It is a priority to install a reference instrument to compare the behaviour of the instruments used on the field with respect to what is certified by the manufacturer;
3. It is necessary to define the value of the coefficient of thermal reduction k and its distribution k_i evaluated over a sufficiently long time period;

4. The position of the temperature sensors has to allow the exact comparison of the temperature values and of the deformation of the measured parameters;
5. In this study the applied procedures to reduce the effect of the temperature excursions have been: (a) evaluation of the correlation between displacement and temperature; (b) correct evaluation of the displacements displaying it as a function of the temperature; (c) extrapolation of the reduction coefficient k; (d) statistical evaluation of the value k_i with and without deformation; (e) re-elaboration of the measured curves with the reduction of the effects connected to the thermal excursion and to the instrumental anomalies to highlight sub-millimeter displacements;
6. The proposed method cleans up the displacement curves allowing to detect also minimal variations in the measured trends.

REFERENCES

Duncliff, J. 1988. Geotechnical instrumentation for monitoring field performance. New York: John Wiley & Sons.
Hanna, T.H. 1985. Field Instrumentation in Geotechnical Engineering. Trans Tech Publications. Clausthal-Zellerfeld, Germany.
Košt'ák, B. 2002. Cycles, trends, impulses in rock movement monitoring. In Rybář Stemberk & Wegner (eds), *Landslide; Proc. of the first European Conference on Landslides, Prague, 24–26 June 2002*: 603–609. Rotterdam: Balkema.
Monma, K., Kojima, S., Takaki, T. & Uekita, K. 2000. One case study of temperature correction on data of rock slope monitoring. *Proc. The Japan society of erosion control engineering. N°. 33*: 16–17.
Youn, H., Fukuoka, H., Greif, V., Tamari, Y. & Sassa, K. 2002. Estimation of Temperature Change Component in Monitoring Data of Rock Slope Movement. *Proc. of Landslide Risk Mitigation and Protection of Cultural and Natural Heritage, Kyoto, 21–25 January 2002*: 459–468.

Geotechnical Measurements and Modelling, Natau, Fecker & Pimentel (eds)
© 2003 Swets & Zeitlinger, Lisse, ISBN 90 5809 603 3

A modular triaxial testing device for unsaturated soils

J. Engel & C. Lauer
Institute of Geotechnical Engineering, TU Dresden, Germany

M. Pietsch
Federal Waterways Engineering and Research Institute (BAW), Karlsruhe, Germany

ABSTRACT: Soil-mechanics experimentation on partly saturated specimens requires special equipment to measure forces and deformations. A new triaxial testing device has been developed to determine the soil-water characteristic curve, the unsaturated hydraulic conductivity, and the shear strength. The volume change of the unsaturated specimen is measured by the flow of the cell fluid into or out of the triaxial cell. To allow accurate measurements of the volume change, a double walled triaxial cell has been designed. Ruggedly designed tie bars prevent vertical deformation. To prevent any falsification of the volume change measurement silicon oil is used as cell fluid. Matric suction ($u_s = u_a - u_w$) is applied at the top and bottom of the soil specimen via the axis-translation technique. The aim is a fast and nearly uniform distribution of matric suction in the specimen. The unsaturated hydraulic conductivity is determined by subjecting different matric suctions at the top and bottom of the specimen. The presented test procedure avoids many of the known problems which are connected with the direct and indirect measurement from the unsaturated hydraulic conductivity; by taking the total and pore-water volume change of the unsaturated specimens into account.

1 INTRODUCTION

The soil mechanical concepts have mostly been developed for saturated soils. Especially the effective stress has been introduced into the constitutive equations for saturated soils. These concepts clearly describe the interaction between total-stress, pore-water pressure and intergranular stress. The experimental investigations have eversince been planned and evaluated and the test equipment has been designed on the basis of these constitutive concepts.

In practical geotechnical applications, the problem of partly saturated soils is mostly considered by empirical relations or with an adaption of the models for saturated soils. Fig. 1 shows the results of triaxial tests on different soils. The capillary pressure ($u_a - u_w$) was measured. This is an example for an interpretation of tests on the basis of adapted models from saturated conditions. It is assumed, that an effective stress exists which can expressed in terms of σ, u_a and u_w.

Another method to consider the behaviour of unsaturated soils are empirical relations. Fig. 2 shows a graphical representation of the phase composition in a (n_w, $1 - n$)-coordinate system. The volumetric water content n_w is defined to $n_w = V_w/V = w\rho_d = \rho_w (V_w -$ volume of water within the specimen, V – volume of

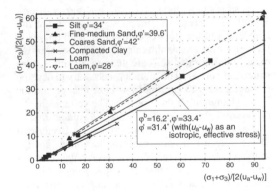

Figure 1. Results of triaxial compression tests on unsaturated soils.

the specimen, ρ_d – dry density, $\rho_w = 1\ \text{g/cm}^3$ – density of water) and the solid content to $(1 - n) = VS/V = 1 - \rho_d/\rho_S$ (V_S – volume of solids, ρ_S – density of solids). Compression tests on unsaturated samples, starting from a low density, leads to lines of equal consolidation pressure. These relationships between phase composition and consolidation pressure is used like an equivalent pressure to calculate the change in the soil mechanical behaviour.

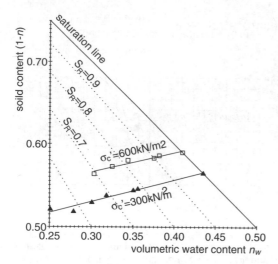

Figure 2. Phase composition of unsaturated soils and results of compression tests.

To link the hydraulic conductivity with soil mechanical stress–strain relations, it is essential to incorporate the soil-water-characteristic curve. On the other hand, the relation between suction and phase composition is also a function of the stress–strain state, the stress history and the fabric of the soil.

To understand the mechanical and hydraulical behaviour of unsaturated soils, it is necessary to check all assumptions about the stress–strain relations. Therefore the test equipments should allow the full control about the stresses and strains of all phases in the soil. However, this aim can only be reached approximately. Unsaturated soils are mixtures of solids, pore water and pore air. Therefore it is necessary to develop test equipment to control the volume change and the stress change of all three phases. The volume change of the solid phase is negligible and the volume change of the pore air is very difficult to measure. Because of this reasons it is usual to measure the volume change of the whole specimen and of the pore water. The volume of the pore air is than calculable from the initial phase composition. However the measurement of all forces causes a great difficulty.

In the following sections, a new triaxial equipment will presented that avoids many of the known problems according to tests on unsaturated specimens.

2 MODULAR EXPERIMENTAL EQUIPMENT FOR TESTING UNSATURATED SAND

2.1 Introduction

The total volume change of an unsaturated soil specimen is the sum of the pore-air and pore-water components of volume change. There are different possibilities for determining the total volume change: (1) measuring pore-water and air-volume change; (2) measuring total volume change in real double-wall triaxial cell or in modified triaxial cells with an insert; (3) contacting or non-contacting internal instrumentation for axial and radial strain measurements; (4) determining of the total volume change by imaging technique (X-ray radiography, Digital imaging processing and analysis techniques) or Laser measurement.

With the help of the internal instrumentation, it is possible to measure ε_1 and ε_3 very accurately. However, most of them are manufactured for investigation in the small strain stiffness of soils or the prefailure behaviour. The resolution of the imaging technique is poor (Scholey et al. 1995), and even the air-volume change is difficult to measure (compressibility of air, environmental effects e.g. temperature) (Wulfsohn et al. 1998).

For future tests with axial strains up to 20% from specimen height, a double wall cell has been manufactured to measure the total volume change of the unsaturated specimen.

2.2 Triaxial cell

Over the past years several double-wall triaxial cells have been built and have been described (Anderson et al. 1997; Aversa and Nicotera 2002; Wheeler 1988). Main focus was on the measurement of the volume change of the cell fluid. The whole constructive design of this new cell has been tuned to this aspect. The goals are (1) to reduce axial deformations of the cell; (2) to avoid clearance volume e.g. threads, quick connects; (3) ensure optimal airing of the cell; (4) small volume of the inner cell; (5) undisturbed specimen preparation by air pluviation; and (6) design of an optimal measuring system for unsaturated sand specimen.

The cell has been designed for specimens with a diameter of 50 mm and 125 mm height (height to diameter ratio of 2.5). The radial deformation of the specimen is measured by monitoring the flow of the cell fluid into or out of the triaxial cell. Errors due to the deformation of the cell wall were avoided by applying the same pressure in the outer and inner cell. The cell consists of a head plate, a base plate, and a basis plate made of stainless steel. The head plate is manufactured in one piece without any splices. The loading ram is guided by a linear roll bearing which is installed in the upper part. It has the same diameter like the specimen. The sealing between loading ram and head plate can be solved by using a rod seal with a high sealing efficiency, low friction, and a minimal stick-slip effect. Therefore the loading ram must have a hard-chromium plated surface with special demands on the surface profile.

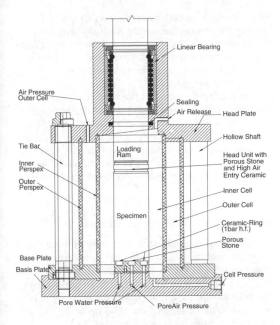

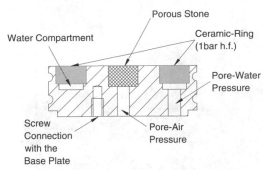

Figure 4. Base and head unit, cross section.

Figure 3. Double wall triaxial cell.

During cell filling, it is possible to bleed air with an lateral air release installed at the highest point of two inclined levels. For the cell walls, acrylic pipes are used. The inner and outer cells have a diameter of 9 cm and 11.5 cm respectively. Thus, the diameter of the inner cell enables radial deformation of the specimen or even the development of shear bands by compression tests with large axial deformation (20% from specimen height). The inner cell, with a relative small volume, is filled with silicon oil of a low viscosity.

The head plate, the perspex walls, and the base plate are held together by hollow shafts. The perspex walls are collected in fluts with o-ring seals which lie in a second flut. When the cell is put together the flut is completely filled with the o-ring sealing. There is no place left for the cell liquid. The cell is connected by five 16 mm diameter stainless steel tie bars with the basis plate. To prevent the flow of silicone oil between the two plates an o-ring sealing is installed. The diameter of the o-ring sealing is nearly the same as the diameter from the inner cell. The cell is positioned and centred by a recess in the basis plate.

There are five connections for the supply with the different media. The tubes are connected with the cell by quick connectors. When they are disconnected, there is a minimal spillage (0.3 cm³), and when they are coupled there is an air inclusion of 0.3 cm³. Like the basis plate, the quick connectors are made in stainless steel. A stainless steel ball valve without any clearance volume is installed in front of every connection. At three connections the pressure will be measured. Therefore, the pressure transducers are durably screwed in T-pieces. They point to the bottom to prevent the air from becoming trapped. There are two connections for the cell pressure and the pore-water pressure, and one for the pore-air pressure.

To separate pore-water from pore-air pressure a base pedestal is connected with the basis plate (Fig. 4). Leakage is prevented by an o-ring seal. Even the tubes for the pore-water pressure and the pore-air pressure are sealed with o-rings. Two tubes for pore-water pressure end in a water compartment under a high-air-entry ceramic-ring (1 bar high flow). It is possible to measure the pore-water pressure and flush the diffused air. The pore-air pressure is applied to the specimen through a porous stone. At the end of the loading ram (top of the specimen), the same unit is installed.

Changing the units to apply or to measure higher matric suction is simply possible. With two ceramic discs it is possible to apply the same or different matric suction(s) at the top and the base of the specimen. This ensures a shorter equalization stage (Romero et al. 1997), a nearly uniform distribution of matric suction about specimen height, and a defined matric suction gradient.

During pluviation the cell is run up and driven backwards on profil rail guides without any vibration. When the specimen is prepared the cell is driven to the starting position and dropped down for connection with the basis plate.

The absorption of water by the acrylic cell wall (Anderson et al. 1997; Wheeler 1988) and even the diffusion from water and air through the membrane (Leroueil et al. 1988) are well-known. Thus, a silicon oil is used as the cell medium. Another advantage is, that oil has a lower surface tension than water. The surfaces in the inner cell will be well wetted and the risk of any air bubbles in the tubes is much lower than using deaired water. Because of the greater compressibility of oil the volume change due to different cell pressure stages is considered by the bulk modulus.

343

Figure 5. Modular loading frame and triaxial testing equipment.

2.3 Loading frame and axial load

The cell is surrounded by a loading frame. The axial force is produced by an hydraulic cylinder or a servo motor with a gearbox. The rotation of the electric motor is changed to linear motion using of a recirculating roller screw. The fastest vertical deformation speed is at 25 mm/min and the slowest is less than at 0.0001 mm/min. At the lowest speed the motor rotates 1.2 times in every second. The vertical feed is still continuous.

2.4 Axial stress

The vertical tensile or compression force is measured out of the double-walled cell by a 10 kN load cell. The load cell is installed between the loading ram and the recirculating roller screw without any fetch. Lateral forces can not be excluded because of the length of the loading ram. However, preliminary tests with conventional load cells have shown that lateral forces have an enormous effect on the measurement. To solve this problem, a special load cell with an integrated lateral force compensator has been used. The load cell can easily be installed on the construction with flanged connections on both sides.

The relative deviation of the characteristic value of the load cell is ±0,2% from the ultimate value. Considering the maximal mistake due to a lateral force (10% from the nominal force), the maximal deviation for measuring the axial stress is smaller than 1.53 kPa.

The axial displacement of the specimen is measured externally to the cell with an opto-electronical displacement transducer.

2.5 Pressure control and measurement

The air pressure is adjusted manually with precision pressure control units. For the cell pressure, a pressure regulating valve with an adjustable range from 0.15 up to 10 bar and for the pore water-pressure and the pore air-pressure four pressure regulating valves with a adjustable range from 0.03 up to 2.1 bar are used. The sensitivity is less than 0.032 kPa.

The accuracy of matric suction measurement or control by the axis-translation technique depends on the accuracy and resolution of pore-air and pore-water pressure control and measurements. For tests with unsaturated sand, the precise control of the matric suction is very important. The essential effects happen at very low values of matric suction (e.g. air entry value: $u_s = 1$ to 2 kPa). So, it must be possible to apply very small matric suction increments to the unsaturated sand specimen.

Hence, the pore-water and the pore-air pressure are measured with pressure transducers with an effective range from 0 to 3.5 bar. The accuracy is less than 0.04% from the final value (0.14 kPa). The cell pressure can be measured with an accuracy less than 0.25% from the final value.

2.6 Volume change of the specimen and the porewater

The volumetric deformation of the unsaturated specimen is measured by monitoring the flow of the cell fluid into or out of the double-wall cell with double-walled burettes. Also, the flow of the pore-water into or out of the specimen can be observed with burettes.

The amount of the volume change depends on the tests which will be carried out. By the determination of the SWWC, small overall volume change and high pore-water volume change will be observed. During triaxial compression high radial deformation and poorer changes in pore-water volume can be monitored. During hydrostatic compression tests both the pore-water and the overall volume change are small.

Table 1. Accuracy and resolution from the measurement system.

Parameter	Accuracy	Resolution
Axial stress (σ_1)	$< \pm 1.528$ kPa	0.155 kPa
Cell pressure (σ_3)	$< \pm 2.5$ kPa	0.031 kPa
Pore-water pressure (u_w)	$< \pm 0.14$ Pa	0.011 kPa
Pore-air pressure(u_a)	$< \pm 0.14$ kPa	0.011 kPa
Axial strain (ε_1)	$< \pm 0.0008\%$	0.001 mm
Radial strain ($2*\varepsilon_3$)		
$d_1 = 6,7$ mm	$< \pm 0.025\%$	0.0002%
$d_2 = 16$ mm	$< \pm 0.145\%$	0.001%
$d_3 = 26$ mm	$< \pm 0.382\%$	0.0027%
Pore-water volume		
$d_4 = 12$ mm	$< \pm 0.285$ cm^3	0.0017 cm^3
$d_5 = 21$ mm	$< \pm 0.874$ cm^3	0.0053 cm^3
$d_6 = 26$ mm	$< \pm 1.339$ cm^3	0.0082 cm^3

Therefore, modular burettes-units consisting of three burettes have been built. The height of the oil respectively the water column is measured by differential pressure transducers.

The different burettes are designed to ensure a sufficient accuracy and resolution for each possible test.

2.7 Data logging

All transducers are connected to a pc with a 16 channel-measurement card. The developed software allows a continuous recording of the data. A high resolution of the measurement is ensured by using a 16-bit card. The output voltage for the system is in the range of ± 5 V.

3 OUTLOOK

The planned tests will be carried out in different stages. At first functional tests are intended for checking the apparatus and the instrumentation. This stage includes (1) tests (triaxial compression, hydrostatic compression) on saturated specimens and the comparison with other test results (e.g. literature, other triaxial testing apparatus); and (2) the comparison of test results (shear strength) acquired with a conventional and the new equipment. These tests will be carried out on unsaturated sand specimens by constant water content.

Thereafter a scientific testing program for the determination of the (1) relationship between saturation and shear strength; (2) soil-water-characteristic curve; and (3) the unsaturated conductivity will be started. The SWCC and the unsaturated conductivity will be examined for different isotropic stresses with recording the total volume change. The aim is to develop and to test models, which describe the SWCC and the unsaturated conductivity for deformable unsaturated soils.

Most of the well-known problems associated with the direct or indirect measurement of unsaturated hydraulic conductivity (Leong and Rahardjo 1997) should be avoided with the new apparatus. Shrinkage caused by matric suction is measured as total volume change.

ACKNOWLEDGEMENTS

The presented developments have been carried out during the work in the subproject 2 "Experimental and theoretical investigation of partially saturated granular material" of the DFG Research Group – Mechanics of Partially Saturated Soils.

REFERENCES

Anderson, W.F., A.K. Goodwin, I.C. Pyrah, and T.H. Salman (1997). Equipment for one-dimensional compression and triaxial testing of unsaturated granular soils at low stress levels. *Geotechnical Testing Journal* 20(1), 74–89.

Aversa, S. and M.V. Nicotera (2002). A triaxial and oedometer apparatus for testing unsaturated soils. *Geotechnical Testing Journal, GTJODJ* 25(1), 3–15.

Leong, E.C. and H. Rahardjo (1997). Permeability functions for unsaturated soils. *Journal of Geotechnical and Geoenvironmental Engineering* 123(12), 1118–1126.

Leroueil, S., F. Tavenas, P.L. Rochelle, and M. Tremblay (1988). Influence of filter paper and leakage on triaxial testing. In R.T. Donaghe, R.C. Chaney, and M.L. Silver (Eds.), *Advanced Triaxial Testing of Soil and Rock, ASTM STP 977*, Philadelphia, pp. 189–215. ASTM.

Romero, E., J. Facio, A. Lloret, A. Gens, and E.E. Alonso (1997). A new suction and temperature controlled triaxial apparatus. In *Proceedings of the Fourthteenth International Conference on Soil Mechanics and Foundation Engineering*, Hamburg, pp. 185–188.

Scholey, G.K., J.D. Frost, L.C. F. Presti, and M. Jamiolkowski (1995). A review of instrumentation for measuring small strains during triaxial testing of soil specimens. *Geotechnical Testing Journal, GTJODJ, 18*(2), 137–156.

Wheeler, S.J. (1988). The undrained shear strength of soils containing large gas bubbles. *Géotechnique* 38(3), 399–413.

Wulfsohn, D., B.A. Adams, and D. Fredlund (1998). Triaxial testing of unsaturated agricultural soils. *J. Agric. Engng Res.* 69, 317–330.

Geotechnical Measurements and Modelling, Natau, Fecker & Pimentel (eds)
© 2003 Swets & Zeitlinger, Lisse, ISBN 90 5809 603 3

Long-term monitoring of 4500 kN rock anchors in the Eder gravity dam using fibre-optic sensors

W.R. Habel
Federal Institute for Materials Research and Testing (BAM), Berlin, Germany

ABSTRACT: Steel anchors were installed in a gravity-dam, in order to increase the stability of the dam. Quasi-distributed fibre-optic strain sensors are centrally positioned in the complete length of these anchors to detect the long-term behaviour of the tendon bonded to the rock foundation. Because the sensor rod containing the measuring lengths is directly bonded to the grout in the fixed anchor length area, it is able to measure the longitudinal strain distribution in the anchor bond area. Such sensors help to evaluate the bonding of the fixed anchor bond length in a rock or in difficult ground areas. Design of the anchors, manufacturing and onsite installation are shortly described. The measurement principle, its characteristics and specific features are presented. Measurement results obtained over a number of years show the ability of sensor-equipped anchors to monitor the long-term tendon bond behaviour.

1 INTRODUCTION

The EDER-dam, with a storage capacity of $202.4 \times 10^6 \, m^3$, is one of the largest dams in Germany. This dam and the smaller DIEMEL-dam are owned and operated by the German government (Federal Republic of Germany), while more than 250 other dams in Germany belong to regional water and hydro power plant companies. Like many other dams, which were built at the turn of the century, the stability of the EDER-dam did not satisfy today's technical standards. Therefore, an improvement was necessary and several methods to improve the stability were discussed. The method finally chosen for enlargement of the vertical forces was pre-stressing the dam with 4500 kN-anchors against the bedrock. The anchor heads were located in a newly built gallery inside the dam crown. Every tenth of the 104 anchors was equipped with a special type of a segmented fibre-optic sensor for monitoring the pre-stressing process as well as the long-term behaviour of the pre-stressed anchors. Following, a short report is given on the construction itself and its stability problems, on design of the anchors and the fibre sensors, the tests carried out and the behaviour of the anchors up till now.

2 THE STRUCTURE

The EDER-dam, located in the northern part of the German Federal State, Hesse, about 35 km southwest of the town Kassel, was erected between 1908 and 1914. At that time, the main reason of such a construction was to supply the Mittellandkanal. Other reasons were to prevent floods and to improve the shipping conditions on the rivers Fulda and Weser. The dam is curved in the ground plan with a radius of 305 m and constructed with quarry stone, mainly greywacke in lime-trass-mortar. The bedrock consists of greywacke and claystone. The height of the dam from the foundation to the top is 47 m, the length being 400 m on the top and 270 m in the lowest foundation line. Before World War II no essential maintenance and repairs were necessary.

The building has a tragic story attached to it since the British Airforce successfully damaged it in the

Figure 1. The EDER-dam.

night from 16th to 17th of May 1943 by using a bomb especially developed for this purpose. A large part of the dam – 22 m depth by 60 m width – was damaged. $160 \times 10^6\,m^3$ of water was lost at a maximum rate of $8500\,m^3/s$. The hole was closed within approximately 4 months, the total time taken for repairs standing at 13 months. Grouting with cement injections was necessary and a new drainage system was installed. At the same time, a longitudinal drainage, service and inspection galleries were excavated. In 1961/62 a grout curtain was created near the waterside by cement injections because the seepage was still relatively high since World War II damage. At the same time, 42 pressure gauges, 15 temperature gauges and 4 inclinometers were installed, all using the principle of vibrating wire.

3 REHABILITATION IN 1992–1994

3.1 Evaluating the state of the dam and its monitoring system

In 1983, the Federal Waterways Engineering and Research Institute (BAW) Karlsruhe evaluated the stability of the dam, the state of the rock beneath the dam and the quality of the measuring systems. For this purpose, the existing measuring instrumentation has been completed and different measurements have been carried out. More details can be found in Feddersen (1986), Köhler & Feddersen (1991) and Habel et al. (2000).

It could be concluded from the investigations that the dam material is still in good condition. The existing seepage was judged as being less than tolerable. However, the uplift pressures in and beneath the dam, which had not been taken into account in the original static calculation, were of such a magnitude that its overall stability was insufficient. It should be mentioned here that the waterside of the wall was not designed for vertical tensile stresses. There were no reserves because the stress is about zero, not taking any uplift pressure into account. Therefore constructive remedial measures had to be taken in order to increase the stability.

3.2 Measures to enlarge the vertical dam forces

In order to improve the stability of the dam, six fundamentally different proposals were made. From an economic point of view, and following, the interests of fishery and tourism as well as considering aspects of the architectural monument protection, a vertical anchoring of the dam was favoured (see Figure 2).

With the evaluation of the wall, the maximum water discharge rate (HQ_{1000}) also had to be calculated. A design flood rate of $680\,m^3/s$ was taken into account up to 1985, this rate rose to $1100\,m^3/s$ after a new calculation. This fact required extensive structural alterations at the top, including the bridge over the dam.

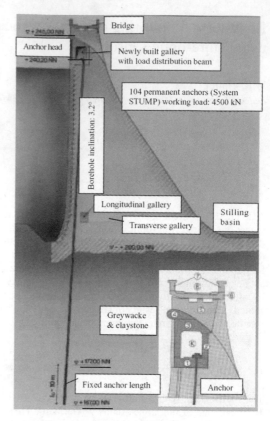

Figure 2. Cross-section of the dam with anchor position (Edertalsperre, 1994). The small picture shows a detail of the upper part of the dam with the anchor gallery. 1–4: newly built upper gallery, 5: pillar, 6–8: bridge; NN – above sea level.

4 DESIGN OF THE ANCHORS AND INSTALLATION OF THE SENSOR ROD

In order to compensate for the mass deficit of the dam, 104 anchors with an average length of 70 m, each with a load of 4500 kN, were installed in the dam (producer: SUSPA-Stump GmbH). Figure 3 shows the structure of an anchor. Each tendon consists of 34 steel strands. Because there is no general permission for the use of anchors with such high forces and because such anchors had not yet been built in Germany either for short-term or long-term use, the building owner decided to carry out pull-out tests in order to establish the necessary fixed anchor length. In these tests, it was not possible to make any of the anchors fail. They were stressed almost to the yielding point of steel approximately 7600 kN (see Section 8). The loading was carried out in accordance to the German Standard DIN 4125. Then, a special permission was given to use these anchors for this single case.

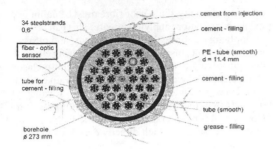

Figure 3. Cross-section of an anchor in the range of the free anchor length.

On the other hand, additional monitoring of the fixed anchor length was proposed. For this purpose, an aramid rod which contains fibre-optic sensors was placed in the centre of one test anchor and in ten of the 104 anchors (photo in Figure 4). The fibre-optic sensors were to measure the strain of the anchor along the free anchor length and especially the non-linear strain distribution in the fixed anchor length inside the rock. The second important aspect was the long-term monitoring of the bond in the fixed anchor length (in the grouted rock boring). The sensor-equipped aramid rod was pre-fabricated onsite on a specially developed production unit in the anchor producer's factory (SUSPA-Stump GmbH). After providing proof of the ability of the sensor fibre's function, it was attached to the aramid rod (diameter: 7.5 mm) at definite points. For reasons of redundancy, each sensor rod contains two separate sensor fibres over the complete anchor length. Each sensor fibre is connected to a separate returning fibre (sensor function is described later in this paper). The sensor rod was then completed by splicing the connectors to the upper fibre ends, and mounting the splicing box on the rock-oriented fibre ends, where the two sensor fibres are connected with their returning fibres. The completed sensor rods were protected for transfer to the anchor production places on-site. Care was taken that the aramid rods containing the fibre sensors did not snap off and that the bend radius of the sensor rod did not fall below the minimum value of 1500 mm. The fibres were tested before putting the sensor rod into the anchor and again before grouting the sensor-equipped fixed anchor length. The sensor rod runs now parallel to the steel strands over their complete anchor length. After hardening the grout inside the PE-tube in the fixed anchor length, the fibre sensor function was checked for one last time before installation.

Two different sections over the anchor length are distinguished: The fixed anchor length section at the bottom end to be bonded to the rock by grout. This length bonded to the rock is 10 m. In this anchor part the fibre sensor is divided into several measuring

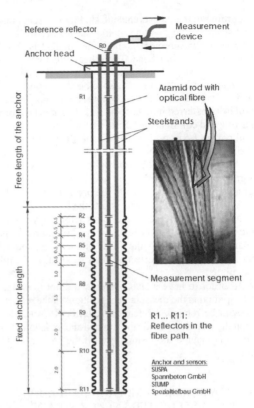

Figure 4. Simplified description of the sensor segments position inside the anchor (each anchor contains two identical sensors; all measures in m).

segments of different length (see Figure 4). The other part of the anchor is the free anchor length above the fixed area up to the anchor head in the gallery. Along the free anchor length (about 60 m) the sensor fibre is non-segmented. Only at the beginning and at the end measuring markers are set.

5 SENSOR STRUCTURE

Along an optical glass fibre with a diameter in the range of 125 μm to 140 μm, reflecting elements were inserted at regular intervals. Two neighbouring markers form a measuring section in the optical fibre. Its spacing can be determined by using an optical time domain sensing method (see next Section). In order to create a long-gauge-length sensor fibre for quasi-distributed (segmented) strain sensing, the optical fibre was physically segmented into discrete sections.

The reflectors consist of two glued fibre endfaces which partially reflect light transmitted through the fibre. These reflectors can be designed and positioned

349

according to the measurement task. Figure 4 schematically shows the position of the sensors R1 to R11. The first measurement point (reflector R1) is fixed beneath the anchor head. The other reflectors R2 to R11 are positioned with relatively small spaces in the fixed anchor length area. The spacing in the upper part of the fixed anchor length, where the anchor forces will be transferred into the rock, was chosen to be more narrow and equidistant (0.5 m). The spacing in its second part is 1.5 m or 2 m. The distance from R1 to R2 is 58 m and 63 m respectively.

A small metallic tube (the outer diameter is about 1 mm) protects the reflectors and is fixed in a small groove of the aramid rod. On the opposite side of the rod, the same sensor type is attached by reason of redundancy. When the aramid rod is deformed straight, the distance between the reflectors will vary. From this change, the strain in the fibre segment, related to the spacing of the reflectors, can be measured and calculated.

In order to have a stable reference point to which all variation in the measurement positions of the reflectors can be related, a reference reflector R0 was additionally installed outside the strained anchor length. R0 does not change its position even if the anchor strain varies.

6 MEASUREMENT METHOD TO READOUT THE DISTRIBUTED ANCHOR STRAIN

The measurement of the distance between two neighbouring reflectors is based on the measurement of the flight time of a monochromatic laser pulse takes to travel into the sensing fibre, then be transmitted to the markers (reflectors) positioned along the fibre, partially reflected there and, finally, received at the photodetector. This time taken to be travel to the markers is measured for relevant markers. A change in the pulse propagation time can be interpreted as a shift of the reflectors, i.e., an increase in the time difference of two neighbouring reflectors can be interpreted as an increase of the measuring segment due to mechanical loads, as well as vice versa in case of shortening. The length Δl of the measuring segment, that means, the spacing between two reflectors l_1 and l_2 is calculated from two separate measurements to determine t_1 and t_2:

$$\Delta l = \frac{c}{2 \cdot n} \, \Delta t \cdot \left(\frac{1}{1+a} \right), \tag{1}$$

where c is the speed of light in vacuum, n is the group refractive index of the fibre, a is the photo-elastic coefficient. This coefficient takes into consideration changes in the optically relevant glass parameters, when the fibre is strained. In order to evaluate the influence of photo-elasticity, the sensor fibre must be calibrated in laboratory before applying on-site. This can be done by applying a definite fibre elongation Δl and measuring the resultant pulse delay Δt. Thus, by using equation (1) and solving it with respect to a, the photo-elastic coefficient can be determined from the strain calibration curve. Zimmermann (1990) gave a value of $a = -0.1839$ for a polyimide-coated multimode fibre. However, it is more practical to use a K-factor as calibration coefficient, which directly describes the relation between Δl and Δt:

$$K = \frac{\Delta l}{\Delta t} = \frac{c}{2n} \cdot \left(\frac{1}{1+a} \right) \tag{2}$$

From equations (1) and (2) also follow that the refractive index must be known and should be constant over the complete fibre length. This cannot necessarily be assumed in such cases, where different fibre sections (may be from different cable drums) are connected or when significantly different temperatures occur along the fibre sections. A variation of the temperature changes the refractive index of the glass fibre, and thus the pulse propagation time. Assuming a temperature difference ΔT in a sensor segment that has the length l and is not experiencing any stress-induced elongation, the pulse propagation time difference is given by

$$\Delta t = 2 \cdot l \cdot \Delta T \, \frac{\alpha n + \beta}{c}. \tag{3}$$

α is the coefficient of thermal expansion and β is the fibre's thermo-optic coefficient which describes the thermo-induced change in the refractive index. Assuming the values $n = 1.485$, $\alpha = 5.5 \times 10^{-7} \mathrm{K}^{-1}$ and $\beta = -6.3 \times 10^{-6} \mathrm{K}^{-1}$ for an all-silica fibre (Zimmermann (1990) used a quite large value of $\beta = -20 \times 10^{-6} \mathrm{K}^{-1}$), it follows for a temperature change of $\Delta T = 15$ K a change in the pulse propagation time of $\Delta t = -0.55$ ps in a sensor segment of the length $l = 1$ m (it follows $\Delta t = -1.92$ ps according to Zimmermann's assumption). Taking into consideration our calculated temperature-induced time delay of -0.55 ps for equation (2), it follows for an assumed K-factor of 100 mm/ns a temperature-induced length change of approximately $\delta(\Delta l) \cong -0.055$ mm. By using an OTDR-device with ps-resolution and with a minimum sampling interval of 2.5 picoseconds, a flight time resolution of around 2.5 picoseconds should theoretically be possible. According to equation (1), a displacement resolution of 0.25 mm is expected. System drift, triggering uncertainties and other factors limit the actual resolution of the system. For a strain resolution of 0.25 mm, a temperature-induced strain of -0.055 mm is not significant. It is, indeed, necessary to choose optical fibres with a small thermo-optic

Figure 5. ps-OTDR device connected to the sensor rod at the anchor head.

coefficient. In the case of Zimmermann's value, the temperature-induced strain would be approximately -0.2 mm. Thus, a temperature change in the fibre increases the uncertainty of measurement considerably.

In the same way, a constant photo-elastic coefficient cannot be assumed, when in the sensor system fibres with different coatings were used. In such cases, the uncertainty of measurement increases and only relative changes from one measurement to the next can be determined. An absolute value of the segment length cannot be given exactly due to the influences described.

In order to get the pulse propagation time down to and back from the reflectors, commercial devices (OTDR[1]-devices) with a resolution in the picosecond range are available. Such a system consists of a high-speed, low-jitter, pulsed semiconductor laser, a highly accurate pulse delay generator and a super-fast rise-time photodetector. In addition, a signal average processor is used to reduce the scatter of the measuring result. The data acquisition is carried out automatically by a GPIB interface. The measuring equipment is shown in Figure 5. The measured data and other information such as temperature in the gallery, anchor force from load cells and level of back water are recorded and saved automatically on a PC.

7 INSTALLATION OF THE ANCHORS

7.1 Installation in the boreholes

The first anchor was installed in the dam in December 1992, the last one in November 1993. The nearly 70 m long anchors – ten of them containing the described sensors – were put into the boreholes using a special device with the support of an autocrane.

The minimum spacing between the anchors was 2.25 m. Special considerations were necessary concerning the installation of these long anchors, e.g., the

[1] OTDR – Optical time domain reflectometry.

small spacing and the accuracy of the position of the anchor boreholes (1% deviation from the required inclination of 3.2°). Because of the magnitude of the anchors (diameter and length), problems resulted which required special consideration, e.g., the strain in the fibres during storing, transportation and installation must not damage the reflectors. Fibre sensors were indeed subjected to particularly rough conditions.

7.2 Fixing the anchors by grouting

After the anchor was fully positioned in the borehole, it was filled with cement grout. Whereas the aramid rod with the segmented fibre sensor in the fixed anchor length area is directly embedded in the grout, the rod part above this fixed anchor length is only protected by a PE tube. In this way, the cement filling between the steel strands does not reach this part of the sensor rod. Therefore, the sensor fibre measures the integral deformation between the fixed anchor length area and the anchor head in the gallery.

In order to reach the sensor rod, the steelstrands must be separated from it. The upper end of the aramid rod was stuck inside a stainless steel tube that has a thread on its upper end.

7.3 Tensioning the anchors and the sensors

The installation of the tensioning jack could be detrimental to the fibre sensor rods. Therefore, care was necessary to avoid any damage. No measurement was carried out during the process of tensioning. Measurements were taken only before tensioning and after reaching the current tensioning steps. By these measurements, only the elongation in the fixed anchor area could be found out. The sensor rod in the free anchor length area did not yet extend. After the working load was reached, the tensioning jack was removed and the pulled out steel strands were cut (see Figure 6). For this reason, the measuring device was removed, and a steel tube protected the sensor rod. The sensor fibre (aramid rod) in the free anchor length area was then tensioned separately and a further measurement was taken.

After finishing the tensioning procedure, all parts were filled with grout and the upper part with Vaseline to avoid corrosion. Finally, the anchor heads were closed by a hat (see Figure 7). From those anchors that contain fibre sensors, a stainless steel tube with the fibre ends protrudes. Figure 7 shows the connectors in the sensor cable terminal for linking the measurement device to the sensor fibers. The sensor cable terminal is closed by a separate cylindrical hat (see Figure 5, right).

More details concerning the anchoring with such 4500 kN-anchors are described by Feddersen et al. (1992).

351

Figure 6.   Anchor head with protruding sensor rod.

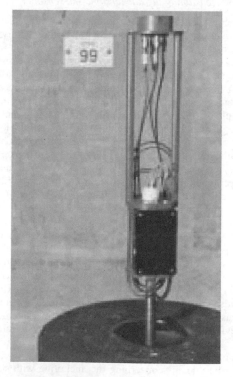

Figure 7.   Fibre sensor cable terminal above the anchor hat (the black box houses the reference reflector R0).

8   MEASUREMENTS, RESULTS AND EXPERIENCE

Installation and tensioning of sensor-equipped anchors is, in principle, a very difficult task. Despite of very careful handling of the sensor-equipped anchors, two of the ten sensor-equipped anchors were damaged. In one case, the damaged leading fibre for connecting the measurement device could be repaired. In the other case, the sensor rod could not be saved. From today's point of view, a critical part of the sensor system is the egress of the optical fibres from the stainless steel tube at the anchor head as well as the return area in the splicing box at the bottom end of the anchor (inside the rock).

Another experience with consequence to the measurement uncertainty concerns the problems of ghost signals, occurring by destructive interference of signals from several equidistant neighbouring reflectors. By superposition of the measuring reflection and the ghost reflection(s) coming from the next reflector(s), the position of the measured reflector cannot exactly be calculated by the measurement device.

The first real measurement of strain variations in the anchors was carried out after tensioning. It could be confirmed that the higher the induced load is the longer will the length of the fixed anchor length area be. This behaviour confirmed the anchor bond in the rock. Figure 8 shows the measured anchor deformation during a suitability test of an additionally

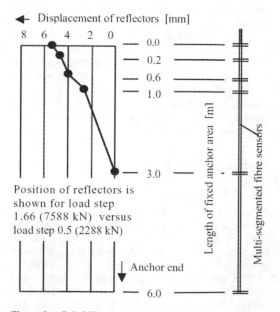

Figure 8.   Suitability test of a 7400 kN anchor in the test area. The displacement of 0 mm corresponds to a load of 2288 kN (Edertalsperre, 1994).

352

produced 7400 kN test anchor installed in the spilling basin before the installation in the wall. The elongation in the fixed anchor length area is shown versus the load steps induced into the anchor. It had been evident that for this load induced, only the upper 2.5 m to 3.0 m of the complete tendon bond length of 10 m makes a contribution to the bond. Thus, the anchor design provides a large reserve of load capacity.

No regular measurements were carried out between 1994 and 1998. In 1998, the actual state of the sensors was evaluated. It could be found out that the sensor technique is able to deliver values for a long-term monitoring of the anchors. In order to achieve a measurement uncertainty for strain measurements in the range of 0.3 mm until 0.4 mm, an additional reference reflector had to be installed. On the basis of this stable reference reflector, Figure 9 shows the strain development between reflectors R2 and R6 measured in anchor no. 11 over a period of 3 years from spring 2000 (S-00) till autumn 2002 (A-02). No significant changes in the bond of the anchor to the rock could be observed apart from the fact that the anchor strain in the bonding area seems to be increased very slightly. In this case, the strain development should be observed over the next years.

The OTDR measurement method revealed two convincing advantages:

- If the parameters n, α, β, a (see equations 1 to 3) of the optical fibre are known, each measurement delivers an absolute value of the actual length of the measurement segment. Any variations in attenuation or bending of the leading fibre as well as an exchange of optical components do not directly influence the measurement signal because the measurement segment is positioned out of the area of possible influences.
- The determination of the position of all reflectors is referred to the (stable) position of the reference reflector. Therefore, there is no propagation of

error during the measurements propagating from one reflector to the next one.

On the other hand, there are a few specific features that have also to be considered:

- In order to determine the position of a reflector, the device sends out a number of very small laser pulses and, following, scans the back-reflections. Depending on the number of pulses for each measurement (mostly 512 or 1024), there is time needed to integrate all back-reflected signals. One scanning run takes between a few and some ten seconds depending on the required precision. For this reason, only static measurements are possible.
- Because OTDR-devices with transit-time resolution in the ps-range are very expensive and provide only one channel for the connection of a sensor line, off-line measurement will be preferred. This off-line measurement method might cause changes in the launching conditions whenever a measurement is taken. Different launching conditions results in a different mode distribution and following in an enlargement of the uncertainty of measurement. Those condition changes can become relevant if the length of the leading cable between device and anchor is too short.
- The index of refraction n is influenced by temperature. Inside the anchor, the temperature cannot exactly be measured, however, a temperature difference of up to 15 K could occur over the whole sensor fibre length. Such a non-linear temperature distribution might slightly enhance the measurement uncertainty.
- The quality of back-reflected pulses depends on the quality of the reflectors itselves as well as on their long-term stability. In a few cases, the precision of the measured reflector position changed due to ageing effects.
- Analysis and interpretation of the recorded data are very time-consuming. For this reason, software-supported data evaluation have to be applied.

9 CONCLUSIONS

The ability to monitor rock anchors by means of structure-integrated fibre-optic sensors is an important contribution to improve the reliability and safety of highly stressed anchor systems with anchor forces of >1000 kN. This method has been used for several 4500 kN rock anchors installed in an large gravity dam during rehabilitation to enlarge the vertical dam forces. Ten of 104 anchors of a length of around 71 m have been equipped with fibre-optic quasi-distributed strain sensors. Such strain sensors are able to measure the strain distribution along the fixed anchor length during the tensioning procedure as well as later under

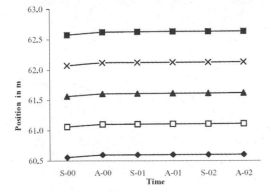

Figure 9. Strain development in the fixed bonding area of anchor no. 11 over a period of 3 years.

service conditions. From regular measurements during the service life, the long-term loading capacity can be evaluated and beginning failure could be detected.

Design, manufacture and installation of the anchors have been described in this paper. A few specific conditions, which are necessary to obtain reliable measurement results, were discussed. Experience from measurement during the past 3 years were given. It can be stated that the measuring system is able to deliver results in such an harsh environment with an measurement uncertainty of about 0.35 mm, if the optical components are well-matched and a stable reference position is available. The measurement uncertainty depends strongly on the quality of the reflectors positioned in the fibre path. Temperature-induced influence and long-term changes of the reflectors do not significantly influence the strain results. Because of its ability to deliver absolute and cable-independent, that means, line-neutral measurement signals, this method seems to be made for long-term monitoring of large structures, especially then, if off-line measurements are to be carried over a number of years.

REFERENCES

Habel, W.R. et al. 2000. Embedded Quasi-distributed Fiber-Optic Sensors for the Long-term Monitoring of the Grouting Area of Rock Anchors in a Large Gravity Dam. *J. of Intelligent Material Systems and Structures* 10(4): 330–339.

Edertalsperre 1994. *Wasser- und Schiffahrtsdirektion Mitte.* Hannover.

Feddersen, I. 1986. Untersuchungen bei der Standsicherheitsüberprüfung der Eder- und Diemeltalsperre. *Geotechnik, Sonderheft 1987 zum 7. Nationalen Felsmechanik-Symposium*, Aachen, p. 221–228, ISSN 0172-6145.

Feddersen, I. et al. 1992. Instandsetzung der Staumauer der Edertalsperre (Rehabilitation of the Eder-dam). *Wasserwirtschaft* 82(2): 62–72, ISSN 0043-0978.

Feddersen, I. 1997. Improvement of the Overall Stability of a Gravity-dam with 4500 kN-anchors. *Proceedings of the International conference on Ground Anchorages and Anchored Structures*. London, UK, Edited by G.S. Littlejohn. Thomas Telford, ISBN: 0 7277 2607 2.

Köhler, H.-J. & Feddersen, I. 1991. *Proceedings of the 3rd International Symposium on Field Measurements in Geo-mechanics*. Oslo, Norway. Edited by Geraldine Sorum, NGI. Balkema, Rotterdam, p. 107–116, ISBN 90 5410 0265.

Messung der LWL-Sensoren an den Ankern in der Eder-staumauer … (Measurement of Fibre Sensor's Strain Stage in the anchors of the Eder dam …). Technical Report BAM-S.12. 1998 (unpublished).

Wolff, R. & Mießeler, H.-J. 1992. Novel optical fibre Sensors for Monitoring Structures, Rock and Earth Bolts and measured Sections in Rock. *Felsbau* 10(5): 223–227.

Zimmermann, B.D. et al. 1990. Fibre-Optic Sensors Using High-Resolution Optical Time Domain Instrumentation Systems. *J. of Lightwave Technology* 8(9): 1273–1277.

Geotechnical Measurements and Modelling, Natau, Fecker & Pimentel (eds)
© 2003 Swets & Zeitlinger, Lisse, ISBN 90 5809 603 3

Stress measurements in salt mines using a special hydraulic fracturing borehole tool

G. Manthei & J. Eisenblätter
Gesellschaft für Materialprüfung und Geophysik mbH, Ober-Mörlen, Germany

P. Kamlot
Institute for Geomechanics, Leipzig, Germany

ABSTRACT: Hydraulic fracturing tests are often utilized in order to measure stresses in rock. In most cases, the fracture forms perpendicular to the orientation of the minimum principal stress whose magnitude can be determined by the so-called shut-in pressure. In order to find the orientation of the minimum principal stress the fracture orientation has to be measured. This can be done by Acoustic Emission (AE) measurements. Usually, for this purpose AE transducers are positioned in separate boreholes around the central borehole where the hydraulic fracturing takes place. Recently, we have developed a new borehole tool which is able to do the same job utilizing only one borehole. The said tool includes the hydraulic pressurization unit with the AE sensors. Due to the small distance between the injection interval and the AE sensors, this tool is much more sensitive than the conventional AE method. In addition, the sensitivity is the same for all fractures independent of borehole depth. During the last six years, we have performed about 100 hydraulic fracturing tests on rock salt and anhydrite in different salt mines. The paper presents some results of these tests which demonstrate the capabilities of the new borehole tool.

1 INTRODUCTION

The hydraulic fracturing method is able to measure the stress in rock directly. During a hydraulic fracturing experiment the hydraulic pressure in a sealed volume of a borehole is increased up to fracture initiation in the rock. At the end of the forties of the last century, the hydraulic fracturing technique was applied first to raise the oil or gas output by increasing the permeability of rock. Later on, in the fifties, a lot of attempts were made to find out the correlation of in-situ stress state and results of hydraulic fracturing tests. Significant in this field was the work of Hubbert & Willis (1957). They were the first to demonstrate the influence of local tectonic stresses on the orientation of hydraulically induced fractures. They assumed that the fracture propagates in the direction of least resistance and that the pumping pressure to merely keep an induced fracture open is equal to the minimum principal stress. Kehle's (1964) determinations of tectonic stress through analysis of hydraulic well fracturing agree closely with those calculated using Hubbert's and Willis' model. Kehle concluded that the minimum principal stress corresponds to the so-called instantaneous shut-in pressure, i.e. the pressure reached some

time after stopping pumping at the end of the test. Fairhurst (1964) was the first who suggested to use hydraulic fracturing for stress measurements in rock.

In all applications dimension, shape and orientation of the fractures are of upmost importance for determination of the in-situ stress state. The usually applied methods to determine the orientation of the fracture plane like overcoring, fissure forming with packer or visual inspection using a borehole camera are limited in borehole diameter and very expensive, in particular for hydraulic fracturing tests in greater borehole depths. All these methods need clearly discernible fissures at the borehole wall and assume that the direction of fracture propagation is not influenced by changes of the stress state close to the borehole. With other words, a change of the fracture direction due to stress redistribution in further distance from the borehole can not be observed.

In order to determine the orientation of the principal stresses, the fracture orientation must be known. A lot of papers show that acoustic emission (AE) is suitable to measure crack orientation and extension utilizing three-dimensional source location. Lockner and Byerlee (1977) located AE events during hydraulic fracturing experiments in small Weber sandstone specimens. Eisenblätter (1988) located microcracks during hydraulic fracturing tests in a salt mine. Manthei et al. (1989) applied AE measurements during a hydraulic fracturing test on a large specimen of rock salt. Ohtsu 1991 used the moment tensor analysis to study crack types and orientations of AE sources detected during in-situ hydraulic fracturing tests in siliceous sandstone. Manthei et al. (1998a) demonstrated that macroscopic crack orientations depend on the tectonic stress field in the rock. They showed that the orientation of the fracture planes as measured by acoustic emission agrees remarkably well with the orientation of the calculated principal stresses.

In order to measure the crack orientation and extension, AE transducers are usually positioned in separate boreholes around a central borehole (Manthei et al. 1998b) where the fracturing tests take place. Recently, we have developed a new borehole tool which is able to do the same job utilizing only one borehole. This borehole tool includes the hydraulic pressurization unit with the AE sensors. Due to the same distance between injection interval and sensor arrays the sensitivity of AE registration is always the same independent of the borehole depth. It is possible to trace back the realistic fracture propagation in distances up to 15 to 20 times of the borehole diameter. Other expensive inspection methods are not needed.

During the last six years, we have performed about 100 fracturing tests on rock salt and anhydrite in horizontal as well as in vertical boreholes in different salt mines. In the following, after a short description of the experimental set-up some examples will illustrate the capabilities of the new borehole tool.

2 EXPERIMENTAL SET-UP

The new borehole tool which was developed by IfG Leipzig and GMuG Ober-Mörlen (Manthei et al. 1989)

is shown schematically in Figure 1. It consists of two parts – the hydraulic pressurization unit in the middle and two AE sensor arrays at both ends. It is applicable to borehole diameters between 98 mm and 104 mm. The overall length of the borehole tool is about 2 m. Each sensor array includes four AE transducers in a cross section perpendicular to the borehole axis. The distance between the AE arrays is approximately 1.5 m. The transducers with integrated preamplifiers are placed in a common housing which is screwed to the pressurization unit. The transducers are pressed pneumatically against the borehole wall. The preamplified signals are supplied to an 8-channel transient recorder card which is controlled by a portable personal computer. The transient recorder card (sampling rate 1.25 MHz, resolution 12 bit) is read each time a signal passes the trigger threshold. The borehole pressure and the pressure which is applied to the packer are measured using pressure cells. The signals of the pressure cells are digitized usually each second and stored on the hard disk of a notebook.

The evaluation of signals stored in digital form was made first in the laboratory. At present, a quasi online location is possible onsite. For locating the events the onsets of the compressional and/or shear waves are picked automatically. A special localization algorithm has been developed in order to estimate the distance, inclination (to the borehole axis) and the azimuth of the AE source. The test measurements showed that the accuracy of distance and azimuth estimation was about 0.15 m and ±10°, respectively.

After a tightness test a high pressurization rate is applied for fracturing the rock. In the shut-in phase the pressure is observed up to several hours. After each hydraulic fracturing test a refracturing test is normally performed in order to confirm the observed shut-in pressure. During a fracturing test several hundreds of milliliter oil are injected.

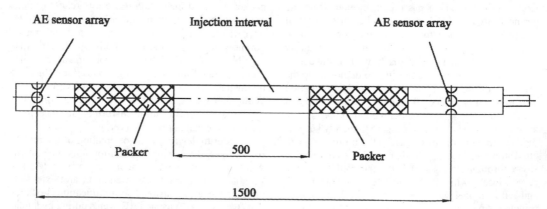

Figure 1.   Hydraulic fracturing borehole tool (schematically).

3 RESULTS

3.1 *Hydraulic fracturing test series at the 500-m level in the salt mine Bernburg*

A hydraulic fracturing test series at the 500-m level in the salt mine Bernburg was performed in an area where the rock stresses are influenced by the propagation of an excavation front in direction to the test site. Figure 2 shows the location of the AE events which were detected during a hydraulic fracturing test (including the fracturing and the refracturing test) in 4 m depth in a horizontal borehole in projection to three coordinate planes (x–y-plane: top view; x–z-plane and y–z-plane: two lateral views). The location of the transducer arrays and the injection interval is indicated by means of circles and rectangles, respectively. The y-axis is parallel to the injection well.

All together, more than 5,100 AE events could be located during the test. The located events mark a clearly discernible fracture plane. The fracture initiated in the middle of the injection interval and propagated in radial direction transverse to the injection well. In order to get the orientation of the fracture plane the volume of a parallelepiped was minimized through rotation in perpendicular axes. The edge lengths of the parallelepiped are the mean deviations of the events from the centre of gravity in each coordinate axis. Two rotations (first with azimuth angle around the z-axis and second with inclination angle around the x′-axis) are necessary to minimize the volume. After rotation of the coordinate system a nearly perfect plain fracture appears in the x′–y′-plane in Figure 3. Most events are located at the crack tip (x′–z′-plane). This is due to the fact that during fast crack propagation at the beginning of each fracture phase AE events are emitted so frequently that they overlap each other and, therefore, cannot be located. Figure 4 shows separately the events of the fracturing (left-hand side) and refracturing tests (right-hand side) in 4 m borehole depth in a rotated coordinate system where the fracture plane lies within the x′–z′-coordinate plane. During the refracturing test the crack slightly enlarged from 2.1 m to 2.5 m (measured in the largest diameter). The AE activity of the refracturing test starts after reaching the crack extension of the previous performed fracturing test.

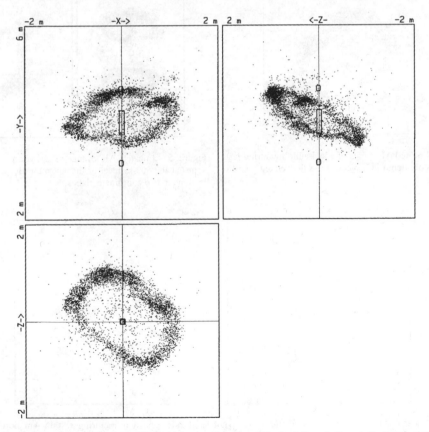

Figure 2. Located AE events of a the hydraulic fracturing test in 4 m borehole depth in projections to the three coordinate planes.

Figure 5 gives an overview on all hydraulic fracturing tests performed in this borehole at depths of 2 m, 4 m, 7 m, and 10.4 m. Approximately 15,000 located events are shown in a top view (at the left-hand side) and in a lateral view (at the right-hand side). Most of the events (11,216) could be localized during the fracturing test in 2 m and 4 m borehole depths. On the contrary, in larger borehole depths much less events (3696) were located in spite of the fact that the same oil volume was injected. This observation may be explained by larger deviatoric stresses close to the contour of a gallery. The extension of the fractures is nearly independent of borehole depth.

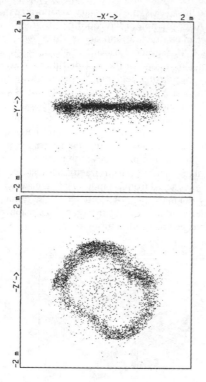

Figure 3. Located AE events of a hydraulic fracturing test in 4 m borehole depth in projections to the rotated coordinate planes.

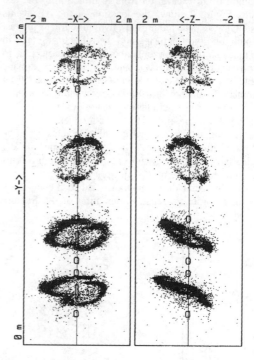

Figure 5. Located AE events of a hydraulic fracturing series at the 500-m level of the salt mine Bernburg in projections to two coordinate planes.

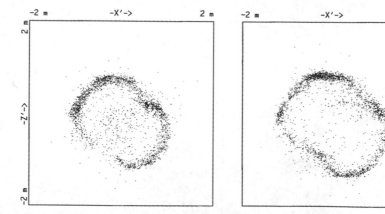

Figure 4. Located AE events during a fracturing test (left-hand side) and a refracturing test in 4 m borehole depth (right-hand side) in projection to the rotated coordinate system.

Figure 6 displays the orientation of the normal vector of the fracture planes (filled dots) in a so-called Schmidt-net diagramme. In this diagramme the cutting points of the normal vectors which correspond to the direction of the minimum principal stress are plotted in an equal-area projection to the lower hemisphere for each fracturing test. The injection well runs

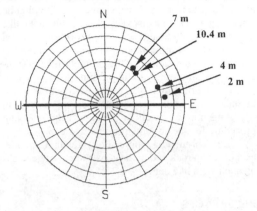

Figure 6. Orientation of the normal vector of fracture planes as seen in Figure 5 in the Schmidt-net diagramme.

west-east (marked by a thick line). The orientations of the fracture planes change with borehole depth. In distances up to 4 m from the contour of the gallery, the fracture planes are inclined nearly vertical and orientated in west-east direction. The azimuth angle (measured clockwise from north) ranges from 83° (at 2 m) to 72° (at 4 m). In greater borehole depths the inclination of the fracture planes decreases up to a value of approximately 45° and the fracture planes turn north.

3.2 Hydraulic fracturing test series at the 800-m level in the salt mine Asse

The fracturing series in the salt mine Asse was carried out in a pillar of 10 m width between two drifts. The horizontal injection well of 35 m length was located in the centre of the pillar and was drilled parallel to the drifts. In this injection well four hydraulic fracturing tests took place in 3.5 m, 7 m, 10 m, and 13 m borehole depths. Figure 7 shows at the left-hand side the located AE events (in total 6125 events) in projection to two coordinate planes. In the top view (x–y-plane) it can be seen that all fracture planes have nearly the same orientation independent of borehole depth. The Schmidt-net diagramme at the right-hand side indicates the same.

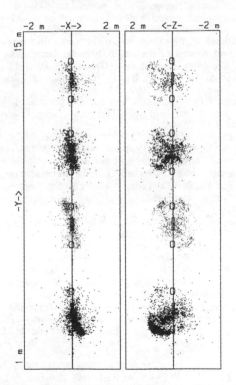

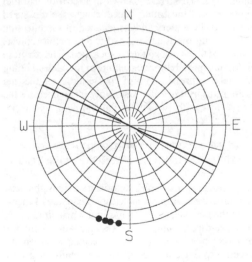

Figure 7. Left-hand side: Located AE events of a hydraulic fracturing series at the 800-n level of the salt mine Asse in projections to two coordinate planes. Right-hand side: Orientation of the normal vector of the fracture planes in the Schmidt-net diagramme.

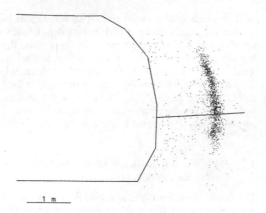

1 m

Figure 8.   Location of AE events in projection parallel to the gallery axis together with the profile of the gallery and the location of the injection well.

All fracture planes are vertical and extend between the transducer arrays parallel to the borehole axis.

Another example of a hydraulic fracturing test with remarkable results is given in Figure 8. The test was carried out in approximately 2 m borehole depth in a horizontal injection well at the same test site in the salt mine Asse. This figure shows a projection parallel to the wall of the gallery together with the measured profile of the gallery and the location of the injection well. During the fracturing and two refracturing tests about 1560 events could be localized. The fracture plane runs parallel to the contour of the wall imaging perfectly its profile. This of course indicates that the stress state in this zone is strongly influenced by the free surface at the wall of the gallery.

4   CONCLUSIONS AND FURTHER APPLICATIONS OF THE BOREHOLE TOOL

The presented examples demonstrate that the new hydraulic fracturing borehole tool is capable of measuring the orientation of the minimum principal stress by three-dimensional AE source location utilizing only one borehole. The short distance between the injection point and the AE sensors leads to a high sensitivity of AE event registration. Therefore, the macroscopic fracture plane will be imaged by a great number of AE locations. Due to the fast progress of registration hardware and computer technique the number of located events could be raised in the last years by orders of magnitude. During the first fracturing tests with the said borehole tool, only a view hundred events could be localized. Now it is possible to locate up to ten thousand events during a single test.

Generally, the borehole tool is applicable in all rocks which show spontaneous and fast crack formation. Difficulties will occur in layered or multiply jointed rocks. Such rock types have a high attenuation and strong absorption of the elastic waves.

The borehole tool was developed and applied in underground waste disposal research projects where knowledge about the stress state in rocks was needed to characterize strength, tightness, and deformation behaviour of the host rock which has to isolate hazardous radioactive or chemical wastes from the biosphere for a long time. On the other hand, hydraulic fracturing measurements deliver valuable data like absolute magnitude and orientation of the minimum principal stress for the validation of structural models which are used to calculate the geomechanical evolution of the long-term stability of mines.

The borehole tool is also useful where stress measurements are needed to evaluate safety aspects in operating mines as well as to support the licensing procedure for sites of underground waste disposals.

Further promising applications relate to tunnel excavation in rock and to the construction of geotechnical barriers like dams.

REFERENCES

Eisenblätter, J. 1988. Localisation of Fracture Planes During Hydraulic Fracturing Experiments in a Salt Mine. In *Acoustic Emission*, Deutsche Gesellschaft für Metallkunde e.V., *Oberursel*: 291–303.
Fairhurst, C. 1964. Measurement of In-situ Rock Stresses with Particular Reference to Hydraulic Fracturing. *Rock Mechanics and Engineering Geology* 2(3–4): 129.
Hubbert, M.K. & Willis, D.G. 1957. Mechanics of Hydraulic Fracturing. *Trans. AIME* 210: 153–168.
Kehle, R.O. 1964. The Determination of Tectonic Stresses through Analysis of Hydraulic Well Fracturing. *J. Geophy. Res.* 69: 259–273.
Lockner, D. & Byerlee, J.D. 1977. Hydrofracture in Weber Sandstone at High Confining Pressure and Differential Stress. *J. Geophy. Res.* 82: 2018–2026.
Manthei, G. & Eisenblätter, J. 1989. Untersuchung der Rißbildung in Salzgestein mit Hilfe der Schallemissionsanalyse. *Materialwissenschaft und Werkstofftechnik* 20: 240–249.
Manthei, G., Eisenblätter, J. & Salzer, K. 1998a. Acoustic Emission Studies on Thermally and Mechanically Induced Cracking in Salt Rock. In H.R. Hardy Jr. (ed.), *Acoustic Emission/Microseismic Activity in Geologic Structures and Materials; Proc. 6th Conference, Pennsylvania, State University, 11–13 June 1996*: 245–265. Clausthal: Trans Tech Publications.
Manthei, G., Eisenblätter, J., Kamlot, P. & Heusermann, S. 1998b. AE Measurements During Hydraulic Fracturing Tests in a Salt Mine Using a Special Borehole Probe. In M.A. Hamstad, T. Kishi & K. Ono (eds.), *Progress in Acoustic Emission IX, Proc. 14th International Acoustic Emission Conference & 5th Acoustic Emission World Meeting, Big Island, Hawaii, USA, 9–14 August 1998*: II60–II69.
Ohtsu, M. 1991. Simplified Moment Tensor Analysis and Unified Decomposition of Acoustic Emission Source: Application to in situ Hydrofracturing Test. *J. Geophy. Res.* 96: 6211–6221.

Experimental and model studies on the Modified Tension Test (MTT) – a new and simple testing method for direct tension tests

R.J. Plinninger
Dipl.-Ing. Bernd Gebauer Ingenieur GmbH, Munich, Germany

K. Wolski & G. Spaun
TU München, Lehrstuhl für Allgemeine, Angewandte und Ingenieur-Geologie, Munich, Germany

B. Thomée & K. Schikora
TU München, Fachgebiet Baustatik, Munich, Germany

ABSTRACT: The "Modified Tension Test" represents a new and innovative approach to the laboratory research of the uniaxial tensile strength. The test features a cylindrical specimen of special geometry, so a unidirectional, direct tensile stress field is created in the sample. The test may easily be carried out in any standard testing machine to test the Unconfined Compressive Strength (UCS). The presented results evaluate the MTT as an easy-to-carry-out laboratory testing method, which on the one hand shows a good ratio of the required testing equipment and demands for the testing material. On the other hand, it provides a realistic value for the direct tensile strength of a rock or concrete sample. From the experience of this program, some practical suggestions are also made on testing circumstances such as sample geometry, sample preparation and documentation of the test result.

1 PROCEDURES FOR TESTING THE TENSILE STRENGTH OF ROCK AND BUILDING MATERIAL

In addition to the unconfined or triaxial compressive strength and deformability, the tensile strength is one of the most important parameters for the mechanical description of a rock or building material.

Unfortunately testing of direct tensile strength is a rather difficult task with a lot of technical problems: if mechanical clamps are used to fix the sample, problems of point loads and uneven stress distribution in the sample may arise. Especially in hardrock testing, the use of adhesives is a problem. And even when these problems are solved, complex bending tensile stresses can occur during failure, when an initial crack on the one side of the sample is propagated to the other side. Additionally, testing systems that can be used for direct tensile tests are not as widely available as standard systems for testing compressive strength. Therefore direct tensile tests are used rather infrequently in the field of rock mechanics and geotechnical engineering (Fecker & Reik, 1996, p. 269f; Prinz, 1997, p. 49).

In contrast to this, indirect testing procedures, such as the Brazilian, point load or bending tests are widespread throughout the world. A number of standards such as the DIN 1048 German standard and testing recommendations such as DGEG (1982, 1985) and ISRM (1978, 1985) deal with these tests and provide a good background for comparable test results. Nevertheless, comparisons between direct and indirect tension tests are difficult and empirical equations have to be used for such purposes.

The presented paper summarizes results of a dissertation at TU München, Lehrstuhl für Allgemeine, Angewandte und Ingenieur-Geologie (chair for general, applied and engineering geology; Wolski, 2002) and finite element studies that were carried out at the TU München, Fachgebiet für Baustatik (Professorship for the Analysis of Civil Engineering Structures, Faculty for Civil Engineering and Surveying) in the course of a PhD thesis on the modelling of steel fiber reinforced concrete in structural mechanics (Thomée, in preparation).

2 INTRODUCING THE MODIFIED TENSION TEST (MTT)

The "Modified Tension Test" (MTT) dealt with in this paper was developed at the institute for rock mechanics and tunneling at the TU Graz, Austria. Basics of the

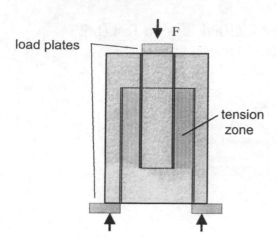

load plates

F

tension zone

Figure 1. General testing layout and sample geometry for the MTT.

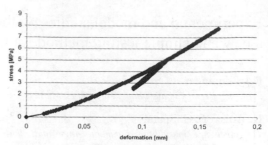

Figure 2. Example for a stress deformation curve of an MTT on rhyolithe.

Table 1. Testing results of rhyolithe samples.

MTT	3.8 ± 0.97 MPa
Brazilian test	7.2 ± 1.6 MPa
(acc. to DGEG 1985, ISRM 1978)	
Point-load test	8.0 ± 1 MPa
(acc. to DGEG 1982, ISRM 1985)	
Bending test	13.5 ± 1.5 MPa
(acc. to DIN 1048)	
Unconfined Compressive Strength	
(UCS)	102.3 ± 9.1 MPa
Young's modulus	25.8 ± 2.1 GPa
Destruction work W_Z	204.1 ± 21.8 kJ/m^3

testing principle and testing requirements were presented at the EUROCK 2000 symposium in Aachen, Germany by Blümel (2000). The test uses a simple, cylindrical specimen that is over cored from the top and bottom by two axial core drill holes with different diameters (Figure 1). After placing a load plate (top) and load ring (bottom), the sample is then loaded in a standard testing device for compressive testing. Failure occurs by direct tension in the area in between the both overlapping core drill holes ("tension zone").

The MTT tensile strength σ_{MTT} is calculated from the maximum compressive load F_{max} and the area of the tension zone A_{TZ} which depends on the radius r_1 and r_2 of the core holes (Equation 1):

$$\sigma_{MTT} = \frac{F_{max}}{A_{TZ}} = \frac{F_{max}}{r_1^2 \cdot \pi - r_2^2 \cdot \pi} \qquad (1)$$

where:

σ_{MTT} MTT tension strength [MPa]
F_{max} failure load [N]
r_1 radius of the larger core hole [mm]
r_2 radius of the smaller core hole [mm]

Suggestions made by Blümel to format samples include a sample diameter of >100 mm, a length-diameter ratio of about 1.5:1 and special formatting of the sample faces according to UCS testing standards.

3   COMPARING DIFFERENT TENSION TESTS

As an example for the wide range of values that can be obtained from different testing procedures, this chapter presents results from a series of tests on a homogenous and isotropic rhyolithe from the Rennsteig tunnel project at Oberhof in Thuringia, Germany (Wolski, 2002; see Figure 2, Table 1). The samples from this fine to medium grained and medium weathered rock are characterized by hypidiomorphic feldspar minerals that are embedded in a very fine grained quartz matrix.

As further explained in chapter 5, the tensile strength values obtained from the MTT are very close to the theoretical tensile strength of the rock material. In comparison with the MTT, other test results are up to about 90% (Brazilian test), 110% (point-load test) or even about 260% (bending test) higher than the direct uniaxial tensile strength of the rock material.

4   DEFORMATION MEASUREMENTS

In order to get an idea of probable bending moments in the sample, axial and lateral deformation was monitored during some tests. Realistic deformation measurements were also useful for adapting the finite element model presented in chapter 5.

The measured deformations were very low, with a maximum axial deformation of about 0.2 mm, +0.004 mm lateral deformation at the bottom and −0.001 mm lateral deformation at the top of the sample (Figure 3), which is only a little above the measuring inaccuracy of the used system.

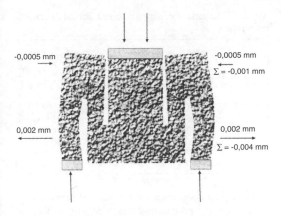

-0,0005 mm

-0,0005 mm
Σ = -0,001 mm

0,002 mm

0,002 mm
Σ = -0,004 mm

Figure 3. Scheme of the measured lateral and axial deformations of MTT sample under load. Note that lateral deformations are shown with 300× magnification.

5 RESULTS FROM THE FINITE ELEMENT STUDIES

In order to further investigate stress distribution during testing, the MTT was simulated with a non-linear finite element calculation at the Professorship for the Analysis of Civil Engineering Structures at the TU München. The sample was modeled using four-node axisymmetric 2d solid elements (Figure 4, upper right).

The material model is based upon the incremental flow theory within the framework of the theory of plasticity and was originally developed for the calculation of concrete and steel fiber reinforced concrete structures. The yield surface is composed of two partial areas, in order to be able to model different material behavior under compression and tension. Tensional failure is described by the Rankine criterion with linear, isotropic softening and a fracture energy concept. Under compression the Drucker-Prager criterion is

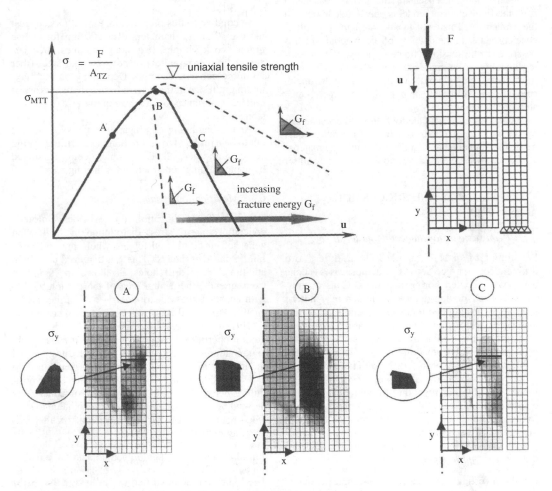

Figure 4. Models for and results from the finite element analysis.

363

used which shows both isotropic hardening and isotropic softening.

The qualitative results of the finite element calculation are illustrated in the upper left part of Figure 4. The diagram shows the correlation between the mean tensional stress σ_m (which is calculated from the load F and the area of the tension zone A_{TZ}) and the axial deformation u.

In the lower part of Figure 4, the distribution of stresses in y-direction is illustrated in cross sections through one half of the sample for 3 different stages of the test: (A) shows the stress distribution in the pre-failure area, (B) is at maximum load (failure point) and (C) shows the post-failure situation. The calculations give rise to the supposition, that the variable stress field in the pre-failure area (A) becomes more or less equally distributed in the tension zone when the failure point is reached (B). This effect shows to be largely influenced by the ductility of the material.

The maximum mean tensile strength σ_{MTT} calculated from the finite element model is only a little lower than the implemented material tensile strength. The difference correlates to the ductility of the material which is described by the tensional fracture energy. With increasing ductility of the material, the calculated maximum mean tensile strength σ_{MTT} comes closer to the implemented material tensile strength due to a more equal stress distribution.

As a result of the calculations, the authors state that the tensile strength obtained in the MTT is rather equal to the theoretical uniaxial tensile strength of a material with respect to the normal variation of testing results.

6 EXPERIENCES AND SUGGESTIONS FOR MTT TESTING

6.1 Requirements for the testing material

Preparation of an MTT sample includes at least two different coring processes which in most cases is done using water-cooled boring machines. Consequently, jointed, weak or non-durable materials may not be suitable for testing due to a lack of stability that may lead to the destruction of the sample at the formatting stage. As a simple guide for judging whether a material is suitable for formatting, a classification system was developed by Wolski (2002) from his experiences on samples from bunter sandstone, lower muschelkalk limestone, phyllite, rhyolithe, granite mylonite and quartz conglomerate (Table 2). A material is suitable for preparation if all influencing factors are located in the "good" areas of Table 2.

6.2 Preparation of MTT samples

For the investigations on hardrock samples a water-cooled boring machine was first used with a 120 mm

Table 2. Criteria for material suited for MTT sample preparation.

Durability in water	non durable	considerable change	neglectable change	durable

Spacing of joints and bedding planes [cm]	< 0,6	0,6-2	2-6	6-20	20-60	60-200	> 200

Grain binding	none	poor	low	medium	good	very good

Rock strength (UCS)	< 1 MPa	1 - 5 MPa	5 - 25 MPa	25 - 50 MPa	50 - 100 MPa	> 100 MPa

Legend: probability for gain of intact MTT sample 0% — 50% — 100%

diameter diamond core bit to core cylinders out of blocks. Further preparation was done with a diamond rock saw and a disc grinding machine for preparation of the sample faces. The cylinder was then over cored from both ends with core bits of 79 mm and 47 mm diameter. The length-diameter ratios varied between 0.5 and 1.4.

In contrast to this preparation, the authors suggest the use of larger diameters over 200 mm for coarse grained rock samples (e.g. conglomerates, breccias, coarse grained granites) or concrete and steel fiber reinforce concrete samples. Depending on the maximum grain size of the material, such diameters are of crucial importance for a representative size of the tension zone.

At the beginning of preparation works, it turned out to be a problem to assure both core drillings being centered and vertical. This problem could be solved by using guiding construction for drilling.

6.3 Modification of sample geometry

Especially for investigations on steel fiber reinforced concrete, constant stress distribution in the tension zone had to be assured for the whole pre and post-failure phase of the test. This was achieved using two additional core drill holes as shown in Figure 5. Consequently, the central area of the tension zone – with constant stress distribution – was further weakened and the initial crack was forced to propagate here.

This testing setup has proved to deliver very good properties for steel fiber reinforced concrete. In combination with deformation-controlled testing and monitoring of the whole stress–strain path, it also allowed detailed and realistic investigation of the post-failure behavior, which for this type of concrete is defined by distinctive post-failure strength due to steel fibers being pulled out of the concrete matrix after failure (Figure 5).

6.4 Testing setup and control

The MTT tests were carried out following the ISRM suggestions for uniaxial compressive strength tests

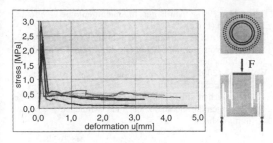

Figure 5. MTT testing results for deformation-controlled tests on steel fiber reinforced concrete samples with modified geometry. Besides the pre-failure behavior such MTT tests are able to deliver properties for the post-failure behavior of ductile materials.

(ISRM, 1978b). Depending on testing material and investigation aims, the tests were either simple stress-controlled tests to the failure point with a constant loading rate of 0.05 MPa/s or deformation-controlled tests including the whole stress–strain path. Forces were applied using 5 mm V2A steel plates and rings in the size of the sample geometry.

Good experiences were also made with the testing system, a ToniNorm UCS testing machine with a testing frame for 200 kN maximum load. Sample deformation was measured using three different measuring devices: Total axial deformation was monitored via three inductive displacement transducers between the load plates. Lateral and axial deformation were measured via a device with several displacement transducers that was mounted onto the sample surface.

Logging and analysis of the collected data were done by a HBM Spider 8 data logger and a PC system with HBM CatMan 2.0 and Microsoft Excel software.

It appeared useful to conduct these tests without a ball joint at the loading plates. If an initial crack formed in one side of the tension zone, a ball joint will further propagate only this crack, which may in the worst case lead to asymmetric stress distribution and inclination of the inner core.

6.5 Parameters from the MTT

During all tests, total axial deformation and applied force were logged and plotted in a force deformation diagram like those shown in Figures 2 and 5. The MTT tensile strength is calculated for the failure point using Equation 1.

Calculation of deformation modules (e.g. a kind of Young's modulus) from this plot does not appear useful since complicated load transfers take place in the sample during testing and thus a calculated deformation modulus would not be very significant for describing any compression or tension behavior of the material.

For significant ductile behavior (like the tested steel fiber reinforced concrete) calculation of a post-failure tensile strength is possible.

7 CONCLUSIONS

Judging from their experience, the authors evaluate the Modified Tension Tests as an innovative and easy-to-carry out testing procedure for determining the uniaxial tensile strength of hardrock and building materials. The MTT shows a good ratio for required testing equipment and demands for the testing material.

In detail, the MTT is characterized by the following positive features:

1. The tensile strength determined with the MTT comes very near to the real tensile strength of a tested material or rather equals the tensile strength with respect to the normal variation of testing results due to material differences.
2. The MTT provides good possibilities for monitoring material behavior in the post-failure area of ductile materials.
3. In comparison with standard UCS tests, the MTT needs no or only little extra expenses with regard to time, costs and required equipment.
4. The MTT is very well suited for materials with high strength (especially hardrock), where the use of adhesives is no longer possible.

REFERENCES

Blümel, M. (2000): Neue Laborversuchstechniken für felsmechanische Versuche (Improved Procedures for Laboratory Rock testing) – DGGT (ed.): Proceedings of the EUROCK 2000 Symposium, Aachen, 27–31. März 2000: 573–578, Essen (Glückauf).

DGEG – Deutsche Gesellschaft für Erd- und Grundbau e.V. (1982): Empfehlung Nr. 5 des Arbeitskreises 19 Versuchstechnik Fels der DGEG: Punktlastversuche an Gesteinsproben – Bautechnik, 1982, 1: 13–15, Berlin (Ernst).

DGEG – Deutsche Gesellschaft für Erd- und Grundbau e.V. (1985): Empfehlung Nr. 10 des Arbeitskreises 19 Versuchstechnik Fels der DGEG: Indirekter Zugversuch an Gesteinsproben – Spaltzugversuch – Bautechnik, 1985, 6: 197–199, Berlin (Ernst).

DIN – Deutsches Institut für Normung (1991): DIN 1048, Teil 5: Prüfverfahren für Beton. Festbeton, gesondert hergestellte Probenkörper – 8 p. Berlin (Beuth)

Fecker, E. & Reik, G. (1996): Baugeologie – 429 p., Stuttgart (Enke).

ISRM – International Society for Rock Mechanics (1978): Suggested methods for determining tensile strength of rock materials. Commission on Standardization of Laboratory and Field Tests – International Journal of Rock Mechanics, Mining Sciences and Geomechanical Abstracts, 15: 99–103.

ISRM – International Society for Rock mechanics (1978b): Suggested methods for determining the uniaxial compressive strength and deformability of rock materials. Commission on Standardization of Laboratory and Field Tests – International Journal of Rock Mechanics, Mining Sciences and Geomechanical Abstracts, 16: 135–140.

ISRM – International Society for Rock mechanics (1985): Suggested method for determining point load strength. Commission on Testing Methods, Working Group on Revision of the Point Load Test method – International Journal of Rock Mechanics, Mining Sciences and Geomechanical Abstracts, 22: 51–60.

Prinz, H. (1997): Abriß der Ingenieurgeologie – 546 p., Stuttgart (Enke).

Thomée, B. (in preparation): Strukturmechanische Modellierung von Stahlfaserbeton (working title) – PhD thesis TU München.

Wolski, K. (2002): Der Modified Tension Test – Aussagekraft und Durchführbarkeit eines neuen Zugversuches – Diplomarbeit am Lehrstuhl für Allgemeine, Angewandte und Ingenieur-Geologie der TU München – 52 p. (unpublished).

Geotechnical Measurements and Modelling, Natau, Fecker & Pimentel (eds)
© 2003 Swets & Zeitlinger, Lisse, ISBN 90 5809 603 3

Swelling behaviour of sedimentary rocks under consideration of micromechanical aspects and its consequences on structure design

E. Pimentel

Institute of Soil and Rock Mechanics, University of Karlsruhe, Germany

ABSTRACT: The adsorption of water is a characteristic of some sedimentary rocks. Although the consequences of this phenomena are known since a long time it has not been complete studied until now. Depending on the kind of rock the swelling process will follow different swelling mechanism. The respectives conceptual models considering microstructural aspects are presented. For quantifying the swelling potential new swelling testing procedures are proposed. Based on the conceptual models and on laboratory test results, structure design aspects were derived.

1 INTRODUCTION

The adsorption of water is a characteristic of clay, argillaceous and anhydrite rock, known since a long time. This phenomena called swelling of rock not only can cause an increase of volume or in case of a volume constriction an increase of pressure, but also a significant reduction of the shear strength and stiffness. An underestimation of the swelling potential during design can lead to engineering problems after construction, such as breakdown of tunnel linings, heave of foundations or roads surfaces or slope instabilities. In the past this type of damages were observed at many strutures. On the other hand, the overestimation of the swelling potential can increment the project costs considerably.

Swelling of rock is a complex process. The amount of adsorbed water, i.e. the swelling potential of the rock, depends on the material composition and structure and on the boundary conditions as for example stress level (before, during and after construction) including potential negative pore pressure due to unloading and water availability, their salinity and acidity. The swelling process can, depending on the rock type activate different chemical or physical mechanism, i.e. with or without inducing changes in the rock structure.

2 SWELLING PROCESSES – CONCEPTUAL MODELS AND MECHANISM

2.1 Physical swelling

In the mineralogy the clay minerals belong to the group of silicaminerals and to the subgroup of sheet silicates.

The typical structure consists on sheets of tetraeder and octaeder crystals (Fig. 1). A clay particle consists on several of these sheetspackages with a more or less parallel arrangement. Due to temperature and weathering, some of the central cations of the clay minerals will be exchanged with other of lower electrical charge, so that the particle will have a positive charge deficit. Additionally and due to the typical sheet structure of clay particles, their surface will show a concentration of negative electrical charge. This two factors will cause the adsorption of water molecules and cations, since both have an electrical charge.

Since the water molecules are electrical dipols, their taking up will be oriented, so that the outer surface after each layer of took up dipols will show also a negative charged surface, but with lesser charge concentration. Beyond the absorbed layer, water will fill the pores between particles, or it will be pressed out depending on changes on the pore pressure, specially due to mechanical loads. So it can be distinguished between three zones (Fig. 2), the intracrystalline, where water has a high density and behaves like a solid,

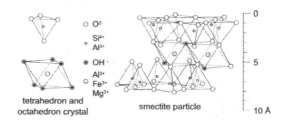

Figure 1. Structure and electrical charges of an elemental clay layer.

followed by a diffuse zone, where water molecules are distributed in an atmospheric likely way, i.e. the density of adsorbed water is closer to the particle higher and at least the zone of water filling the remaining pore

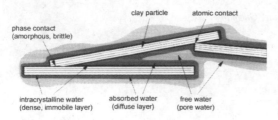

Figure 2. Clay particle and water types.

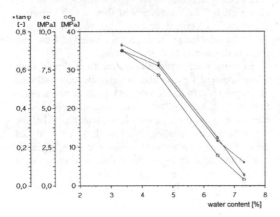

Figure 3. Structure of a diagenerically consolidated claystone (arrows represent normal vectors to particle surface).

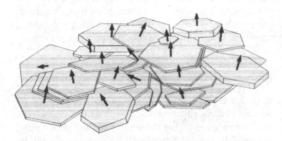

Figure 4. Strength loss in dependance of water content.

volume, the free water. The bonding energy and behaviour of the water for each zone is different, and therefore it can be distinguished between intracrystalline, osmotic and mechanical swelling. In Table 1 theoretical maximal values of swelling strains and stresses corresponding to the different mechanisms are presented.

The argillaceous rocks are sedimentary rocks and their raw material consists on clay particles, which were separated from the origin rock due to weathering. Since this happpens with a surplus of water and cations, the intracrystalline and the osmotic swelling processes must have been finished during the long transport or at least during the sedimentation. The overburden pressure of the subsequent thick deposits of sediments or ice, acting over long time periods, caused a consolidation. The pores were getting closer and (free) water was pressed out. If the pressure increased and by having enough time, the clay particles would show a markedly alignment perpendicular to the pressure direction and also an increase of stiffness. During this process the borders and surfaces of the particles were still getting closer, also by pressing out absorbed water. The result will be either a stiff clay or an argillaceous rock, diagenetically consolidated, depending on the duration and load levels of the process.

Due to the above mentioned markedly alignment of the clay particles during the diagenetic process and due to their sheet like structure (Fig. 3) a transversal isotropic swelling behaviour must be expected.

The long time swelling behaviour of claystone is comparable with diffusion processes, i.e. the closer the system is to a new thermodynamical equilibrium, the slower will be its development. The swelling strains will cause damages in the diagenetic bounds between clay particles, so that a strength and stiffness loss associated with the swelling process must be expected. In Figure 4 results of triaxial tests on Opalinus clay samples after different swelling stages, i.e. after the samples have absorbed different amounts of water, are presented. They show a decrease of shear strength of more than 90% due to swelling. Therefore the swelling process can induce squeezing of rock.

Each unloading of saturated argillaceous rock or clay in presence of water will cause negative pore pressure. The equalization of this pore pressure will induces the took-up of water molecules and causes a mechanical swelling, which usually will not have the

Table 1. Physical swelling mechanism of argillaceous rock and their maximal theoretical stresses and strains.

Type of took-up water	Material behaviour	Swelling mechanism	σ_{max} [MPa]	ε_{max} [%]
Intra-crystalline	Solid	Intra-crystalline	100	100
Absorbed	Diffuse layer	Osmotic	10	20
Free	Liquid	Mechanical	$<<1$	<1

368

same relevance for engineering purposes, as the other swelling processes (Table 1).

As indicated above, the swelling process happened during the sedimentation is partially reversed, so that the clay particle will have again a swelling potential. Thereby no intracrystalline water is being pressed out. The required pressure would correspond to an overburden of h > 3000 m, which will cause metamorphic changes to the sedimentary deposit and so loss swelling potential. In case a rock will have considerable osmotic and mechanical swelling potential, latest can be neglected.

2.2 Chemical swelling

Differing from the physical swelling processes, the chemical swelling is under moderate energetic levels a non reversible process, associated with changes in the mineralogy of the swelled rock. The most observed process is the swelling of anhydrite in gypsum:

$$CaSO_4 + 2H_2O \Rightarrow CaSO_4 \cdot 2H_2O \qquad (1)$$

Theoretically it would produce a volume increase of $\Delta V = 60,8\%$ and in case of volume constriction a swelling pressure similar as in case of intracrystalline swelling, i.e. $\Delta\sigma > 50$ MPa. Under for engineering purposes typical temperatue conditions (T < 40°C) anhydrite is metastable, i.e. due to this high saturation concentration and in contact with water it will turn into solution and precipitate to the more stable gypsum. The solution phase is necessary since anhydrite has a tricline and gypsum a monocline crystal structure, so that the incorporation of water molecules cannot happened in the purely solid phase as by clay particles.

In laboratory it has been observed, that in case a thick anhydrite block will be in contact to a low amount of water, only the outer surface will transform into gypsum and it will impermeabilized the rest of the material. In case the amount of water will be increased, some gypsum will solute so that the process can continue. In Central Europe the most important geological formation for engineering purposes containing anhydrite is the 230 mio years old Keuper formation corresponding to the Triassic. Typical for this formation is an intercalation of thin anhydrite and clayey layers and clay embedded in carbonate matrix. Laboratory tests with this material have shown, that the swelling process is much slower but with a higher swelling potential as by the physical swelling of clays. A swelling test can take many years up to some decades. It has also been observed, that in case of an increasing of water salinity will reduce the swelling process.

The swelling of this type of rock is an alternating process. Since the anhydrite does not have any electrical field it is not able to attract water dipols or cations

from far away. This is done by the clay particles as far as the boundary conditions are adequate, i.e. the rock is nearly saturated and an unloading has happened. In the contact between clay and anhydrite layers the anhydrite crystal will, due to this high potential energy, caught the took-up water molecule from the clay particle, get in solution and precipitate as gypsum. So the osmotic swelling of the clay particle must be repeated. Due to the sedimentation conditions in the past, the contact zone between these materials is not smooth but it contains many imperfections. Therefore subsequent crystal layers of anhydrite will swell, as far as the clay particles will provide them with water. During swelling these imperfections form weak planes, which can rise up and so increase the water permeability and reduce the rock mass strength and stiffness. Since the water transport along the clay layers is a diffusion process, mostly along the diffuse double layer, i.e. across zones with low water mobility, the whole swelling process will take long time. At the end of the process and in case even less water circulation is allowed the gypsum crystals will dilute, be eroded and some holes will remain. This new pore volume can, depending on the stress level collapsed and cause instabilities.

Another chemical swelling mechanism that has been observed is a weathering process due to oxidation, as for example in case of pyrit crystals embedded in a clayey matrix:

$$FeS + H_2O + 7 O \Rightarrow FeSO_4 + H_2SO_4 \qquad (2)$$

In case the clayey layer will contain carbonates, as it is the case for the about 190 mio years old Posidonienschiefer formation from the Jurassic, the sulfuric acid will quickly react and produce gypsum and carbonate acid:

$$H_2SO_4 + CaCO_3 + 2H_2O \Rightarrow CaSO_4 \cdot 2H_2O + H_2CO_3$$
$$(3)$$

Equation (2) shows, that for this swelling process, not only water but also oxygen is needed. The activation and development of this swelling process will be discussed below.

2.3 Activation of the swelling process

As indicated in the conceptual models above, in nearly all the cases the rock will be in equilibrium before construction. Unless the project is situated in arid zones, in many cases it can be assumed, that the rock will be saturated. The activation of the swelling process in these rocks with swelling potential will happened during the driving the tunnel itself, firstly due to unloading. Often the rock stratification is more or less horizontal, so that we will have in floor and

369

roof area an unloading normal to the strata and a loading parallel to it. For the sidewalls areas the loading and unloading direction relative the strata is just the opposite. Due to the anisotropic behaviour, we must expect swelling preferentially in roof and floor, but because of gravitation, water will migrate from the roof to the floor. The consequence will be a deficit of water in the roof area and an excess in the floor area. Therefore, as it has been observed in all tunnels in the past, swelling will occur in the floor area.

The general recommendation, to drive a tunnel in such formations completely dry, is like a wishfull thinking, i.e. an activation of the swelling process cannot be avoid. One of the fundamental problems is, to quantify the extension of the rock area where the swelling potential will be activated. Hereby not only swelling must be considered, but also the softening of rock due to swelling. This process will increase deformations and permeability of the swelled rock and continue unless the rock swells complete out or a new equilibrium can be reached, as for example due to deformation constriction induced from a closed lining. It must be pointed out, that neither shear strength and stiffness (Pimentel, 1996) nor permeability are material constants (Pimentel, 1999). On the other hand, it is obvious, that the swelling process can be increased if groundwater from other rock stratas is not beeing drained adequately during construction, so that a considerable excess of water will, as indicated above, dramatically increase deformations as consequence of plastification due to shear strength loss, i.e. due to squeezzing.

In the past is was incorrectly assumed by Wittke & Rissler (1976), that during the driving of a tunnel and before lining was placed, the swelling process will be activated on the surfaces exposed to air circulation with high but not saturated relative humidty. This concept does not consider unloading and gravitation. On the other hand, exposed clayey rock will in contact with circulating wet air, although its hygroscopic properties, dry out and shrink but not swell. The circulating air will evaporate the free water and some molecules in the diffuse layer of the exposed surfaces. Due to this evaporation, the salinity in the diffuse layer will increase and the thickness of the diffuse layer will shrink losing water molecules to the free water, which will be also evaporated. Additionally it must be considered, that due to unloading, shrinkage of the diffuse layer and due to the surface tension of water, negative pore pressure will be generated, producing water meniscus and tensile stresses. This tensile stresses can open some latent cracks. In summary, the exposition of rock surfaces with wet air with relative humidity less than 90% will contribute to a weathering process but not activate a swelling process.

In case of anhydrite interlayered with argillaceous rock the activation of the swelling process in the clay particles will occur, as indicate above, during tunnel driving. As explained in the conceptual models, the swelling of anhydrite will continue furthermore after tunnel construction, unless a new equilibrium is reached. It must be pointed out, that in comparison with clay swelling, usually very high pressures are needed to achieve a new thermodynamic equilibrium. In the next chapter the pressures involved by achieving a new equilibrium are discussed.

For the oxidation swelling mechanism water works as transport vehicle for the oxygen. Since groundwater does not content enough oxygen, this process cannot be activated, although stress unloading in saturated rock happens. This swelling phenomena has been observed on the cellar of buildings founded on this kind of rock and with pour thermical isolation. The heating of the cellars in winter will dry part of the rock, produce shrinkage and increase cracks along weak planes or imperfections. With beginning of the raining season during spring time the open cracks and other rock surfaces will get in contact with "fresh" water, i.e. with oxygen. So the process will be activate again. Due to the shrinkage during the drying process negative pore pressure will be developed. The mechanical and osmotical swelling potential will be initially increased again. With further dry-wet cycles the rock will increase cracks openings and pores. After some cycles and as far as the pore volumes are enough to allow expansion without considerable pressure changes, the rock will definitely loss its swelling potential.

3 NEW SWELLING TESTS

Another fundamental aspect for the design of structures in or above rock with swelling potential is to determine or assume realistic pressures and deformations under stationary conditions, i.e. after achieving a new equilibrium.

As indicated above, the swelling potential depends on several conditions as composition, microstructure, diagenesis and structure specific boundary conditions. Tests which only determine the mineralogical composition as x-ray diffraction, DTA-tests or water absortion tests are not suitable for this purpose and have at most an index character.

The only adequate way to quantify the swelling potential of rock mass is by performing swelling tests on undisturbed and representative rock samples and also under representative boundary conditions. In the literature several swelling tests are proposed, as by Huder & Amberg (1970), Kaiser & Henke in DGGT (1986) or Madsen in ISRM (1999). They are supposed to be performed with the classical apparatus for consolidation tests and have as common characteristic, that they include a considerable loading phase of

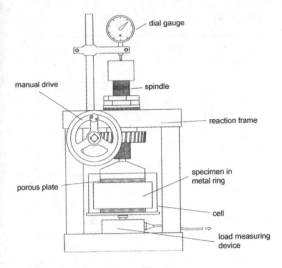

Figure 5. Apparatus for swelling tests.

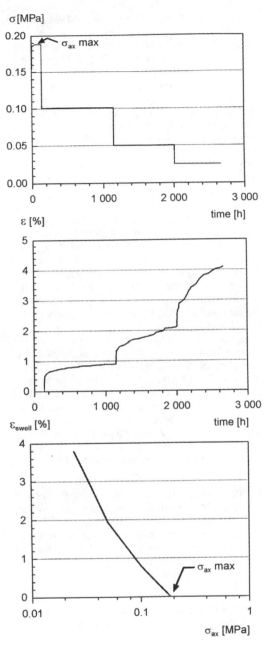

Figure 6. Results of a swelling test on argillaceous rock with weak diagenetics bounds.

the specimen. This load value is comparable to the maximal swelling pressure and therefore it will cause damages in the microstructure and falsify the tests results. It must be pointed out, that the mentioned apparatus have limitations. Tests can only be driven stress-controlled and the maximal applied load is $P < 1$ to, i.e. for a sample with a diameter $D = 80\,mm$, $\sigma_{max} < 2\,MPa$.

An alternative apparatus for swelling tests was developed at the Chair of Rock Mechanics of the University of Karlsruhe (Fig. 5). The testing device consist of a 4-columns frame. Load or deformation is applied respectively induced manually over a spindle, so that tests can be performed stress-, as well as strain-controlled. The maximal applied load is $P < 5$ to, i.e. for a sample with a diameter $D = 80\,mm$, $\sigma_{max} \leq 10\,MPa$.

For the investigation of the swelling behaviour of argillaceous rocks following new multistage swelling test is proposed by the author:

A cylindrical rock sample is being cut on a lathe, fitted in a stiff metal ring, mounted in the testing frame and prestressed with a minimal load of about $\sigma_{preload} < 0.02\,MPa$. In order to avoid structure damage in the specimen no external loads should be applied, i.e. the loads will be generated by the specimen itself due to swelling.

In the first stage of the test, the swelling process is activated by watering the sample under volume constriction, i.e. $\varepsilon_{vol} = 0\%$. The increase over the time of the swelling pressure is being recorded until the maximal swelling pressure σ_{max} is reached, i.e. a new equilibrium is being reached.

After this the sample will be subsequently unloaded. This can happens strain- or stress-controlled. Both pro-

cedures will take similar time. Since the swelling strains can often hardly be estimate, it is more suitable to perform a uniform stress-controlled unloading. The load values can be chosen as fractions of the determined maximal swelling pressure σ_{max}, for example with σ = 50%, 25%, 10%, 5% of σ_{max}. The unloading should be continued until a minimal stress is reached. For each load step the load will be kept constant until a new equilibrium has been reached. After that the next unloading should be applied.

If time is running out, a non-linear extrapolation can be applied in some load steps for estimation of the equilibrium values (Pimentel, 1996). Finally the equilibrium values of the strains without elastic component, i.e. ε_{swell} with their corresponding stresses are plotted in a semilogarithmical scale. For argillaceous rock in the uniaxial swelling case a linear relation should be expected (Grob, 1978). In Figure 6 the curves of a swelling test are presented.

For the case of anhydrite interlayered with argillaceous rock principally the same testing procedure should be applied. According to the conceptual model presented above, a considerably higher σ_{max} and longer testing duration will be expected. In case that the swelling pressure during the firts step reaches the capacity load of the testing device the load should be kept constant at this level (begin of 2nd load step) and the maximal swelling pressure σ_{max} should be determined with a non-linear extrapolation.

In the past swelling tests on samples from the Keuper formation in southwest Germany with a duration of about 20 years were performed in our laboratories. In Figure 7 the results of one of these tests is presented. By this test several unloadings and reloadings happens, so that for evaluation purposes only the first 6 years have been considered, whereby only the absolute strains are considered and the corresponding stresses have been extrapolated and plotted (Fig. 7 bottom). In this diagram we can recognize three swelling phases:

I. development of a very high swelling pressure, with relative small deformations and without considerable microstructural changes
II. transition phase with high swelling pressures, considerable increment of deformation and microstructural changes
III. increment of deformation by realtive reduction of swelling pressure

It must be pointed out, that for engineering purposes the swelling pressures are also in the last phase high.

Since the oxidation swelling mechanism can only be activated by water with oxygen, another testing procedure must be choosed, i.e. with dry-wet cycles. Depending on the project conditions there are principally two different possibilities:

I. wet-dry cycles under volume constriction (ε_{vol} = 0) in order to determined σ_{max}
II. wet-dry cycles under a constant load in order to determine the maximal heave ε_{max}

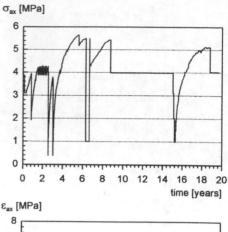

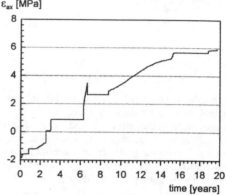

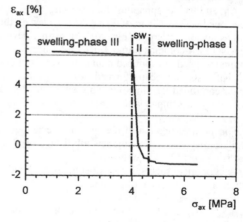

Figure 7. Results of a swelling test on anhydrite interlayered with argillaceous rock with strong diagenetics bounds.

The results of a swelling test on a sample of the Posidonienschiefer formation performed after modus I are shown in Figure 8. From the first to the 8th wet-dry cycle the swelling steady increased and reaches a maximum value of about σ_{max} = 1 MPa. In the subsequent wet-dry cycles a continous decay of the

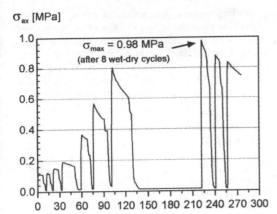

σ_ax [MPa]

$\sigma_{max} = 0.98$ MPa
(after 8 wet-dry cycles)

time [days]

Figure 8.   Results of a swelling test with wet–dry cycles.

swelling pressure can be observed, probably due to solution and increaisng of pores.

4   DESIGN CONSIDERATIONS

As indicated above in the conceptual models, the swelling mechanism are, due to the microstructural considerations too complicated. Therefore a purely theoretical derivation of constitutive laws will not be possible. The empirical derivation, i.e. based on laboratory tests results for example, is only suitable, if the samples and boundary conditions of the tests are representative for the project, otherwise all numerical calculations are more or less useless. A deterrently example for the last case is Kiehl (1990), who derivated a constitutive law for both main swelling materials, i.e. argillaceous and anhydrite rock, based on swelling tests with remolded clay. These tests were performed by Pregl et al. (1980).

For argillaceous rock considering only a uniaxial case the linear relationship between the swelling strain and the logarithm of the stress proposed by Grob (1978) is often suitable. For the the 3D case, the author derivated a constitutive law, based on triaxial swelling tests (Pimentel, 1996). It must be pointed out, that, since the swelling process is activated during construction, the argillaceous rock will be able to dissipitate part of the swelling potential, i.e. the maximal swelling pressure determined with the swelling tests are not necessary the design pressure for tunnel lining. This value must corrected by allowing a deformation usually in an order of magnitude of about 0.5% to 1.0%. Figuratively it means, that the x-axis of the semilogarthmical stress strain diagram must be

move upwards. In case that geological conditions are those, that due to construction some aquifers will get in contact with the swelling rock, a considerable loss of strength and stiffness of the rock must be considered in the design calculations. The same will occur due to poor drainage of underground water.

Until now there is no constitutive law, which satisfactorily describes the swelling process of anhydrite interlayered with argillaceous rock. Since the swelling process will be much slower as in case of pure argillaceous rock, a reduction of the swelling pressures determined with laboratory tests due to construction time is not suitable. On the other hand, the swelling pressures are considerably higher and much more problematic. Principally there are two tunnel design philosophies for this type of rock:

I.   stiff tunnel design (Widerstandsprinzip)
II.   ductile tunnel design (Ausweichprinzip)

In the first case the tunnel lining is supposed to support a high swelling pressure, hoping that this pressure will stop the swelling process itself. It can be achieved with an extreme thickness of the tunnel lining, for example in the Engelbergbasis-Tunnel in southwest Germany of about t = 3 m. Until now we can construct tunnels considering a swelling pressure not much higher than 2 MPa. In the Keuper formation swelling pressures can be a multiple of this value (Fig. 7 – swelling phase I), so that the assumption that the activation of the resistance of the tunnel lining will induce a new stable equilibrium is a dangerous wishfull thinking. Additionally, since these resistance is smaller than the maximal swelling pressures (in Fig. 7, $\sigma_{max} > 6$ MPa), the permeability cannot be reduced in such a way that a new thermodynamical equilibrium can be reached. The great disadvantage of these design is its brittle behaviour, i.e. the structure will be loaded up to his shear resistance and collapse without warning. This process can take many years to decades. A remediation of the structure will be difficult.

In the case of ductile tunnel design a package of high compressible material (Knautschzone) is build in the floor area under the tunnel lining. Compared to the stiff design, in this case the tunnel lining will have a usual dimensions. This compressible material, for example expanded clay, has the function to absorb the greatest part of the swelling strains. After this compression stage, the remaining swelling potential will induce a moderate swelling pressure on the tunnel lining. If some extensometers are placed during construction in the floor area, the development of the compaction phase can be monitored, and remediation measures can be taken before the structure collapses, as for example horizontal boreholes under the tunnel. Due to the ductil behaviour of the system the observational method can be applied in this case.

373

For rock with oxidation swelling potential the only reasonable design consideration is to avoid wet-dry cycles, i.e. good thermal isolation and surface impermeabilisation.

5 CONCLUSIONS

The most important swelling mechanism were identified and discussed above. Their corresponding conceptual models together with results of swelling tests performed with appropriate boundary conditions and with resentative samples allows us to derivate the following conclusions:

- It is necessary to distinguish between the different swelling mechanism.
- The only way to evaluate the swelling potential and determine representative material parameters is by performing swelling tests.
- Swelling tests as oedometric tests can be performed, but in order to conserve the microstructure of the material, a loading of the specimen must be avoid.
- Depending on the material, i.e. swelling mechanism, the duration of a swelling test can take months to decades. This must be contemplate in the project.
- In case of rocks with oxidation swelling potential, dry–wet cycles must be avoid.
- In case of osmotic swelling, the maximal swelling pressure determined in laboratory can be reduced for design purposes, according to project conditions, i.e. due to preswelling during construction. If groundwater will increase considerable the water supply, a substantial shear strength and stiffness reduction must be considered even during the construction phase.
- Anhydrite interlayered with argillaceous rock is the most problematic swelling rock. A reduction of the swelling pressure due to project considerations is not applicable here. The maximal swelling pressures can exceeds considerable the design pressures of the tunnel lining. Due to the slow development of the swelling process, the activation of the maximum swelling potential in form of strains or stresses will take years to decades after construction. Due to lack of knowledge about this phenomena, the engineering design concept must be flexible, including possible elements of remediation. Therefore the only reasonable design philosophy is a ductile tunnel design.

REFERENCES

DGEG 1986. Empfehlung Nr. 11 des Arbeitskreises 19 – Versuchstechnik Fels – der DGEG. Quellversuche an Gesteinsproben. *Bautechnik 3 (1986), 100–104*

Grob, H. 1978. Schwelldruck im Belchentunnel, *Proc. Int. Symp. Für Untertagebau, 99–119, Luzern, 1972*

Huder, J. & Amberg, G. 1970. Quellung in Mergel, Opalinuston und Anhydrit. *Schweizerische Bauzeitung Jg. 83, Heft 43, 975–980*

ISRM 1999. International Society for Rock Mechanics – Commission on Swelling Rocks and Commission on Testing methods – Suggested Methods for Laboratory Testing of Swelling Rocks. *Intern. Jour. of Rock Mechanics and Mining Sciences 36 (1999), 291–306.* Pergamon

Kiehl, J.R. 1990. Ein dreidimensionales Quellgesetz und seine Anwendung auf den Felshohlraumbau, *Proc. 9. Nat. Felsmechanik Symposium Aachen Nov. 1992, in Sonderheft Geotechnik 1993.* Essen: Verlag Glückauf

Pimentel, E. 1996. Quellverhalten diagenetsich verfestigtem Tonstein. *Veröffentlichungen des Institutes für Bodenmechanik und Felsmechanik der Universität (TH) Fridericiana in Karlsruhe, Heft 139*

Pimentel, E. 1999. Quelldurchlässigkeitsverusche an hochverdichteten Bentoniten und Mischungen aus Bentonit und Sand. *Proc. 9th International Congress on Rock Mechanics ISRM, Paris 1999.* Rotterdam: Balkema

Pregl, O., Fuchs, M., Müller, H., Petschl, G., Riedmüller, G. & Schwaighofer, B. 1980. Dreiaxiale Schwellversuche anTonsteinen. *Geotechnik 1 (1980), 1–7*

Wittke, W. & Rißler, P. 1976. Bemessung der Auskleidung von Hohlräumen in quellendem Gebirge nach der Finite Element Methode. *Veröffentlichung des Inst. für Grundbau, Bodenmechanik, Felsmechanik und Verkehrswasserbau der RWTH Aachen, Heft 2, 1976*

Geotechnical Measurements and Modelling, Natau, Fecker & Pimentel (eds)
© 2003 Swets & Zeitlinger, Lisse, ISBN 90 5809 603 3

Presentation of a new soil investigation technique: the in situ triaxial test

Ph. Reiffsteck
Laboratoire Central des Ponts et Chaussées, Paris, France

G. Reverdy
Centre d'Étude et de Conception des Prototypes, CETE Normandie Centre, Le grand Quevilly, France

ABSTRACT: The in situ triaxial testing device is an apparatus whose objective is to reproduce the testing conditions of the triaxial apparatus, in situ. The apparatus is a sampler of which the inner tube is equipped with a membrane and whose head comprises a jack. During testing, it creates a homogeneous stress field and allows to impose a stress path to the sample and to study the behaviour of soils at a small deformation level.

For two decades, the developments in the field of measurement, acquisition and data processing have made it possible to multiply measurements during the same test. This evolution contributed to the sophistication of the tests and allowed the appearance of innovations. However, these innovations have principally led to device creating inhomogeneous stress field. For this reason the parameters needed by numerical models are obtained using correlation or empirical laws. So there is still a need for apparatus able to propose to numerical methods, rheological parameters directly measured in situ.

We present in this paper the principles of a new in situ test called "triaxial in situ" and we more particularly will illustrate the techniques used for the design and the manufacture of this testing device. The triaxial in situ testing device is an apparatus whose objective is to reproduce the testing conditions of the laboratory triaxial test, in situ. It consists of a thin-walled sampler where one substituted for the liner, a membrane equipped with transducers and whose head comprises a piston. This device is forced into the soil by pushing. Progressively with the penetration in the ground, a jetting tool desegregates in the driving module the soil that penetrates inside the device. It acts in this phase as a self-boring pressuremeter. Once the apparatus reached the wanted depth, the stages of the test takes as a starting point the traditional triaxial test: application of horizontal pressure using a measuring cell then application of vertical stress using a piston, the horizontal stress being generally maintained constant. It creates a homogeneous stress field. Thanks to the control of the vertical piston and measuring cell, the apparatus makes it possible to impose various stress paths.

Since special attention is paid to the characterization of deformability, local measurements of displacements inside the apparatus are made thus allowing studying the behaviour of soils at a small deformation level. We will in a first part compare this new technology to more classical devices used in situ and in laboratory. Some preliminary results are also presented in the end.

1 INTRODUCTION

Recently, in situ triaxial testing devices were developed simultaneously in Japan and in France. They were born from the need for realizing in situ, tests closer to the laboratory tests, to free themselves from the sampling phase causing disturbance and with an aim of direct comparison of the results between laboratory and in situ tests. These tests try for that to recreate in place a homogeneous stress field similar to the one applied on the sample during triaxial test in laboratory.

The idea is simple: one seeks to confine the level to be tested in a cell constituted of a flexible membrane filled with a fluid, which will apply to the core a horizontal pressure. The vertical load from the piston is transmitted to the core through the top loading cap, which requires a reaction force. During the test, the radial pressure is maintained constant and the axial stress grows until rupture of the core is obtained by shearing. The absence of base pedestal is compensated by the resistance of the subjacent ground which can be estimated by the cone resistance a static penetrometer. Due to their design, these apparatuses can impose stress paths, thanks to the control of the vertical force and the radial pressure by for example a pressure volume controller.

The tests developed in Japan are intended mainly for rocks and rise from the overcoring test (Tani, 1999) (figure 1). The two first are carried out starting from ground surface on core of respectively 35 mm and 100 mm diameter. The vertical force developed by the confining pressure in the first test requires a reaction by anchoring. This reaction force disappears in the second test, where a cylinder is enough to confine the cell pressure. A reaction force is still necessary to counterbalance the vertical force. With this geometry it is difficult to measure the vertical deformation of the core.

The last test proposed by Tani is made on a hollow cylindrical specimen prepared in the bottom of the borehole by rotary drilling. A center hole of a small diameter is added. Hydraulic pressures are provided to the inner and outer cells which are equipped with rubber membrane. This enables to impose an internal pressure different from the external pressure. One can imagine that the designer wants in a further step to apply a torsion to carry out a shearing test in place, as on hollow cylindrical specimen for laboratory torsional shear tests. The dimensions used at the time of the first tests are an internal diameter of 85 mm, a external diameter of 400 mm and a height of 1000 mm. These dimensions allow to easily set up a displacements measuring system. The tests without rigid cell have been tested for the moment only in soft rocks and hard soils, because of the problems related to the expansion of the membrane towards outside in the surrounding ground. This configuration limits the use of stress paths. The first and third tests cannot be easily made in soil without disturbance as rotary drilling is needed to perform the test specimen. Furthermore the rubber membrane can not be placed without trapping air bubble.

2 IN SITU TRIAXIAL TEST DEVICE DEVELOPED AT LCPC

The test developed at LCPC is related to the need to carry out an in situ test which confines the ground by avoiding to the maximum the disturbance applied to the core and by removing the drilling phase. The idea which have been patented is to have a rigid confining cell able to provide axial and radial loading and which can be jacked into the ground (LCPC, 1999). The objective being to study the behavior of soils in small deformations, the absence of lower base pedestal is not then any more a real problem. It is also possible after test to retrieve the core for visual analysis and/or tests of physical characterization in laboratory.

2.1 Principle

The apparatus consists of a thin walled sampler which "internal" sides include measurement or testing devices (figure 2). It is set up in the ground by pushing. To reach the field of the small deformations, the realization of local measurements are carried out while placing the transducers in contact with the soil: radial and axial displacement and pore pressure transducers are fixed on the membrane.

In its self-boring configuration, the device can be jacked steadily into the ground and there is no need to retrieve the probe from the borehole between the tests. As in a self-boring pressuremeter a jetting nozzle placed in the driving module desegregates the soil that penetrates inside the device above the measuring cell (Baguelin et al., 1978; Benoît et al., 1995).

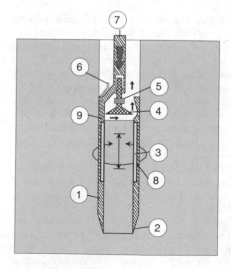

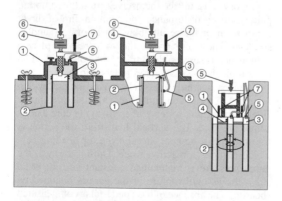

Figure 1. In situ triaxial test: Ishibashi patent, Fukushima patent, CRIEPI patent (according to Tani, 1999) (1: cell, 2: membrane, 3: cap, 4: force transducer, 5: connecting lines, 6: axial load, 7: displacement transducer).

Figure 2. In situ triaxial test: LCPC patent (1: cell, 2: cutting shoe, 3: membrane, 4: cap, 5: force transducer, 6: connecting line, 7: axial load, 8: displacement transducer, 9: jetting tool).

2.2 Comparison with the traditional equipment

The comparison of this equipment with the laboratory triaxial test reveals many similarities with the triaxial cell developed by Bishop and Wesley (1975). The cylindrical core is placed in a pressure chamber called "triaxial cell", which is connected to various systems of pressurization and measurement (figure 3). The load application and the follow-up of the deformations are very close. In both cases, local measurements are carried out on the core and measures are made of the vertical deformation on the central third, of the entire height of the core, and of the radial deformation (figures 3 and 4).

On another side, the test takes as a starting point the self-boring pressuremeter. Indeed, the probe is inserted in the ground and penetrates with a cutting shoe and can be self-boring (figure 3). The measuring system is fixed rigidly on the body of the probe and follows the radial displacement of the membrane. The apparatus is also able to measure the in situ horizontal stress after the installation.

However, there are two notable differences: the soil tested is contained inside the probe and the jetting tool is placed in the head of the apparatus. The loading configuration is not more the expansion of a cylindrical cavity in an infinite solid mass but the compression of a cylindrical sample in principal stresses (Baguelin et al., 1978; AFNOR, 1994a) (figure 4).

2.3 Design

2.3.1 Geometry of the probe

The design of the probe was carried out to achieve the above mentioned goals and to answer some requirements related to the sampling techniques (Horvslev, 1949; Bigot et al., 1996; Bat et al., 2000). These requirements are gathered in the French standard NF-P 94-202 which impose the respect of a diameter higher than 75 mm. Obtaining the least disturbed possible core also requires to respect of an area ratio of relatively small value and if possible in agreement with standard NF-P 94-202. The probe does not have inside clearance or offset in order to avoid imposing deformations of extension and with its internal diameter of 100 mm and its 15 mm thickness, it presents an area ratio of 69% for a value of 15% asked. In these requirements, the internal diameter does not seem to be influential but reality shows well that beyond these indices, this dimension is dominating (Hight, 2000). This high diameter is thus necessary to limit the remoulding. Furthermore, the installation of the instrumentation also imposes the

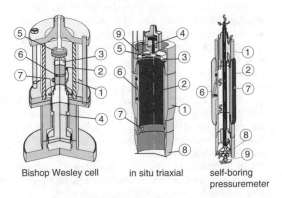

Bishop Wesley cell   in situ triaxial   self-boring pressuremeter

Figure 3. Comparison of the equipment (1: cell or body, 2: membrane, 3: top cap, 4: piston, 5: force transducer, 6: displacement transducer, 7: pressure transducer, 8: cutting shoe, 9: tool for disintegration).

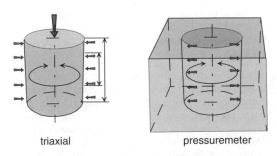

triaxial   pressuremeter

Figure 4. Comparison of equipment.

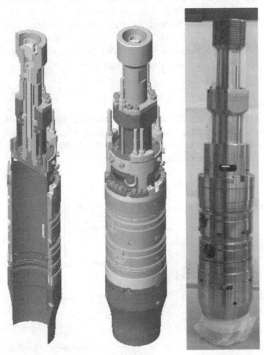

Figure 5. In situ triaxial probe.

377

size of the apparatus. On the other hand, the use of a large diameter increases the force to insert the probe in the soil. For example, the pushing effort is estimated at 5 tons for a cohesion Cu > 200 kPa, case of Flandres clay (clay present in the North of France and similar to Boom clay).

The slenderness of two was selected to be close to the laboratory triaxial apparatus geometry; consequently the active membrane is 200 mm long. The displacement of the piston is to the maximum of 30 mm. The piston has conical grooves in order to purge the chamber at the time of the setting in contact with the core and has a washer to prevent a blocking due to the intrusion of particles.

2.3.2 Structure of equipment

The physical architecture of the machine is as follows: the actuators of the probe (membrane and jack) are connected to two pressure volume controllers being able to be controlled in pressure and in volume, the various transducers are connected to two data-acquisition unit to collect measurements (figure 6).

The device developed is thus mainly made up of parts (data-acquisition card, data-acquisition units, transducers ...) and of software (Labwindows from National Instrument) available on the market, in order to reduce the costs. The data-acquisition units (power supply of the transducers, acquisition of the data) and the pressure controller are connected to the computer by RS232 interface. A specific software analyzes the measurements, calculates the test parameters and sends the instructions to the controllers. The software is also programmed to carry out the various preliminary and posterior phases to the test.

2.4 Measurement technique

Because of the reduced place available, measurements of displacement use recent technologies like Hall effect semiconductors for vertical displacement on the membrane, and proximity transducers for radial displacement. Measurements of force and pressure are taken in a traditional way. The other displacement transducers used for validation, are placed outside of the membrane to provide a reference in which must fit the arriving signal of the longitudinal transducers.

2.4.1 Displacement measurement with Differential Variable Reluctance Transducers (DVRT)

Three non-contact DVRT are used for the measurement of radial displacements of the membrane. The head of the transducer is the sensitive part that comprises two coils excited in frequency producing an electromagnetic field. This field is modified by the presence (approach or distance) of a metallic object thus changing the impedance of the coil. This metal object is in our case the membrane.

2.4.2 Displacement measurement with Hall effect semiconductors

The use of Hall effect semiconductors was retained to measure vertical relative displacements. The principal interest of this type of transducer is to allow side displacement or measures of location through a nonferromagnetic wall separating from the probe the support object of the magnet. The Hall effect displacement transducer is a semiconductor plate through which a current is flowing, and subjected by a magnet to a magnetic field whose flux lines are perpendicular both to the plate and the current flow. Any displacement of the magnet in the direction of the current generates a voltage (Asch et al., 1998; Clayton et al., 1989). The sensitivity to displacement depends on the magnetic circuit; this one is generally carried out using several rare earths permanent magnets associated in order to present an area where the gradient of induction is important (0,1 to 1 T/mm) and appreciably constant at a distance of a few mm. The transducer that is visible on figure 7, has a measuring zone of 3,86 mm length.

The objective of a measurement of displacement with accuracy of 10^{-6} is achieved with a stable and precise measuring equipment and a current transformer.

3 PROCEDURE OF TEST

3.1 Preparation of the test

Before the test, the general set up and procedure are the preparation and saturation of the system: controllers, tubing, jack and measuring cell, gauging and calibration of the measuring cell and saturation of the pore water pressure measurement circuit. One proceeds to the installation of the porous stone, saturated and protected by a film of clay to prevent desaturation.

In the first phase of our study, we do not use the self boring module. Our first objective is to validate the

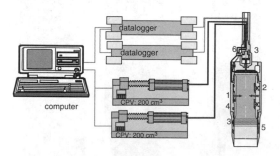

Figure 6. Architecture of equipment (transducers: (1) radial, (2) local axial, (3) external axial, (4) cell pressure, (5) pore pressure, (6) axial force).

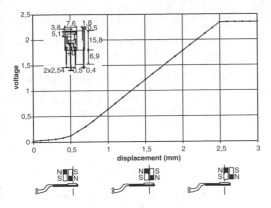

Figure 7. Transducer and variation of the Hall voltage according to the relative position of the magnet and the semiconductor.

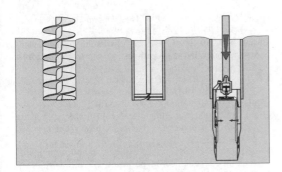

Figure 8. Principle of installation.

concept of the probe. So our borehole is made in soils with a helical auger. On the level, which one wishes to test, a flat bottom auger is used to clean the borehole immediately before inserting the probe (figure 8).

3.2 Procedure of test

The probe is then inserted by pushing at the speed of a traditional self-boring pressuremeter on a predetermined depth defines as the distance between the edge of the cutting shoe and the vents. During this phase, the local transducers move upward until they are against abutment (figure 9). One can then put in contact the piston with the surface of the ground.

After that, it is necessary to proceed to a phase of relaxation: measure of pressures at constant volume. It is then possible to carry out an isotropic or anisotropic consolidation by controlling the vertical force and the pressure in the measuring cell (figure 10). The phase of shearing can then be carried out according to the preset stress path (AFNOR, 1994b).

Lastly, one can apply the procedure of end of test described below or after unloading and adjusting the

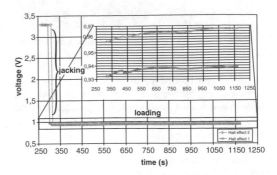

Figure 9. Evolution of vertical displacements during installation.

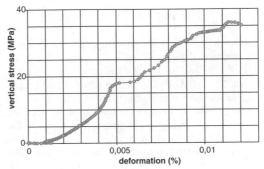

Figure 10. Results of loading in oedometric condition on clay.

membrane in the alignment of the cutting shoe, to continue the sounding while descending the probe by self boring.

3.3 Procedure of end of test

Several operations different from those of a pressuremeter bring closer the probe to a sampler: the probe can be extracted of the borehole after test like a traditional sampler, the piston still in contact with the ground, thus blocking the vents. Once retrieved from the borehole, the probe lay out on a specific frame, the double effect action of the jack is used to retract the piston. After disconnection of tubings and cables of the transducers, and disassembling of the head containing the jack, one can eject the core by taking support with the thread of the cutting case.

4 ANALYZING THE RESULTS

This apparatus does not imply, in its concept, of development specific of theory like empirical relations. This aspect is interesting because it will be able to follow the testing method of the triaxial apparatus (AFNOR, 1994b).

Since the major and minor stresses are applied, Mohr-Coulomb and relationship between deviatoric stress and strain can be drawn precisely. The Ko parameter can be deduce from the relation between horizontal pressure and vertical stress during tests made in oedometric conditions. The knowledge of the pore pressure allows to make drained or undrained analysis.

It is however necessary to notice that the techniques of tests described have a limited field of investigation. Certain boundary conditions cannot be applied and consequently, certain states of stress cannot be obtained such as, for example, pure extension.

5 CONCLUSIONS

A new in situ test method is proposed to investigate strength and deformation of soil masses. It's characterized as a field loading test realized in the bottom of borehole, placed by jacking. The advance along a soil profile can be performed using a jetting system housed inside the head of the device. At the opposite of borehole expansion tests, the soil is tested in the probe, which looks like a sampler.

The actual version including 12 transducers for a validation purpose can be simplified for practical purpose of site investigation. For research purpose, it can be also imagined to construct a square probe able to test cubic samples and then performing true triaxial tests in situ.

The design phase of the apparatus, including the validation of the technology of test and measurement on models, is finished. A prototype has been build and we plan a validation on a real test site in 2003.

REFERENCES

AFNOR 1994a. Norme NF P 94-070 Essai à l'appareil triaxial de révolution – Généralités Définitions.
AFNOR 1994b. Norme NF P 94-074 Essai à l'appareil triaxial de révolution – Appareillage, Préparation des éprouvettes Essais UU, CU + u, CD.
AFNOR 1995. Norme NF P 94-202 Sols: reconnaissance et essais – Prélèvement des sols et des roches – Méthodologie et procedures.
Asch, G. 1998. *Les capteurs en instrumentation industrielle*, Dunod Ed.
Baguelin, F., Jézéquel, J.-F., Shield, D.H. 1978. *The pressuremeter*, Transtech publications.
Bat, A., Blivet, J.-C., Levacher, D. 2000. Incidence de la procédure de prélèvement des sols fins sur les caractéristiques géotechniques mesurées en laboratoire, *RFG* 91: 3–12.
Benoît, J., Atwood, M.J., Findlay, R.C., Hilliard, B.D. 1995. Evaluation of jetting insertion for the self boring pressuremeter, *Can. Geotech. J.* 32: 22–39.
Bigot, G., Blivet, J.C. 1996. Prélèvement des sols et des roches, *Bull. LPC* 204: 113–117.
Bishop, A.W., Wesley, L.D. 1975. A hydraulic triaxial apparatus for controlled stress path testing, *Géotechnique* 25(4): 657–670.
Clayton, C.R.I., Khatrush, S.A., Bica, A.V.D., Siddique, A. 1989. The use of Hall effect semiconductors in geotechnical instrumentation, *GTJODJ ASTM* 12(1): 69–76.
Hight, D.W. 2000. Sampling effects in soft clay: an update, *The 4th Int. Geotec. Eng. Conf., Faculty of Engineering, Le Caire, Egypt.*
Hvorslev, J. 1949. Subsurface exploration and sampling of soils for civil engineering purposes, *ASCE Report.*
LCPC 1999. Procédé et dispositif d'essai triaxial in situ, Patent N°99.137.92, 9 pages.
Tani, K. 1999. Proposal of new in-situ test methods to investigate strength and deformation characteristics of rock masses, In Jamiolkowski, Lancellotta & Lo Presti (eds), *Prefailure deformation of geomaterials*: 357–364. Rotterdam: Balkema.

Geotechnical Measurements and Modelling, Natau, Fecker & Pimentel (eds)
© 2003 Swets & Zeitlinger, Lisse, ISBN 90 5809 603 3

Resonant column tests with cohesive and non-cohesive granular materials

S. Richter & G. Huber
Institute for Soil Mechanics and Rock Mechanics, University of Karlsruhe, Karlsruhe, Germany

ABSTRACT: Resonant column (RC) experiments are used for the determination of shear stiffness and energy dissipation during cyclic shearing of granular materials. We give an introduction into the principle of the resonant column device. The RC device used at the University of Karlsruhe is introduced along with recent improvements and developments. Typical experimental results for non-cohesive and cohesive model materials consisting of granular α-Al_2O_3 will be presented. Effects such as the degradation of shear stiffness with strain and the appearance of harmonics in the system response are indicators of nonlinear system behavior. The RC device is suited for the testing of cohesive and non-cohesive granular material and allows the investigation of the influence from particle surface forces.

1 INTRODUCTION

The issue of soil dynamics has become a substantial part of soil mechanics where soil behavior under shear cycles and wave propagation are of interest. Typical examples are foundations of dynamically loaded structures, the wide field of earthquake engineering, dynamic loads resulting from traffic or construction processes, wave propagation in conjunction with soil exploration or the effect of vibration on people and human health.

For the prediction of soil behavior under a dynamic load the corresponding soil parameters are required. In the case of shear deformation the shear modulus G is usually considered the governing parameter. Strictly speaking, there is not one shear modulus for granular material but rather a state-dependent shear stiffness, which is a function of shear strain γ, void ratio e, stress state, history, etc. Therefore we use the term "shear stiffness" whose symbol will still be G here.

In order to describe the dissipation of energy during cyclic loading the equivalent viscous damping D is widely used.

The resonant column (RC) device can be employed to determine G and D for the small and medium strain range, implying a linearized system. By small strain range we refer to values of $\gamma \leqslant 10^{-5}$ while the medium strain range ends at $\gamma \leqslant 10^{-3}$.

Important systematic investigations on the dynamic behavior of non-cohesive soils have been started several decades ago, e.g. Hardin & Richart (1963), Hardin & Drnevich (1972a), Hardin & Drnevich (1972b).

Due to the complex nature of cohesive soils their dynamic behavior is more difficult to describe. Additional influencing parameters are the plasticity index (Vucetic & Dobry 1991, Ishibashi & Zhang 1993), overconsolidation ratio (Hardin & Drnevich 1972b) and time (Hardin & Black 1968, Anderson & Woods 1976, Anderson & Stokoe 1978).

2 PRINCIPLE OF RESONANT COLUMN DEVICE

The RC device is commonly used in soil dynamics for the determination of shear modulus and damping characteristics of a soil sample. The principle of the device is based on a cylinder, which is set into torsional vibration. Figure 1 represents the model.

We consider an element of a cylinder with radius r and height dx. Since we are referring to a torsional mode of deformation we have to equilibrate the moments acting at the representative cross section. For a first approximation we neglect energy dissipation. Therefore we only consider moments at the boundaries $M(x)$ and $M(x + dx)$ as well as the moment resulting from inertia, M_i, which can be written as

$$M_i = -I\frac{\partial^2\vartheta}{\partial t^2} = -I_A \cdot \rho \cdot dx\,\frac{\partial^2\vartheta}{\partial t^2} \tag{1}$$

with I = polar mass moment of intertia of the cylinder element with respect to the x-axis; ρ = bulk

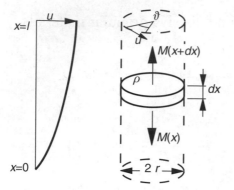

Figure 1.   Element of cylindrical sample and tangential displacement u for fundamental mode of deformation (sample fixed at $x = 0$, harmonic excitation).

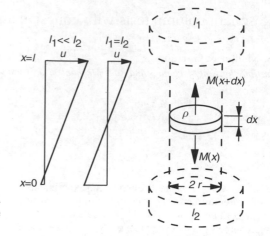

Figure 2.   Element of cylindrical sample with additional mass polar inertia at top and bottom and displacement u for fundamental mode of deformation (both ends free; $I_1, I_2 \gg I$; I = polar mass moment of intertia of entire sample).

density; ϑ = rotation of the cross section; I_A = geometrical polar moment of intertia.

Under the condition of equilibrium the sum of all moments has to be zero, hence

$$M + \frac{\partial M}{\partial x} dx - M - I_A \cdot \rho \cdot dx \, \frac{\partial^2 \vartheta}{\partial t^2} = 0 \qquad (2)$$

which eventually results in

$$\frac{\partial^2 \vartheta}{\partial t^2} = \frac{G}{\rho} \frac{\partial^2 \vartheta}{\partial x^2} \text{ or } \frac{\partial^2 u}{\partial t^2} = c_s^2 \, \frac{\partial^2 u}{\partial x^2} \qquad (3)$$

where u = tangential displacement; c_s = shear wave velocity; $G/\rho = c_s^2$.
The displacement u is a function of x and t. A general solution of Equation 3 is

$$u(x,t) = \left(A \cos \frac{\omega}{c_s} x + B \sin \frac{\omega}{c_s} x \right) \\ + (C \cos \omega t + D \sin \omega t) \qquad (4)$$

where ω = angular frequency.
With the boundary condition of a cylinder being fixed at $x = 0$ and free at $x = l$ we obtain $A = 0$ and

$$\frac{\omega l}{c_s} = \frac{(2n - 1)\pi}{2}; \quad n = 1, 2, 3, \dots.$$

Hence, the wavelength of the fundamental resonance frequency of a fixed-free cylinder is $\lambda = 4 \cdot l$.

For several reasons the arrangement of a fixed-free sample is unfavorable when testing soil by the resonant column method:

- Even the resonance in the fundamental mode of deformation leads to frequencies that are far off the frequencies encountered in situ. Thus, if viscous effects exist, they would be overestimated in the experiment.
- The fundamental mode of deformation being a quarter sine wave the shear strain γ at a given radius varies from its maximum at the fixed end to zero at the free end (Fig. 1). This highly inhomogeneous shear strain distribution is not desirable.
- The realization of a fixed end requires a support of infinite stiffness. Since this is in reality not achievable there will always be a "leakage" of elastic energy from the system due to wave propagation out of the testing device. This effect results in an overestimation of damping. Results of other researchers confirm this issue (Avramidis & Saxena 1990).

Therefore an improved RC device has been developed at the Institute for Soil Mechanics and Rock Mechanics, University of Karlsruhe starting in 1982 (Prange 1983). The device is of the free-free type and features additional masses at top and bottom of the soil sample. These additional masses (Fig. 2) lead to the following boundary conditions.

$$\tau = G \frac{\partial u}{\partial x} \bigg|_{x=0} = \frac{I_2 \cdot \rho \cdot l}{I} \cdot \frac{\partial^2 u}{\partial t^2} \qquad (5)$$

$$\tau = G \frac{\partial u}{\partial x}\bigg|_{x=l} = \frac{I_1 \cdot \rho \cdot l}{I} \cdot \frac{\partial^2 u}{\partial t^2} \qquad (6)$$

Neglecting the intermediate arithmetics the boundary conditions result together with Equation 4 in the following implicit eigenvalue equation

$$1 + \frac{I_1}{I_2} - \frac{I_1}{I}\beta \tan\beta + \frac{I}{I_2} \cdot \frac{1}{\beta} \tan\beta = 0 \qquad (7)$$

where I = polar mass moment of intertia of sample; $\beta = \omega l/c_s = 2\pi l/\lambda$. Equation 7 has an infinite number of solutions. Each solution yields a value for β, which represents the ratio of sample length l over the wavelength λ in the corresponding mode of deformation. For $I_1, I_2 \gg I$ γ is quasi-constant over the sample height.

Knowing the solution of β for the fundamental deformation mode of a given arrangement the shear stiffness of the sample can be determined from the resonance frequency f_R to

$$G = \rho c_s^2 = \rho \left(2\pi f_R \frac{l}{\beta}\right)^2 \qquad (8)$$

Equation 7 can easily be adapted for different boundary conditions. The corresponding moments of inertia are $I_{1/2} = \infty$ for a fixed end or $I_{1/2} = 0$ for a free end.

In order to describe the energy dissipation in a resonant column sample the equivalent viscous damping ratio[1] D is common. D can be derived from a single degree of freedom system with mass, spring and viscous damper described by m, c and r, respectively.

$$D = \frac{r}{r_{cr}} = \frac{r}{2\sqrt{cm}} \qquad (9)$$

In the field of rheology an alternative approach is used to describe stiffness and energy dissipation of a viscoelastic material. Stiffness and damping can be summarized in the complex valued shear modulus G^*. The elastic properties of the system are represented by the storage modulus G' (real part); the viscous properties are described by the loss modulus G'' (imaginary part). The complex modulus is given by

$$G^* = G'' + G'' = G'\left(1 + i\frac{\mu\omega}{G'}\right) = G'(1 + 2iD) \qquad (10)$$

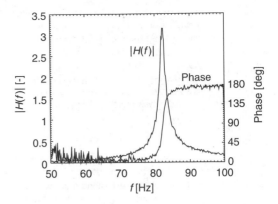

Figure 3.   Magnitude and phase of transfer function $H(f)$ for typical α-Al$_2$O$_3$ sample ($\gamma \approx 8 \cdot 10^{-8}$).

when μ = viscosity (corresponds to real component of complex viscosity η^*); $i = \sqrt{-1}$. The ratio of G'' over G' is the loss factor $\tan\delta$, which is related to D by

$$\tan\delta = \frac{G''}{G'} = 2D \qquad (11)$$

The value of G from Equation 8 corresponds to G'.

3   DETERMINATION OF DYNAMIC PARAMETERS

In order to calculate the shear stiffness G of the sample the resonance frequency of the RC system has to be determined. Two ways are widely used:

- The RC system can be excited by a harmonic signal. A frequency sweep and a simultaneous observation of the phase between input and output or the amplitude of the output signal indicates the resonance frequency.
- An excitation with a random noise signal in connection with a digital signal analyzer can be used to determine the resonance frequency from the transfer function $H(f)$.

The transfer function is restricted to linear, time-invariant systems. Its advantage is the possibility to measure the resonance frequency at very small strain ($\gamma < 10^{-7}$) since the signal-to-noise ratio can be improved by averaging. This requires long measurement periods and therefore a time-invariant behavior.

Figure 3 shows the course of magnitude and phase of the transfer function $H(f)$ of a typical α-Al$_2$O$_3$ sample in the vicinity of a resonance frequency.

---

[1] It has to be pointed out here that even though the behavior of a soil sample can be described by a viscoelastic model, its use does not mean that damping in granular materials is indeed of viscous, but rather of hysteretic nature (Hardin 1965, Ellis et al. 2000).

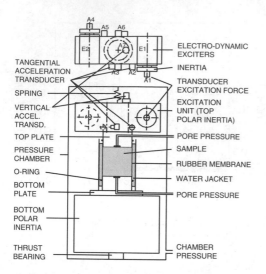

A4 A5 A6
E2 A0 E1
ELECTRO-DYNAMIC EXCITERS
A3 A2 A1
INERTIA
TANGENTIAL ACCELERATION TRANSDUCER
TRANSDUCER EXCITATION FORCE
SPRING
EXCITATION UNIT (TOP POLAR INERTIA)
VERTICAL ACCEL. TRANSD.
TOP PLATE
PORE PRESSURE
PRESSURE CHAMBER
SAMPLE
RUBBER MEMBRANE
O-RING
WATER JACKET
BOTTOM PLATE
PORE PRESSURE
BOTTOM POLAR INERTIA
THRUST BEARING
CHAMBER PRESSURE

Figure 4. Resonant column device at the Institute for Soil Mechanics and Rock Mechanics, University of Karlsruhe.

There are various methods for the determination of D for a linear system:

- Logarithmic decrement from free vibration decay.
- Half-power points in the vicinity of a resonance frequency.
- Magnification factor at resonance frequency f_R.

In the so-called nonlinear strain range, where G decreases with γ, D becomes also a function of γ. Therefore the above methods are only approximations of D.

4 THE RESONANT COLUMN DEVICE

4.1 General configuration

Figure 4 is a detailed sketch of the RC device used at the Institute for Soil Mechanics and Rock Mechanics, University of Karlsruhe. The sample has a diameter and a height of 100 mm, each. It is placed in a pressure chamber where it can be exposed to an isotropic state of stress. The top mass is free to rotate and its static vertical load is compensated by a soft spring. The bottom mass is carried by a thrust bearing and is therefore also free to rotate. For an average soil sample this arrangement leads to a ratio of wavelength over sample height of $\lambda/l \approx 35$. Thus the strain γ can be assumed to be constant over the sample height (see Fig. 2) but it is still a function of radial distance.

The arrangement of additional masses and sample is driven at the top by means of two electro-dynamic shakers with attached masses. (Brüel & Kjær Vibration

Exciter 4810). The signal can be chosen to be harmonic (HP Function/Arbitrary Wave Generator 33120A) or random noise (HP Digital Signal Analyzer 35650). The force of excitation is determined by two acceleration transducers attached to the masses of the shakers (Brüel & Kjær, type 4366).

The setup of free ends and the decoupled driving system lead to low energy losses from radiation of elastic energy out of the system. A test of the RC device with a sample dummy made of an aluminum tube yields damping values of $D < 0.2\%$.

4.2 Measurement of tangential sample deformation

The tangential displacement of the RC sample is determined from acceleration signals, which are measured at the top end (Brüel & Kjær, tranducer types 4366 and 4334; Fig. 4). The tangential displacement u is obtained from the integration of the signal.

The arrangement of four acceleration transducers allows the verification of a torsional sample deformation. By comparing the phase between transducers A2, A3, A5 and A6 (see Fig. 4) we can ensure that the top mass is in a torsional mode. Thus a flexural mode can be identified and excluded.

The signals of the acceleration transducers are transmitted by double-shielded cables and fed into charge amplifiers (Kistler type 5001). The signals can then be visualized on a two-channel digital storage oscilloscope or processed and analyzed by a Digital Signal Analyzer. Figure 5 shows the layout of signal generation and processing.

4.3 Measurement of vertical sample deformation

Due to the dilatant behavior of granular materials during shear a vertical deformation of the RC sample can be expected. In order to determine this deformation a vertical acceleration transducer has been installed at the top mass (Brüel & Kjær, type 4370; Fig. 4).

4.4 Air migration through rubber membrane

Most rubber membranes have the tendency to allow gas molecules to pass through them. Depending on material and type of gas the amount of gas passing through the membrane can turn a saturated granular sample into the unsaturated state. Natural rubber without any additives can feature a permeability coefficient of 0.3 cm³mm/h/m²/kPa for air (v. Amerongen 1954). At an effective pressure of 100 kPa the rubber membrane around the RC sample would pass an air volume of about 3 cm³/h. Thus for longer experiments the leakage of air into the sample cannot be neglected. Compared to natural rubber, butyl rubber exhibits a

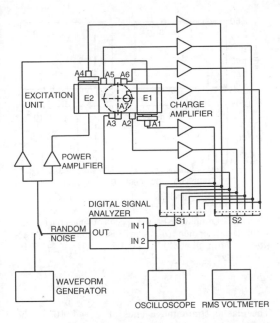

Figure 5. Components and signal flow chart for signal analysis and data storage.

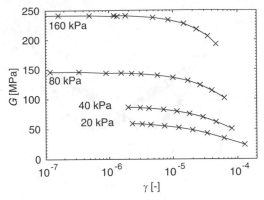

Figure 6. Results of G vs. γ for dry α-Al$_2$O$_3$ with $d_{50} = 0.5$ mm at varying effective pressures.

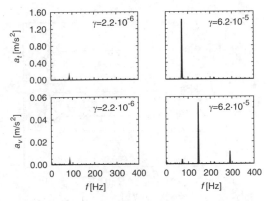

Figure 7. Amplitude Fourier spectra of tangential (a_t) and vertical acceleration (a_v) signal for dry α-Al$_2$O$_3$ ($d_{50} = 0.5$ mm, $p' = 80$ kPa, $e = 0.68$).

significantly reduced gas permeability. Unfortunately, butyl rubber membranes turned out to be difficult to handle due to a reduced resilience.

Our solution to the problem of air migration is the installation of a water jacket around the RC sample (de-aired water). This prevents the contact of air with the membrane. In order to reduce the re-aeration of the water a film of silicone oil is applied. This film of silicone oil also avoids an increased humidity inside the pressure chamber, which can cause problems with the isolation of transducers and cables.

Other authors (Marcuson & Wahls 1972, Marcuson & Wahls 1978) have used mercury jackets around the sample, which was avoided here due to reasons of safety and health.

In addition to a water jacket a gas with a reduced permeation of natural rubber could be used. Compared to air, nitrogen gas (N$_2$) reduces the permeability of natural rubber by 25%.

5 EXPERIMENTAL RESULTS

In this section we report about results from RC experiments with granular α-Al$_2$O$_3$ samples. Experimental results from two different materials with mean particle diameters of $d_{50} = 0.5$ mm (non-cohesive) and 0.7 μm (cohesive) are presented.

5.1 Non-cohesive material

Figure 6 represents typical results for a dry granular α-Al$_2$O$_3$ sample with $d_{50} = 0.5$ mm. G increases with the effective pressure p' and starts to decrease slowly with γ once the linear strain range is exceeded. The limit of the linear strain range for this material is $\gamma_{el} \approx 3 \cdot 10^{-6}$.

Nonlinear material behavior is not only characterized by a decrease in stiffness with increasing strain but can also result in frequencies that differ from those in the input signal of a system.

Figure 7 shows amplitude Fourier spectra of the tangential and vertical acceleration signal for $\gamma = 2.2 \cdot 10^{-6}$ and $6.2 \cdot 10^{-5}$ (e.g. transducers A2 and A7). At strains below γ_{el} the signal contains only the input frequency ($f_R = 86.1$ Hz). At larger strains the resonance frequency decreases ($f_R = 72.5$ Hz) and harmonics of

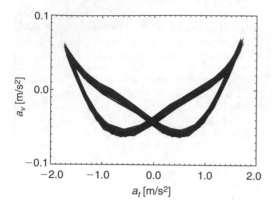

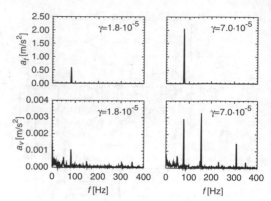

Figure 8. Lissajous figure from tangential (a_t) and vertical acceleration (a_v) signal for dry α-Al$_2$O$_3$ ($d_{50} = 0.5$ mm, $p' = 80$ kPa, $e = 0.68$, $\gamma = 6.2 \cdot 10^{-5}$).

Figure 10. Amplitude Fourier spectra of tangential (a_t) and vertical acceleration (a_v) signal for saturated α-Al$_2$O$_3$ ($d_{50} = 0.7$ μm, $p' = 80$ kPa, $e = 1.21$); noise at $f = 50$ Hz and 150 Hz.

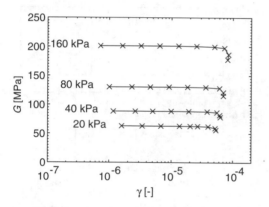

Figure 9. Results of G vs. γ for saturated α-Al$_2$O$_3$ with $d_{50} = 0.7$ μm at varying effective pressures.

the input signal can be observed in the vertical direction. The twofold and fourfold frequencies are apparent, which are additional indicators of non-linear system behavior.

If the vertical acceleration signal is drawn vs. the tangential acceleration, integer frequency ratios can be visualized. Figure 8 shows the resulting Lissajous figure for $\gamma = 6.2 \cdot 10^{-5}$. The dominant frequency ratio of 1:2 is obvious. Similar results have been found by Huber (1998) for large scale RC tests with railroad ballast.

5.2 Cohesive material

Figure 9 shows the course of G over γ for a fine-grained saturated α-Al$_2$O$_3$ sample with $d_{50} = 0.7$ μm.

The elastic strain range ends at $\gamma_{el} \approx 4 \cdot 10^{-5}$. The stiffness decreases rapidly with γ at $\gamma > \gamma_{el}$ and also becomes dependent on the number of shear cycles.

The much finer granular material also exhibits nonlinear behavior with a twofold frequency in the vertical direction. Figure 10 shows the spectra of the tangential and vertical acceleration signal for the linear ($\gamma = 1.8 \cdot 10^{-5}$) and nonlinear ($\gamma = 7.0 \cdot 10^{-5}$) strain range. However, the finer material shows a much smaller vertical amplitude. Possible reasons are its saturation, which counteracts volumetric changes, and also the lower density of the sample.

6 CONCLUSIONS

A resonant column device is suitable for testing samples of granular material at small and medium shear strain.

Air migration through rubber membranes has to be considered if complete saturation of the sample is required. A water jacket of de-aired water was used to reduce air migration.

Cohesive and non-cohesive model materials show a typical decrease of G with γ at $\gamma > \gamma_{el}$ but the specific course of G vs. γ depends on the particle size. Another indicator of nonlinear behavior are harmonics of the excitation frequency in the system response, which result from dilatant soil behavior.

The mean particle size of a granular material influences its behavior during small shear cycles. It can be shown that for small particles with a large surface-to-volume ratio the influence of particle surface forces affects the macroscopic behavior (Richter & Huber 2003).

ACKNOWLEDGEMENTS

The experimental results shown in this paper were sponsored by the German Research Council (DFG), Research Group FOR 371–2. The support is gratefully appreciated.

REFERENCES

Anderson, D. G. & Woods, R. D. 1976. Time-Dependent Increase in Shear Modulus of Clay. *Journal of the Geotechnical Engineering* 102 (GT5): 525–537.

Anderson, D. G. & Stokoe II, K. H. 1978. Shear Modulus: A Time-Dependent Soil Property. In: *Dynamic Geotechnical Testing,* ASTM STP 654.

Avramidis, A. S. & Saxena, S. K. 1990. The Modified "Stiffened" Drnevich Resonant Column Apparatus. *Soils and Foundations* 30(3): 53–68.

Ellis, E. A., Soga, K., Bransby, M. F. & Sato, M. 2000. Resonant Column Testing of Sands with Different Viscosity Pore Fluids. *Journal of Geotechnical and Geoenvironmental Engineering* 126 (1): 10–17.

Hardin, B. O. & Richart, Jr., F. E. 1963. Elastic Wave Velocities in Granular Soils. *Journal of the Soil Mechanics and Foundations Division* 89 (SM1): 33–65.

Hardin, B. O. 1965. The Nature of Damping in Sands. *Journal of the Soil Mechanics and Foundations Division* 91(1): 63–97.

Hardin, B. O. & Black, W. L. 1968. Vibration Modulus of Normally Consolidated Clay. *Journal of the Soil Mechanics and Foundations Division* 94 (SM2): 353–369.

Hardin, B. O. & Drnevich, V. P. 1972a. Shear Modulus and Damping in Soils: Measurement and Parameter Effects. *Journal of the Soil Mechanics and Foundations Division* 98 (SM6): 603–624.

Hardin, B. O. & Drnevich, V. P. 1972b. Shear Modulus and Damping in Soils: Design Equations and Curves. *Journal of the Soil Mechanics and Foundations Division* 98 (SM7): 667–692.

Huber, G. 1998. Dynamisches Schotterverhalten im Resonanzsäulenversuch. *VDEI-Bahn-Bau-Fachtagung, 28–30. Oktober, Berlin:* 209–213.

Ishibashi, I. & Zhang, X. 1993. Unified Dynamic Shear Moduli and Damping Ratios of Sand and Clay. *Soils and Foundations* 33(1): 182–191.

Marcuson III, W. F. & Wahls, H. E. 1972. Time Effects on Dynamic Shear Modulus of Clays. *Journal of the Soil Mechanics and Foundations Division* 98 (12): 1359–1373.

Marcuson III, W. F. & Wahls, H. E. 1978. Effects of Time on Damping Ratio of Clays. In: *Dynamic Geotechnical Testing,* ASTM STP 654.

Prange, B. 1983. Der Resonant-Column-Versuch: Theorie und Experiment. In: *Symposium Messtechnik im Erd-und Grundbau, DGEG. München, 23–24. November 1983. Essen: Deutsche Gesellschaft für Erd-und Grundbau e.V.:* 99–104.

Richter, S. & Huber, G. 2003. Resonant Column Experiments with Fine-Grained Model Material – Evidence of Particle Surface Forces. *In preparation.*

van Amerongen, G. J. 1954. Der Einfluß von Füllstoffen auf die Gasdurchlässigkeit von Kautschuk. *Kautschuk und Gummi* 7(6): WT 132–WT137.

Vucetic, M. & Dobry, R. 1991. Effect of Soil Plasticity on Cyclic Response. *Journal of Geotechnical Engineering* 117(1): 89–107.

Geotechnical Measurements and Modelling, Natau, Fecker & Pimentel (eds)
© 2003 Swets & Zeitlinger, Lisse, ISBN 90 5809 603 3

Water percolation inducing micro-seismicity: evidences from small scale in-situ experiment

V. Saetta
LaEGO-INPL-INERIS, School of Mines of Nancy, France;
also at Politecnico of Turin, Department of Structural and Geotechnical Engineering, Turin, Italy

D. Amitrano & G. Senfaute
LaEGO-INPL-INERIS, School of Mines of Nancy, France

Y. Guglielmi
Geoscience-Azur, University of Nice-Sophia Antipolis, France

ABSTRACT: The aim of the paper is to investigate the correlation between the hydro-mechanical behaviour and the seismicity induced by hydraulic loading in a small fractured rock mass (about $1.5 \cdot 10^3 \, \text{m}^3$) located in the Alpes–Maritimes at Coaraze (France). Hydro-mechanical studies have pointed out different types of behaviour for rock joints thanks to punctual measurements of water pressure and deformations. The results of seismic monitoring show that the most permeable joints are aseismic: the seismic activity is quasi-absent even though a high pressure increment is applied. The seismic activity starts to occur when water percolates within low permeability joints. We suggested that the seismicity may results from the rupture of crystallised zone, the formation of new micro-fractures or the sudden opening and the shearing of bedding joints. These results reveal that even for low water pressure variation (less than 0.1 MPa) hydraulical loading of a fractured rock mass can induce observable seismicity. These small scale (1–10 m) results could be of interest for understanding induced seismicity at larger scale (100–1000 m like dam, reservoir etc.).

1 INTRODUCTION

Elastic waves are generated whenever a transient stress imbalance is produced within or on the surface of an elastic medium. Almost any sudden deformation or movement of a portion of the medium results in such a source. A great variety of physical phenomena in the Earth involve rapid motions that excite detectable seismic waves. As internal source, it is possible to consider earthquake faulting, buried explosions, abrupt phase change, magma movements, hydrological circulation (Lay & Wallace 1995). Seismic activity can be triggered by human actions, including underground mining, reservoir impoundment, fluid injection and fluid extraction.

The following discussion is focussed on stimulated seismicity due to human activity with regard to reservoir impoundment, fluid injection and extraction.

The analysis of documented case histories of seismicity at reservoirs allows to distinguish two primary types of induced seismicity. The first type is a rapid response, in which the seismicity onsets immediately after the first loading of the reservoir. The seismic activity consists primarily of low–magnitude swarm–like activity, confined to the immediate reservoir area, and closely time correlated with changes in the water level in the reservoir. Classical examples are Nurek and Kabira for large reservoirs, and Monticello and Manic 3 for relatively small reservoirs (Simpson et al. 1988). The second type is a delayed response (depending on the hydraulic properties of the rock), in which the seismicity onsets after a significant delay after the first filling. It is often associated with large magnitude earthquakes, which may extend significantly beyond the confines of the reservoir, and which may not show an immediate correlation with major changes in the reservoir level (Niitsuma et al. 1998; Brandt 2001). Aswan, Koyna and Oroville are well known examples where the reservoirs have undergone a number of similar cycles of water level changes before the major earthquakes occur. In all identified cases of delayed response, the highest level of seismicity appears to be triggered by short term changes associated with the seasonal maximum of water level (Gupta 2002).

The causal mechanisms responsible for reservoir seismicity can involve increases in both shear and normal stress due to the surface load, pore pressure increases due to compaction or water pressure diffusion, remaining in the elastic domain. The rapid response following the filling of the reservoir is mainly related to elastic deformation and associated pore pressure changes, whereas the delayed response is related to diffusion or flow from the reservoir (McGarr et al. 1997).

An important aspect of all cases of seismicity caused by fluid injection is that the injection takes place in a geological formation having hydraulic connection to the fault zone on which the earthquakes occur. The earthquake activity is usually concentrated, in both lateral and vertical extent, near the bottom of the injection well. Thus the anomalous pressures are transmitted directly to the depths at which the earthquakes occur. There is also evidence of hysteresis and time dependency in injection related seismicity. Initial stages appears to be concentrated near the injection point and respond rapidly to changes in injection pressure or volume. As injection proceeds, the zone of influence increases, earthquake magnitudes increase and the direct response to input pressures is less obvious. For events close to the injection point, activity usually stops immediately after the injection ceases; whereas events farther from the injection well may continue for some time after pumping stops (McGarr et al. 1997). Some examples are Hijiori (Sasaki 1998) and Fjällbacka hot dry rock research site (Wallroth et al. 1996). The seismicity results from a decrease in the effective normal stress caused by increased pressure, which may cause shear rupture of existing fractures.

Seismicity associated with fluid extraction was first recognized in the oil field at Goose Creek, Texas, in 1925. In the meantime, numerous other cases have been recognized related to pore compaction, horizontal deformation and attendant earthquakes. The earthquakes tend to occur either immediately above or below the reservoir and are most prevalent in brittle strata than in ductile ones. Seismicity in the immediate environs of a reservoir onsets when the pore pressure reduction reaches a low level and is stimulated by stress changes associated with differential compaction of the producing formation (McGarr et al. 1997). In some cases, pore pressure changes and mass transfers leading to incremental deviatoric stresses $\ll 1\,MPa$ are sufficient to trigger seismic instabilities in otherwise historically aseismic areas. Once triggered, less stress variations (0.01–0.1 MPa) are enough to sustain seismic activity (Grasso & Sornette 1998).

The aim of the report is to investigate the correlation between the hydro-mechanical behaviour and the seismicity induced by hydraulic loading in a superficial fractured rock structure, subjected to low variations of hydrostatic water pressure.

2 DESCRIPTION OF COARAZE SITE AND EXPERIMENTAL PROCEDURE

2.1 Description of the experimental site of Coaraze

The experimental site is located in south eastern France, in the Alpes–Maritimes area about 30 km north of Nice. It is a small reservoir composed of a 17 m thick pile of Lower Cretaceous fractured limestone limited at the top and the bottom by a 2 m thick, impervious, glauconious marl layer. The joint network is made of three main types of discontinuities (Guglielmi 1998; Fénart et al. 2001):

1. The bedding surface S_0 with a N40–45E strike;
2. Fault set: N140 with a dip of 75 NE;
3. Fault set: from N60 to N90 with a dip of 70 to 80 SE.

Schematically, the reservoir network consist in 8 planar macro-faults extending up to the reservoir boundaries and intersecting the numerous remaining joints composed of S_0 joints and of small extension fractures (less than 2 m). The reservoir is drained by a spring of annual average yield of $12\,L\cdot s^{-1}$. A water-gate has been installed on the spring in order to permit an artificial variation of the water table level (up to 10 m) in the rock mass (Fig. 1A, B and C).

To prevent water leakage on the massif surface from the rock joints, concrete was used to clog the outcropping part of the reservoir. Stable water table level in the massif is reached when water overflows at the upper spring T1. The total volume of rock mass involved by water table variation is $19\cdot10^3\,m^3$. The volume of instrumented part of the reservoir is about $1.5\cdot10^3\,m^3$.

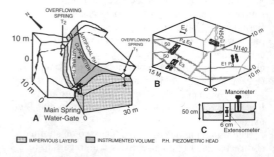

Figure 1. Measurement points into instrumented rock volume; A and B: 3D representation of the site and partial scheme of instrumented rock volume; C: scheme of drilled holes normal to the joint (Y. Guglielmi 1999).

Hydraulic conductivity measurements of isolated joints from the different families were taken. Main result is that there is an important conductivity contrast between the macro-faults with values ranging from 10^{-3} to $10^{-5}\,m^2\,s^{-1}$ and the low conductive joints with values ranging from 10^{-4} to $10^{-8}\,m^2\,s^{-1}$ (Fénart et al. 2003). The embedding rock matrix is considered by contrast as impervious.

2.2 *Hydro-mechanical experimental device*

Hydro-mechanical instruments have been located into corings (6 cm in diameter). This low size drilling is supposed not to significantly modify the stress state into the reservoir. For each monitored hydro-mechanical point two cores were drilled normally to the monitored joint 0.2 m from each other (Fig. 1C). In the first hole a chamber allows hydrostatic pressure monitoring in the joint (Pi). In the second hole, a vibrating-wire extensometer is centered on the joint in order to monitor the deformation normal to the joint (E_i). The extensometer is encased in concrete and the deformation measurement is that of a concrete cylinder (6 cm in diameter) whose long axis coincides with the extensometer. Strain measurements are made with GLOZL® 15 cm gauges (accuracy of $0.5 \cdot 10^{-6}$). Pressure measurements are made with GLOZL® vibrating-wire pressure gauges (accuracy of 0.001 MPa) (Fig. 1A, B and C). All the measurements are automatically recorded every 60 s with a CAMPBELL® data acquisition station. 10 pressure deformation measurements are taken along the reservoir for the three different joint families.

2.3 *Seismic monitoring system*

In order to monitor the seismic activity during the experiment, four mono-directional accelerometers (A2, A3, A4, A5) and one tri-directional geophone (G1) are employed (Fig. 2). Five cores has been drilled at a depth between 0.4 and 0.6 m to lodge the sensors. This configuration is supposed not to disturb the state of stress inside the reservoir. Sensors G1, A4 and A5 are located near faults (straight lines in Fig. 2), while A2 and A3 are located in low conductive zones (dotted lines in Fig. 2).

The accelerometers (Wilcoxon, Model 793L) have a flat response in the range of 1 to 300 Hz, with 500 mV/g sensitivity. The geophone (GEO SPACE Corporation, Model GD-20 DH 600) have a passing band in the range of 10 to 1000 Hz, with a natural undamped frequency of 40 Hz (peak sensitivity of 27.55 V/(m/s)). The analog signals are transmitted to an analog–digital (A–D) interface (Interface AD-v2.0, INERIS, Nancy). The A–D converter has a 16 bit resolution card, with a range of input signals up to ±15 V, 0.1 to 2300 Hz bandwidth on −3 dB.

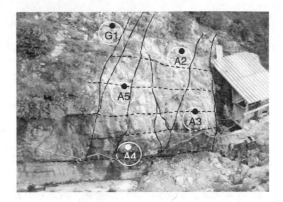

Figure 2. Instrumented rock volume: seismic sensors location and displaying of faults (straight lines) and bedding joints (dotted lines).

Signals amplification is 50 dB for the accelerometers and 60 dB for the geophone. Sampling frequency is 20 kHz, record duration is 0.5 s and trigger threshold is 300 digit.

3 HYDROMECHANICAL AND SEISMOLOGICAL EXPERIMENTAL RESULTS

3.1 *Experimental procedure*

Before starting the experiment, the massif is fully saturated. The experiment consits in a 30 minutes opening of water gate. During this time, the water level in the macro-faults is reduced to a minimum level. After that, the water gate is closed again and the water level increases until reaching the T1 spring overflowing. Depending on the hydraulic condition, this can take more or less time (Guglielmi 1998). In the following described experiment, the level is reached after 10 minutes.

3.2 *Hydro-mechanical experimental results*

A volume of about 2000 L is stored in the rock mass at the end of the experiment. 500 to 800 L are stored in the macro-faults and the remaining volume is stored in the bedding joints. When the water-gate is closed, measured deformations on different points range from 1 to 500. Deformations can be contractions or extensions that vary with water pressure elevation and with time. When the water-gate is opened, there is an instantaneous variation of deformation followed by a residual deformation with a slow return (10 to 20 mn depending on the points) to the initial state. Concerning the joints, it is possible to distinguish three different behaviour related to macro-faults and bedding joints (Kadiri et al. 2003).

1. In all macro-faults, there is an opening of the joint linked to the water pressure elevation in the joint. At the end of the experiment, deformation values range from $20 \cdot 10^{-6}$ (E1, P3 in Fig. 1) to $250 \cdot 10^{-6}$ (E7, P5 in Fig. 1). Water pressure reaches values of 0.04–0.06 MPa. In those coupled measurements, deformation versus time and pressure versus time curves show a similar non linear variation.

2. In some S_0 joints (coupled measurements E8, P2 in Fig. 1), it is possible to observe a tiny opening of the joint of $0.5 \cdot 10^{-6}$. Water pressure values are very variable and range from 0.003 to 0.1 MPa (Charmoille et al. 2003). Deformation varies independently from pressure elevations in the joints.

3. In other S_0 joints (coupled measurements E3, P4 in Fig. 1), there is a closure of $3 \cdot 10^{-6}$. Water pressure reaches values of 0.04–0.06 MPa. Deformation varies independantly from pressure elevations in the joints.

Results show that hydro-mechanical behavior is very different from one point to the other. In particular, macro-faults have a coupled hydro-mechanical behaviour and deformation values are important; on the other hand, hydro-mechanical behaviour of low conductive joints is not only linked to pressure increases in the joints but can also be induced by macro-fault hydro-mechanical deformations. Deformation values remain much lower than in the macrofaults.

3.3 Seismic observations during hydraulic loading

Figure 3 shows the cumulate of signal energy of all seismic sensors and water pressure measured by 4 vibrating wire manometers (Fig. 1B) during the fifteen minutes closing of the water gate. The recorded high initial and final seismic activity (Fig. 3) are related to the noise caused by the water gate closing and opening. This is accredited by the fact that the number of recorded signals is very low (11) and they are concentrated in few seconds. After that, the activity becomes stable at the beginning of the experiment and stops definitely at the end of the experiment when the spring flow rate comes back to the normal value. During the experiment, seismic events are generally recorded only by one sensor. Hence, no spatially location is possible.

It is possible to distinguish four periods during the experiment.

1. (0–3rd minute) water fills the more permeable joints in the rock mass as revealed by increasing of pressure measured by P3, P4 and P5. Indeed P2, located in low permeability zone, does not record any pressure increase. The seismicity is low even though a pressure increment in the joints occurs.

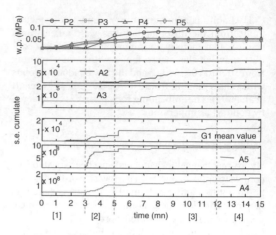

Figure 3. G1, A2, A3, A4 and A5 signal cumulated energy and interstitial pressure measured by vibrating wire manometers during the fifteen minute closing of the water gate. P2 is situated in a rock block, P3, P4 and P5 are situated in rock joints.

2. (3rd–5th minute) a steep increase of the seismic activity for G1, A4 and A5 sensors is recorded, meanwhile a strong increase in water pressure in low permeability zone is found. During this stage the A2 and A3 don't show any seismic activity increase.

3. (5th–12th minute) G1, A4 and A5 sensors show a discontinuous seismic activity, whereas A2 and A3 sensors begin to record seismic activity. Water pressure starts stabilizing in macro-faults and increasing in low conductivity zone.

4. (12th–15th minute) during this stage seismic activity is low and continuous. It is possible to consider the steady state in the rock mass has been reached.

4 DISCUSSION

On the basis of the previous bibliographic research, it can be stated that the fluid injection in a reservoir triggers a rapid seismicity in the immediate environs of the injection due to the increase of loading and a delayed seismicity shown beyond the confines of the reservoir due to the permeation of water. In applying these considerations to Coaraze site, it must be considered that the above results refer to reservoirs located at high depth in the earth subjected to high water pressure. Basing on results obtained during the experiment carried out on Coaraze site, we observed that a seismic activity is triggered by a little hydraulic loading (0.1 MPa) in a superficial fractured rock mass. In particular, as depicted in Figure 3, the most permeable joints are aseismic (no seismic activity

392

even though a high pressure increment in the joints occurs). Seismic activity is correlated to water permeation into low permeability zones.

Considering that seismic events are generally recorded only by one sensor, this means the recording is only local around each sensor. Seismic activity recorded by each sensor is representative of a small zone surrounding the sensor. Furthermore, source energy is very low and can only reach the nearest sensor and not all the seismic network.

It is possible to distinguish a delay between the seismic activity recorded by G1, A4, A5 sensors and the one recorded by A2, A3 sensors. The G1, A4 and A5 are located in the neighbourhood of a permeable joint, whereas A2 and A3 are located in low permeability zones far from permeable joints (see Fig. 2). Pressure variations are higher in the neighbourhood of macro-faults and occur with smaller delay. Pressure variations decrease with the distance to macro-faults and the delay increase. So, we propose that seismicity is related to water pressure increase in the low permeability joint (percolation). Mechanisms candidate for creating instabilities able to produce seismic activity are:

- failure of crystallizated calcite,
- new microfracturing,
- stratification joints aperture and sliding due to water diffusion.

5 CONCLUSION

Precedent studies (Guglielmi 1998) have shown correlation between hydraulic loading and rock mass deformation. Results show that the hydraulic injections tend to activate existing fractures rather than create new ones due to the low variation in effective stresses.

The results of seismic monitoring show that the most permeable joints are aseismic: the seismic activity is quasi-absent even though a high pressure increment in the faults occurs. Then, the seismic activity begins when water percolates into low permeability joints. Water percolation seems to induce instability phenomena like rupture of crystallised zone, formation of new micro-fractures and sudden opening and the shearing of bedding joints.

These results reveal that even for low water pressure variation (up to 0.1 MPa), hydraulic loading of a fractured rock masses can induce observable seismicity. These low scale (1–10 m) results could be of interest for understanding induced seismicity at larger scale (100–1000 m like dam, reservoir etc.).

ACKNOWLEDGMENT

This work was supported by the french program ACI-Catnat and the INERIS Nancy.

REFERENCES

Brandt M. B. C. 2001. *A review of the reservoir-induced seismicity at the Katse Dam, Kingdom of Lesotho – November 1995 to March 1999.* Rockbursts and Seismicity in Mines – RaSiM5, South African Institute of Mining and Metallurgy, pp. 119–132.

Charmoille A., Cappa F., Guglielmi Y. 2003. *Spacialization of fluid flow laws in a fractured reservoir by in-situ multiparametric measurements and numerical modelin.* Accepted to C. R. Géosciences.

Fenart P., Guglielmi Y., Henry J. P. 2003. *Validation of a discontinuity network characterisation to model hydraulic behaviour of fissured aquifer.* Accepted to Groundwater.

Grasso J-R., Sornette D. 1998. *Testing self-organized criticality by induced seismicity.* J. Geophys. Res. 103 (B12), 29965–29987.

Guglielmi Y. 1999. *Apport de la mesure des couplages hydromécaniques à la connaissance hydrogéologique des reservoir fissurés: approche sur site expérimental.* Habilitation à diriger la recherche, Université de Franche – Comté.

Guglielmi Y. 1998. *Hydro-mechanics of fractured rock masses: results from an experimental site in limestone.* Mechanics of Jointed and Faulted Rock, ed. A. A. Balkema, Rotterdam, pp. 621–624.

Gupta H. K. 2002. *A review of recent study of triggered earthquakes by artificial water reservoirs with special emphasis on earthquakes in Konya, India.* Earth Science Reviews, article in press.

Kadiri I., Merrien-Soukatchoff V., Guglielmi Y. 2003. *Interpretation and modelling of a hydro-mechanical in situ experiment within a fractured calcareous rock mass.* ISRM 2003-Technology roadmap for rock mechanics, South African Institute of Mining and Metallurgy, 2003.

Lay T., Wallace T. C. 1995. *Modern global seismology,* ed. Academic Press, San Diego.

McGarr A., Simpson D. 1997. *Keynote lecture: a broad look at induced and triggered seismicity.* Rockbursts and Seismicity in Mines, eds Gibowicz & Lasocki, Balkema Rotterdam, pp. 385 – 395.

Niitsuma H., Nakatsuka K., Chubachi N., Yokoyama H., Takanohashi M. 1998. *Downhole AE measurement of a geothermal reservoir and its application to reservoir control.* Fourth Conference on Acoustic Emission/ Microseismic Activity in Geologic Structures and Materials, ed. Trans Tech Publications, Clausthal – Zellerfeld, F. R. Germany, pp. 475–489.

Sasaki S. 1998. *Characteristic of microseismic events induced during hydraulic fracturing experiments at the Hijiori hot dry rock geothermal energy site, Yamagata, Japan.* Tectonophysics, Vol. 289, ed. Elsevier Science Ltd, pp. 171–188.

Simpson D. W., Leith W. S., Scholz C. H. 1988. Two types of reservoir-induced seismicity. *Bulletin of the Seismological Society of America,* Vol. 78, n° 6, pp. 2025–2040.

Wallroth T., Jupe A. J., Jones R. H. 1996. *Characterisation of a fractured reservoir using microearthquake induced by hydraulic injections.* Marine and Petroleum Geology, Vol. 13, n° 4, ed. Elsevier Science Ltd, pp. 447–455.

Measurement of the spatially distributed water content in geotechnics

A. Scheuermann & A. Bieberstein
University of Karlsruhe, Institute of Soil Mechanics and Rock Mechanics

R. Becker & W. Schädel
University of Karlsruhe, Institute of Water Resources Management, Hydraulic and Rural Engineering

S. Schlaeger & R. Nüesch
Research Center Karlsruhe, Institute of Technical Chemistry, Water and Geotechnology Section

C. Hübner
University of Cooperative Education, Heidenheim

R. Schuhmann
Research Center Karlsruhe, Institute of Meteorology and Climate Research

ABSTRACT: The physical properties of soil such as water content or hydraulic conductivity are very import-ant in many applications in soil mechanics and other geosciences. Due to a lack of adequate measurement methods the Soil Moisture Group (SMG) at the University of Karlsruhe has developed a new soil moisture measurement technology. This measurement system consists of moisture sensitive transmission lines which are embedded in the soil. Electromagnetic pulses are applied to the sensor and the reflections are recorded with a time domain reflectometer (TDR). With a new inversion algorithm which uses the full information content of the TDR signal, the spatial variability of the dielectric properties of the soil which are correlated to the water content can be determined. Experimental results obtained in various projects dealing with measurement and monitoring of water content demonstrate the applicability of the proposed method for practical problems.

1 INTRODUCTION

Many applications in hydrology, agriculture and civil engineering require the moisture profile or hydraulic conductivity. Most of the measurement techniques known to date are unable, however, to meet the user requirements in terms of accuracy, measurement volume, reproducibility, automation capability, and spatial resolution. The propagation evaluation of an electromagnetic pulse on unshielded transmission lines makes it possible to measure soil moisture distribution. Up to now TDR was restricted for measuring the integral or very coarse resolved water content along the transmission line. A new sensor technology and an inversion algorithm has been developed which uses the full information content of the TDR data of an embedded sensor. For the first time soil moisture measurements are possible with a high resolution both in space and time using this new method (Spatial TDR).

2 SOIL MOISTURE MEASUREMENTS WITH TDR

2.1 TDR moisture sensor

The typical TDR sensor is a metallic fork, which is inserted in the porous material. The maximum length is limited, because the electromagnetic pulse is attenu-ated and it disappears on longer lines. For longer transmission lines insulated forks or flat band cables arc proposed. They show much less pulse attenuation than non-insulated metallic forks in the same media. The flat band cable used in the experiments in chapter 3 and 4 is shown in Figure 1. In chapter 5 an applica-tion of insulated forks is presented.

The electrical field concentrates around the con-ductors and defines a sensitive area of 3 to 5 cm around the cable. The electric properties of the flat band cable used in this study can be calculated and measured using the capacitance model from Figure 2.

Figure 1. Insulated flat band cable (short section with bare conductors to visualize the geometry and the electrical connection of the cable).

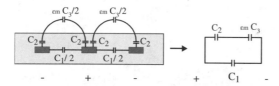

Figure 2. Capacitance model for the insulated flat band cable.

The total capacitance C may be replaced by three capacitances C_1, C_2, and $\varepsilon_m C_3$ and can be transformed into a direct relation between the dielectric coefficient ε_m of the surrounding material and the total capacitance:

$$C(\varepsilon_m) = C_1 + \frac{C_2 \varepsilon_m C_3}{C_2 + \varepsilon_m C_3} \qquad (1)$$

The three unknown capacitances were derived from calibration measurements of three different materials with well known dielectric properties, cf. Becker et al. (2002).

2.2 Reconstruction algorithm

A three step algorithm for reconstructing the water content profile along the flat band cable was developed. In the first step the transmission line parameters $C(x)$ and $G(x)$ were reconstructed with two independent time domain measurements from both sides of the flat band cable. In the second step $C(x)$ was transformed into dielectric coefficient $\varepsilon_m(x)$ by (1). In the third step $\varepsilon_m(x)$ was converted into a water content profile by standard transformations (Topp et al. 1980) or material specific calibration functions. The first step as a key component for the reconstruction will be discussed in detail (see also Schlaeger et al. 2001).

In order to reconstruct $C(x)$ and $G(x)$, two independent measurements are needed. Consequently, the problem is divided into two parts, one part that deals with an incident wave from the left and another part dealing with an incident wave from the right of the transmission

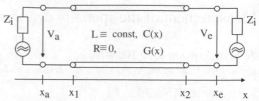

Figure 3. The uniform flat band cable, situated between x_1 and x_2, is connected to two lossfree uniform coaxial cables with matched impedances Z_i at their endpoints.

line. Figure 3 shows an experimental setup for receiving the reflection data from both sides of the unknown material. Two separate measurements must be carried out with the external current $F_{ex}^1 = \delta(x - x_a)f(t)$ and $F_{ex}^2 = \delta(x - x_e)f(t)$, respectively.

The telegraph equations describe the variation of the voltage $V(x,t)$ and the current $I(x,t)$ whilst it runs along the transmission line.

$$\frac{\partial}{\partial x}V(x,t) = -R(x)I(x,t) - L(x)\frac{\partial}{\partial t}I(x,t)$$
$$\frac{\partial}{\partial x}I(x,t) = -G(x)V(x,t) - C(x)\frac{\partial}{\partial t}V(x,t) \qquad (2)$$

The inverse method for solving these equations is based on an iterative search for the electrical parameters of the nonuniform flat band cable with full wave solution $V_1(x,t)$ and $V_2(x,t)$ of both direct problems. During the search, the solutions of the line need to be calculated repetitively. It is therefore important to use a technique that is computationally efficient, and which provides simple mapping from the parameters to the direct solution, to guarantee a fast convergence.

The input data $f(t)$, which describes the incident pulse of the time domain reflectometer, can be easily calculated from the measurements $V_a(t)$ of the coaxial cable between x_a an x_1. The output data $\lambda_i(t)$, i = 1,2, can be measured from the two separate experiments with the incident wave from the left and from the right side, respectively.

A cost function $J(C,G)$ is defined here which measures the difference, in the L_2-norm, between the solution of the direct problem corresponding to the parameters $C(x)$ and $G(x)$, and the given measurements,

$$J(C,G) = \sum_{i=1}^{2} \int_0^T [V_i(x_i,t) - \lambda_i(t)]^2 \, dt \qquad (3)$$

with $T > 2\tau(x_1,x_2)$, where $\tau(x_1,x_2)$ is the traveltime between x_1 and x_2. The cost function refers to the error in the solution for left-sided and right-sided incidence, respectively. The concept of the method is to find the parameters that minimize the cost function J. One more reason for choosing the L_2-norm is that it makes it possible to derive exact expressions for the gradient of J (Schlaeger 2002).

Only reflection data are used to determine uniquely the parameters $C(x)$ and $G(x)$ (He et al. 1994). The optimisation is achieved with a conjugate gradient method using Fletcher-Reeves update formula. As the result of the minimisation of (3) we obtain the spatial distribution of the total capacitance C and total conductance G, which are transformed into $\varepsilon_m(x)$ and volumetric water content $\theta_v(x)$.

The following examples will demonstrate the use and the effectiveness of this new method "Spatial TDR".

3 MEASURING OF WATER CONTENT ON A FULL-SCALE DIKE MODEL

In order to investigate transient hydraulic processes in river dikes during a flood, a full-scale dike model at the Federal Waterways and Research Institute was equipped vertically with 12 sensor cables between

Figure 4. Full-scale dike model at the Federal Waterways and Research Institute (BAW) in Karlsruhe during a flood simulation test in December 2000 (steady state of seepage condition).

0.7 m and 3.0 m in length. Both sides of the sensor cables were connected to the TDR device. The model is built up homogeneously with uniform sand (grain size 0.2 to 2 mm) and it is based on a waterproof sealing of plastic. As a result of this construction, the water infiltrating into the dike will flow to a drain at the toe of the land side slope and directed to a measuring device (cf. Figure 4).

In the course of the investigation mentioned above, flood simulation tests and sprinkling tests were carried out on the dike model (cf. Scheuermann and Brauns 2002, Scheuermann et al. 2001). Figure 5 shows the steady state of seepage condition during a flood simulation test in December 2000. The measurements of water content were reconstructed spatially along the sensor cable (cf. chapter 2) and interpolated as distribution of saturation in the considered cross section (assumed porosity of 37%). The dots represent the position of the sensor cables together with the saturation at these points. On the water side the water level in the basin is shown, and the position of the phreatic line within the body of the dike is given, as estimated from the pore water pressure measurements in the base of the dike. It can be seen that the measurements of the water content correspond very well with the position of the phreatic line. During the flood simulation tests on the dike model it was discovered that the percolation area below the phreatic line does not become fully saturated. Up to 15% of the pore space remained filled with air. This observation was verified by independent measurements during the steady state condition.

With this system it is possible to record the distribution of water content in a cross section within 5 minutes. This way, measurements of the water content with a spatial resolution of 3 cm are possible. The profiles of saturation can be determined with an average uncertainty of ±4% compared to other independent field measurements.

With this technique the observation of transient hydraulic processes within a dike was made possible for

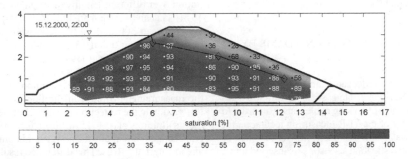

Figure 5. Interpolated distribution of saturation (from measured volumetric water content with an assumed porosity of 37%) in the dike model received from the cable sensor measurements during a flood simulation test in December 2000. The points represent the positions of the cable sensors, and the numbers next to the points show the corresponding values. The phreatic line was derived from the piezometer measurements in the base of the model.

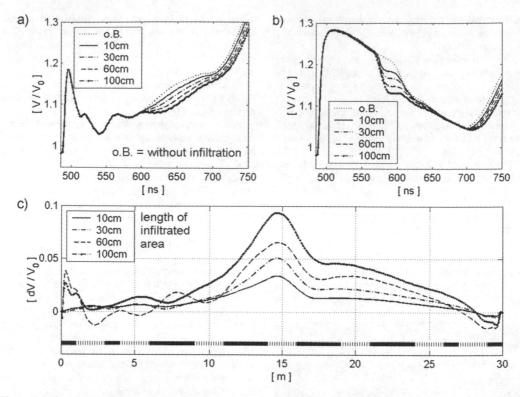

Figure 6.   Time domain reflection measurements from x = 0 m (a) and x = 30 m (b) of a 30 m long flat band sensor cable during an infiltration at x = 15 m (i.e. in the middle of the cable) using different lengths of infiltrated areas (10, 30, 60 and 100 cm). The measurements from (a) and (b) can be mapped to a spatial moisture estimations quite simple (c).

the first time with this spatial resolution. However, it takes several hours on a common desktop PC to reconstruct the water content profile for a cross section. This means that with this reconstruction algorithm real time monitoring is not possible at the moment. But direct evaluation of the measured signals can lead to an improvement by means of localization of water content anomalies (Chapter 4).

4   MONITORING OF DIKES AND DAMS BY MEANS OF TDR

At the Institute of Soil Mechanics and Rock Mechanics, University of Karlsruhe, large scale infiltration tests were carried out on a 30 m long cable sensor, in order to test the suitability of the TDR measuring technique for assessing the water side sealing elements of dikes or dams. For this investigation a sensor was inserted into a sand-filled pipe, which had been slit open at the top and perforated at the bottom. Water was sprinkled over the sensor at different positions and at varying widths. At the same time reflection measurements were carried out

from both ends of the sensor, in order to improve the resolution accuracy. The aim is not to reconstruct the water content along the cable, as in the dike model at the BAW (cf. chapter 3), but to directly evaluate the measuring signals to localize any water content anomalies.

In Figure 6a) and b) the reflection measurements are shown, which are carried out from both ends for infiltration over different lengths of irrigated areas at x = 15 m . As can be seen from the curve in Figure 6a) the area between 500 and 600 ns (corresponds to sensor positons between 0 and 15 m) is "disturbed" by prior infiltrations at x = 2 m (510 ns), x = 5 m (540 ns) and x = 10 m (570 ns). The soil between x = 16 m and x = 30 m retained its original dry condition. As opposed to the measurement without infiltration, the changes in this section in the reflection signals originate only from the infiltrations at x = 15 m. In Figure 6a) these several changes start at 600 ns, however, in Figure 6b) they already start at 570 ns. This difference results from the slower velocity of the signal in a medium with a higher dielectric constant – i.e. with a higher soil water content. An exact localization of the soil water anomaly from a one-sided reflection measurement is thus hardly possible. However, if we use the information from both reflection measurements we have to

- subtract the initial measurements (i.e. without infiltration) from the test measurements,
- reverse the corresponding curve from fig. 6b) and superimpose it with the one from fig. 6a) (by addition of normalized curves with unit travel time) and
- transform these normalized resulting curve to the cable length,

to get a very good estimation of the position of the water content anomaly (cf. Figure 6c). In comparison to a reconstruction of the water content, this estimation can be calculated quickly and without any great effort. Thus it is very well suited as an evaluation algorithm for a monitoring system. This monitoring system will be first used at the river Rhine in a zoned dike with mineral sealing on the water side and a drain at the toe of the land side slope, Scheuermann et al. (2002).

5 MONITORING OF SOIL SATURATION FOR AN IMPROVED FLOOD WARNING SYSTEM

The sustained protection of rivers, their stream banks and their inhabitants in many cases requires environmentally sound flood protection. Especially in catchments up to a few hundred square kilometers severe damage often occurs because high floods are recognized too late. In areas of that size critical flood discharge builds up in only some hours. If these floods are observed with conventional runoff gauges, the remaining advances warning time does not allow effective protection to lower damage.

The flood peaks are mostly formed by fast runoff components. The high dynamic runoff processes of these components take place in the upper soil horizon. The state of the soil is mainly influenced by soil moisture, which decides if a precipitation event leads to critical runoff.

Figure 7 shows the runoff discharge dependency upon the actual soil moisture, observed in the Dürreych catchment ($7 \, km^2$, northern Black Forest, Germany). Therefore a conventional TDR-probe is installed horizontally in the depth of 35 cm. In the observation period of 2 years higher runoff discharge is only observed after exceeding a measurement site typical threshold. As a consequence a distinction between two different catchment states can be made.

The new "Spatial TDR" technique offers the possibility to detect several different saturation states. Hence it serves as base of an innovative flood warning system with spatially distributed continuous observation of soil moisture saturation at representative indicator sites of the catchment.

A soil moisture measurement cluster is therefore realized with 46 twin rod probes. With only a slight

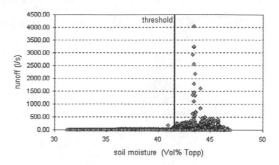

Figure 7. Runoff dependant upon the soil moisture, 15000 measurements, observation period: 2 years.

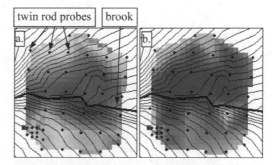

Figure 8. interpolated soil moisture distribution at two different measurement dates registered at 46 twin rod probes. Dark grey: wet, light grey: dry, uncalibrated.

disturbance of the surrounding soil matrix the probes (insulated metallic forks) with a length of 60 cm are installed vertically in the Goldersbach catchment (Tübingen, Germany). As part of a flood warning system the cluster is connected via modem for data update. The cluster site is a shallow floodplain around a small brook. The dominant runoff formation process is saturation excess.

Figure 8 shows the top view of two measurement dates of the cluster. The difference in size of the zone of high saturation is obvious. In both images the zone spreads along the brook but it blurs in the wetter image (b). The capillary fringe is also displayed as transition between the high saturated lower parts and the low saturated upper parts of the cluster. The height is illustrated by contour lines.

This time dependant information of the local saturation state of the cluster must have a representative character for various catchment states. As one of the differentiating monitors of the catchment state, sites with high soil moisture dynamics are allocated using satellite imagery.

Five Landsat-TM satellite images with different antecedent precipitation conditions and different runoff

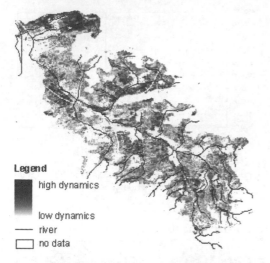

Legend

■ high dynamics

□ low dynamics

—— river

☐ no data

Figure 9.   Soil moisture dynamics derived of Landsat-TM.

reactions are chosen. The six spectral bands of each image are processed by the tasseled cap transformation. This linear combination of all bands with fixed coefficients leads to five different soil moisture states for each pixel. The standard deviation of these five different states is base of an index of the soil moisture dynamic (figure 9).

The inhomogeneity of precipitation events and the different runoff formation processes lead to several soil moisture measurement clusters for an operating flood warning system. The resulting different local saturation states must be transferred to catchment size using regionalisation approaches. In this fashion the catchment state can be determined and combined with the precipitation, thus allowing the comprehensive risk calculation of critical floods.

6   CONCLUSION

Advanced measurement methods in time domain reflectometry are presented. A new transmission line sensor technology has been developed which is based on insulated flat band cables and extends the maximum length of TDR transmission line sensors up to 10 m or more. The electrical parameters of the transmission line sensors have been investigated with numerical field calculations and incorporated into an electrical equivalent circuit. A new algorithm for determining the water content distribution along transmission lines has been developed. It is based on TDR measurements from both sides of the transmission line and an optimization approach for determining the parameters of the telegraph equations. On this way the full information content of TDR data of an embedded sensor can be

analised. For the first time soil moisture measurements are possible with a high resolution both in space and time using this new method (Spatial TDR).

The new reconstruction algorithm has been integrated into a soil moisture measurement system for investigating the water transport processes in a full-scale dike model. For this purpose 12 sensor cables were installed in one cross section of the dike and connected to a TDR and data acquisition equipment. A simulation of a flood event was carried out. The spatial resolution had an accuracy of about 3 cm and the saturation profiles along the cables were determined with an average uncertainty of $\pm 4\%$ compared to independent measurements. Due to the data acquisition time for the TDR data of about 5 min. for the complete cross section of the dike, fast running water transport processes could be monitored for the first time in this spatial resolution.

Future improvements of the measurement system will include an optimized processing of the TDR data and a better soil specific calibration.

REFERENCES

Becker, B., Brandelik, A., Hübner, C., Schädel, W., Scheuermann, A., Schlaeger, S. (2002) "Soil and snow moisture measurement system with subsurface transmission lines for remote sensing and environmental application" URSI Commission-F.

He, S., Romanov, V.G., Ström, S. (1994) "Analysis of the Green function approach to one-dimensional inverse problems. Part II: Simultaneous reconstruction of two parameters" J. Math. Phys. 35(5): 2315–2335.

Scheuermann, A., Schlaeger, S., Hübner, C., Brandelik, A., Brauns, J. (2001) "Monitoring of the spatial soil water distribution on a full-scale dike model" In Kupfer, K. and C. Hübner: Fourth International Conference on Electromagnetic Wave Interaction with Water and Moist Substances, Weimar, Germany, May 13–16, S. 343–350.

Scheuermann, A., Brauns, J. (2002) "Die Durchfeuchtung von Deichen – Modellversuche und Analyse" In: DGGT 12. Donau-Europäische-Konferenz 2002, Proceedings, Passau, 197–200.

Scheuermann, A., Brauns, J., Schlaeger, S., Becker, R., Hübner, C. (2002) "Monitoring von Dämmen und Deichen mittels TDR" In Kupfer, K. et al.: 11. Feuchtetag 2002, Proceedings, Weimar, 187–196.

Schlaeger, S. (2002) "Inversion von TDR-Messungen zur Rekonstruktion räumlich verteilter bodenphysikalischer Parameter" Veröffentlichungen des IBF, Universität Karlsruhe, Heft 156.

Schlaeger, S., Hübner, C., Weber, K. (2001) "Moisture profile determination with TDR" In Cam Ngyen: Subsurface sensing technilogies and applications, Kluwer Academic/ Plenum Publishers, Submitted.

Topp, G.C., Davis, J.L., Annan, A.P. (1980) "Electromagnetic determination of soil water content: Measurements in coaxial transmission lines", Water Resources Research, 16, 579–582.

Fiber optic sensor systems for static and dynamic strain measurements of geotechnical structures

C. Schmidt-Hattenberger, M. Naumann & G. Borm
GeoForschungsZentrum, Potsdam, Germany

ABSTRACT: We assess the feasibility of fiber optic sensors as structurally integrated components for short- and long-term strain monitoring under rough environmental conditions. Geotechnical testing procedures at different scales and in various materials have been carried out with these sensors. Case studies include a static axial borepile test, borehole inclinometer measurements, and uniaxial compression tests at laboratory scale.

1 INTRODUCTION

1.1 General background

For the determination of the stability, safety and lifetime of geotechnical structures, intelligent measurement systems can deliver reliable and precise information on the most relevant quantities such as strain, stress, temperature, and chemical properties. Monitoring needs an optimized technology during the construction process and in long-term controlling during operation.

For several years, fiber optic sensors have been exploited for harsh environmental conditions that generally prevail during geotechnical and civil-engineering monitoring programs. Because of their compact size they can be designed as structurally integrated systems, so-called "smart sensors" (Udd 1995).

Three successful geotechnical applications are described below:

(I) in-situ strain monitoring of a large diameter drilled concrete pile during quasi-static cycling loading
(II) permanent captors instrumentation at inclinometer tubes for borehole strain measurements
(III) fiber Bragg grating strain sensors for laboratory rock mechanical testing.

For all cases, the developed measurement technique is outlined, and experimental results are benchmarked to conventional reference sensors. The advantages of the new technologies and their great potential for the requirements in geoengineering is discussed.

1.2 Fiber Bragg grating technology for strain and temperature sensor systems

Fiber Bragg gratings (FBG) are intrinsic sensor elements written onto optical fibers by an UV laser-inscription process, usually at a wavelength of about 245 nm. The sensing element comprises a periodic modulation of the refractive index n of the fiber core with a spatial frequency Λ. The grating acts as a very selective spectral reflector at the characteristic wavelength λ_B given by

$$\lambda_B = 2 \cdot n \cdot \Lambda \qquad (1)$$

Any change of n or Λ has a proportional response on the reflected wavelength as depicted in Figure 1.

Therefore, in principle, FBG sensors can be used for measurements of strain, temperature, and pressure in static and dynamic regimes. The wavelength shift $\Delta\lambda$ corresponding to a change of strain $\Delta\varepsilon$ is given by

$$\Delta\lambda = \Delta\varepsilon \cdot (1 - \rho_a) \cdot \lambda_B \qquad (2)$$

where $\rho_a = 0.22$ is the photoelastic coefficient of the fiber.

For the employed silica glass fiber, the wavelength-strain sensitivity of a 1550 nm FBG is 1.15 pm/$\mu\varepsilon$, and its temperature sensitivity is 1.3 pm/°C (Kersey 1997). For strain measurements under large temperature variations, an interrogation technique is required which can distinguish between both effects.

For measuring static strain, a fiber Fabry-Perot tunable filter system (FFP-TF) is used as readout unit

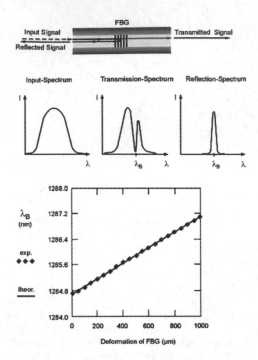

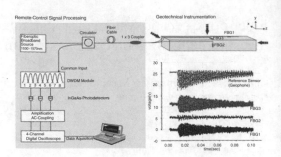

Figure 3. FBG interrogation system for dynamic strain detection, based on Dense-Wavelength-Demultiplexing (DWDM) filter technique.

Figure 1. Top: measurement principle of fiber Bragg grating sensor, bottom: linear relationship between Bragg wavelength shift and deformation of the fiber with intrinsic grating.

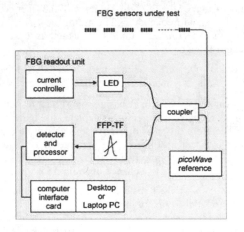

Figure 2. FBG interrogation system for static strain detection.

for the FBG sensors (Fig. 2). This interrogation technique allows the simultaneous detection of several sensors in series, the total number of them depending on the expected range of strain. The system has a temperature-stabilized fixed Fabry-Perot multi-wavelength reference to achieve stability and accuracy

of ± 5 pm which corresponds to a static strain resolution in the order of 10^{-6} mm/mm.

For detecting dynamic strain signals in the frequency band of 10 to 10 000 Hz, a prototype has been developed on the principle of ratiometric intensity modulation (e.g. small-band filters) as depicted in Figure 3 which offers the capability for field applications (Schmidt-Hattenberger 2002, Liu et al. 2002).

2 STRAIN MONITORING OF A LARGE DIAMETER BOREPILE

2.1 Testing conditions

A network of fiber Bragg grating sensors has been installed into a large diameter concrete pile for deformation monitoring during several quasi-static loading cycles. The FBG sensors have been assembled into the steel reinforcement cage of the pile, where they are embedded as structural integrated sensors by cementation (Fig. 4). The sensor setup was designed to meet the rough conditions at the construction site. The technical procedure of the tests was carried out according to the recommendations of the German Society of Geotechniques (DGGT 1998).

Electronic concrete strain gages (CSG) have been installed in the pile for discrete strain registration to verify the fiber optic measurements. These conventional sensors were implemented in separate measurement bars of 1 meter in length, which are also mounted directly at the pile reinforcement cage. The arrangement of the strain gages in pairs or triples and at various measurement locations allows the determination of the deformation distribution.

The tested bore pile of a multistory building has a length of 19 m and a diameter of 1.20 m. The FBG sensors were installed close to the conventional CGS sensors. The FBG sensor elements – all of them having the same central wavelength of about 1550 nm – are embedded into glass fiber reinforced plastic (GRP)

402

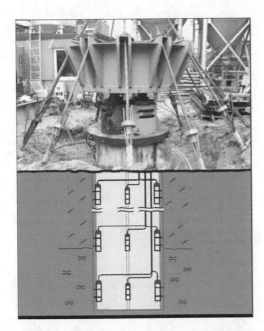

Figure 4. Experimental setup with FBG sensor rods for axial pile load testing at levels 1.50 m, 9.20 m, and 15.50 m below surface. Load crown with hydraulic jacks on top of the pile.

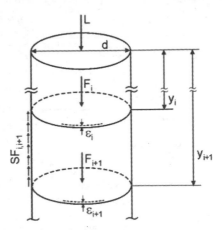

Figure 5. Schematic of the pile geometry.

Table 1. Abbreviations used in the model calculation.

Symbol	Denotation
i $(i = 1, 2, 3)$	Number of measurement cross section (MCS)
ε_i	Deformation at position i
L	Load on the top of pile
d	Diameter of the pile
y_i	Depth of MCS i
σ_i	Axial stress at position i
F_i	Axial force at position i
A	Cross-sectional area (constant)
$SF_{i,i+1}$	Skin friction between MCS$_i$ and MCS$_{i+1}$

rods of 1 m length. This matrix material has already been proven itself in several geotechnical static and dynamic strain measurements (Borm et al. 1998, Schmidt-Hattenberger et al. 1999, Nellen et al. 1999).

The testing procedure was carried out according to the following scheme: starting from a pre-loaded stage of 500 kN, subsequent loading increments of approximately 1/8 of the desired ultimate load of 10 MN were applied. These increments have been executed stepwise after the time-dependent settlement had ceased below a certain threshold. The ultimate load during the test was twice the intended working load of the structure.

Skin friction and pile tip resistance depending on load and depth have to be taken into consideration for a more sophisticated analysis of the pile behavior. The skin friction can be estimated using a linear model based on Hooke's law under specific simplifying assumptions:

$$SF_{i,i+1} = \frac{L \cdot (\varepsilon_i - \varepsilon_{i+1})}{\pi \cdot d \cdot \varepsilon_1 \cdot (y_{i+1} - y_i)} \qquad (3)$$

where abbreviations and symbols are given in Figure 5 and Table 1.

The results in Figure 6 demonstrate the excellent agreement between both measurement systems. For the deepest level (MCS3), only very small strains

occur, while for the uppermost level (MCS1) a final compression of about 0.035% was recorded. This observation leads to the conclusion of a proper pile working due to sufficient load decrease by skin friction (Fig. 7). The minor deviation between both measurement systems (FBG vs. CSG) is within the variance of the three individual sensors at each level.

3 PERMANENT CAPTORS FOR DOWNHOLE MONITORING

3.1 Concept of instrumentation

A new measurement technique based on FBG sensors has been proposed for long-term downhole monitoring of deformation occurring e.g. in active tectonic fault zones. For such applications, one can utilize the advantages of fiber optic sensing systems, such as optimal integration of sensor elements, high data-rates on the fiber optic links and long life time under corrosive conditions. Generally, the downhole strain measurement

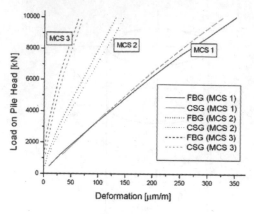

Figure 6. Static load-deformation behavior of concrete pile after ceasing of time-dependent settlements, averaged for each measurement cross section.

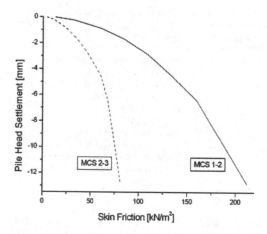

Figure 7. Plot of skin friction versus pile top settlement according to equation (3).

setup consists of a long optical fiber cable with an array of highly sensitive intrinsic FBG elements. The optical fiber string is guided along the entire casing. It is protected by a buffer coating outside its sensitive area. The casing segments are connected by coupling elements. This cascade design allows a proper arrangement of the sensors along the fault (Fig. 8). Simultaneously to the fiber optic measurements, the casing deflection caused by the deformation of the fault will be observed by an inclinometer probe as reference instrument. Essential criteria for the borehole application of the sensor elements of the newly developed fiber optic strain cable are: reliability, reproducibility, packaging, and fixing on the Acrylonitril-Butadiene-Styrole-(ABS) casing tubes. Two optical fiber cables at 0° and 180° azimuthal position of the casing cross section consist of a sensing part

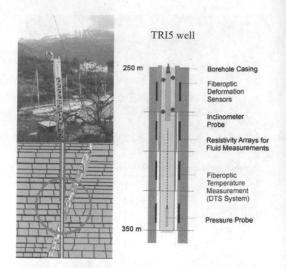

Figure 8. Fiber Bragg gratings as permanent captors on a borehole casing for long-term strain detection at Trizonia island, Greece.

covering the critical monitoring interval of about 100 m and a transmission line part to acquire data from the surface.

Over the 100 m intersection with the fault, a FBG sensor is placed in the middle of each 3 m-tube section. These individual fiber lines are fused together with crimp-splices. Each FBG element has been embedded in special protection material possessing an elastic behavior comparable to the ABS tubes. The sensor properties have been optimized to achieve a linear hysteresis-free performance.

3.2 Field test

A first field experiment has been carried out to evaluate the newly developed technology. Over a length of 15 m, casing tubes were cascaded along the outer wall of an office building to simulate the vertical installation conditions of a borehole (Fig. 9).

The installation tools were tested together with two prototypes of the fiber optic strain cable. Each meter a cantilever with metallic bracket was used to simulate a distinct deflection at this position. In the experiment, several defined deflection profiles have been applied along the casing. Simultaneously, the position of each cantilever element was determined by a conventional measurement system (tachymetry). A typical "log" of the fiber Bragg grating strain cable for a selected deflection profile of the casing is drawn in Figure 10. The curvature of the cascade was calculated from the recorded tachymeter data and compared with the results of the FBG system.

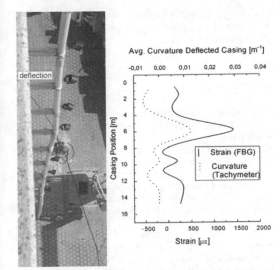

Figure 9. Experimental setup for simulating downhole installation conditions of the fiber Bragg grating strain cable prototype installed along 15 m casing (left side). Strain profile measured by fiber optic permanent captors in comparison with tachymeter (right side).

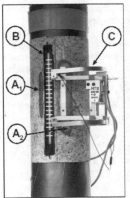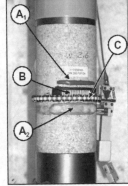

Figure 10. Sensor instrumentation A1: epoxy-glued FBG sensor, A2: salol-bonded FBG sensor, B: barcode for Laserextensometer measurement, C: mechanical extensometer for axial strain measurements (left side) and radial strain measurements (right side).

4 STRAIN MEASUREMENT TECHNIQUES FOR LABORATORY ROCK MECHANICAL TESTING

4.1 Strain measurement in uniaxial compression tests

Fiber Bragg gratings (FBG) can be used as strain transducers for strain measurements in uniaxial compression (UC) tests. The elastic Young's modulus and Poisson's ratio can be derived from the measured strain and from the applied loads using Hooke's law for linear elastic solids.

A systematic comparison of the optical fiber Bragg grating (FBG) sensors against conventional strain gages has been conducted. The results cover mechanical extensometers based on cross-flexure strain gages, surface mounted FBG sensors, and a non-contacting laser extensometer measuring system as additional optical reference. The test equipment consists of a rock mechanical testing load frame with a load cell capacity up to 2600 kN. The experimental setup and measurement conditions are in agreement with the standards of UC tests on rock (Fairhurst & Hudson, 1999). In the tests, the maximum applied stress (115 MPa) corresponds to about 70% of the uniaxial compressive strength of the material.

In Figure 10, the complete instrumentation is shown. In the immediate proximity of the mechanical extensometers (C), the FBGs are mounted on the surface of the sample (diameter: 75 mm) by two different types of bonding materials. First, a two-component epoxy resin for optimum adhesion was used (A1). The second one (A2) was Phenylsalicylate (Salol) with crystalline consistence at room temperature and a melting point of about 40°C (Kamioka 1993). For the FBG elements, the Salol bonding provides the great advantage of detachability by applying a heat source and reutilize them in further experiments.

4.2 Results

The FBG sensor elements mounted on the surface of the sample can well compete with mechanical extensometers with respect to accuracy, preparation, and handling. In the case of compressional strain the use of Salol as fixing agent cannot be recommended due to the larger differences to the benchmark. This effect could be caused by buckling effects of the fiber under compressive strains. Adequately pre-stressed sensors can help to overcome this difficulty.

In the case of radial strain measurements, when extensional circumferential strains are present under the loading conditions, we found that Salol performs as good as conventional fixing. The measured absolute strain values were larger than for the mechanical extensometer system. It is assumed that a transversal strain is transmitted by the bonding material in addition to the extensional strain caused by the Poisson effect due to the compression acting perpendicularly to the fully bonded FBG. To reduce this interfering effect, the FBG element was applied over a range of about 3 cm on the sample surface without fixing, and bonded on both sides with the fixing agent. This technique avoids squeezing of the glue matrix as in the

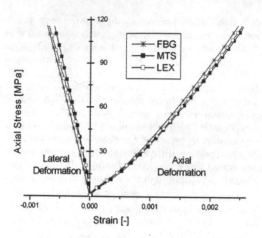

Figure 11. Comparison of strain measurements obtained from the various measurement systems.

case of completely fixed FBG sensors and proved to be applicable (Fig. 11).

5 CONCLUSIONS AND OUTLOOK

FBG sensors integrated in various objects and structures can be used for short- and long-term deformation monitoring. The experimental setup was developed in a robust and easy installable laboratory or field test system, which fits the application requirements. Very good results have been obtained for the measured strain values by the FBG sensors in comparison to conventional reference systems. A further development is directed to attempt the simultaneous exploitation of the FBG sensor networks for both static and dynamic tests by means of a modular experimental setup with combined interrogation systems.

REFERENCES

Borm, G. & Schmidt-Hattenberger, C. 1998. Anchoring Device with Strain Measurement System, *patent* WO 98/19044.

German Society for Geotechnical Engineering (DGGT), 1998. Recommendations for static and dynamic pile tests, Technical University Braunschweig, Institute of Foundation Engineering & Soil Mechanics, 1–32.

Udd, E. 1995. *Fiber Optic Smart Structures*, New York: Wiley.

Kamioka, H. 1993. Elastic behaviour of salol during melting and solidification processes, *Japanese Journal of Applied Physics*, vol. 32: 2216–2219.

Kersey, A. D. 1997. Fiber grating sensors, *Journal of Lightwave Technology*, vol. 15: 1442–1463.

Liu, J.-G. et al. 2002. Dynamic strain measurement with a fibre Bragg grating sensor system, *Measurement*, vol. 15: 151–161.

Fairhurst, C. E. & Hudson, J. A. 1999. Draft ISRM suggested method for the complete stress-strain curve for intact rock in uniaxial compression, *International Journal of Rock Mechanics & Mining Science,* vol. 36: 279–289.

Nellen, Ph. M. et al. 1999. Structurally embedded fiber Bragg gratings: civil engineering applications, In M. M. Marcus & B. Culshaw (eds.), *Proc. SPIE Fiber Optic Sensor Technology and Applications*, vol. 3860: 44–53.

Schmidt-Hattenberger, C. et al. 1999. Bragg grating seismic monitoring system, In M. M. Marcus & B. Culshaw (eds.), *Proc. SPIE Fiber Optic Sensor Technology and Applications*, vol. 3860: 417–424.

Schmidt-Hattenberger, C. 2002. Wavelength detection with fiber Bragg grating sensors, *patent* 9826784.2.

Laboratory swell tests on overconsolidated clay and diagenetic solidified clay rocks

P.-A. von Wolffersdorff & S. Fritzsche
Baugrund dresden, Ingenieurgesellschaft mbH, Germany

ABSTRACT: In order to the expected swell heaves due to excavation of the cuttings at the high-speed railway Nürnberg–Ingolstadt extensive swell tests on diagentic solidified clay rocks with different weathering grades and on overconsolidated clays were performed. The tests, carried out in different devices and under different loading conditions, are described. The results are shown, assessed and briefly summarized. The determined final swell heaves and their accompanying time–heave plots are described and analysed on the basis of an one-dimensional swelling law. The advantages and disadvantages of the different test types are shown in order to provide realistic swell behaviour in situ.

1 INTRODUCTION

The high-speed railway from Nürnberg to Ingolstadt passes from north to south the Nürnberg Depression Area, the northern Alp Foreland, the Frankish Alp and the Ingolstadt Basin. Because of the considerable relief varieties and high requirements along the alignment, numerous embankments, cuttings, bridges and tunnels are necessary, at which complicated geological and hydrogeological conditions must be accounted.

The alignment traverses regional in the northern Alp Foreland diagenetic solidified clay rock and in the Frankish Alp respectively in the Ingolstadt Basin sedimentations of tertiary, predominantly overconsolidated clays. Because of the potential swellability of these rocks and soils, swell heaves occur in the cutting subgrades due to excavation.

The high-speed railway is constructed as a slab track system, hence the tolerable deformation during the rail traffic is very low. Swell heaves after the installation of the concrete track must not exceed the correction value for the adjustability of the fastening system.

For this reason it is necessary to predict reliable swell heaves to control the swelling for the construction of the high-speed railway (von Wolffersdorff et al. 2002). These predictions of swell heaves include the final swell heaves and the time–heave plots. Both are founded on laboratory tests and on swell heave measurements in situ.

2 SWELL BEHAVIOUR

2.1 *Tertiary clay*

Swell heaves caused by excavation in tertiary clays and diagenetic solidified clay rocks originate mainly from osmotic and mechanical swell processes (Fritzsche 2002). The swellability of the tertiary clay depends not only on the composition of the clay minerals, but also on the loading history, which usually results in an overconsolidation.

2.2 *Diagenetic solidified clay rocks*

Three different kinds of diagenetic solidified clay rock with distinctive swellability were found: Feuerletten, Opalinuston and Amaltheenton. The diagenesis in the clay rocks results, in difference to tertiary clays, in a solidification which counteracts to the swellability. The diagentic consolidation decreases with increasing weathering. A clear relationship between diagenesis, weathering grade and swellability could not be proved yet. For the swell characteristics of solidified clay rocks the following relationships were assumed at first.

– Higher weathering grade causes in higher swell values due to unloading.
– Higher weathering grade causes in lower swell potential, i.e. lower swell pressures at completely restrained volumetric expansion.

– Higher weathering grade causes in faster swell processes.

In the framework of the geological modelling the following 4 homogeneous areas with different swell characteristics have been defined depending on weathering grades:

– weathering grade w_2 and lower weathered
– weathering grade w_2–w_3
– weathering grade w_3
– weathering grade w_3–w_4 and stronger weathered.

3 LABORATORY TESTS

3.1 Type of swell tests

For the investigation of the swelling in the cuttings all together 169 tests in 4 different laboratories has been carried out with the diagenetic solidified clay rocks and with the tertiary clay. With regard to the loading regime the tests differ as follows:

– combined swell pressure–swell heave tests,
– multi-stepped swell heave tests,
– Huder Amberg swell tests.

Table 1 gives an outline of all carried out swell tests ordered by the kind of soil and rock, laboratories and loading regimes.

In the following sections the test procedures and the loading regimes respectively the stress–strain curves are described.

3.2 Experimental procedures

Usually the swell behaviour is investigated with one-axial deformation tests (oedometric tests). The

Table 1. Swell tests.

	Feuerletten (kmF)	Amaltheen clay (lv)	Opalinus clay (al1)	Tertiary clay (tt, tk)
SP–SH test[1]				
TU Karlsruhe	17	13	6	22
Swell heave test				
GH Kassel	20	7	4	
Baugrund Dresden		3		7
TU München				7
Huder Amberg test				
GH Kassel	3	3		
Baugrund Dresden	19	6	15	
TU München				17

[1] Combined swell pressure–swell heave test.

predictions of swell heaves, founded on an one-dimensional swelling law (see Section 4) are supported mainly by one-axial deformation tests.

A test is called swell heave test, if the axial swell expansion ε_z^q is determined by a predetermined axial compressive stress σ_z (see Fig. 1). It is force controlled and can be carried out with conventional devices (oedometer).

In contrast a test is called swell pressure test, if the maximum stress σ_{z0} is determined by a fixed boundary in axial direction (see Fig. 2). It requires a deformation control system, which is generally not available in conventional devices.

Experimental conditions of the combined swell pressure–swell heave tests are much more complicated. Therefore an electronic control system is required to predetermine both, axial forces and axial displacements.

3.3 Sample preparation and stress–strain curves

The swell characteristics in situ of diagenetic solidified clay rock and predominantly overconsolidated clays are significantly different from these, which are ascertained in laboratory tests on prepared sampling material. Therefore it is necessary to place in the samples widely undisturbed, before the swelling process starts. The origin material has not to be exposed to a loading as possible. Otherwise the diagentic bonds will be destroyed or the overconsolidated stresses will be changed.

The combined swell pressure–swell heave test is characterized by a fixed boundary in axial direction after the widely undisturbed samples is placed in.

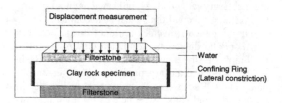

Figure 1. Force-controlled swell heave test.

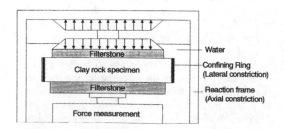

Figure 2. Deformation-controlled swell pressure test.

Subsequently water is admitted and the swelling starts, i.e. the compressive stress σ_z increases until the swelling pressure σ_{z0} is reached. This increasing of stress describes a horizontal line in a stress–strain chart (see Fig. 3).

During the following heaving stage, the axial stress is reduced to a defined value, whereas a spontaneous heave occurs. Afterwards the stress is kept constantly and a swell heave develops. The swelling process is observed until the swell heave fades widely away, thus the final swell heave is reached. The stress–strain curve for this stage has the shape of a step (increasing curve with decreasing stress and subsequently a vertical line). The heaving stage of the combined swell pressure–swell heave test, shown in Figure 3, includes those 5 steps.

In Figures 3–5 it is obvious that in all 3 kinds of tests the 5 stepped heaving stages are the result of the same experimental procedure. The experimental procedure of the swell heave test, the Huder Amberg swell test and the combined swell pressure–swell heave test differ only in the initial state before the swelling starts.

Before admitting water and heaving start, the conventional swell heave test is characterized by an one-axial compression (oedometric loading) until a defined compressive stress is reached. This compressive stress is assumed empirical and should approximate the swelling pressure σ_{z0}.

The Huder Amberg test differs only from the conventional swell heave test in a complete loading–unloading–reloading cycle before the swelling starts.

Both the conventional swell heave test and the Huder Amberg swell test have the following disadvantages:

– Besides the inevitable structural changes as the result of sampling and placement the specimen gets additional disturbed by the preloading to a more than less arbitrarily selected compressive stress.
– The swelling pressure σ_{z0}, which compensates the swell heave completely, cannot be experimentally determined.

The analysis of the swell tests is based on an one-dimensional swelling law, which is briefly described in the following section.

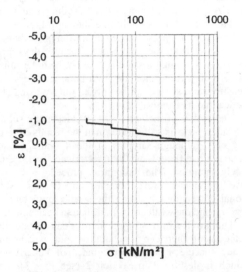

Figure 3. Stress–strain curve of the combined swell pressure–swell heave test (several steps of heaves).

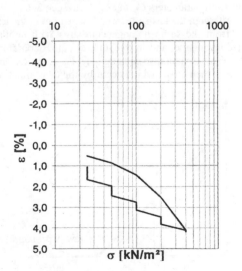

Figure 4. Stress–strain curve of the swell heave test (several steps of heaves).

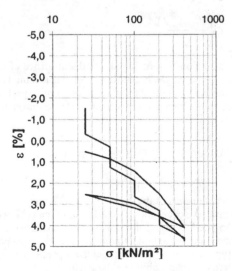

Figure 5. Stress–strain curve of the Huder Amberg test (several steps of heaves).

4 1D SWELLING MODEL

4.1 Final swell heave

The logarithmic approach according to Grob is used here for the presented one-dimensional swelling model (Grob 1972). It is shown in Figure 6.

The stress σ_{z0} of Figure 6 is a maximum value, which over presses the swelling completely. The stress σ_c is a minimum value. The swelling does not increase for lower compressive stresses than σ_c. The swell heave coefficient C_b describes the dependence between the swelling and the vertical stress σ_z. The both parameters C_b and σ_{z0} of the swelling model are determined by laboratory tests.

4.2 Time plot of the swell expansion

Kiehl proposed a time-depended extension of the swelling model that is shown in Figure 6 (Kiehl 1990). With assumption of constant unloading stress values the following equation can be written.

$$\varepsilon_z^q(\hat{t}) = -C_b \cdot \ln\left(\frac{\sigma_{z\infty}}{\sigma_{z0}}\right)\left[1 - \exp\left(-\frac{\hat{t}}{\eta_q}\right)\right] \tag{1}$$

The parameter η_q describes a time reference unit and is determined by using the time–heave curves of the swell tests.

In Equation (1) the new variable $\hat{t}$ is introduced, which defines a modified time to obtain time-plots independent of the layer thickness. The following power approach describes the relationship between the modified time, the real time and the thickness of swelling layer as well as of the sample,

$$\hat{t} = t \cdot \left(\frac{d_{\text{specimen}}}{D_{\text{layer}}}\right)^n \tag{2}$$

The exponent n of (2) adjusts the influence of the thickness dependence on time-swell behaviour. This exponent is calibrated by an approximation to preliminary heave measurements in situ, e.g. extensometer measurements.

5 ANALYSIS OF THE TEST DATA

5.1 Purpose of the analysis

Both parameters of the swelling model C_b and σ_{z0} and the time reference unit η_q according to (1) and (2) were determined by using the above named extensive test series. Various calibration procedures were developed and applied to determine representative parameters and to verify the applicability of the swelling model.

Furthermore the mentioned relationships between weathering grade and swellability were quantified

and checked by using the test series on the diagenetic solidified clay rocks.

5.2 Final swell heave

A few basic results are selected from the extensive analysis of the test series and are presented in suitable charts. Figure 7 shows the results of a series of conventional swell heave tests on tertiary clay in an ε–log σ chart.

In Figure 7 is shown, that the test results scatter significantly. But the stress dependence of the swell heaves relating to every individual test is described very well by means of the above named swelling model.

The swell-strain straight line shown in Figure 7 is the result of a representative calibration procedure of C_b and σ_{z0} from the test results. Because the representative swelling pressure σ_{z0} cannot be determined directly, the calibration for both parameters is very sensitive and should not be carried out by using statistical methods only.

The scatter of the test results of diagenetic solidified clay rocks is still greater than the results of tertiary clay, although samples of the 4 homogeneous areas with different swell characteristics were separately analysed. The calibration procedures of the swell parameters were more difficult because additional correlations between the swell parameters C_b and σ_{z0} and the weathering grade had to be accounted.

The swell-strain straight lines demonstrate the relationship between swellability and weathering grade for the 4 weathering grades of Feuerletten, which is described in Section 2 (see Fig. 8). The represented 4 swell strain straight-lines are the results of the predominantly empiric calibration procedure of the swell parameters C_b and σ_{z0} (Fritzsche 2002).

Although the test series were carried out on material from the same homogeneous area with similar loading regime, and in one and the same laboratory the results scatter also considerably. These scatters are greater than these, which occur by the determination

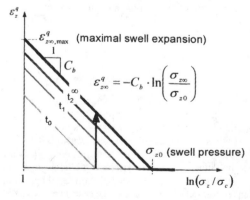

Figure 6. 1D swelling model.

of shear strength parameters φ and c using a comparable data basis.

5.3 Time–heave plots

The time–heave plots of the swell tests are analysed both to determine the reference time unit η_q and to separate the swell heaves from other heave parts.

Figure 9 shows the time–heave plot and the time–swell heave plot of a conventional swell heave test with 5 steps. The separation of the swell heave from the total heaves is more than less arbitrarily.

The time at which spontaneous heaves due to unloading are faded away depends on the material structure and the previous loading history. For instance a pore water pressure balance can occur in terms of a reverse consolidation after the maximum compressive stress is applied. Thus arising heaves occur already within a very short time.

These effects were considered in Figure 9 by the separation of the time–heave plot in a swell part and a further heave part.

The magnitude of the spontaneous heaves due to unloading can also be estimated on the basis of the unloading in oedometer tests on the same material. Figure 10 shows the stress–swell strain curve of a 5-stepped swell heave test and the unloading of an oedometer test. The determined slopes of the spontaneous heave steps should agree with the tendency of the unloading curve of a conventional consolidation test.

6 INFLUENCE OF THE TEST TYPE ON TEST RESULTS

6.1 Diagenetic solidified clay rocks

Swell tests with different loading regimes were carried out on same material to investigate the influence of the loading regime on the swell value before the swell process starts. An analysis is exemplarily given for the Feuerletten. On sampling material of the

homogeneous area "weathering grade w3" combined swell pressure–swell heave tests, conventional multi-stepped swell heave tests and Huder Amberg swell tests were carried out. Figure 11 shows the results of the 3 test series and the accompanying representative swell-heave straight lines. The outcome is a strong dependence between swellability and test type. Hence the following qualitative relationship results.

– Combined swell pressure–swell heave tests
⇒ low swellability (low C_b and σ_{z0} values),
– Conventional swell heave tests
⇒ middle swellability (middle C_b and σ_{z0} values),
– Huder Amberg swell tests
⇒ high swellability (high C_b and σ_{z0} values).

The results of the actual heave predictions and heave measurements in situ prove that the calculated swell heaves on the basis of the swell parameters C_b and σ_{z0}, determined by combined swell pressure–swell heave tests, are close to reality. Against it, the swellability determined on the basis of the other two test types is too high in order to specify the swell behaviour of diagenetic solidified clay rocks.

Conventional swell heave tests and Huder Amberg swell tests are not suitable to assess the swell behav-

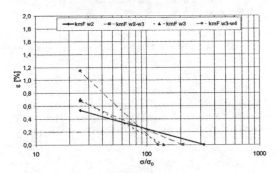

Figure 8. Swell-strain straight-lines for the different weathering grades of Feuerletten as the result of a representative analysis of the swell parameter C_b and σ_{z0}.

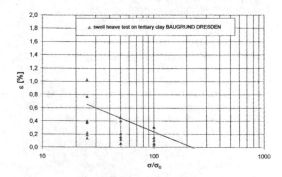

Figure 7. Representative swell-strain straight-line as the result of swell heave tests on tertiary clay.

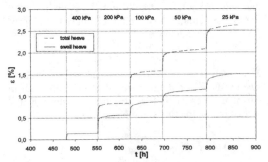

Figure 9. Time–heave plot and time–swell heave plot of a conventional 5-stepped swell heave test.

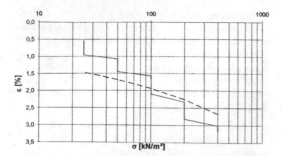

Figure 10. Stress–swell strain curve of a 5-stepped swell heave test and stress–strain curve of a conventional oedometer test.

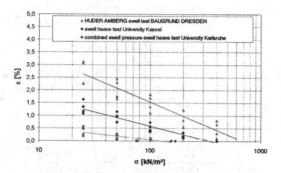

Figure 11. Results of 3 different test series on Feuerletten and corresponding representative swell-strain straight-lines.

iour in situ of diagenetic solidified clay rocks under laboratory conditions. Due to the pre-loading before swelling starts the diagenetic bonds were destroyed, even if widely undisturbed samples were placed in.

6.2 Tertiary clay

All 3 test types were also carried out on tertiary clay. Figure 12 shows the results of the combined swell pressure–swell heave tests and the results of the conventional swell heave tests as well as the accompanying swell-strain straight lines. It is obvious that the experimental determined swellability of the tertiary clays depends on the test type as well as the diagenetic solidified clay rocks do.

The swellability of tertiary clay determined with conventional swell heave tests and Huder Amberg swell tests does not agree with the swell behaviour in situ at all, because the clay structure gets strongly disturbed by the loading regime.

7 CONCLUSIONS

This paper shows that combined swell pressure–swell heave tests are more suitable to describe the in situ

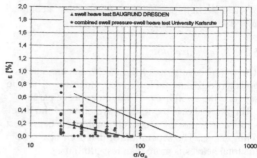

Figure 12. Results of 2 different test series on tertiary clay and corresponding representative swell-strain straight-lines.

swell behaviour of diagenetic solidified clay rocks and tertiary clay under laboratory conditions than the conventional swell heave test and the Huder Amberg swell test.

It is recommended to force the technical development of combined swell pressure–swell heave tests and emphasize the advantages in according technical standards for increased use.

REFERENCES

Fritzsche, S. 2002. Untersuchung und bautechnische Beherrschung des Quellverhaltens von Tonsteinen am Beispiel der Einschnitte der NBS Nürnberg – Ingolstadt. *Diplomarbeit am Institut für Geotechnik der Technischen Universität Bergakademie Freiberg.*

Grob, H. 1972. Schwelldruck im Belchentunnel. *Proc. Int. Symp. für Untertagebau, Luzern*, 99–119.

Huder, J. & Amberg, G. 1975. Quellung in Mergel, Opalinuston und Anhydrit. *Schweizerische Bauzeitung* 83: 975–980.

Kiel, J.R. 1990. Ein dreidimensionales Quellgesetz und seine Anwendung auf den Felshohlraumbau. *Sonderheft der Zeitschrift Geotechnik, Vorträge zum 9. Nationalen Felsmechanik Symposium.*

Paul, A. 1986. Empfehlung Nr.11 des Arbeitskreises 19 – Versuchstechnik Fels – der Deutschen Gesellschaft für Erdund Grundbau e.V., Quellversuche an Gesteinsproben. *Bautechnik* H63(3): 100–104.

Pimentel, E. 1996. Quellverhalten von diagenetisch verfestigten Tonstein. *Veröffentlichung des Inst. für Bodenmechanik Universität Karlsruhe, Heft*, 139.

von Wolffersdorff, P.-A., Hempel, M. & Raithel, M. 2002. Bau einer Hochgeschwindigkeitsstrecke auf quellfähigen Untergrund. *Proc. 12. Donau-Europäische Konferenz, Passau*, 407–410.

Wittke-Gattermann, P. 1998. Verfahren zur Berechnung von Tunnels in quellfähigen Gebirge und Kalibrierung an einem Versuchsbauwerk. *WBI Geotechnik in Forschung und Praxis.* Verlag Glückauf: Essen.

Geotechnical Measurements and Modelling, Natau, Fecker & Pimentel (eds)
© 2003 Swets & Zeitlinger, Lisse, ISBN 90 5809 603 3

Damage evolution during uniaxial compressive tests evaluated by acoustic emission monitoring

J. Wassermann, D. Amitrano, G. Senfaute & F. Homand
LaEGO-INPL-INERIS, Nancy, France

ABSTRACT: Experiments of fracturation have been performed under uniaxial compression tests on three arenite samples from iron mine of Tressange (Lorraine, France). In addition to usual mechanical parameters such as strains and axial stress, acoustical parameters such as P-wave velocity, maximum signal amplitude received (a kind of measure of the attenuation) and AE activity, have been investigated. These two independent types of measurements involve coherent results suggesting a progressive anisotropic damage leading to the failure.

1 INTRODUCTION

Damage in rocks under stress is related to cracks nucleation and propagation. Acoustic emissions (AE) are transient elastic waves that are primarily generated by quick microcracks growth (Lockner 1993). Hence the analysis of AE produced during mechanical tests is a non-destructive tool for the study of the rock damage evolution. Microcracks nucleation and growth as they generate new voids, affect elastic waves propagation by changing their velocity and their amplitude (Chow et al. 1995, Yukutake 1997, and Zang et al. 2000). Thus, waves velocity analysis is a complementary tool for the study of the damage state in rocks.

In order to analyse the damage evolution of pillars in Lorraine iron mines, a small scale approach has been carried out through laboratory investigations. Uniaxial compressive tests were performed on argillaceous arenite, micro-arenite and ferriferous arenite from mine of Tressange (Lorraine, France). Uniaxial stress state is considered as a good model of stress conditions existing within a mine pillar.

In this paper we described experiments of rock deformation under uniaxial stress up to the failure on arenite samples. The applied stress and the strains in central part of the sample have been measured. AE activity has been registered, and P-wave velocity in three directions and maximum amplitude of the signal received, have been determined during the tests. We have searched for relations between mechanical parameters such as permanent strain, dilatancy, Young's Modulus and lateral strain modulus, and acoustical parameters previously described. In the first place, in order to evaluate the capability of this two types of

parameters to provide a better understanding of rock damage process under uniaxial conditions.

2 EXPERIMENTAL PROCEDURE

2.1 Rocks tested and samples instrumentation

The tests were performed on ferriferous arenite (S1), micro-arenite (S2), and argillaceous arenite (S3) samples from the same borehole in iron mine of Tressange (Lorraine, France). The samples were cut into cylinders of 140 mm in length and 70 mm in diameter.

In order to measure the local strains in the central part of the sample, six strain gauges (20 mm in length) were cemented to the sample surface, three in axial direction and three in transversal direction (Fig. 1).

Eleven piezoelectric transducers (PZT from Physical Acoustic Corporation, PAC) were coupled to the surface of the sample (Fig. 1). They are characterized by a wide spectral band-pass (from 100 kHz to 1 MHz) and by a major resonant frequency at 300 kHz. Three PZT functioned as transmitters, and were dedicated to P-wave velocity measurements. Eight worked as receivers of acoustic signals and constituted a mini-seismic network.

2.2 Experimental devices

Uniaxial compressive tests have been performed with an hydraulical press of 1000 kN capacity. The press is enslaved by an hydraulical pump controlled by a computer. A pressure sensor between the pump and the press piston allows to calculate the load during the test.

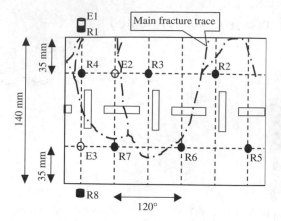

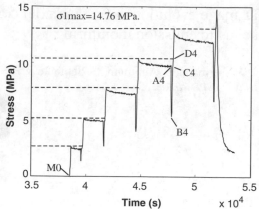

Figure 1. Samples instrumentation schema with 6 strain gauges (rectangles in the central part) and eleven PZT transducers. Length of the sample is 140 mm, its diameter is 70 mm. Transducers E1, E2, and E3 are transmitters. R1 to R8 transducers are receivers. Sensors E1, R1, and R8 are located on the ends of the sample. The dashed-dotted line is the main fracture trace observed after the test performed on the ferriferous arenite (S1).

Figure 2. Loading history with 5 loading–relaxation–unloading cycles for the test performed on the ferriferous arenite (S1). Velocity measurements (in 3 directions) have been performed at points M0, Ai, Bi, Ci, and Di (i = 1 to 5). Ai and Ci correspond to the same stress value. Strength is about 14.76 MPa.

Two LVDT (Linear Variable Differential Transformers) located between the press platens measure the displacements during the sample deformation.

The acquisition system of mechanical data is composed by two distinct units. One is dedicated to the strains measurements, the other controls the press via the valve of the hydraulical pump and registers the load and the displacements. The measure precision for the strains is about 5 μm/m and 1% for the displacements and load.

The acquisition system of the acoustical data consists of two chains. The first includes the receivers connected to 40 dB preamplifiers (PAC) with 50 kHz – 1.8 MHz spectral band-pass, the Analogic/Digital card (Engineering Seismology Group Canada Inc., Hyperion system) which digitizes the signals after preamplification at a rate of 10 MHz and with 14-bit vertical resolution along an amplitude interval of ±2.5 V for each channel. This chain performs also the AE counting. The second acoustical chain allows P-wave propagation velocity measurements, it consists of a pulser (PAC) connected to the three transmitters via a switching box. The pulser generates a negative pulse with maximum amplitude about −360 V and duration inferior to 10 ns. The acoustical data are stored on the hard drive of a computer.

2.3 *Measurements procedure*

Uniaxial tests have been conducted at a $10^{-5} s^{-1}$ constant strain rate. Loading has consisted on succesive

loading–relaxation–unloading cycles with increasing axial load until failure (Fig. 2). The relaxation phase allows to dissipate delayed elastic strain and so to determine properly a Young's Modulus during the following unloading phase with reduced hysteresis.

After the determination of the noise maximum amplitude, the signal recording trigger has been set to 50 mV for each channel.

P-wave propagation velocity measurements have been performed in three directions, one axial and two transversals at 4 points of each cycle (Figs 1, 2). The P-wave arrival times have been manually picked on the signal traces digitized by the A/D card.

The recording of the AE activity was performed during the load excepted while P-wave measurements were realized.

3 EXPERIMENTAL RESULTS

Here, we present detailed experimental results from the uniaxial test performed on the ferriferous arenite sample (S1). The results from the two others tests realized on micro-arenite (S2) and on argillaceous arenite (S3) are presented in Table 1 (for the mechanical results) and in Table 2 (for the acoustical results).

3.1 *Mechanical results*

Figure 3 shows the axial stress plotted against lateral (ε_L), axial (ε_A) and volumetric (ε_V) strains. The envelopes of the $\sigma(\varepsilon_L)$ and $\sigma(\varepsilon_V)$ curves become non-linear

Table 1. Mechanical results for ferriferous arenite (S1), micro-arenite (S2) and argillaceous arenite (S3). E is the Young's Modulus and M_L the lateral strain modulus. S1, S2 and S3 strengths are respectively 14.76 MPa, 54.32 MPa and 22.81 MPa. M_L for the last cycle of the S3 test has been calculated within a shorter stress domain (10.54–16.87 MPa) than it is indicated here.

Sample	Stress domain (MPa)	E (MPa)	M_L (MPa)
S1	1.31–2.41	10183 ± 592	69940 ± 3231
	2.63–4.78	11374 ± 901	60196 ± 4637
	3.93–7.18	12132 ± 455	55303 ± 3215
	5.29–9.78	13432 ± 645	55480 ± 4963
	6.62–11.76	13146 ± 397	45450 ± 3688
S2	2.68–5.10	29163 ± 895	180462 ± 2008
	5.31–10.09	29106 ± 658	168954 ± 686
	7.97–15.09	30279 ± 773	163933 ± 2484
	10.55–20.19	30774 ± 505	158948 ± 1215
	13.23–25.34	31950 ± 572	155556 ± 1737
	16.01–30.52	31651 ± 587	149394 ± 2138
	19.85–37.74	31565 ± 530	140942 ± 2235
S3	2.69–4.52	20509 ± 668	164433 ± 2483
	3.99–6.98	21650 ± 630	161351 ± 2646
	5.29–9.32	21570 ± 561	148914 ± 2904
	6.62–11.87	22186 ± 401	145020 ± 2101
	9.21–16.26	22645 ± 391	131075 ± 1793
	10.54–18.01	23054 ± 229	93962 ± 1557

as soon as dilatancy appears, defined as the beginning of the nonlinear part of the axial stress–volumetric strain curve. It is not the case for $\sigma(\varepsilon_A)$ curve which remains linear until larger stress value. Permanent strain

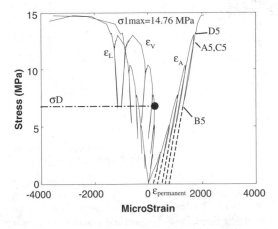

Figure 3. Stress–strain curves. ε_A, ε_V and ε_L are respectively the axial, volumetric and lateral strains. σD is the stress level corresponding to the onset of dilatancy. A5, B5, C5, and D5 correspond to the velocity measurements performed during the fifth cycle.

Table 2. Acoustical results for ferriferous arenite (S1), micro-arenite (S2) and argillaceous arenite (S3). $V_P(A)$ is the P-wave velocity at point A, $A_m(C)$ is the signal maximum amplitude at point C. Axial, Lat.1 and Lat.2, are the acoustical parameters respectively in axial, lateral (or transversal) directions at the top (ray E2R2, fig. 1) and at the bottom (ray E3R6) of the sample. When $A_m = 2.499$ V, the signal is saturated. Maximum cumulative AE activity for S1, S2 and S3 are respectively about 1992, 2730 and 1659 shocks.

Sample	Stress (Mpa)	$V_P(A)$ (m/s) Axial	Lat.1	Lat.2	$V_P(C)$ (m/s) Axial	Lat.1	Lat.2	$A_m(A)$ (V) Axial	Lat.1	Lat.2	$A_m(C)$ (V) Axial	Lat.1	Lat.2
S1	0	2907	2450	2202	2907	2450	2202	0.007	0.013	1.347	0.007	0.013	1.347
	0	2901	2467	2181	2901	2467	2181	0.006	0.014	1.349	0.006	0.014	1.349
	2.39	3295	2828	4117	3325	2828	4141	0.028	0.027	1.418	0.026	0.023	1.394
	4.86	3380	2783	4217	3333	2794	4242	0.042	0.030	1.395	0.039	0.025	1.329
	7.30	3421	2750	4166	3437	2485	4191	0.061	0.017	1.394	0.056	0.015	1.374
	9.74	3470	2216	4191	3462	2216	4217	0.068	0.010	1.343	0.064	0.009	1.345
	11.85	3462	2168	4141	3421	1775	4191	0.079	0.016	0.941	0.076	0.016	0.824
S2	0	4282	5016	3592	4282	5016	3592	0.607	2.331	2.499	0.607	2.331	2.499
	5.12	4359	5016	5282	4333	4980	5203	0.529	2.499	2.499	0.529	2.499	2.499
	10.14	4625	4875	5242	4373	4980	5282	0.608	2.499	2.499	0.602	2.499	2.499
	15.05	4731	4980	5203	4747	4910	5127	0.732	2.499	2.499	0.711	2.499	2.499
	20.15	4655	4841	5242	4731	4875	5242	0.819	2.499	2.499	0.795	2.499	2.499
	25.06	4731	4875	5242	4778	4742	5203	0.911	2.094	2.499	0.881	1.313	2.499
	30.42	4794	4808	5165	4794	4775	5203	1.032	1.260	2.499	0.999	0.626	1.949
	37.68	4747	4356	5203	4794	3915	5089	1.106	0.262	1.386	1.089	0.207	1.324
S3	0	2964	3964	2857	2964	3964	2857	0.024	1.133	2.499	0.024	1.133	2.499
	4.49	3859	3897	4653	3911	3854	4622	0.143	0.044	2.499	0.138	0.038	2.499
	7.01	3932	3941	4561	3922	4009	4622	0.206	0.059	2.499	0.202	0.071	2.499
	9.38	3964	3986	4622	3964	3941	4531	0.246	0.071	2.499	0.237	0.053	2.499
	11.74	3964	3941	4531	3997	3941	4622	0.285	0.020	2.499	0.256	0.014	2.499
	16.29	4019	1627	4502	4041	1627	4502	0.266	0.006	2.499	0.239	0.006	2.499
	18.05	4041	1012	4561	4030	941	4531	0.271	0.002	2.155	0.260	0.002	1.607

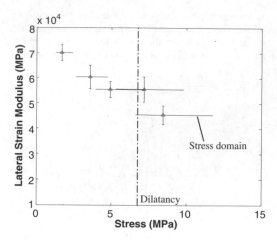

Figure 4. Lateral strain modulus (M_L) values plotted against stress. Each horizontal line represents the stress domain of each unloading phase where linear regression has been done to determine M_L.

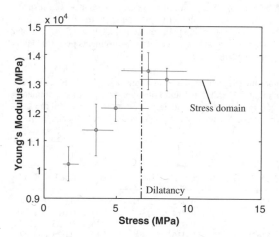

Figure 5. Young's Modulus (E) values plotted against stress.

manifestation is also graphically observed (linear extensions of the unloading phases) even before the onset of the dilatancy at about 46% of the strength. Unloading and reloading phases appear to be linear and reversible (very slight hysteresis).

Lateral strain modulus (M_L) and Young's Modulus (E) variations have been determined by linear regression realized for each unloading phase of the cycle ($M_L = -\sigma/\varepsilon_L$, and $E = \sigma/\varepsilon_A$, where σ is the axial stress). It is observed that M_L decreases and that E increases (excepted for the last cycle) as stress increases (Fig. 4 and Fig. 5 respectively).

Mechanical results from the two others tests performed on micro-arenite (S2) and on argillaceous

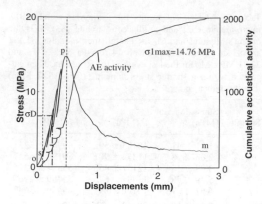

Figure 6. Three domains (os, sp, and pm) distinguished in the acoustical and mechanical behaviour. Stress and cumulative acoustical activity (AE counting) are plotted against displacements. Maximum cumulative AE activity is about 1992 shocks.

arenite (S3) are qualitatively equivalent to these results (see the Table 1).

3.2 Acoustical results

Figure 6 displays results obtained from stress and displacements measurements, and from AE activity monitoring (AE counting) during uniaxial test performed on S1. AE activity is, on the whole, observed from the beginning to the end of our test. We can distinguish three domains in the acoustical and mechanical behaviour of the sample:

- os domain, where the displacement–stress curve is concave upwards. This non-linearity of the curve is due primarily to the closing of pre-existing cracks orthogonal to the major principal stress direction. This domain is also characterized by a low AE activity.
- sp domain is defined by linear and non-linear successive parts of the curve until the maximum stress value ($\sigma1_{max} = 14.76$ MPa) is reached. AE activity increases continuously with a drastic increase just before the peak stress.
- pm is the falling stress domain where AE activity is high. This domain corresponds typically to the growth of a macroscopic discontinuity and hence to the macrorupture of the sample.

Figure 7 shows that AE is observed since the first loading phase of the cyclic loading. The subsequent relaxation, unloading and reloading phases are aseismic. AE activity begins again when the previous maximum stress (the end of the previous loading phase) is exceeded. This well known acoustical behaviour is called "Kaiser effect" (Holcomb 1993). As regards the

416

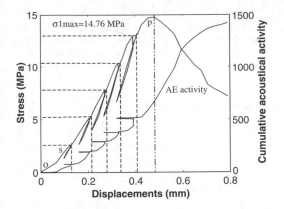

Figure 7. Magnifying of the figure 6 showing aseismicity of the relaxation, unloading, and reloading phases, and Kaiser effect manifestation.

P-wave velocity (V_P) and the maximum amplitude (A_{max}) of the signal received during velocity measurements, their evolutions during the test are plotted in Figures 8 and 9 respectively. It is firstly observed that the whole of V_P and A_{max} values increases during the test until the first cycle (see the os domain in section 3.2 for explanations). Axial V_P and A_{max} increases, while, on the whole, transversal V_P and A_{max} at the bottom of the sample (along the ray E3R6) are constant. Transversal V_P and A_{max} at the top of the sample (ray E2R2) decrease during the test. Moreover V_P values at the end of relaxation phases (point A, see the loading history on Fig. 2) and at a similar stress level during subsequent reloading phase (point C) for each cycle are equivalent. It is also the case for A_{max} values. Concerning the decrease of transversal V_P and A_{max} along the ray localized at the top of the sample (E2R2, Fig. 1), it seems to begin at the onset of the dilatancy.

Acoustical results from uniaxial tests performed on S2 and S3 presented in Table 2 are qualitatively equivalent to the previously exposed results.

4 DISCUSSION

We observe that AE activity begins before any appreciable dilatancy, and increases as stress increases. It is in good agreement with previous study of AE during mechanical tests (Ohnaka & Mogi 1982, Jouniaux et al. 2001, and Amitrano 2003). AE involves locally inelastic processes such as the nucleation and growth of microcracks. At a macroscopic scale, we observe permanent strains and dilatancy. Hence, we suggest that the ferriferous arenite sample has an inelastic behaviour. In parallel, the linearity and the reversibility of the stress–strain curves, the constant velocity and maximum signal amplitude (a kind of measure of the

attenuation), and the aseismicity (absence of AE activity), during unloading and reloading phases of each cycle, reveal a linear elastic behaviour. Thus the rock sample behaviour appears as a superposition of elastic and inelastic processes.

The decrease of the lateral strain modulus (M_L) signifies a lateral expansion of the sample. It corroborates the interpretation of the dilatancy manifestation also observed and which can be attributed to the formation and extension of microcracks parallel to the $\sigma 1$ direction (Jaeger & Cook 1979). The increase of the Young's Modulus involves an axial hardening which can be ascribed to the closing of intergranular voids. Hence, these mechanical observations suggest an anisotropic damage. The acoustical results, in particular the P-wave velocity (V_P) and maximum signal amplitude (A_{max}) variations, reinforce this hypothesis. The increase of the axial velocity and A_{max} is in agreement with axial hardening. The decrease of the transversal V_P and A_{max} (at the top of the sample along the ray E2R2) is consistent with lateral expansion, dilatancy manifestation, and with the ultimate location of the main fractures (Fig. 1). Thus anisotropic damage could result from nucleation and coalescence of extensive axial cracks. This is in agreement with Chow et al. (1995) interpretations of anisotropic velocity field observed from uniaxial test performed on granite.

The difference between the two transversal velocity and A_{max}, id est, the first at the top of the sample (along the ray E2R2, Fig. 1) which decrease as stress increases, and the second at the bottom (along the ray E3R6) which remain constant excepted for the last cycle near the peak stress, is well correlated with localization of the main fractures (more numerous at the top than at the bottom of the sample, Fig. 1). It suggests that the macroscopic discontinuity has grew up from the top to the bottom. The macrofracture would reach the location of the E3R6 segment when transversal V_P and A_{max} along the associated ray begin to decrease (Figs 8, 9, and Table 2). Thus, heterogeneity within the sample is highlighted.

5 CONCLUSION

Mechanical and acoustical parameters that are independent, in term of the measurements, show however an high coherence in the observation of rock damage evolution during uniaxial tests performed on arenite samples. Mechanical and acoustical results suggest an anisotropic damage by nucleation of extensive axial cracks and coalescence. In addition, different mechanical behaviour depending on stress–strain path have been identified, one seems to be linear and elastic during unloading–reloading cycles, the other involves inelastic processes.

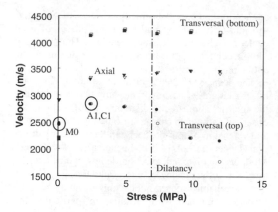

Figure 8. P-wave velocity during the test: triangles, squares and circles correspond to axial velocity, transversal velocity at the bottom and at the top of the sample respectively. Blacks markers are the velocity measured at points Ai. White markers correspond to points Ci – velocity.

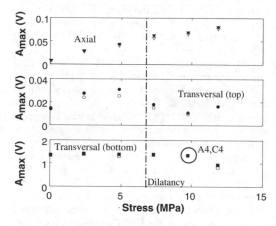

Figure 9. Maximum amplitude of the signal received during velocity measurements and plotted against stress.

Mechanical measurements are only performed locally in the central part of the sample. Thus that are the acoustical measurements, in particular the P-wave velocity and signal maximum amplitude measurements, at two different locations (at the top and at the bottom) which allow to reveal heterogeneity.

Finally, because of their origin, AE provide a tool for the study of microscopic processes involving damage also observed at a macroscopic scale. Macroscopic manifestation of the damage can be observed via P-wave velocity analysis and mechanical measurements.

ACKNOWLEDGMENTS

These researches have been carried out thanks to subsidies from the Ministries for Industry and Research within the GISOS framework. The authors express their gratitude to these organizations for the useful support provided.

REFERENCES

Amitrano, D. 2003. Brittle-ductile transition and associated seismicity: Experimental and numerical studies and relationship with the b value, J. Geophys. Res. 108, B1, 2044.

Chow, T.M., I.L. Meglis, and R.P. Young, 1995. Progressive microcrack development in tests on Lac du Bonnet granite II. Ultrasonic tomographic imaging, Int. J. Rock Mech. Min. Sci. & Geomech. Abstr. 32, 8, 751–761.

Holcomb, D.J. 1993. General theory of the Kaiser Effect, Int. J. Rock Mech. Min. Sci. & Geomech. Abstr. 30, 7, 929–935.

Jaeger, J.C. and N. G. W. Cook, 1979. Fundamentals of rock mechanics, Chapman & Hall, London.

Jouniaux, L., K. Masuda, X.-L. Lei, O. Nishizawa, K. Kusunose, L. Liu, and W. Ma, 2001. Comparison of the microfracture localization in granite between fracturation and slip of a pre-existing macroscopic healed joint by acoustic emission measurements, J. Geophys. Res. 106, B5, 8687–8698.

Lockner, D. 1993. The role of acoustic emission in the study of rock fracture, Int. J. Rock Mech. Min. Sci. & Geomech. Abstr. 30, 7, 883–899.

Ohnaka, M. and K. Mogi, 1982. Frequency characteristics of acoustic emission in rocks under uniaxial compression and its relation to the fracturing process to failure, J. Geophys. Res. 87, B5, 3873–3884.

Yukutake, H. 1989. Fracturing process of granite inferred from measurements of spatial and temporal variations in velocity during triaxial deformations, J. Geophys. Res. 94, B11, 15,639–15,651.

Zang, A., F.C. Wagner, S. Stanchits, C. Janssen, and G. Dresen, 2000. Fracture process zone in granite, J. Geophys. Res. 105, B10, 23,651–23,661.

Geotechnical Measurements and Modelling, Natau, Fecker & Pimentel (eds)
© 2003 Swets & Zeitlinger, Lisse, ISBN 90 5809 603 3

Assessment of a loading test on highly sensitive deformable soil

M.M. Zimmerer, M.D. Datcheva & T. Schanz
Laboratory of Soil Mechanics, Bauhaus-Universität Weimar, Germany

ABSTRACT: One of the main questions in soil mechanics is the relevance and accuracy of soil characteristic parameters determined in laboratory and in field tests. Inaccuracy in soil parameters affects significantly the afterward calculated mechanical response and the theoretically predicted settlement. Very often a significant difference is observed when comparing settlements measured *in situ* during geotechnical works with those obtained after calculations based on the laboratory tests. That is why it is of interest to evaluate soil parameters from *in situ* data by back calculations using a proper mathematical model and to correlate the results with parameters determined from standard laboratory oedometer tests. The aim of this article is to give an example for such procedure. The application is related to the waste disposal site 'Cröbern' which is one of the most modern in Europe. It has been placed in the late open cast mining area 'Espenhain' in the southern area of Leipzig (Germany). Horizontal inclinometer and extensometer have been installed in advance to measure the time and load dependent vertical displacements. These measurements provide data for comparing the actual soil behavior and the results from the mathematical modelling of the process. The mathematical model also takes into account the measured vertical displacements during site building, exploration drillings and the results of laboratory tests as well. The corresponding mathematical problem is solved analytically for the 1D case and numerically using FEM for the 2D case.

1   INTRODUCTION

After nearly 150 years of operation numerous of open dumps in Central Germany have been closed, Figure 1. The large areas need to be now renaturated and prepared for upcoming build-up. The expectation is the restored areas to be used for public recreation and for industrial parks. The opportunity for some traffic routes and service pipes to cross the restabilized land should also be taken into account. Such a programme for renaturation needs extensive ground explorations and

investigations in order to guarantee the required characteristics for safety and suitability of the land.

Before building any construction principle ground investigations are made including a number of laboratory and field testings. The main purpose of measurements and investigations is to obtain a detailed knowledge about the properties of the soil and its behavior under loading.

In this paper we investigate the embankment founded in a former open cast mining area 'Espenhain' with the purpose to evaluate settlement behavior of the highly

Figure 1.   The open cast mining area 'Espenhain'.

sensitive soils encountered there. A part of the carbon open mining 'Espenhain' in the south area Leipzig was covered to serve the large central waste site 'Croebern' (ZDC) in Markkeeberg. A main requirement for doing this was the deformation stability of the ground. With high and mainly different settlements large strain can occur for the sealing basis, which will form a separation of the embankment soil from the sealing body, eventually resulting in cracking of the sealing.

2 CONCEPTUAL MODEL

According to DIN 4022 classification dump material is generally considered as a silt and fine sand mixture. The granular mixture, according to DIN 18196 is classified as SU*. Grain size distribution curves give information over the soil type and also over the drained/ undrained behavior of the given soil. For the investigation of the behavior under compression mainly oedometer test results are used. The difficulty here consists usually in accomplishing useful tests. Problems occur in assembling of undisturbed samples and in insuring initial void ratio of the sample which corresponds to the void ratio of the soil in the field. When oedometer data is available the stiffness parameters can be obtained. In our calculations we determine the normalized tangent stiffness modules using the method of Ohde (Ohde 1939). Following this method we compute the derivative of the regression-function for the relationship $\varepsilon(\sigma)$ and in this way we define the oedometer stiffness. Compared to oedometer tests the cone penetration tests (CPTs) are more profitable in expenses. By such field tests a knowledge about soil undisturbed and natural behavior can be obtained (Figure 2). When both laboratory and field tests are available the correlation between them can be established by determining correlation parameter factors. The procedure for doing this is presented hereafter. We use the stiffnesses obtained from the oedometer tests to establish a procedure for correlating them to the CPT's readings. During the correlation procedure we take into account that several soil layers exist where the material is the same but in different states of compaction (I_D) and therefore having different stiffnesses and different settlement behavior. Using this calibrating procedure we can determine the soil stiffness $(E_{oed}^{ref} (e_0))$ for each particular layer based on the CPTs readings. In this way a more detailed modelling of the soil is achieved. Therefore the relevant soil parameters can be derived from the laboratory and field tests reported in the investigator's documentation. The determined parameters then are used in extended calculations according to DIN V 4019-100 and advanced computations using 2D-FEM-modelling. The computed deformations are compared with the measured deformations of the test embankment obtained using horizontally placed inclinometers and

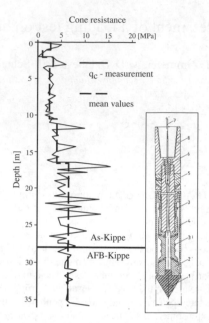

Figure 2.   CPTs readings.

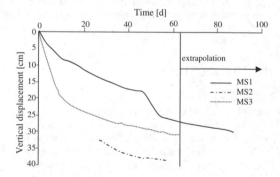

Figure 3.   Readings of inclinometers.

extensometers (see Figure 3, (Jolas 1991)). Then a back analysis is done correcting the model parameters to achieve better agreement with measurements. The optimization possibilities lie mainly in the variation of the stiffness parameters of the ground and their adjustment after CPTs profiles. The result of such an optimization is a general and practice relevant prescription for the correlation between laboratory and field tests.

3 EXAMPLE APPLICATION

3.1 Description of the mining area 'Espenhain'

The Table 1 gives the technical details about the waste site 'Croebern' (ZDC) in Markkeeberg. The natural

420

former open cast mining-bed is located approximately 70 m below the actual ground level. Over the natural mining-bed material there is approximately 50 m thick dump landfill, which has been dumped by overburden conveyor bridges. This layer is called hereafter AFB dump. The range up to the ground level was filled up by a packing apparatus for lignite open cast mines and for this the notation used is AS dump. The groundwater level in the examined area during the measurements was more than 60 m below the ground level and had no influence on the determined soil parameters. Figure 4 shows schematically the dumpsite construction.

Description of AFB dump

The production and fall of predominantly evenly deposited marine sediments, mainly laminated vertically from fine sand to silt, were made by the AFB with multi-bucket excavators. Thus relatively evenly more mixed and an almost homogeneous dump soil arised. The material of the AFB-dump consists of virgin marine cover mountain layers, which are to be assigned in the range from silt to sand (Jessberger 1992).

Description of AS dump

The settler dumps possess an inhomogeneous structure. Due to the used filling technique the AS dump has bigger settlement potential. The depth of this dump layer varies and reaches values between 15 m and 30 m below the ground level. The AS dump contents materials coming from the areas between the lignite seams as well as from the natural mining-bed. The former mainly include tertiary sand and silt. In addition there is a small quantity of filter ash from the power station in Thierbach. The different technical handling during the dump filling as well as the varying material properties lead to the fact that the AS dump layer exhibits a clearly more heterogeneous structure as well as more sensitive behavior concerning settlements compared to the AFB dump layer.

The mining-bed masses consist of silts and clay and shows predominantly cohesive behavior. Table 2 contains soil characteristic values representative for dump soils.

3.2 Oedometer tests

In this section a typical data from the oedometer tests is used for determining the mechanical parameters of the soils composing the dampsite schematically shown in Figure 4. Because during the oedometer test no elastic relation between stress and strain occurs from the oedometer data we determine the coefficient of compressibility noted by E_{oed} and not the module of elasticity.

When calculating settlements according to DIN we use tangent stiffness modules instead of commonly used secant stiffness modules. The equation for the

Table 1. Technical details of the waste site.

Waste area	48.85 ha
Waste height	48 m
Total capacity	$10^7 \, \text{m}^3$
Commissioning date	1995
Running time (expected)	2023

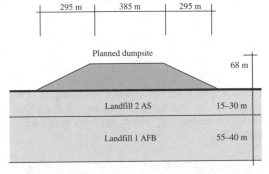

Figure 4. Scheme of the dumpsite.

Table 2. Characteristic soil parameters values.

Friction angle	$\phi = 25°$
Cohesion	$c = 0 \, \text{kN/m}^2$
Specific weight	$\gamma = 17.5 \, \text{kN/m}^3$
Sorting coefficient	$U = d_{60}/d_{10} = 100$

determination of the stiffness E_{oed} according to Ohde (Ohde 1939) is:

$$E_{oed}(\sigma) = E_{oed}^{ref} \, (\sigma/\sigma_{ref})^m \tag{1}$$

Equation (1) corresponds to the derivative of the $\sigma\varepsilon$ – regression function by measured stress strain values from oedometer tests (Figure 5). With the regression function

$$\ln \varepsilon = \alpha_{reg} \, \ln(\sigma_z/\sigma_{ref}) + \beta_{reg} \tag{2}$$

and the regression coefficients α_{reg} and β_{reg} it results that:

$$E_{oed}^{red} = \sigma_{ref} \, e^{-\beta_{reg}/\alpha_{reg}} \tag{3}$$

and $m = 1 \, \alpha_{reg}$. We take $\sigma_{ref} = 100 \, \text{kPa}$ and from the four oedometer test measurements given on Figure 5 we determine $\alpha_{reg} = 0.45$ and $\beta_{reg} = 3.19$. From relation (3) it follows that $E_{oed}^{ref} = 5320 \, \text{kPa}$. The results of the evaluations of all oedometer tests are presented on

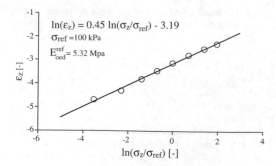

$$\ln(\varepsilon_z) = 0.45 \ln(\sigma_z/\sigma_{ref}) - 3.19$$
$$\sigma_{ref} = 100 \text{ kPa}$$
$$E_{oed}^{ref} = 5.32 \text{ Mpa}$$

Figure 5.   Oedometer test data.

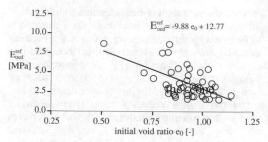

$$E_{oed}^{ref} = -9.88 \, e_0 + 12.77$$

Figure 6.   Reference stiffness vs. initial void ratio.

Figure 6. The reference stiffness is a function of the initial void ratio and this functional dependence is fitted by the linear function with coefficients presented on Figure 6.

3.3   Assessment of CPTs readings

The mechanical properties of dump soils are influenced by the mechanical characteristics of the source rocks as well as by the mining process. Depending on the mining method, assistant transport and dumping technique different mixing occurs and different mechanical loading is applied to the natural loose components (Bennewitz 2002). That is why in the area of the performed sample embankment (pilot block) several cone penetration testing installations were established (DS 2.1−DS 2.5 in Figure 7). In our investigation we use the readings of the CPT device DS 2.3 since it was installed in the middle of the horizontally placed inclinometer MS 2. This particular location of the CPT device gives the opportunity to correlate the measured settlements to the measured stiffness. The average values for the tip–pressure values for several layers are shown together with the DS 2.3 readings on Figure 2.

The oedometer stiffness E_{oed} and the cone resistance are proportionally related and according to DIN EN 1997-3 we have:

$$E_{oed} = \alpha \, q_c \tag{4}$$

The empirical correlation factor α is guessed based on engineering experience. For dump soils it is given that values of α can vary between 2 and 4, see (Philipp 2002). In the appendix B.3 of the DIN EN 1997-3 it is given that for sand $\alpha = 2$ and this value is applicable for $q_c < 5$ MPa. In (Lunne et al. 1997) the value of α is specified to be between 4 and 5 with the range of applicability 2.5 MPa $< q_c < 5$ MPa.

In our 2D-FEM numerical calculations we use the pattern of the embankment shown on Figure 2. Under

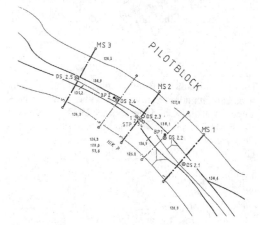

Figure 7.   Test measurement installations.

knowledge of the prevailing dead weight stress the determined stiffness were converted into E_{oed}^{ref} reference stiffnesses with $\sigma_{ref} = 100$ kPa and the equation:

$$E_{oed}^{ref} = \alpha \, q_c \, (\sigma_z/\sigma_{ref})^{-m} \tag{5}$$

As σ_z in each layer the average stress is taken.

Table 3 contains the derived characteristic values. Using the CPTs data we provide the best fit for I_D as a function of q_c as it follows:

$$I_D = 0.98 + 0.66 \log(q_c/\sigma_z) \tag{6}$$

and then using (6) we can determine the initial void ratio e_0 from the known I_D with:

$$e = e_{max} - I_D(e_{max} - e_{min}) \tag{7}$$

The values used for the minimum and maximum void ratios are $e_{min} = 0.5$ and $e_{max} = 1.15$.

Taking $\alpha = 1$ in equation (4) and the minimum and maximum values for the initial void ratios from the

Table 3. Reference stiffness with depth for $\alpha = 3$.

Layer no.	z [m]	E_{oed}^{ref} [MPa]
1	−1	35.8
2	−2	7.3
3	−3.3	15.6
4	−10.5	6.9
5	−16.5	7.5
6	−35.8	8.1

Table 4. Dimensions of the embankment.

Designation	Width [m]	Length [m]	Height [m]
Pilot embankment	50	100	10
Final waste site	975	975	68

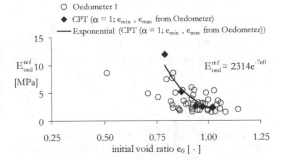

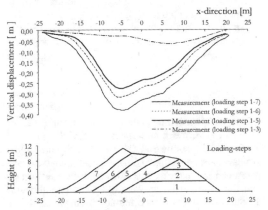

Figure 8. Correlation between CPTs and oedometer.

Figure 9. Loading steps and displacement measurements.

oedometer tests a good agreement for the void-ratio-stiffness relation from oedometer and CP testing can be found (Figure 8).

Table 5. Vertical displacements (extrapolation – extr.).

Point	Max displ. (17.01.1991) s [cm]	Extr. incr. Δs [cm]	Final displ. for back calcul. s + Δs [cm]
MS 1	27.7	5.3	33
MS 2	38.2	2.8	41
MS 3	31.2	2.3	33.5

3.4 Measurements

Using the measured vertical displacements under the embankment body, the settlements in the final state of the construction with planned waste site height of 68 m should be predicted. The dimensions of the embankment and the planned waste site can be seen in Table 4. That staged dumping process of the embankment is shown on Figure 9.

The maximum settlement amount up to 38.2 cm for MS 2 after two months. If the deformations are shown as a function of time, it is obvious that the displacement decreases after short time. For the later modelling and calculation the results were extrapolated according to Figure 3 and Table 5.

The staged construction of the pilot embankment is modelled with the help of the finite element discretization and the mathematical problem is solved with the FEM. The measured values can be compared with the results of the computations.

3.5 Manual calculation

The deformations can be determined according to DIN V 4019-100 by means of indirect settlement calculation. For this it is needed first to determine the lateral stresses in the soil. Depending on the value of the stiffness parameters E_{oed} the stress will induce corresponding deformation of the considered soil layer. The stress increasing and/or the decreasing over the depth can be assessed using different formulations. In this work the solution proposed by (Kezdi 1964) is used. The stresses under any arbitrarily point loaded by a trapezoidal load can be expressed with

$$\sigma_z = \frac{q}{\pi A}\left[A\beta + \left(A + \frac{B}{2}\right)(\alpha_1 + \alpha_2)\right] - \frac{q}{\pi A}[x(\alpha_1 - \alpha_2)]$$

(8)

The geometry of the problem and the meanings of the variables and parameters in (8), namely A, B, x, α_1, α_2, β are given on Figure 10. The unit for the angles is

423

radian. The size of the angles has to be determined with geometrical relations. For the increment of the displacements we have:

$$\Delta s = \int_0^{\Delta z} \int_{\sigma_{z,0}}^{\sigma_{z,0}+\sigma'_{z,q}} \frac{d\sigma' dz}{E_{oed}(\sigma')} \tag{9}$$

The specified depth was determined to the level where $\sigma_{z,0} = 10\,\sigma_{z,q}$ (in contrast to the prescription from the German code of 20%).

Since this area is composed mainly of AS material one single layer is considered. The calculations have been done with values for the input parameters given on Table 6. The results are shown on Figure 12.

3.6 FEM Calculation

For the FEM modelling of the embankment the program PLAXIS, (Vermeer and Brinkgreve 1995) and the advanced 'Hardening Soil' (HS) model ((Schanz 1998)) are used. As a first approximation a computation with parameters from the manual calculation were performed and the different layers were not taken into account. The deformation are maximum of 70 cm vastly higher than the values determined by manual calculation. The next step was modelling of a geometry which corresponds to the penetration testing profiles shown on Figure 2. Reference stiffnesses are determined for reference stress $\sigma_{ref} = 100\,\text{kPa}$. These values are the HS model.

Additionally the phases of the embankment construction are modelled. The aim is to examine whether a reduction of the void ratio and a coupled increase of

the stiffness will cause refinement of the results during the loading procedure. The relation between void ratio and reference stiffness (Figure 8) is included in the computations in the following way. The procedure for doing this is explained hereafter. First with the stiffnesses and the initial void ratios e_0 from the CPT results ($\alpha = 3$) the loading/construction phases 1–3 (production of the right loading cone (Figure 9)) are implemented.

The vertical strain ε_z which corresponds to the increments Δ_z of the displacement under the loading is computed. The problem is considered symmetric and the coordinate axis coincide with the axis of symmetry of the embankment. After that the new, decreased void ratios e of each of the soil layers are determined. With $\alpha = 3$ and the equations (4) and (5) the exponential relationship between void ratio and stiffness is

$$E_{oed}^{ref} = 6941\,e^{-7e_0} \tag{10}$$

For the following computation steps new reference stiffnesses corresponding to the updated void ratios are implemented. These are assigned to the soil layers starting from computation phase 4. The results shows a refinement of the agreement between measurements and calculations, see Figure 12.

3.7 Comparison of results

The results of the FEM computations (70 cm) are nearly as twice as high as those from the computation according to DIN V 4019-100 (38 cm).

Table 7. 1st input parameters for FE-calculation.

Symbol	Unit	Landfill 1	Landfill 2
ϕ	[°]	24	24
c	[kPa]	1	1
v	[-]	0.2	0.2
γ_d	[kN/m³]	16	16
γ_r	[kN/m³]	17.5	17.5
ψ	[°]	0	0
E_s^{ref}	[MPa]	5	5
m	[ ]	0.7	0.7

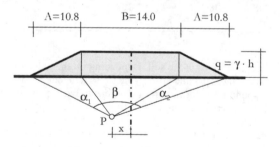

Figure 10. Geometry for the manual calculations.

Table 6. Soil parameters.

	Symbol	Unit	Value
Reference stiffness	E_{oed}^{ref}	[MPa]	5.0
	m	[–]	0.7
Specific weight/ landfill	γ_k	[kN/m³]	17.5
Dam	γ_P	[kN/m³]	16.0

Table 8. Final input parameters for FE-modelling.

Layer no.	Depth [m]	γ_r [kN/m³]	E_s^{ref} [MPa]
1	−1	17.5	35.8
2	−2	17.5	7.3
3	−3.3	17.5	15.6
4	−10.5	17.5	6.9
5	−16.5	17.5	7.5
6	−35	17.5	8.1
7	−70	18.5	16.8

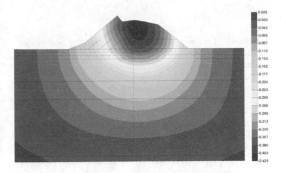

Figure 11. Vertical displacement of FEM-calculation.

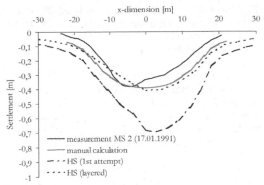

Figure 12. Comparison of results.

An explanation for this is that in the computations according to DIN the considered calculation depth was 35 m, while in the FEM computations there are considerably large deformations in depth lower than 35 m and these deformations have obviously an influence on the total deformation.

A more detailed modelling of the soil lead to better agreement of the measured and computed values. Therefore a back calculation of stiffnesses from the CPTs and the following prescription for the correlation factor α are reasonable and have to be performed.

4 ANALYSIS

In the first part the settlement calculation is arranged according to DIN V 4019-100. The dump soil is accepted and simplified as homogeneous but with increasing stiffness with growing depth. A good agreement with the measurements can be obtained.

For the computations according to DIN V 4019-100 the measured settlement values for a homogeneous soil can be relatively well reconstructed up to a specified depth. The comparison of the results can be seen in Figure 12. The maximum of the measurements is with 38 (+3) cm, for final setting condition, while the computated maximum deformation is 39 cm. The form of the settlement basins agrees with calculated one.

The measured settlements show no symmetry because the real embankment was nonsymmetric. In the computation a trapezoidal dam cross section was selected, while the actual outline over the measured maximum exhibited an extremum. The computation was accomplished with a stress-dependent stiffness (see Figure 13). The soil layering derived in section 3.6 from the CPTs is included into the FEM-computation. After a comparison of measurements and computations the parameter variation supplies for the correlation factor between cone resistance and stiffness of the soil in the field the following value: $\alpha = 3$. This result is in agreement with the indicated values given

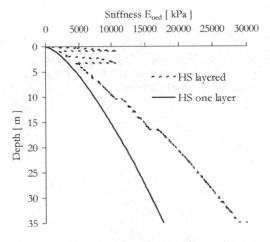

Figure 13. Comparison of the stiffnesses calculated for layered and non-layered models.

in the literature. The computations using the HS model with layered dump soil shows good agreement with the measurements.

It is obvious that by good knowledge about the ground and with close-to-reality derived soil characteristic values the predicted by calculations deformations are suitable and comparative with measurements.

5 CONCLUSIONS

From the performed investigation it can be concluded that the main difficulty and the most important thing in numerical predictions of settlements in highly sensitive soils is the suitable choice of the the soil parameters. Maximum of the deformations is very sensitive to the differences in the selected stiffness of the soil.

That is why stiffness modules defined from dis turbed samples used in the oedometer tests must be brought in correlation with *in situ* tests readings to insure that the actual soil behavior is properly considered.

Extensometer measurements are of a big importance for the determination of the stiffness modules and therefore are needed to predict maximum settlements. Although extensometers were placed inside and below the embankment the readings were not used completely in the calculations presented here because of damages in the devices caused by the settlement. For the future work it is of interest to consider the problem of minimum and maximum void ratios in order to simulate the real conditions throughout the whole embankment.

ACKNOWLEDGEMENT

Some of the results presented in this paper are based on the calculations made by Dipl.- Ing. Christine Glissmann, who is greatly acknowledged.

REFERENCES

Bennewitz, T. (2002). Fachbüro für Consulting und Bodenmechanik GmbH, Espenhain, Kurzfassung: Querung der Tagebaukippen beim Bau der BAB 38. Chemnitz. Bautex 2002.

Jessberger, H. (1992). Verformungsnachweis für die mineralische Basisabdichtung der Zentraldeponie Cröbern. Technical report, Bochum.

Jolas, P. (1991). Zentraldeponie Cröbern, Neu bearbeitete Setzungsauswertung Pilotblock. Technical report, MIBTRAG, BKW Borna, BT Bodenemchanik, Gaschwitz.

Kezdi, A. (1964). *Bodenmechanik*. Berlin: VEB Verlag für Bauwesen.

Lunne, H.-J., P. Robertson, and J. Powell (1997). *Cone Penetration Testing in Geotechnical Practice*. London: E & FN Spon.

Ohde, J. (1939). Zur Theorie der Druckverteilung im Baugrund. *Bauingenieur 20*, 451–459.

Philipp, H. (2002). Bebauung einer Tagebaukippe mit einer Reststoff-Verwertungsanlage. Technical report, Leipzig.

Schanz, T. (1998). *Zur Modellierung des mechanischen Verhaltens von Reibungsmaterialien*. Universität Stuttgart: Mitteilung 45 des Instituts für Geotechnik.

Vermeer, P. and R. Brinkgreve (1995). *PLAXIS: Finite element code for soil and rock analyses (Version 6.3)*. Rotterdam: Balkema.

Recent developments in modelling (Workshop)

A non-local elasto-plastic model to simulate localizations of deformations in quasi-brittle materials

J. Bobinski & J. Tejchman

Gdansk University of Technology, Gdansk, Poland

ABSTRACT: The paper presents results of FE-calculations of the behaviour of quasi-brittle materials during a biaxial compression test. An elasto-plastic material model with a linear Drucker-Prager criterion using isotropic hardening and softening and non-associated flow rule was used. A non-local approach was applied to capture realistically localization of deformations. A characteristic length was incorporated via a weighting function.

1 INTRODUCTION

Failure in most engineering materials like metals, soils, rocks, polymers and concrete is proceeded by the occurrence of narrow zones of intense deformation. Due to these zones, a degradation of the material strength develops. The localizations of deformation can occur as cracks or shear zones. An understanding of the mechanism of the formation of localizations is of a crucial importance since they act as a precursor to ultimate fracture and failure. In order to properly analyze the failure behavior of materials, the phase of the localization of deformation has to be modeled in a physically consistent and mathematically correct manner. Classical FE-analyses within a continuum mechanics are not able to describe properly both the thickness of localizations zones and distance between them since they do not include a characteristic length of microstructure. Thus, they suffer from strong mesh sensitivity (its size and orientation) because differential equations of motion change their type and the rate boundary value problem becomes mathematically ill-posed. Thus, they require an extension by a characteristic length to accurately capture the formation of localizations.

The intention of the paper is to analyze localizations of deformations in quasi-brittle materials (e.g. concrete-like materials). The analyses are performed using a finite element method and a non-local elasto-plastic law with hardening and softening according to Drucker-Prager. Due to the presence of a characteristic length, the law can describe the formation of localizations of deformations with a certain thickness and spacing. The FE-results converge to a finite size of a shear zone via mesh refinement, and initial and

boundary value problems become mathematically well-posed at the onset of localization.

The FE-calculations are carried out with a specimen subjected to uniaxial plane strain compression. Attention is laid on the influence of element size and orientation of the mesh, characteristic length and non-local parameter on the spontaneous shear localization in the specimen.

2 MATERIAL BEHAVIOUR

The behaviour of quasi-brittle materials is very complex due to their heterogeneous structure, strong anisotropy, non-linearity and interaction between friction and cracking. Usual quasi-brittle materials are characterised by the following properties:

– the compressive strength is several times higher than the tensile one,
– the strength increases with increasing confining pressure,
– the shape of the failure surface in a principle stress space is close to parabolic,
– the shape of the failure surface in deviatoric planes changes from a curvilinear triangle with smoothly rounded corners to nearly circular with increasing pressure,
– the initial volume change in the form of compaction is almost linear up to the critical stress, later volumetric expansion (dilatancy) occurs,
– the curvature of the failure surface is produced by interaction between dilatancy and crushing, .
– the curved meridians in planes containing the hydrostatic axis are different for compression and extension,

- the non-associated flow rule prevails,
- the ductile damage occurs at high pressures and brittle (unstable) one at low pressures,
- localizations of deformation can occur in the form of shear zones (if friction dominates) and cracks (if cohesion is dominant),
- the stiffness degradation due to localizations occurs.

To describe the behaviour of quasi-brittle materials, different models were developed: continuum ones within fracture mechanics (Bazant & Cedolin 1979, Hillerborg 1986), damage mechanics (Dragon & Mroz 1979, Chen 1999, Ragueneau et al. 2000) and softening plasticity (Willam & Warnke 1975, de Borst 1986, Pietruszczak et al. 1988), and discrete ones using a lattice approach (Herrmann et al. 1989, Vervuurt et al. 1994) and DEM (Sakaguchi & Mühlhaus 1997, Donze et al. 1999, Place & Mora 2001).

To take into account localizations of deformations with a certain thickness and spacing, a characteristic length has to be included. It can be introduced with micro-polar (Mühlhaus 1989, Tejchman and Wu 1993, Tejchman et al. 1999), gradient (Zbib & Aifantis 1989, Mühlhaus & Aifantis 1991), viscous (Loret & Prevost 1990, Sluys 1989) and non-local (Bazant 1986, Bazant & Lin 1988, Brinkgreve 1994, Strömberg Ristinmaa 1996, Chen 1999, Marcher & Vermeer 2001) theories. Otherwise, the mesh-dependent behaviour is observed, localizations approach zero-volume zones and the dissipative energy decreases to zero (de Borst et al. 1992).

3 CONSTITUTIVE LAW

To simulate the behaviour of a quasi-brittle material, an elasto-plastic material model with a linear Drucker-Prager criterion using isotropic hardening and softening and non-associated flow rule was used. The yield function f was defined as:

$$f = q + p \tan \beta - \sigma_y(\kappa), \qquad (1)$$

where q – von Mises equivalent stress, p – mean stress, β – friction angle, σ_y – yield function and κ – hardening/softening parameter. The equivalent stress q is defined as:

$$q = \sqrt{\frac{3}{2} s_{ij} s_{ij}}, \qquad (2)$$

where s_{ij} is the stress deviatoric tensor. The mean stress p is equal to:

$$p = \frac{\sigma_{ii}}{3}. \qquad (3)$$

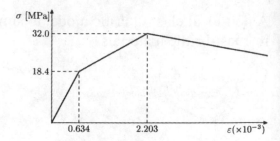

Figure 1. Stress–strain relationship for a material in a compressive regime.

The parameter κ is equivalent to the equivalent plastic strain ε_p:

$$d\varepsilon_p = \sqrt{\frac{2}{3} de_{ij}^p de_{ij}^p}, \qquad (4)$$

wherein e_{ij} – deformation deviatoric tensor.

The model uses non-associated flow rule. The potential function g was chosen as:

$$g = q + p \tan \varphi, \qquad (5)$$

where φ is the dilatancy angle. For the parameters $\beta = 0$ and $\varphi = 0$, a von Mises yield criterion is obtained. In an elasto-plastic constitutive law by Drucker-Prager, the Young modulus was taken as $E = 29$ GPa and Poisson's ratio as $\nu = 0.18$ (softening modulus $H = 0.8$ GPa), Figure 1. The compressive strength of the specimen was 32 MPa at $\varepsilon = 0.22\%$. The internal friction angle β was equal to $25°$ and the dilatancy angle φ was equal to $10°$.

4 NON-LOCAL MODEL

A non-local approach is used to obtain a well-posed boundary value problem, to ensure a mesh-independent solution and to promote convergence of a numerical calculation in a softening regime.

A full non-local model assumes a relation between average stresses and average strains:

$$\sigma_{ij}^* = \frac{1}{A} \iiint w(x_n') \sigma_{ij}(x_n + x_n') \, dx_1' dx_2' dx_3', \qquad (6)$$

$$\varepsilon_{ij}^* = \frac{1}{A} \iiint w(x_n') \varepsilon_{ij}(x_n + x_n') \, dx_1' dx_2' dx_3', \qquad (7)$$

wherein a superimposed star denotes a non-local approach, x_n is a global coordinate, x_n' is a local coordinate with $n = 1, 2, 3$, $w(x'_n)$ denotes a weighting function and A stands for a weighted volume:

$$A = \iiint w(x'_n)\,dx'_1 dx'_2 dx'_3. \tag{8}$$

In homogeneous materials, Eqs. 1 and 2 become

$$\sigma^*_{ij} = \frac{1}{A}\int w(r)\sigma\,dV, \tag{9}$$

$$\varepsilon^*_{ij} = \frac{1}{A}\int w(r)\varepsilon\,dV, \tag{10}$$

where r is the distance from the material point considered to other points of the body. Usually, the error function is used as a weighting function:

$$w(r) = \frac{1}{l\sqrt{\pi}}e^{-(r/l)^2}, \tag{11}$$

$$\int_{-\infty}^{\infty} w(r)\,dr = 1. \tag{12}$$

The parameter l denotes a characteristic length. At the distance of a few times the length l, the function w is equal to zero. In the model used, the non-locality is related to an equivalent plastic strain measure in the softening regime as (Bringreve 1994)

$$\varepsilon^*_p(x) = (1-\alpha)\varepsilon_p(x) + \frac{\alpha}{A}\int_{-\infty}^{\infty} w(r)\varepsilon_p(x+r)\,dV, \tag{13}$$

where α is a non-local parameter. For $\alpha = 0$, a local model is obtained, and for $\alpha = 1$, a classical non-local model (Bazant & Lin 1988) is recovered. To simplify calculations, plastic strain rates are approximated by total strain rates:

$$d\varepsilon^*_p(x) \approx d\varepsilon_p(x)$$
$$+ \alpha\left(\frac{1}{A}\int_V w(r)d\varepsilon(x+r)dV - d\varepsilon(x)\right), \tag{14}$$

where $d\varepsilon_p$ is the plastic and $d\varepsilon$ is the total increment of the local effective strain.

5 FE-RESULTS

Plane strain FE-calculations were performed with a specimen 4 cm wide and 14 cm high subject to uniaxial compression (Fig. 2). All nodes at the lower edge were fixed in a vertical direction. To preserve the stability of the specimen, the node in the middle of the

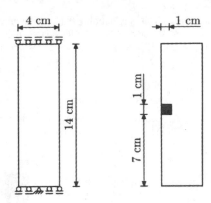

Figure 2. Geometry of the element and location of the imperfection.

lower edge was kept fixed. The deformations were initiated through constant vertical displacement increments prescribed to nodes along the upper edge of the specimen. The lower and upper edges were smooth. The localization was induced by a small material imperfection at mid-height of the specimen (where the maximum strength of the material was diminished by 2%).

To investigate the effect of the mesh size on the results, various discretisations were used: coarse (8 × 28), medium (16 × 56) and fine (24 × 84) where each quadrilateral is composed of four diagonally crossed triangular elements with linear shape functions. To examine the influence of the mesh alignment on the results, two types of meshes were used: with an inclination smaller than $\theta < 45°$ (8 × 56 − θ = 26.6°, 12 × 56 − θ = 36.9°, 16 × 84 − θ = 33.7° and 20 × 84 − θ = 39.8°) and with an inclination greater than $\theta > 45°$ (12 × 28 − θ = 56.3°, 16 × 28 − θ = 63.5°, 20 × 56 − θ = 51.3° and 24 × 56 − θ = 56.3°).

First, a local analysis was carried out (with $\theta = 45°$ and $\alpha = 0$). Figure 3 shows deformed meshes for various discretisations.

The deformations and strains localize in one element wide shear band with an inclination of 45°. The load–displacement curves depend upon the mesh size.

Figures 4–6 demonstrate the FE-results with a non-local model (with $\theta = 45°$). Two additional constants were taken into account: the parameter α ($\alpha = 2$) and characteristic length l ($l = 7.5$ mm) both influencing strongly the width of a localization zone (Brinkgreve 1994). The deformed meshes and calculated equivalent plastic strains are shown in Figure 4. The load–displacement diagrams are presented in Figure 5.

Deformations localize in a band wider than one finite element (Fig. 4). The thickness of the shear zone is approximately 2.4 cm and does not depend upon the mesh size. The inclination of the shear zone is equal to 44.8°, 44.8° and 46.8° with a coarse, medium and fine mesh, respectively. These values are in a good agreement with an inclination obtained from an analytical formula ($\Theta = 45.5°$) based on a bifurcation theory. The evolution of the vertical force along the top after the peak is the same for all discretisations

(Fig. 5). The maximum vertical force is equal to $1.95 \cdot 10^3$ kN at the vertical displacement of 0.4 mm. The results of non-local equivalent plastic strains in a section perpendicular to a shear zone conform mesh-independency (Fig. 6).

The FE-calculations were also performed with a mesh alignment lower than 45°. The width of the shear zone was equal to 2.6 cm for all meshes except for the coarsest one (8×56) where the localization zone was equal to 3.6 cm. The inclination of the shear zone was: 44.0° (8×56, $\theta = 26.6°$), 43.9° (12×56, $\theta = 36.9°$), 47.6° (16×84, $\theta = 33.7°$) and 45.2° (20×84, $\theta = 39.8°$). The load–displacement diagrams were similar using all mesh discretisations.

In the FE-studies assuming a mesh inclination greater than 45°, the width of a shear zone was 2.4 cm with meshes 20×56 and 24×56, and 2.9 cm with meshes 12×28 and 16×28. The inclination of the

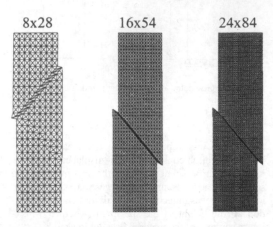

Figure 3.   Deformed meshes (local model $\alpha = 0$).

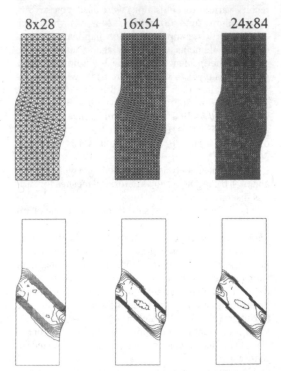

Figure 4.   Deformed meshes and contours of equivalent plastic strains (non-local model, $\alpha = 2$ and $l = 7.5$ mm).

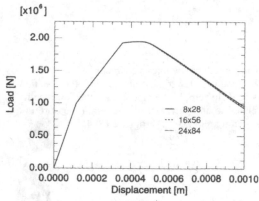

Figure 5.   Load–displacement curves (non-local model, $\alpha = 2$ and $l = 7.5$ mm).

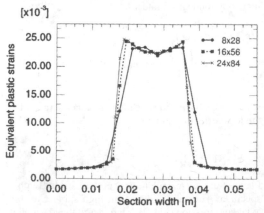

Figure 6.   Equivalent plastic strains in a section perpendicular to a shear zone (non-local model, $\alpha = 2$ and $l = 7.5$ mm).

shear zone was 50.3° (12 × 28, $\theta = 56.3°$), 47.2° (16 × 28, $\theta = 47.2°$), 50.1° (20 × 56, $\theta = 50.1°$) and 49.1° (24 × 56, $\theta = 56.3°$). The evolution of the vertical force along the top was the same for all discretisations.

The FE-results with a classical non-local model ($\alpha = 1$) are demonstrated in Figures 7–9 (using a mesh inclination lower than $\theta < 45°$). The calculations show that the results are only partly mesh-independent. The evolution of the vertical force on the top edge is slightly different in a softening regime. In turn, the width of a shear zone is equal to 2.7 cm for meshes 12 × 56 and 16 × 56, and 1.9 cm for 20 × 84 mesh. The shear zone using a mesh 8 × 56 is very wide and has a different direction. The inclination of a shear zone is 43.6° (12 × 56, $\theta = 36.9°$), 43.8° (16 × 84, $\theta = 33.7°$) and 45.5° (20 × 84, $\theta = 39.8°$).

In the calculations with mesh inclinations equal to 45°, the width of the shear zone was approximately equal to 1.4 cm. The inclination of the shear zone was 44.8°, 45.7° and 46.2° for a coarse, medium and fine

mesh, respectively. The load–displacement diagrams were similar in all cases.

When the mesh alignment was larger than 45°, the width of a shear zone was equal to 2.7 cm, 3.3 cm, 1.2 cm and 1.7 cm for meshes 12 × 28, 16 × 28, 20 × 56 and 24 × 56 mesh, respectively. The inclination of a shear zone was 55.2° (12 × 28, $\theta = 56.3°$), 47.5° (16 × 28, $\theta = 47.2°$), 51.3° (20 × 56, $\theta = 50.1°$) and 50.6° (24 × 56, $\theta = 56.3°$).

The effect of the parameters α ($\alpha = 2$–5) and l ($l = 2$–12 mm) on the width of shear localization was investigated on the basis of a medium mesh (16 × 56), Table. 1.

The results show that the larger the parameters l and α, the wider the shear zone L. The parameters α and l have no influence on the maximum vertical force. However, they strongly influence the load–displacement curves in a post-peak regime. The larger L, the smaller the drop of the curve after the peak. The width L can be approximately correlated with the parameters α and l with a following relation:

$$L \cong (1.2 - 1.6)\alpha l. \tag{15}$$

8x56   16x84   20x84

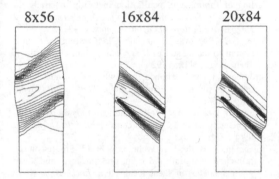

Figure 7. Contours of equivalent plastic strains with various meshes (non-local model, $\alpha = 1$ and $l = 20$ mm).

[x10⁶]

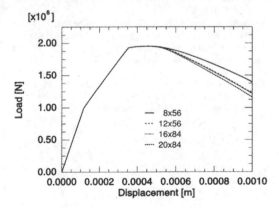

Figure 8. Load–displacement curves (non-local model, $\alpha = 1$ and $l = 20$ mm).

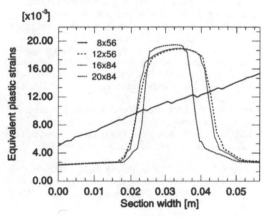

Figure 9. Equivalent plastic strains in a section perpendicular to a shear zone (non-local model, $\alpha = 1$ and $l = 20$ mm).

Table 1. The calculated width of a shear zone L [cm].

l [mm]	α			
	2.0	3.0	4.0	5.0
2	1 element	1.2	1.2	1.5
4	1.4	2.0	2.3	2.5
6	2.1	2.7	3.2	3.6
8	2.7	3.5	4.0	4.7
10	3.2	4.6	diffuse	–
12	3.8	diffuse	–	–

6 CONCLUSIONS

The FE-calculations demonstrate that conventional elasto-plastic models suffer from a mesh-dependency when material softening is included. The thickness and inclination of shear zones inside of a specimen, and load–displacement diagram in a post-peak regime depend strongly upon the mesh discretisation.

A modified non-local model causes a full regularisation of the boundary value problem. Numerical results converge to a finite size of the localization upon the mesh refinement. The thickness of a localized shear zone increases with increasing characteristic length and non-local parameter.

The results of a classical non-local theory suffer partly from the mesh sensivity.

REFERENCES

Bazant, Z. P. & Cedolin, L. 1979. Blunt crackband propagation in finite element analysis. *J. Engrg. Mech. Div. ASCE* 105(2): 297–315.

Bazant, Z. P. 1986. Mechanics of distributed cracking. *Appl. Mech. Rev.* 26: 675–705.

Bazant, Z. P. & Lin, F. B. 1988. Non-local yield limit degradation. *Int. J. Num. Meth. Engrg.* 35: 1805–1823.

de Borst, R. 1986. Non-linear analysis of frictional materials, *PhD Thesis*, Delft University.

de Borst, R. Mühlhaus, H. B., Pamin, J. & Sluys, L. 1992. Computational modelling of localization of deformation. In D. R. J. Owen, H. Onate & E. Hinton (eds), *Proc. of the 3rd Int. Conf. Comp. Plasticity*, Swansea: Pineridge Press, 483–508.

Brinkgreve, R. 1994. Geomaterial models and numerical analysis of softening, *Dissertation*, Delft University, 1–153.

Chen, E. P. 1999. Non-local effects on dynamic damage accumulation in brittle solids. *Int. J. Num. Anal. Meth. Geomech.* 23: 1–21.

Donze, F. V., Magnier, S. A., Daudeville, L., Mariotti, C. & Davenne, L. 1999. Numerical study of compressive behaviour of concrete at high strain rates. *J. Engrg. Mech.* 1154–1163.

Dragon, A. & Mroz, Z. 1979. A continuum model for plastic-brittle behaviour of rock and concrete. *Int. J. Engrg. Science* 17.

Herrmann, H. J., Hansen, A. & Roux, S. 1989. Fracture of disordered elastic lattices in two dimensions. *Phys. Rev. B* 39: 637–647.

Hilleborg, A. 1985. The theoretical basis of a method to determine thre fracture energy of concrete. *Mat. Struct.* 18: 291–296.

Loret, B. & Prevost, J. H. 1990. Dynamic strain localisation in elasto-visco-plastic solids, Part 1. General formulation and one-dimensional examples. *Comp. Appl. Mech. Engrg.* 83: 247–273.

Marcher, T. & Vermeer, P. A. 2001. Macro-modelling of softening in non-cohesive soils. In P. A. Vermeer et al. (eds), *Continuous and discontinuous modelling of cohesive-frictional materials*, Springer-Verlag, 89–110.

Mühlhaus, H. B. 1986. Scherfugenanalyse bei Granularen Material im Rahmen der Cosserat-Theorie. *Ingen. Archiv* 56: 389–399.

Mühlhaus, H. B. & Aifantis, E. C. 1991. A variational principle for gradient plasticity. *Int. J. Solid Struct.* 28: 845–858.

Pietruszczak, S., Jiang, J. & Mirza, F. A. 1988. An elasto-plastic constitutve model for concrete, *Int. J. Solid Struct.* 24(7): 705–722.

Place, D. & Mora, P. 2001. A random lattice solid model for simulation of fault zone dynamics and fracture processes. In H. B. Mühlhaus et al. (eds), *Bifurcation and Localisation Theory in Geomechanics*, 321–333.

Ragueneau, F., Borderie, Ch. & Mazars, J. 2000. Damage model for concrete-like materials coupling cracking and friction. *Int. J. Num. Anal. Meth. Geomech.* 5: 607–625.

Sakaguchi, H. & Mühlhaus, H. B. 1997. Mesh free modelling of failure and localisation in brittle materials. In A. Asaoka, T. Adachi & F. Oka (eds), *Deformation and Progressive Failure in Geomechanics*, 15–21.

Sluys, L. J. 1992. Wave propagation, localization and dispersion in softening solids, *Dissertation*, Delft University of Technology, Delft, 1992.

Strömberg, L. & Ristinmaa, M. 1996. FE-formulation of nonlocal plasticity theory. *Comp. Meth. Appl. Mech. Engrg.* 136: 127–144.

Tejchman, J. & Wu, W. 1993. Numerical study on patterning of shear bands in a Cosserat continuum. *Acta Mechanica* 99: 61–74.

Tejchman, J., Herle, I. & Wehr, J. 1999. FE-Studies on the influence of initial void ratio, pressure level and mean grain diameter on shear localisation. *Int. J. Num. Anal. Meth. Geomech.* 23: 2045–2074.

Willam, K. J. & Warnke, E. P. 1975. Constitutive model for the triaxial behaviour of concrete, *IABSE Seminar on concrete structures subjected to triaxial stress*, Bergamo, Italy, 1–31.

Vervuurt, A., van Mier, J. G. M. & Schlangen, E. 1994. Lattice model for analyzing steel-concrete interactions, In Siriwardane & Zaman (eds), *Comp. Methods and Advances in Geomechanics*: 713–718, Balkema, Rotterdam.

Zbib, H. M. & Aifantis, C. E. 1989. A gradient dependent flow theory of plasticity: application to metal and soil instabilities. *Appl. Mech. Reviews* 42(11): 295–304.

Geotechnical Measurements and Modelling, Natau, Fecker & Pimentel (eds)
© 2003 Swets & Zeitlinger, Lisse, ISBN 90 5809 603 3

Intermediate soils for physical modelling

K. Boussaiad, V. Ferber, J. Garnier & L. Thorel
French Public Works Research Laboratory, LCPC, Nantes, France

ABSTRACT: More realistic soils than pure sand and clay may sometimes be needed in physical models. Special techniques must be developed to prepare and characterise such "intermediate soils". Mixtures of Kaolin Speswhite clay and Fontainebleau sand have been investigated. Physical properties of these mixtures, as Atterberg limits and methylene blue value, depends linearly on the percentage of clay. Results of tests carried out using either static or dynamic loads show the effect of the clay percentage on soil compaction and give information on soil sample making method. The maximum density corresponds to 20% clay. The static compaction becomes more efficient compared to the Proctor procedure when the clay percentage is larger than 40%. The influence of sample density, water content and clay proportion on the sample compressibility and shear strength has been investigated in a modified oedometer apparatus and shear box.

1 INTRODUCTION

Two kinds of soils are generally used in geotechnical physical models (pit and centrifuge tests, calibration chambers): dry or saturated sand and saturated clay. Former is frictional soils usually tested under drained conditions. The latter is cohesive and is more used in undrained loading tests. However, natural soil deposits are often more complex since they can be unsaturated and present both cohesion and friction.

Classification of soil is generally based on the proportion of fines (particles smaller than 80 μm). A soil is considered as sandy if the fines proportion is lower than 20% and its behaviour is usually studied by considering the effective stresses. On the other hand, when the fines proportion is higher than 50%, the soil is considered as a fine soil, which behaviour is very often determined by total stress analysis. Soils containing between 50 and 80% of sandy particles are called intermediate soils, because of their intermediate behaviour between sand and clay (Tanaka, 2001). Other studies (see Cola, 2000 for example) have shown that the soil response is influenced by the fine fraction behaviour if it exceeds 70–80%.

Moreover artificial soils composed of kaolin and inert granular filler simulate soils behaviour in a more realistic way than pure kaolin (Rossato, 1992; Kimura, 1994).

This is the reason why a research programme has been engaged to develop experimental techniques and procedures for preparing "intermediate" soils. The main objective are to propose mixtures of reference

materials, as Fontainebleau Sand (FS) and Speswhite Kaolin (SK) and to determine the best preparation process for homogeneous soil masses as well controlled as possible.

The experimental work includes determination of Atterberg limits and methylene blue values, compaction, compressibility and shear strength characteristics of the mixtures. This program also aims to draw the main tendencies of the behaviour of the compacted mixtures.

2 PHYSICAL PROPERTIES OF MIXTURES

Particle size distribution of the mixtures were calculated from the FS and SK ones (fig.1). Some specific characteristics of the FS and of the SK can be noted here. In particular, the bad graded curve of the sand and the high proportion of clay particles (i.e. particles under 2 μm) in the SK lead to rather discontinuous particle size distributions of the mixtures.

Atterberg limits increase linearly with clay content (fig. 2). At very low clay content, Atterberg limits are difficult to evaluate but plasticity index can be estimated by extrapolating the data linearly. Kimura (1994) and Al-Shayea (2001) obtained similar results. Kimura showed that values obtained for IP on the extrapolated region are larger than the real plasticity index of soils.

The methylene blue value also depends linearly on the percentage of SK (fig. 3).

The linear variation of consistency limits and methylene blue values with fines proportion can be

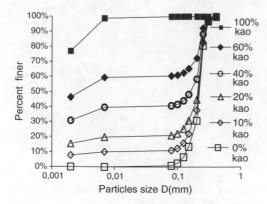

Figure 1. Particle size distributions of mixtures.

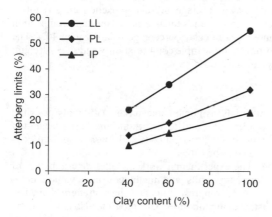

Figure 2. Atterberg limits versus clay content.

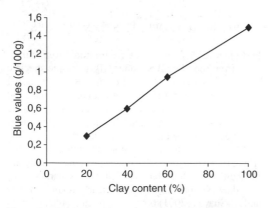

Figure 3. Methylene blue values versus clay content.

explained by the fact that the sand fraction behaves as an inert filler. Thereby, it shows that these two parameters could be considered as physico-chemical parameters.

3 COMPACTION

Dynamic compaction is considered as an efficient process for sandy soils. For clayey soils and more generally for cohesive soils, static compaction is simpler and more adequate (Camapum de Carvalho et al., 1987; Moussai et al., 1993).

In the present program, several mixtures of FS and SK were wetted at various water contents (w). The main objective was to highlight the effect of the particles size distribution and mineralogical properties of the mixtures on their behaviour under compaction. Dynamic and static compactions have been studied distinctly.

3.1 Dynamic compaction

Dynamic compaction was carried out according to the principle of the normal Proctor test with a new material developed recently by the division for Soil Mechanics and Site Investigation of the French Public Works Research Laboratory (LCPC) and described in detail by Boussaid (2002).

Compaction is carried out in an oedometer mould , which implies that the largest size of particles can not exceed 2 mm. This process enables to reduce the amount of materials that have to be prepared. It consists in compacting two successive 12,5 mm thick layers in a 25 mm high oedometer mould and to level the soil in excess. The compaction is carried out by means of a mini-hammer which dimensions (fig. 4) were calculated in order to apply the normal Proctor volumetric energy Ev:

$$Ev = (N \cdot H \cdot m \cdot g)/V = 593 \, kN.m/m^3 \qquad (1)$$

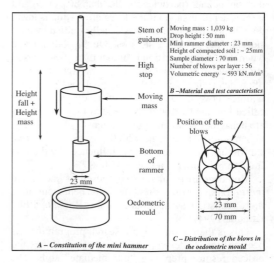

Figure 4. Mini hammer scheme & characteristics.

436

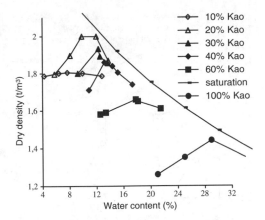

Figure 5. Water content versus max dry density (Proctor).

where N is the total number of blows (112), H the drop height of the hammer ($5 \cdot 10^{-2}$ m), m the mass of the hammer (1,039 kg), g the gravity (9, 81 m.s^{-2}) and V the volume of compacted material ($115 \cdot 10^{-4}$ m^3).

The mini-hammer diameter was designed so that the blows can be distributed as in the conventional CBR test (fig. 4c).

Figure 5 shows that the dynamic compaction behaviour of the mixtures is strongly related to the fines proportion. The maximum dry density increases until an optimal fines proportion of 20%. Beyond this proportion, the maximum dry density decreases steadily to reach the lowest value corresponding to pure SK maximum dry density.

3.2 Static compaction

The sample at the chosen w is has been here introduced in the oedometer mould.

Then, using a hydraulic press, the sample was loaded, 500 kPa by 500 kPa, up to 2500 kPa. Settlement was recorded at each loading stage. For the mixtures containing more than 30% of clay, static compaction was very efficient, at every w. With applied stress about 2000 kPa it enabled to reach density similar to the maximum dry density of the normal Proctor test (fig. 6).

On the other hand, as expected, this process was less appropriate for sandy mixtures because of the development of friction between sandy particles (fig. 7).

The variation of the maximum dry density with the fines proportion is approximately the same as with dynamic compaction and the optimal content of kaolin is about 20% (fig. 8).

3.3 Interpretation

In order to characterize the physical state of the fractions constituting the mixtures, two parameters can be introduced:

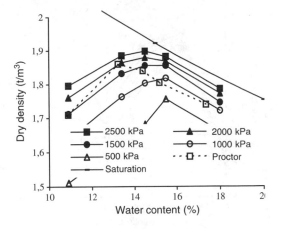

Figure 6. Static and dynamic compaction of 40% SK mixture.

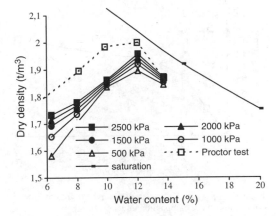

Figure 7. Static and dynamic compaction of 20% SK mixture.

− ρ^*_{sand}: Dry density of the sand grains in the mixture.

$$\rho^*_{sand} = M_{sand}/V_{total} = \rho d\,(1-Pc) \qquad (3)$$

− ρ^*_{clay}: Dry density of the clayey fraction in the void volume.

$$\rho^*_{clay} = M_{clay}/(V_{total}-V_{sand})$$
$$= (\rho d \cdot \rho s \cdot Pc)/[\rho s - \rho d(1-Pc)] \qquad (4)$$

where M_{clay}, M_{sand}, ρd, ρs and Pc represent, respectively, the dry masses of clay and sand fractions, the dry density of the mixture, the grain density and the fines proportion in the mixture. Figure 9 shows that ρ^*_{sand} first increases slightly for low clay ratio, keeps a constant value up to a fines proportion of 20% and then decreases and reach the zero value for pure kaolin. The parameter ρ^*_{clay} increases in two stages, first with a sharp rope and beyond 20% of fines with

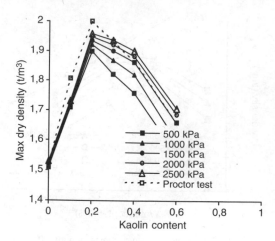

Figure 8. Maximum dry density versus clay content.

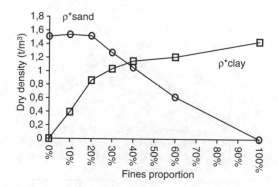

Figure 9. ρ^*_{sand} and ρ^*_{clay} variation on the optimum Proctor conditions (static compaction under 500 kPa).

a softer one. For 60% of fines, the dry density of the fines fraction in the mixture is almost at 90% of the maximum dry density of pure kaolin.

Thus, below 20% of fines, the lubricate role of clay on the sandy particles generates a new distribution of the sandy grains. It results in increasing ρ^*_{sand} with the proportion of SK, and it is one of the sources of increase in mixtures density. The clayey particles occupy larger and larger part of the volume of the void separated by the coarse grains and so ρ^*_{clay} increases in a continuous way between 0 and 20% of SK. Beyond 20% of fines, the decrease of ρ^*_{sand} represents the dispersion of the sandy particles in the mixture.

4 MECHANICAL BEHAVIOUR

Compressibility and direct shear tests have been carried out using mixtures at 20, 40 and 60% of SK representing, respectively, a sandy, an intermediate and a

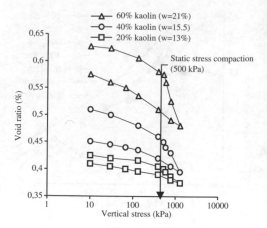

Figure 10. Compressibility curves of different mixtures.

clayey soil. All samples were compacted at the same static pressure of 500 kPa before testing.

4.1 Compressibility

The oedometer test with a constant w has been developed by Delage and Fry (2000). It derives from the conventional oedometer test with a device enabling w of samples to be kept constant. Saturation of the air inside the cell is done by placing a small water container and insulating the test-tube from the ambient air by an impermeable membrane. In these conditions, the loss of w does not exceed 0.75 point for approximately one week (Delage & Fry, 2000).

In these conditions, settlement is almost instantaneous: it reaches 90% of final settlement after only a few minutes (the settlement is considered as completed after 24 hours of application of the load). This phenomenon was explained by the fact that in the unsaturated soils, the air phase is continuous and the settlement corresponds to an expulsion and immediate compression of this air, and local reorganisation of the water meniscus without any transfer of liquid (Delage & Fry, 2000).

For mixtures of 60% and 40% of SK, only small variations of volume are observed for stresses smaller than the compaction stress of 500 kPa (fig. 10). Beyond 500 kPa, strains increase and become irreversible. Thus, as saturated soils, compacted unsaturated mixtures seem to keep a trace of their compaction history. Delage and Fry (2000) obtained similar results when carrying out tests on the silt of Jossigny compacted at a w of 22%. Moreover, it can be observed that larger is the proportion of clay, more the stress compaction appears clearly as a pre-consolidation stress.

This influence of the proportion of clay is also illustrated by the values of the compression index Cc ($-\Delta e/\Delta \log \sigma$). As a matter of fact, Cc decreases from

0.2 to 0.14 when the proportion of SK decreases from 60% to 40%. At 20%, Cc is only 0.06.

4.2 *Shear strength*

The influence on shear strength of w and of the proportion of SK has been investigated through shear box tests. All samples were prepared at the right w and consolidated in the box under a static vertical pressure of 500 kPa. Shear tests were then carried out on these samples under vertical stress of 50, 160 and 260 kPa. OCR were then respectively close to 10, 3.1 and 1.9. The displacement rate during the shear test was 0.11 mm/min. As the tested samples are far from being saturated (fig. 6 and 7), these shear tests may be considered as drained tests.

Figure 11 shows shear tests results obtained on the mixture at 40% of SK: friction angle decreases with the w. The reduction is larger when w goes from 9% to 12% and is less for further w decrease (from 12% to 20%). This reduction in friction angle with increase in w may be explain by the lubrication capacity of the clayey particles when they are wetted. Lubrication is due to the mobility of the absorbed film because of its increasing thickness and its larger surface ion hydration and dissociation (Mitchell, 1993; Al Sheyea, 2001).

The effect of w on the cohesion is more complex (fig. 11). Cohesion first increases (when w goes from 9% to 12%) but then decreases for larger w.

This reduction in cohesion when increasing the amount of water in the mixture beyond a certain limit (sample density and SK proportion being constant) can be explained by the fact that the distance between the clay particles becomes larger and the electrostatic and electromagnetic attractions decrease. In the same way, water in excess decreases adhesion and cementing and when w reaches saturation, suction becomes null. Al shayea (2001) obtained similar results on different materials.

According to figure 6, it seems that the zone of increase in cohesion with w corresponds to the "dry" side of the compaction curve (w less than 15%) and the zone of decrease in cohesion to the "wet" side (w larger than 15%). The rather good coincidence observed between the two optimum values of w is not surprising. The shear strength is indeed expected to be larger in samples at smaller void ratio. But this effect is only observed on the cohesion component and not at all on the friction angle.

To study the effect of the proportion of SK on the shear strength, it was decided to compare shear tests results on mixture prepared at the optimum w (since w has also a strong influence on shear strength as seen above). Results are given in Table 1. At these optimum conditions, the friction angle decreases continuously with the proportion of clay (from 28% to 12%). At the opposite, cohesion is larger for a middle SK proportion of 40%.

5 FIRST RECOMMENDATIONS FOR THE RECONSTITUTION OF SOIL SAMPLES

For the clayey mixtures with SK content exceeding 40%, the routine techniques developed for pure clay may still be applied (mixing soil materials and water followed by static consolidation of the mixture under controlled stress). Without any more results on their behaviour, it could be cautious to keep these samples saturated when using them in geotechnical modelling. It is indeed not yet possible to perfectly control the suction during the whole experiment on a physical

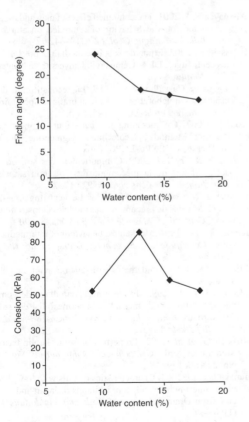

Figure 11. Cohesion and angle of internal friction versus water content for 40% SK mixture.

Table 1. Shear strength of mixtures prepared at the optimum water content and consolidated at 500 kPa.

SK proportion	20%	40%	60%
Friction angle (°)	28	18	12
Cohesion (kPa)	8	57	40

model. Otherwise, as the friction angle of these mixtures is larger than the clay one, it could be needed to reduce the friction between the sample and the mould or container wall as it is sometimes done in centrifuge modelling (latex or neoprene membrane, layer of silicone grease between the container and the membrane to reduce friction).

When the percentage of sand is higher than 60%, the energy needed for compaction becomes rather large and the equipment developed in the different laboratories for static compaction will not be strong enough to apply the required consolidation stress. As an example, when a soil sample of about a square meter has to be prepared (as in the large geotechnical centrifuge facilities), a 2000 kPa pressure on the sample requires a 2 MN hydraulic press and consolidometer. More difficult than the hydraulic press capacity itself will be the design of the piston that will transfer the load to the sample. Its stiffness would be extremely large to keep its own deformation below a given limit value. First tests carried out in LCPC demonstrate that this is absolutely needed to guarantee a good homogeneity of the soil sample over the whole container.

One alternative to the design of these new huge equipment may be the dynamic compaction, layer by layer, in the container. This technique has already been used in LCPC and compaction was done with a pneumatic tamper (Garnier et al., 1988). Again, good homogeneity of the sample is required and the compaction process must be perfectly controlled (energy of compaction at each blow, distribution of the blows over the sample). Automatic systems may be needed based on the same principle as for Proctor compaction tests but at a larger scale.

6 CONCLUSION

Some results have been obtained during this first study on the preparation and characterisation of intermediate soils:

1. Physical properties of sand-clay mixtures as Atterberg limits and methylen blue value depends linearly on the fines proportion (figs 2&3).
2. The ability to compaction is mainly affected by the percentage of clay. The maximum density was obtained for a clay proportion of 20% (fig. 5).
3. When the percentage of clay is more than 30%, static compression under 2000kPa gives samples as dense as Proctor dynamic compaction (fig. 8). For more sandy soils, static compaction is less efficient.
4. The settlement of unsaturated mixtures is almost instantaneous. Mixtures with high clay content seem to keep a trace of the stress compaction since the strain become increasingly important and irreversible beyond the compaction stress value.
5. For a given stress compaction, friction angle of the compacted mixtures decreases with water but cohesion shows a maximum value for middle water content (that corresponds approximately to the optimum for compaction).
6. For samples prepared at the optimum condition, the friction angle seems to decrease continuously with the clay proportion, from the value of pure sand to the value of pure clay.
7. Again at the optimum condition, cohesion reaches a maximum value for mixtures with middle clay proportion.
8. New equipment must be developed to prepare large samples of intermediate soils in well controlled conditions. Additional tests are also needed to determine the best procedures and the clay/sand/water proportions for preparing samples with given characteristics (stiffness and strength).

REFERENCES

Al-shayea N.A. 2001. The combined effect of clay and moisture content on the behavior of remolded unsaturated soils. *Bull. Asso. Engng Geol.* 62 : 319–342.

Boussaid K. 2002. Etude de sols intermédiaires pour modèles centrifugés. DEA. Génie civil. Université de Nantes-*LCPC Nantes*, 44 p.

Camapum de Carvalho *et al.* 1987. La reconstitution des éprouvettes au laboratoire, théorie et pratique opératoire. *Rapport de recherche LCPC*, 145, 54 p.

Cola S. 2002. On modeling the behaviour of melange. Numerical Methods in Geotechnical Engineering, Mestat (ed) Presse de l'ENPC/LCPC, Paris.

Delage P. & Fry J.-J. 2000. Comportement des sols compactés: apport de la mécanique des sols non saturés. *Revue française de géotechnique* 92 : 17–29.

Garnier J. & Cottineau L.M. 1988. La centrifugeuse du LCPC: Moyens de préparation des modèles et instrumentation, *Centrifuge 88, Corté (Ed.)*, Balkema, 83–90.

Kimura T. *et al.* 1994. Mechanical behaviour of intermediate soils. *Centrifuge 94, Singapore, Leung et al. (Ed.), Balkema:* 13–24.

Mitchell J.K. 1993. Fondamentals of soils behavior. 2nd ed. Wiley, New York.

Moussai B. *et al.* 1993. Etude d'un appareillage de compactage statique et de mesure de la perméabilité des sols fins argileux, *Bulletin Liaison Labo. P. et Ch-188-nov dec 1993-Ref 3804* : 15–22.

Rossato G. *et al.* 1992. Properties of some kaolin-based Model Clay Soils, *Geotechnical testing journal*, Vol.15, No. 2 : 166–179.

Tanaka H *et al.* 2001. Characteristics of soils with low plasticity: Intermediate soil from ishinomaki, Japan and lean clay from Drammen, Norway. *Soils and foundations* 41 (1): 83–96.

Geotechnical Measurements and Modelling, Natau, Fecker & Pimentel (eds)
© 2003 Swets & Zeitlinger, Lisse, ISBN 90 5809 603 3

Experimental and numerical investigation of the influence of local site conditions on the ground motion during strong earthquakes

M.M. Bühler, R. Cudmani, V.A. Osinov, A.B. Libreros-Bertini & G. Gudehus
Institute of Soil and Rock Mechanics, University of Karlsruhe, Karlsruhe, Germany

ABSTRACT: The paper presents a numerical method for the analysis of ground response during strong earthquakes validated using laboratory and field data. The method is based on a one-dimensional wave propagation model in which the nonlinear soil behaviour is described by a (visco)hypoplastic constitutive law. Model tests using a novel laminar box mounted on a shaking table have been performed in the laboratory to verify the applicability of the (visco)hypoplastic constitutive relation to the modelling of the soil behaviour under dynamic earthquake-like excitation. To validate the model for real field conditions, two typical soil profiles from the Bucharest Vrancea region are used: soft cohesive soil near INCERC station and non-cohesive soil near EREN station. It is concluded that local soil conditions significantly affect the ground response during strong earthquakes and should be properly considered when performing seismic hazard, vulnerability and risk analyses. As shown by the present study, the proposed numerical model is able to fulfil this requirement.

1   INTRODUCTION

Observations of the damage caused by earthquakes clearly reveal that ground shaking induced at different sites by the same seismic event can be very different depending on the locally existing geological, geotechnical and topographic conditions. Reports of historical earthquakes dating from the last two centuries present clear evidence of the influence of local site conditions on the intensity of ground shaking and earthquake damage. For instance, MacMurdo (1823) reported that during the 1819 earthquake in Cutch, India buildings founded on rock did not suffer as much damage as those whose foundations did not reach the bedrock. Gutenberg (1927) was the first who tried to relate observed seismic response with local site conditions. Based on micro-seismic records at sites with different subsurface conditions, he developed site-dependent amplification factors. In recent years, the interpretation of strong motion data from sites with different subsurface conditions has made possible more reliable quantification of local site effects. The comparison of measured ground motions at Yerba Buena Island (rock station) and Treasure Island (soil station) during the 1989 Loma Prieta earthquake (Seed et al. 1991, Arulanandan et al. 1997), and also the interpretation of the acceleration records at Port Island during the 1995 Kobe earthquake (Ishihara et al. 1996) show how local site conditions may influence all important

characteristics – amplitude, frequency content and duration – of ground motion.

In general, the influence of local conditions is controlled by: (1) geological characteristics of the site (origin of the soils and type of the underlying bedrock, subsurface stratification); (2) granulometric properties and state of the soil layers (stresses, density, fabric, degree of saturation); (3) characteristics of the earthquake input motion (frequency content, duration, amplitude of the bedrock motion); and (4) topography of the site.

Practical applications of many of the available models, especially those based on the theory of elasticity, to strong earthquakes are limited by the inability of the used constitutive laws to model the non-linear soil behaviour under alternating loading.

In this paper, the applicability of the hypoplastic relations to the modelling of the dynamic soil behaviour is investigated in the laboratory using a novel 1-g laminar shake-box. Finite element simulations of the laboratory experiments have been performed to show the adequacy of both the experimental device and the numerical model. Furthermore a one-dimensional nonlinear wave propagation model based on the theory of (visco)hypoplasticity is presented (Osinov 2000, 2001, 2003, Osinov & Gudehus 2003). The model describes the dynamic response of a soil deposit induced by a plane wave coming from below. Applications of the proposed numerical model are exemplified by the evaluation of the ground response to strong earthquakes for two typical soil profiles from the Bucharest Vrancea region.

2 EXPERIMENTS WITH A LAMINAR SHAKE BOX

In the last decades, different model testing devices – so-called shake boxes – have been developed to investigate the dynamic behaviour of soils and embedded geotechnical foundations and structures under seismic excitation in the laboratory. Experiments with shake boxes are usually performed in a normal laboratory environment (so-called 1-g condition) or in a centrifuge where the magnitude of the acceleration field is n times greater than gravity. Shake boxes differ in the construction of their walls. Basically, the walls of a box can be rigid ("rigid box") or consist of movable segments ("laminar box"). Most of the laminar boxes consist of rigid rectangular frames that are allowed to translate relatively to each other. Although such a laminar box is an improvement of a rigid box for the modelling of plane waves, the strain field in it still differs from that of plane wave propagation since shear strains are localized – as in a conventional shear box test – in thin zones between the rigid frames.

In order to investigate the dynamic response of soil under dynamic earthquake-like excitation, model tests have been carried out with the novel 1-g laminar box shown in Figure 1 (length 400 mm, width 300 mm, height 500 mm). To overcome the limitations of existing laminar boxes, the lamellas of our laminar box are allowed to translate and rotate. Opposite lamellas are constrained to undergo exactly the same motion by hinged bar connectors. Parallel to the shaking direction, the box consists of smooth steel walls, which are rigidly connected to the base. The dynamic base excitation is generated by means of springs attached to the base of the box.

The spring forces are activated by enforcing an initial displacement to the base and fixing it in the new position. Shaking is initiated by a manual release mechanism. Displacements of the lamellas, settlements of the surface, and pore pressures at the bottom can be recorded, processed and consequently analyzed.

In most of the experiments fine quartz sand ($d_{50} = 0.25$ mm, $e_{max} = 0.98$, $e_{min} = 0.65$, $U = 2.35$) was used in both dense and loose initial state. A few experiments have also been made with a soft, nearly water saturated silty clay. The lateral wall adhesion was removed by means of a water film produced by electrophoresis.

In the first test series, loose dry sand has been investigated for different intensities of excitation of the base by applying initial deflections of the base of 2, 4, and 8 mm. These deflections cause peak acceleration values of slight, moderate and strong earthquakes, respectively. Figure 2 shows the horizontal displacements of the lamellas over time. As can be seen, damping increases with increasing shaking intensity and decreases with increasing initial density. This demonstrates the hysteretic nature of soil damping, and indicates that the assumption of viscous damping in elastic models is not realistic for moderate and strong earthquakes. During shaking the surface settles due to densification. After about five test series the further settlement of the surface became negligibly small.

Subsequent experiments have been carried out with saturated sand (Fig. 3) where the soil showed higher energy dissipation than in a dry state even for small amplitudes. Liquefaction, which is assumed to occur when the measured excess pore water pressure at the base equaled the initial effective stress, took place already for an initial base deflection of 4 mm. Since the upper half of the soil specimen moved as a rigid body (Fig. 3) we deduce that liquefaction must have occurred underneath. Whereas for an initial deflection of 8 mm skeleton disaggregation must have extended to the whole specimen since the observed dynamic response of the material resembled that of a viscous suspension (Fig. 3).

Excess pore pressure dissipates and subsequent settlements of the surface develop about ten times faster than calculated by means of the conventional consolidation theory taking the permeability of the material into account. Observation of the surface during the experiment reveals a system of fine vertical water channels allowing faster drainage. After repeating the test the sample densifies quickly and thus, the liquefaction susceptibility is reduced.

3 NUMERICAL MODELLING OF THE LAMINAR BOX TESTS

A numerical back-analysis of the laminar box tests was carried out to verify the adequacy of the device and the applicability of the constitutive equation used in the calculation for cohesionless fine-grained soils. The simulation was run with the finite element program

Figure 1.   Laminar shake box.

ABAQUS. The domain was discretized using 8-node continuum elements (Fig. 4).

The mass of the laminar box and the lamellas were modelled with mass elements. Spring and dashpot elements were attached to the bottom of the box to model the shaking mechanism. The nodes were constrained to undergo the same motion in the direction of shaking. No constraint was imposed in the vertical

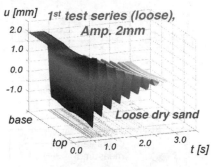

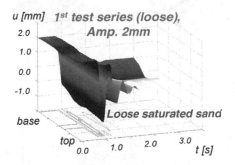

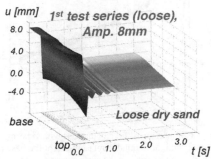

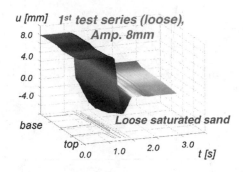

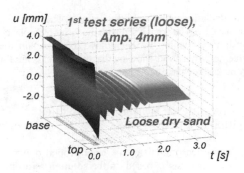

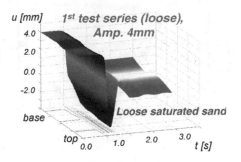

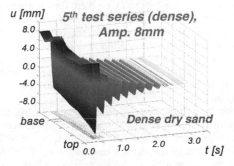

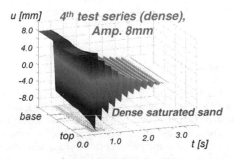

Figure 2.   Laboratory results for dry sand.

Figure 3.   Laboratory results for saturated sand.

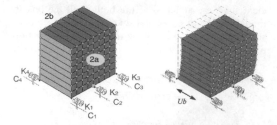

Figure 4. FE model for the calculation of the laminar box experiments.

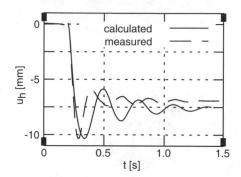

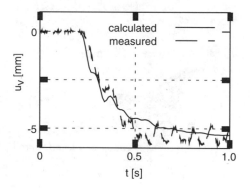

Figure 5. Comparison of FE calculations with experiments for dry sand: horizontal displacements at the base (u_h) and vertical displacements at the surface (u_v) vs. time.

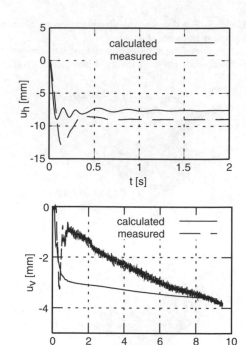

Figure 6. Comparison of FE calculations with experiments for saturated sand: horizontal displacements at the base (u_h) and vertical displacements at the surface (u_v) vs. time.

and calculations show a good agreement. The tests with clay have been backanalysed with the visco-hypoplastic constitutive relation by Niemunis (2003). Again the agreement was good, showing that this model is apt for strong earthquake effects in clay layers.

4 NUMERICAL MODEL FOR SEISMIC RESPONSE ANALYSIS

The numerical model for the level ground dynamic response analysis is based on the solution of a one-dimensional boundary value problem for a horizontal soil layer (Osinov 2000, 2001, 2003, Osinov & Gudehus 2003). In this problem, the unknown variables are functions of depth and time. The material velocity vector has two components (horizontal and vertical), and nonzero components of the stress tensor are σ_1, σ_2, σ_3 and τ_{12} with the x_1-axis being directed vertically.

For the part of the layer above the water table, the governing system of equations consists of the equations of motion, the constitutive equation for the solid skeleton, and the mass balance equation for the void ratio. For the soil below the water table, the stress in the skeleton represents the effective stress, and pore

direction. In the first calculation step the desired initial deflection was enforced to the base.

In the subsequent dynamic step the constraint was released and the system was allowed to oscillate. In the third step the settlements were obtained by allowing the excess pore pressure to dissipate. Calculated and measured horizontal displacements of the base and settlements of the surface are compared in Figure 5 and 6 for dry and saturated sand. As can be seen, experiments

444

pressure is introduced as an additional variable with an appropriate constitutive equation. The constitutive equation for the pore pressure involves the compressibility of the pore fluid, which depends strongly on the degree of saturation. In problems with saturated soil, either locally undrained or drained conditions may be assumed. The problem with undrained conditions is described by Osinov (2000). In the case of drained conditions, the socalled u-p formulation is used for the present study (Zienkiewicz et al. 1980, 1999), the numerical algorithm is described by Osinov 2003. This formulation constitutes an approximation to the dynamic equations for a two-phase medium, and allows seepage of the pore fluid to be taken into account. Whether seepage should be taken into account in the dynamic problem depends on the permeability of the soil. Calculations show that the evolution of the pore pressure during an earthquake calculated with locally drained conditions is practically the same as the one with locally undrained conditions if the permeability of the soil is less than 10^{-5} m/s (Osinov 2003).

The boundary condition at the base of the layer consists of the velocity components as functions of time. In addition, if the soil is saturated, this boundary is assumed to be impermeable. The upper boundary of the layer is free of traction. The pore pressure at the water table is taken to be constant (zero or atmospheric). A finite-difference algorithm with implicit time integration solves the problem numerically.

5 APPLICATION EXAMPLES

Verification of the proposed one-dimensional propagation model using real earthquake data is presented by Cudmani et al. (2003). In the following the application of the model is exemplified by the analysis of two level ground sites from the Bucharest Vrancea region (Balan et al. 1982) for a strong earthquake (Fig. 7).

At the EREN seismic station, northern Bucharest, the ground consists predominantly of cohesionless soil layers. On the other hand, the soil profile at the INCERC seismic station, eastern Bucharest, is a predominantly cohesive soil profile (Lungu & Aldea 2000). For the determination of the constitutive parameters available geotechnical and geological data were used (Ciugudean-Toma 2001, Moldoveanu 2001). For the upper layer down to ca. −20 m results of index and standard laboratory tests were available so that the hypoplastic constitutive parameters could be determined with the procedure given by Herle 1997, Herle & Gudehus 1999. For deeper soil layers, for which laboratory tests were not available, the constitutive parameters were estimated from granulometric properties. The initial relative density in the cohesionless soil layers was estimated from standard penetration test data. The void ratios in the clay layers were

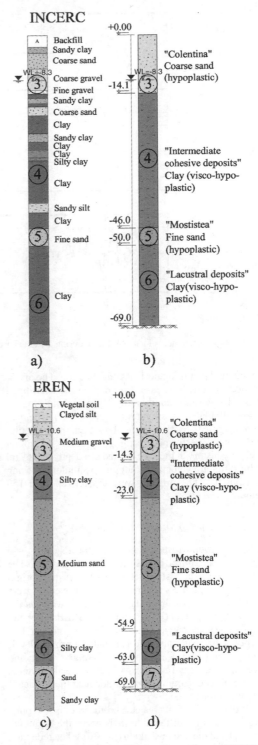

Figure 7. Real and simplified soil profile at INCERC (a, b) and EREN (c, d) sites.

445

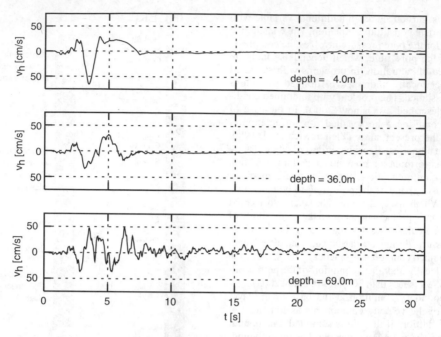

Figure 8. Horizontal velocities vs. time during a strong earthquake at INCERC site.

calculated assuming normal consolidation. The distribution of vertical stresses over depth was determined from the geostatic equilibrium. The horizontal stresses were calculated assuming $K_0 = 1 - \sin(\varphi_c)$, where φ_c is the critical friction angle. The numerical calculations were performed using a synthetic horizontal velocity history at the base located at a depth of -69 m. The synthetic velocity is thought to simulate a strong earthquake. The profiles were discretized with 150 elements.

5.1 INCERC station

For the ground response analysis, the INCERC soil profile was simplified as shown in Figure 7b. The intermediate cohesive deposits of lacustral origin (layer 4) and the lacustral deposits (layer 6) were modelled as visco-hypoplastic material, whereas the "Colentina" gravel (layer 3) and the "Mostistea" bank of sands (layer 5) were modelled as rate independent hypoplastic materials. The calculated horizontal velocity time histories at depths of –4 m and –36 m and the input velocity time history at the base are given in Figure 8. The maximum magnitude near the surface is obtained for $t = 3$ s.

Figure 9 shows the calculated mean effective pressure vs. depth at different times between the beginning ($t = 0$) and the end of shaking ($t = 30$ s). The reduction of the effective stresses concentrates in the "Colentina" and "Mostistea" sand layers. Liquefaction is first

obtained in the lower boundary of the "Colentina" layer after $t = 8$ s. Shortly thereafter the "Mostistea" layer liquefies, thus impeding further passage of shear waves to the upper layers. Shear isolation and the relatively high permeability of the "Colentina" gravel ($k = 10^{-3}$ m/s) lead to an increase in the effective stresses due to drainage and consolidation between $t = 8$ s and 30 s. On the other hand, the clayey layers did not experience substantial reduction of the effective stresses, and can therefore contribute to the amplification of the base shaking before the sand layers become liquefied.

5.2 EREN station

The idealization of the EREN soil profile is depicted in Figure 7d. The soil profiles of both stations are similar, with the difference that the "Mostistea" layer at the EREN site is ca. 8 times thicker.

The calculated horizontal velocity time histories at depths of -4 m and -36 m are given in Figure 10. Amplification of the base shaking is observed at depths of -4 m between $t = 2$ s and $t = 6$ s. Thereafter, the intensity of shaking near the surface reduces considerably. Strong deamplification up to decoupling from the base shaking occurs after $t = 12$ s. On the other hand, amplification of the base shaking at a depth of -36 m is observed between $t = 4$ s and $t = 8$ s. At this depth, the base shaking is deamplified after $t = 12$ s. At both considered depths, deamplification

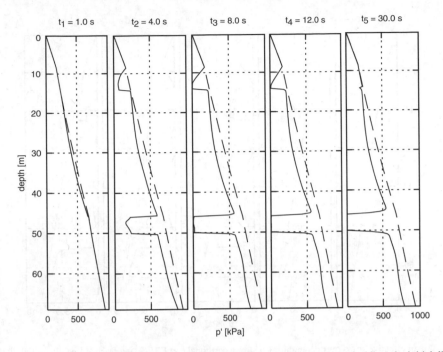

Figure 9.   Mean effective stress over depth during a strong earthquake at INCERC site. Dashed line: the initial distribution.

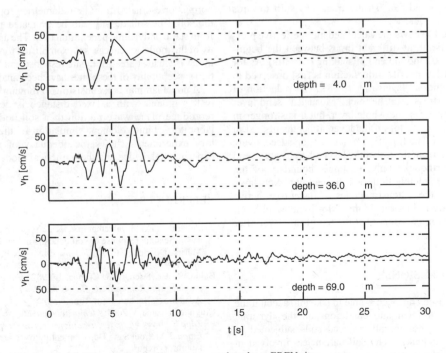

Figure 10.   Horizontal velocities vs. time during a strong earthquake at EREN site.

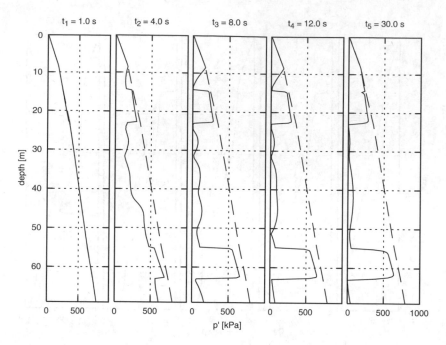

Figure 11. Mean effective stress over depth during a strong earthquake at EREN site. Dashed line: the initial distribution.

is accompanied by a strong frequency shift towards lower frequencies.

Figure 11 shows the calculated mean effective pressure vs. depth at different times between the beginning (t = 0) and the end of shaking ($t = 30$ s). As for the INCERC profile, liquefaction is first observed in the "Colentina" layer after $t = 8$ s. The reduction of effective stresses in the thick "Mostistea" sand layer takes place over the whole layer, but it is stronger in thin zones at the top of the layer, at depths of -31 m and -51 m. The liquefied zones develop successively beginning from the top of the layer whereby at the end of the earthquake only the upper boundary of the layer is completely liquefied. In the sand layer (layer 7) beneath the "Lacustral deposits" (layer 6) effective stresses are reduced as in the "Mostistea" sand layer. Liquefaction is observed at the top of this layer after $t = 30$ s.

6 CONCLUSIONS

The proposed approach is able to take into account the influence of local site conditions on the dynamic response of horizontally layered soils subjected to strong earthquakes. The soil parameters involved in the constitutive equation and required for practical applications of the model can be determined using disturbed samples from conventional exploratory borings or estimated from granulometric properties of the soil. The soil density can be evaluated indirectly from static or dynamic penetration tests. The applicability of the model to real cases is exemplified for two sites in the Bucharest area. Both site response and liquefaction susceptibility of these sites have been analysed.

A novel laminar box consisting of moving opposite wall segments with a fixed distance is devised to reproduce plane wave motion of a soil body in the laboratory. Finite element simulation of the laboratory experiments shows the adequacy of both the experimental device and the numerical model.

REFERENCES

Arulanandan, K., Muraleetharan K. K. & Yoganchandran, C. 1997. Seismic response of soil deposits in San Francisco Marina District. *J. Geot. Geoenv. Eng.*, ASCE, 123(10), 965–974.

Balan, S., Cristescu, V. & Cornea, I. 1982. *Cutremurul de pamint din Romania de la 4 martie 1977*, Editura Academiei, Bucharest (in Romanian).

Ciugudean-Toma, V. 2001. *Memoriu tehnic Mag 4. Racord1, Zona I, Investigatii teren conform cerinte expert BEI, Etapa 2.* Metroul S.A., Departamentul Proiectare, internal laboratory report.

Cudmani, R. O., Osinov, V. A., Bühler, M. M. & Gudehus, G. 2003. A model for the evaluation of liquefaction susceptibility in layered soils due to earthquakes,

The 12th Panamerican Conference on Soil Mechanics and Geotechnical Engineering June 2003, Cambridge, USA, in press.

Gutenberg, B. 1927. *Grundlagen der Erdbebenkunde*, Verlag Borntraeger, Berlin.

Herle, I. 1997. Hypoplastizität und Granulometrie einfacher Korngerüste. *Veröffentlichungen des Instituts für Boden- und Felsmechanik der Universität Karlsruhe.* No. 142.

Herle, I. & Gudehus, G. 1999. Determination of the parameters of a hypoplastic constitutive model from properties of grain assemblies. *Mech. Cohesive-frictional Mater.* 4: 461–486.

Ishihara, K., Yasuda, S. & Nagase, H. 1996. Soil Geotechnical aspects of the January 17, 1995 Hyogoken- Nambu earthquake: soil characteristics and ground damage, *Soils and Foundations, Special Issue,* Jan. 1996, 109–118.

Lungu, D. & Aldea, A. 2000. Near-surface geology and dynamic properties of soil layers in Bucharest. In: *Vrancea Earthquakes: Tectonics, Hazard and Risk Mitigation,* F. Wenzel et al. (eds.), Kluwer Academic Publisher, Germany, 137–148.

MacMurdo, J. 1823. Papers relating to the earthquake which occurred in India in 1819, *Philosophical Magazine, Trans. Literary Soc. Bombay,* 63, 105–177.

Moldoveanu, T., Dinu, C., Negut, A., Lungu, D. & Pascu, R. 2001. *Geological profile and the seismic velocity model at INCERC Bucharest test site,* Institute of Geotechnical and Geophysical Studies (GEOTEC), Bucharest, Romania, internal report.

Niemunis, A. 2003. Extended hypoplastic models for soils. *Monografia nr 34,* Politechnika Gdanska, Poland.

Osinov, V. A. 2000. Wave induced liquefaction of a saturated sand layer. *Continuum Mech. Thermodyn.,* 12(5): 325–339.

Osinov, V. A. 2001. The role of dilatancy in the plastodynamics of granular solids. In: *Powders and Grains 2001,* Y. Kishino (ed.), Swets and Zeitlinger, Lisse, 135–138.

Osinov, V. A. 2003. A numerical model for the site response analysis and liquefaction of soil during earthquakes. *This volume.*

Osinov, V. A. & Gudehus, G. 2003. Dynamics of hypoplastic materials: theory and numerical implementation. In: *Dynamic Response of Granular and Porous Materials under Large and Catastrophic Deformations,* K. Hutter and N. Kirchner (eds.), Springer, Berlin, 265–284.

Seed, R. B., Dickenson, S. E. & Idriss, I. M. 1991. Principal geotechnical aspects of the Loma Prieta Earthquake, *Soils and Foundations,* 31(1): 1–26.

Zienkiewicz, O. C., Chang, A. H. C., Pastor, M., Schrefler, B. A. & Shiomi, T. 1999. *Computational geomechanics with special reference to earthquake engineering.* Chichester, John Wiley.

Zienkiewicz, O. C., Chang, C. T. & Bettess, P. 1980. Drained, undrained, consolidating and dynamic behaviour assumptions in soils. *Géotechnique,* 30(4): 385–395.

Geotechnical Measurements and Modelling, Natau, Fecker & Pimentel (eds)
© *2003 Swets & Zeitlinger, Lisse, ISBN 90 5809 603 3*

New approaches in modelling recent subrosion processes

W. Busch, C. Hanusch, St. Knospe & K. Maas
Institute of Geotechnical Engineering and Mine Surveying (IGMC), Technical University of Clausthal, Clausthal-Zellerfeld, Germany

ABSTRACT: In order to combine the representation of geochemical underground leaching processes in salinar rocks with associated time variant changes of cavern shapes, a temporal Geo-Information System (TGIS) is to be developed involving a database management system and an object-oriented process model based on Volume-NURBS (Non-Uniform Rational B-Splines) supplemented by 4D-visualisation-techniques. Different rock-types and cavities are represented as Geo-objects (free form bodies) and treated uniformly as temporal Volume-NURBS, respecting their geochemical and physical properties. In addition to thermodynamic calculations underground leaching processes in salinar rocks can be specified by solution restricting processes, e.g. recrystallisation and accumulation of insoluble substances. Within the complex system of shape changing processes uncertainties and lacking data are to be considered. For this, a fuzzy-rule based system has to be set up. Process model and data are to be managed within an information system, providing methods for time- and space-continuous representation (TGIS).

1 INTRODUCTION

In the following the concept and the main purpose of the DFG (German Research Foundation) promoted research project *Analysis, modelling and simulation of recent subrosion processes in abandoned mine sites regarding feasible methods of controlling and interference* are presented. In the context of this project a spatio-temporal process-model is developed representing underground leaching processes connected with deformation processes of participating geological bodies and changing structural characteristics of the Geo-system. The development is based upon object-oriented analysis-, design- and modelling-techniques.

Geo-objects are treated as real temporally varying volumes, so called temporal volume NURBS (Non Uniform Rational B-Splines), processes are implemented as methods of the objects. The adjusted data model consists of the central module CORE (combined object relationally data model) with the extension of GeoCORE for the processing of spatial data. The application of the new system serves exemplarily the prognosis of underground leaching process within the abandoned potash mine *Neustaßfurt*. In 1972 the last mine of the Staßfurt mining district was closed and finally flooded in order to keep the mine safe from collapsing. Unfortunately there are still leaching processes taking place at underground exposed carnallitic salt

rock (subrosion). Subsidence and mining sinkholes are the consequences of the resulting destabilisation of overburden layers.

Since any required remedial action means significant capital and temporal investments, sustainability through simulation is essential. Therefore a scientific Geo-Information System is to be developed using existing data, expert knowledge – advanced by data attained in tests (field, laboratory) – as well as physicochemical process- and interaction-modelling in order to enable the spatio-temporal representation of the Geo-system.

2 PROCESS-MODELLING AND SIMULATION IN GEO-SYSTEMS

Process simulations to exemplify temporal behaviour as well as prognostic scenarios in Geo- and Eco-systems are of an immense economical, ecological and safety-engineering importance.

Such systems could be assigned:

– the winning of raw materials (any mining activities)
– the construction and operating underground and surface large-scale facilities
– the founding of power plants, barrages and dams
– the construction of tunnels, caverns, underground storage sites and disposals

– the operating and closing mine sites and industrial facilities.

The difference between modelling and simulating technical systems and Geo-/Eco-systems is that the integrated elements (objects, subjects) and their natures and dynamic interdependencies (structures, processes) are mostly known incompletely. Creating and developing an appropriate model takes a long time and is methodically and according to the intended application a real challenge for every modeller (Bossel 1992). Since the mathematical exact description of the geological-geotechnical system with all its interactions and interdependencies is mostly impossible, one often uses a simulation based on discretisation and abstraction. Reality close simulation of processes in extreme complex Geo-systems does not only help to understand, they are rather important for controlling and optimising geotechnical intentions and interactions within the Geo-system. Crucial is furthermore a most possible holistic use of all relevant data and information that are attainable.

Basic essentials in this context are:

– the spatio-temporal modelling (model-development) of geological – geotechnical systems (Geo-objects with attributes and interdependencies)
– the analysis of process-related temporal changes of the Geo-system
– the modelling (analysis, representation) and simulation of dynamical processes as well as anthropogenic induced events and changes
– the prognosis, its verification and further analysis.

3 TEMPORAL GEO-INFORMATION SYSTEM

Setting up on the concepts of a former developed system (BAGIS, Baugeologisch-Geotechnisches Informations-System; Schöttler et al. 1998) a GIS-technology based system is to be developed to analyse, control, simulate and predict anthropogenic influenced processes of an exemplary Geo-system. The actual project is part of an interdisciplinary research cooperation carried out with the Department of Computer Science, Chair Computer Engineering, at the University of Hamburg.

Fundamental for BAGIS is the conceptual model of the database CORE (Combined Object Relational Database Model) that serves the management of thematic information. Concepts of object-oriented and relational database management systems are combined, while taking the particular systems advantageous qualities (Hörner 1999). CORE was expanded with concepts for representation and concepts for more efficient administration of spatial data (GeoCORE; Kesper 2001). The basis for the management of geometric data is the differentiation of topological

and metric information. Objects are meant to be described extraordinary flexible and expressive through parameterised cubic polynoms (NURBS, Non Uniform Rational B-Splines; Piegl & Tiller 1997).

Presently there are divers extension modules for integration of temporal and versioned information about to be developed, that are essential for spatio-temporal modelling and designated to simulate and analyse.

Spatial subregions of the investigation area are represented as Geo-objects, featuring homogeneous (in the broadest sense) or even single value characteristics and typical process behaviour. The modelling of spatial, temporal and thematic information is made consistently, transparently and flexibly within one uniform database consisting of three closely connected levels, the thematic level, the topology and the metric. The developed conceptual model of the proposed system is shown in Figure 1.

The development of the process-model is based on object-oriented design and modelling techniques (Booch 1996). This approach enables the integration of metric, topological and thematic attributes including processes and interactions and thereby the strong connection of spatial data (static feature) with time-related processes (dynamics and interactions between individual objects) implemented as methods.

The data- and process-adequate discretisation and description of the Geo-system arise during the system analysis. Outcoming is a complex object-class-catalogue describing the structure and therefore the interdependencies within the system. Furthermore it enables the working out of an object-class-model. Geo-objects are represented within the model like real, time-variate volumes as temporal Volume-NURBS (Körber et al. 2003).

4 MODEL SHAFT VI

For the exemplary investigation area (*Neustassfurt Mine*, Shaft VI), there are several datasets existent, such as a spatial model for the date of flooding, sonar measurements and chemical analysis from recently growing cavities in the mine (Schwandt & Seifert 1999).

The data and process adequate description and discretisation of the Geo-system was acquired within the initial phase of generating the model in context with an intensive system analysis (Figure 2). The result is a complex object-class-catalogue, entered in UML (Unified Modelling Language; Booch et al. 1999), showing the structure and therewith the interactions and interdependencies in the Geo-system, as well as a corresponding object-class-model. In this static view there are besides attributes also methods formulated that are representing the dynamic and physicochemical processes of leaching and geometrical changes

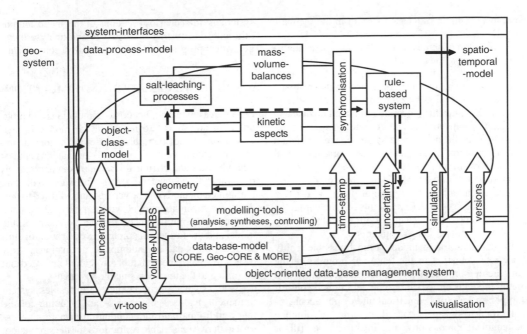

Figure 1. Conceptual model.

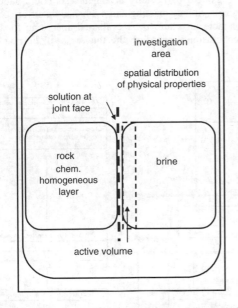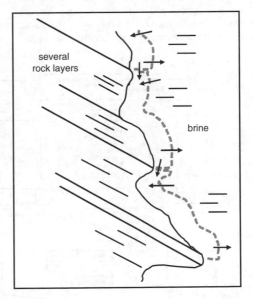

Figure 2. Object-oriented analysis of salt leaching processes.

(Busch et al. 2003). The actual work focuses on the design of the process model, i.e. formulating content and mathematical expressions of the processes and interactions.

Aiming for the application the prognosis of the salt leaching progress in Shaft VI, the calculation of temporal changes through synchronisation of individual processes is indispensable. Geochemical and thermodynamically balances (Herbert & Reichelt 1992), geometric and physical conditioned kinetic shape-changes of the Geo-objects (Grüschow 2001) as well as topological relations and interactions are to be combined within a rule based system in the thematic model.

Since heterogeneities of the data and the available and not always clear defined – or just qualitative describable – interferences between process components, there are besides analytic concepts also heuristic and statistic as well as fuzzy-set based concepts designated. These allow the processing of uncertain data and interrelationships and the integration of expert knowledge assisting the designing and later on the simulation and optimisation of the model. Therefore, the actual existing system and the rule based system (in actual development) are to be expanded with accordant concepts. For calibrating the temporal progress, existing data are to be used at definite points of time within the process. Comparing calculations are feasible by using so called solution-mining-simulators.

In this data-process-model, a most possible reality close representation of the temporal progress of the salt-leaching process as time-referenced changes of a spatial model is aimed. Still to be determined controlling parameters (of geometrical and/or physicochemical properties) can be varied, in order to simulate geotechnical interactions, e.g. injection, backfill or plugging of the shaft. The necessity of simulation derives from the insufficient knowledge of effects of geo-technical safety- and remedial-actions, that are furthermore extraordinary cost-intensive. Different simulations are to be saved persistently within the system as 4d-scenarios (versions of the spatio-temporal-model), to further on visualize and analyse them, i.e. with appropriate VR (Virtual Reality) components.

5 MODEL VERIFICATION AND SIMULATION

For any reasonable application of simulations and prognoses an adjustment to reality of the model through evaluation is indispensable. Therefore a comparison with measured values (empirical evaluation) and/or with physical and chemical laws and boundary conditions (analytic adequateness) has to be made. With that, the applied methods can be verified and adjusted to predominating conditions and with possibly new gained information optimised. For the development of the model and the further on empirical verification of simulations, lots of data and additional information are necessary, that has to be preferably structured managed and according to the intention suitable acquired.

As consistence criteria there are besides system immanent features like mass- and volume-consistency additional and independent information necessary to derive reasonable restrictions for the simulation. This information is to be announced by experts or to be gained in specially arranged tests (field, laboratory). Already existing and published data from comparable tests are to be evaluated according to their conclusions regarding the thickness of the reactive

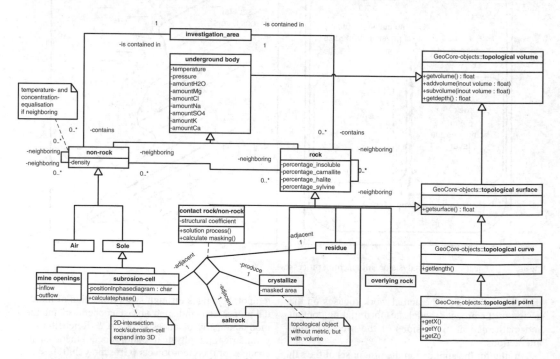

Figure 3.   UML-class-diagram, static view fragment of the thematic model "Subrosion".

layer, the amount of the substituted masses through density flow, respective the area of availability, since these are some of the main critical parameters of the process-model. Furthermore the shape and composition of insoluble components and recrystallisation phases are decisive for the calculation of the active volume.

6 LOOKING FORWARD

After an intense examination of the thematic matters concerning the processes of salt-leaching and geometrical changing and the commencing implementation of the same, further processes and extensions are to be added to the system. With that, mainly convergence laws and calculations and causal correlations to geo-mechanical characteristics are focussed.

The influence of geo-mechanic characteristics like the condition of the rock surface or the overburden pressure as well as the stress conditions in the solid rock are assessed of high importance, e.g. for the breaking of rocks within a cavern making "fresh" rock available for the solution progress. In this context is also to be mentioned the phenomenon of preferred solution in the hanging wall, which is heuristic parameterised in the actual model, but could also be added as a physical determined process.

ACKNOWLEDGEMENT

This project is part of an interdisciplinary research co-operation *Development and application of geo-information science technologies to the analysis of impacts, prognosis and control of anthropogenic influenced processes in Geo-Systems* carried out with the Department of Computer Science, Chair Computer Engineering, at the University of Hamburg founded by the German Research Foundation (DFG). The authors gratefully thank all BAGIS team members.

REFERENCES

Booch, G. 1996. Objektorientierte Analyse und Design: Mit praktischen Anwendungsbeispielen. Addison-Wesley.

Booch, G., Rumbaugh, J. & Jacobson, I. 1999. The Unified Modeling Language, Addison-Wesley.

Bossel, H. 1992. Modellbildung und Simulation: Konzepte, Verfahren und Modelle zum Verhalten dynamischer Systeme. Braunschweig, Wiesbaden: Vieweg-Verlag.

Busch, W., Hanusch, C., Knospe, S. & Maas, K. 2003. A time and space continuous model of salt leaching processes in a flooded potash mine. In Troch, I. & Breitenecker, F. (eds.), MathMod IV; Proceedings IV. IMACS Symposium on Mathematical Modelling, Vienna, 5–7 February 2003. Vienna: ARGESIM-Verlag.

Grüschow, N. 2001. Interpretation und Prognose von Lösungsprozessen im salinaren Gebirge. Exkursionsführer und Veröffentlichungen GGW 211: 6/31–6/47.

Herbert, H.-J. & Reichelt, C. 1992. Sieben Jahre Laugenentwicklung im gefluteten Salzbergwerk Hope – Geochemische Messungen und rechnerische Modellierung. Kali und Steinsalz 11(1/2): 44–48.

Hörner, C. 1999. Erweiterte Modellierungskonzepte in einem objektorientierten Datenbankmodell: Beziehungen, dynamische Klassenzugehörigkeit und Handhabung von Informationsdefiziten. Dissertation, TU Clausthal.

Kesper, B. 2001. Konzeption eines Geo-Datenmodells unter Verwendung von Freiformkörpern auf der Basis von Volume Non Uniform Rational B-Splines. Dissertation, Fachbereich für Informatik, Uni Hamburg.

Körber, Ch., Möller, D. P. F., Kesper, B, & Hansmann, W. 2003. A 4D modelling concept using NURBS applied to geosciences. In Troch, I. & Breitenecker, F. (eds.), Math Mod IV; Proceedings IV. IMACS Symposium on Mathematical Modelling, Vienna, 5–7 February 2003. Vienna: ARGESIM-Verlag.

Piegl, L. & Tiller, W. 1997. The NURBS Book, 2.ed. Springer.

Schöttler, D., Busch, W., Möller, D.P.F. & Reik, G. 1998. Bedeutung der Visualisierung in einem dreidimensionalen baugeologischen und geotechnischen Informationssystem. In DMV (ed.), Das Markscheidewesen in der Rohstoff-, Energie- und Entsorgungswirtschaft. "Neue Technologien und Aufgaben in den Geowissenschaften" 41. Wissenschaftliche Tagung des DMV 10–13 September 1997. Wissenschaftliche Schriftenreihe im Markscheidewesen 17.

Schwandt, A. & Seifert, G. 1999. Natürliche und gelenkte Flutung von Salzbergwerken in Mitteldeutschland. In Kali-, Steinsalz und Kupferschiefer in Mitteldeutschland. Exkursionsführer und Veröffentlichungen GGW 205: 61–72.

Geotechnical Measurements and Modelling, Natau, Fecker & Pimentel (eds)
© 2003 Swets & Zeitlinger, Lisse, ISBN 90 5809 603 3

Modelling of soil behaviour

J. Feda

Institute of Theoretical and Applied Mechanics, Prague, Czech Republic

ABSTRACT: Nonstandard compression and creep of some (geo)materials are commented on. Such a behaviour results from some special features of soil structure and/or specific state parameters (density, water content etc.). The present study is based on laboratory tests showing some examples of nonstandard behaviour. This results mostly from the grain breakage and bonding-debonding mechanism. The study of the nonstandard behaviour is practically important because of showing the pitfalls of the routine approach.

1 INTRODUCTION

The aim of exploring the mechanical behaviour of soils is twofold. Firstly, the soil behaviour should be mathematically modelled for the use in the design procedures. Secondly, the soil-structure mechanical changes as reflected in the constitutive behaviour are explored. Both goals are closely related, though the first approach is more of practical value while the second one of mostly academic interest. The author follows the second way when investigating some specific structures (clay, lime, granular silicagel) intended to model the behaviour of undisturbed samples in oedometric compression, creep and shear (Feda 2002). Standard and nonstandard behaviour are critically confronted (Feda 2003).

2 STANDARD BEHAVIOUR

Figure 1a depicts an array of compression curves (oedometer) generated by the mixture of granulated silicagel with varying proportion of lime (5% to 30%). The addition of lime should model the structural bonds of undisturbed material. The shape of the compression curves is, as expected, exponential which is typical for granular materials. Figure 1b represents a semilogarithmic compression curve of a reconstituted clay as a special case. Figure 2 shows the compression index $C_{c\varepsilon}$ ($= C_c /(1 + e_0)$; $C_c = \Delta e/\Delta \log \sigma$; e_0, initial void ratio; e, void ratio; σ, axial stress) as depending on the content of lime. Bonding is clearly an important structural component.

Compression behaviour in Figure 1 and 2 can be classified as standard behaviour. By some appropriate transformation the effect of debonding can be visualized. Figure 3 shows two compresssion curves from Figure 1a logarithmically transformed. They show the effect of debonding as an increased compression at about $\sigma = 150$ kPa.

Figure 4c presents three creep curves of the same material. One may recognize the common secondary compression (consolidation) $C_\alpha = $ const. ($C_\alpha = \Delta e/ \Delta \log t$; t = time) for $\sigma = 800$ kPa and 30% of lime. This relation can also be labelled as a standard one.

3 NONSTANDARD BEHAVIOUR

In Figures 4a, b creep of two (oedometric) samples is depicted. Arrows indicate the kinks at about 500 min ($\sigma = 500$ kPa, 5% of lime) – Figure 4a, and $\sigma = 1000$ kPa and 10% of lime (Figure 4b). This behaviour is a nonstandard one indicating some structural perturbations produced as the time effect. In some cases, the structural collapse may be more accentuated. Figure 5 shows a collapse of a garlandlike nature at a time of about 800 min (granular clay, grain size 4–8 mm). This is certainly a nonstandard behaviour, which depends on the state parameter "time" while the standard behaviour requires $C_\alpha = $ const. irrespective of time.

The material tested was the granulated clay and (lime)cemented silicagel. Granulated clay belongs to the family of double-porosity geomaterials. They consist of pseudograins (clusters of elementary particles) which act as a bearing skeleton, but finally fail. The change from the reinforcing role into the progressively failing material is deemed to be the reason of the observed collapsible behaviour. Collapse of a smaller

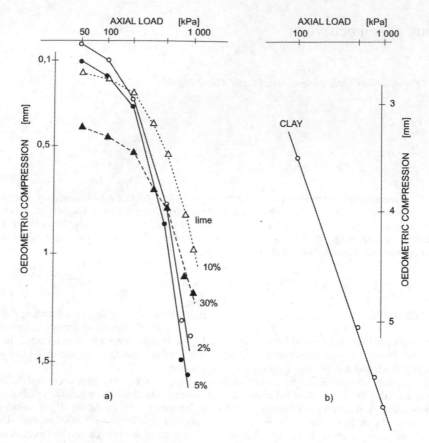

Figure 1. Curves of the oedometric compression: a) mixture of silicagel and lime (2%, 5%, 10%, 30%); b) clay Sokolov.

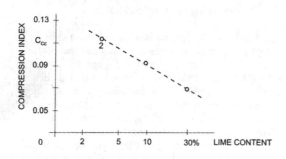

Figure 2. The effect of the lime content on the compression index $C_{c\varepsilon}$ of silicagel according to Figure 1 (2 is the mean of two points).

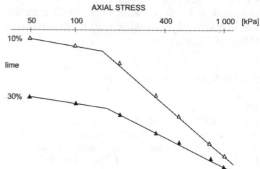

Figure 3. Transformation of two compression curves in Figure 1 (lime content 10% and 30%).

extent may be responsible for the branching-off creep in Figures 4a, b.

In this case, progressive failure of the structural bonds (produced by lime) is considered to be responsible for this nonstandard behaviour.

Figure 6 depicts stress–strain diagrams of shear box tests of silicagel–lime mixture (10% of lime). The behaviour is nonstandard in the low normal stress range ($\sigma_n < 20$ kPa) which is the stress range in the routine laboratory scarcely tested. At the small stress

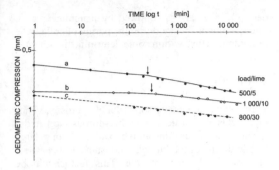

Figure 4. Oedometric creep of three samples: a, b – silicagel with 5% and 10% of lime content at 500 and 1000 kPa; c – 30% and 800 kPa.

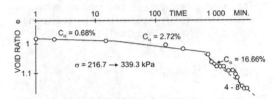

Figure 5. Oedometric creep of a wet specimen 4–8 mm of a fragmentary clay loaded by 339.3 kPa.

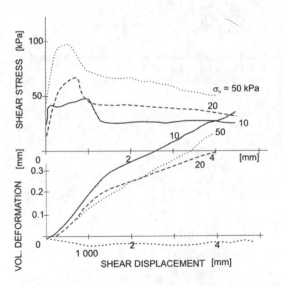

Figure 6. Stress–strain diagrams of a clay with 10% of lime admixture (shear box, porosity of about 70%).

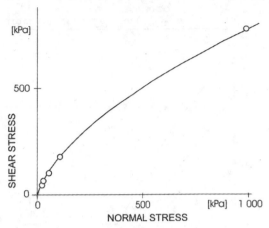

Figure 7. Mohr-Coulomb envelope of tests depicted in Figure 6.

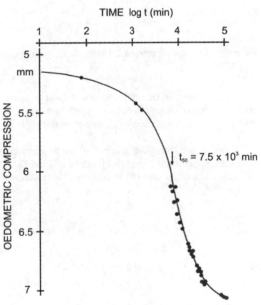

Figure 8. Oedometric creep of a dry granulated clay 2–4 mm at 5.67 MPa.

($\sigma_n = 10$ kPa) progressive failure occurs starting at $\tau = 40$ kPa and ending by the collapse at the shear displacement of about 1 mm (= 1%). The behaviour in the low stress range is strictly dilatant (contractance displayed but at 1000 kPa). The Mohr-Coulomb envelope (Figure 7) is highly nonlinear, due to the severe structural changes. All these behaviourable features indicate collapsible behaviour and the non-linearity of the M-C envelope is the consequence of debonding.

Figure 8 depicts a creep curve of heavily loaded double porosity geomaterial (dry granulated clay 2–4 mm, $\sigma = 5.67$ MPa). S-shape of the creep curve declares that creep is produced as a diffusion process (diffusion of the deformation from the loaded surface).

459

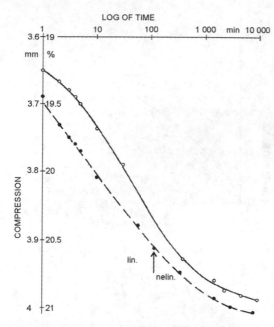

Figure 9. Oedometric creep at 2.5 MPa of angular silicagel (2–4 mm).

Such processes can be observed not only at primary consolidation (dissipation of pore-water pressure) but under some circumstances with dry granular soils as well.

Figure 9 presents two creep curves of identical oedometer specimens. One creep curve (○) indicates diffusion, the other one (●) is complex: initial linear part (C_α = const.) is followed by a nonlinear part. The specimens' behaviour in the period of collapse is of random nature within some deterministic limits.

4 CONCLUSION

With some problematic soils (double porosity, bonding, metastable structure, highly porous etc.) under specific conditions (anisotropic loading, wetting, large load and time of loading etc.) nonstandard behaviour may play an important role. This fact should be respected because of heavy consequences induced by the nonstandard behaviour. Generally one may observe in this field of behaviour intensive nonlinearity up to collapsibility.

The intrinsic regions of nonstandard behaviour should be more deeply explored because of forming a specific domain of geomaterials' behaviour.

ACKNOWLEDGEMENT

The research referred to was supported by the grant No. IAA2111301 of the Academy of Sciences of the Czech Republic which is gratefully acknowledged.

REFERENCES

Feda, J. 2002. Simulace chování neporušených vzorků zemin – in Czech. Geotechnika 5(1): 3–5.
Feda J. 2003. Nelinearita v chování zemin – in Czech. Proc. Conf. Brno – in print.

Geotechnical Measurements and Modelling, Natau, Fecker & Pimentel (eds)
© *2003 Swets & Zeitlinger, Lisse, ISBN 90 5809 603 3*

Speed of elastic waves in inhomogeneous media with initial stresses

A.G. Kolpakov

Novosibirsk, Russia

ABSTRACT: The equations for inhomogeneous elastic body with initial stresses can be written both on the micro-level (for the real inhomogeneous material) and on the macro-level (for the fictitious homogeneous averaged medium corresponding to real one). The correct way for obtaining the macro-equations is the homogenization of the micro-equations as the dimension of inhomogeneity $\varepsilon \to 0$. The theory of the inhomogeneous medium based on the homogenization theory is present.

1 INTRODUCTION

The equations for inhomogeneous elastic body with initial stresses can be written both on the micro-level (for the real inhomogeneous material) and on the macro-level (for the fictitious homogeneous averaged medium corresponding to real one).

In the engineering and geophysics practice the so called "intermediate" homogenization is widely used. It is carried out as follows: one homogenizes the non-homogeneous body having no initial stresses and calculates the stresses in it, and then one compiles an operator that should arise in describing a real homogeneous body having those elastic constants and initial stresses in accordance with the classical theory. The "intermediate" homogenization arises in particular from a phenomenological approach to a non-homogeneous body. In this case the experimentally measured elastic constants are the homogenized ones. It is shown that the "intermediate" homogenization in general leads to an incorrect result. Mathematically, this is due to the fact that the G-limit of a sum is not equal to the sum of G-limits [1]. From the mechanical viewpoint it is explained by the occurrence of a general state of local stress and strain when the uniform homogenized stresses are applied to a non-homogeneous medium [2].

The correct way for obtaining the macro-equations is the homogenization [3] of the micro-equations as the dimension of inhomogeneity $\varepsilon \to 0$.

2 THE HOMOGENIZATION PROCEDURE FOR INHOMOGENEOUS MEDIUM WITH INITIAL STRESSES

The linear problem of elasticity for such a body subjected to a mass forces $\mathbf{f}(\mathbf{x},\mathbf{x}/\varepsilon)$ and clamped on the boundary ∂G can be formulated in the following way

$$\partial \sigma_{ij}^\varepsilon / \partial x_j = f_i(\mathbf{x},\mathbf{x}/\varepsilon) \text{ in G,} \tag{1}$$

$$\mathbf{u}^\varepsilon(\mathbf{x}) = 0 \text{ on } \partial G, \tag{2}$$

$$\sigma_{ij}^\varepsilon = A_{ijkl}(\mathbf{x},\mathbf{x}/\varepsilon)\partial u^\varepsilon / \partial x_l, \tag{3}$$

where

$$A_{ijmn}(\mathbf{x},\mathbf{x}/\varepsilon) = c_{ijmn}(\mathbf{x}/\varepsilon) + \delta_{im}\sigma_{jn}^*(\mathbf{x},\mathbf{x}/\varepsilon). \tag{4}$$

We use the following asymptotic expansions widely used in the literature on the homogenization of 3-D composites [2, 3]:

Asymptotic expansion for displacements

$$\mathbf{u}^\varepsilon = \mathbf{u}^{(0)}(\mathbf{x}) + \varepsilon \mathbf{u}^{(1)}(\mathbf{x},\mathbf{y}) + \cdots$$
$$= \mathbf{u}^{(0)}(\mathbf{x}) + \sum_{k=1}^{\infty} \varepsilon^k \mathbf{u}^{(k)}(\mathbf{x},\mathbf{y}), \tag{5}$$

Asymptotic expansion for stresses

$$\sigma_{ij} = \sum_{k=0}^{\infty} \varepsilon^k \sigma_{ij}^{(k)}(\mathbf{x},\mathbf{y}). \tag{6}$$

Here $\mathbf{x}$ is a "slow" variable and $\mathbf{y} = \mathbf{x}/\varepsilon$ is the "fast" variable. The functions in the right-hand side of (5) and (6) are assumed to be periodic in $\mathbf{y}$ with periodicity cell Y. The term $\mathbf{u}^{(0)}(\mathbf{x})$ in (2.1) depends on the "slow" variable $\mathbf{x}$ only.

Denote

$$\langle \, \rangle = (\text{mes Y})^{-1} \int_Y d\mathbf{y} \tag{7}$$

the average value over the periodicity cell Y.

With the use of two-scale expansion, the differential operators are presented in the form of sum of operators in $\mathbf{x}$ and in $\mathbf{y}$. For the function $Z(\mathbf{x}, \mathbf{y})$ of the arguments $\mathbf{x}$ and $\mathbf{y} = \mathbf{x}/\varepsilon$, as in the right-hand sides of (5) and (6), this representation takes the form

$$\partial Z/\partial x_i = Z_{,ix} + \varepsilon^{-1} Z_{,iy}. \tag{8}$$

The subscript ,ix means $\partial/\partial x_i$ and subscript ,iy means $\partial/\partial y_i$.

Substituting (5) and (6) into equation (4), we obtain with allowance for (8)

$$\sum_{k=0}^{\infty} \varepsilon^k \sigma_{ij}^k = \sum_{k=0}^{\infty} \varepsilon^k c_{ijmn}(\mathbf{y})(u_{m,nx}^{(k)} + \varepsilon^{-1} u_{m,ny}^{(k)})$$
$$k = 0, 1, \ldots. \tag{9}$$

Equating the terms with identical power of ε in (9), we obtain

$$\sigma_{ij}^{(k)} = c_{ijmn}(\mathbf{y})(u)_{(m,nx)}^{(k)} + c_{ijmn}(\mathbf{y}) u_{m,ny}^{(k+1)}$$
$$k = 0, 1, \ldots \tag{10}$$

As it was assumed

$$u^{(k)}(\mathbf{y}) \text{ is periodic in } \mathbf{y} \tag{11}$$

with periodicity cell Y, $k = 0, 1, \ldots$.
Substituting (6) into the equilibrium equation (1), we obtain with allowance for (8)

$$\sum_{k=0}^{\infty} \varepsilon^k \sigma_{ij,jx}^{(x)} + \sum_{k=0}^{\infty} \varepsilon^{k-1} \sigma_{ij,jy}^{(k)} = f_i(\mathbf{x}, \mathbf{y}). \tag{12}$$

Equating the terms with identical power of ε in (12), we obtain an infinite sequence of equations, the first two of which are the following:

$$\sigma_{ij,jy}^{(0)} = 0, \tag{13}$$

$$\sigma_{ij,jy}^{(1)} + \sigma_{ij,jx}^{(0)} = f_i(\mathbf{x}, \mathbf{y}). \tag{14}$$

Proposition 1. *The following relationship takes place: for $\varepsilon \to 0$*

$$\int_G Z(\mathbf{x}, \mathbf{x}/\varepsilon) \, d\mathbf{x} = \int_G \kappa Z \lambda(\mathbf{x}) \, d\mathbf{x}$$

for every function $Z(\mathbf{x}, \mathbf{y})$ periodic in $\mathbf{y}$ with the periodicity cell Y.

Averaging (14) over the periodicity cell Y with allowance for Proposition 1, we obtain an infinite sequence of the homogenized equilibrium equations, the first of which is the following

$$\langle \sigma_{ij}^{(0)} \rangle_{,jx} = \langle f_i \rangle(\mathbf{x}). \tag{15}$$

Here we use equality $\langle \sigma_{ij,jy}^{(1)} \rangle = 0$, which follows from the well-known formula

$$\int_Y \sigma_{ij,jy}^{(1)} \, d\mathbf{y} = \int_{\partial Y} \sigma_{ij}^{(1)} n_j \, d\mathbf{y} = 0$$

and periodicity of $\sigma_{ij}^{(1)}(\mathbf{y})$ on Y.

Let us consider the problem (10, k = 0) and (13). It can be written as

$$(A_{ijmm}(\mathbf{y}) \, u_{m,ny}^{(1)} + A_{ijkl}(\mathbf{y}) u_{m,nx}^{(0)}(\mathbf{x}))_{,jy} = 0 \tag{16}$$

in Y, $\mathbf{u}^{(1)}(\mathbf{y})$ is periodic in $\mathbf{y}$ with periodicity cell Y.

Allowing for the fact that the function of the argument $\mathbf{x}$ plays the role of a parameter in the problems in the variables y and $\mathbf{u}^{(0)}$ depends on $\mathbf{x}$, only, solution of the problem (16) can be found in the form

$$\mathbf{u}^{(1)} = \mathbf{K}^{lkl}(\mathbf{y}) u_{k,lx}^{(0)}(\mathbf{x}) + \mathbf{V}(\mathbf{x}). \tag{17}$$

Here $\mathbf{V}(\mathbf{x})$ is an arbitrary function of the argument $\mathbf{x}$, and the periodic function $\mathbf{K}^{kl}(\mathbf{y})$ represents a solution of the following cellular problem:

$$(A_{ijm}(\mathbf{y}) K_{m,ny}^{kl} + A_{ijkl}(\mathbf{y}))_{,jy} = 0 \tag{18}$$

in Y, $\mathbf{K}^{kl}(\mathbf{y})$ is periodic in $\mathbf{y}$ with the periodicity cell Y.

Substituting (17) into (10, k = 0), we have

$$\sigma_{ij}^{(0)} = (A_{ijmn}(\mathbf{y}) K_{m,ny}^{kl}(\mathbf{y}) + A_{ijkl}(\mathbf{y})) u_{k,lx}^{(0)}(\mathbf{x}). \tag{19}$$

Averaging (19) over the periodicity cell Y, we obtain the following homogenized governing equation

$$\langle \sigma_{ij}^{(0)} \rangle = C_{ijkl}(\sigma) u_{k,lx}^{(0)}(\mathbf{x}), \tag{20}$$

where

$$C_{ijkl}(\sigma) = \langle A_{ijmn}(\mathbf{y}) K_{m,ny}^{kl}(\mathbf{y}) + A_{ijkl}(\mathbf{y}) \rangle. \tag{21}$$

are called the homogenized characteristics of the stressed composite body.

The equilibrium equation (14, k = 0) is

$$\langle \sigma_{ij}^{(0)} \rangle_{,jx} = \langle f_i \rangle(\mathbf{x}). \tag{22}$$

From the boundary condition (2) and expansion (5) we obtain the following boundary conditions

$$\mathbf{u}^{(0)}(\mathbf{x}) = 0 \text{ on } \partial G. \tag{23}$$

The problem (20)–(23) is the homogenized problem for stressed body. The fundamental difference of this problem from the homogenized problem for body having no initial stresses is the dependence of the cellular problem (18) and, as a result, the homogenized coefficients $C_{ijkl}(\sigma)$ on the initial stresses σ_{ij}^*.

3 SMALL INITIAL STRESSES

The formulas obtained above can be used to compute the homogenized constants numerically. The problems related with the numerical solution of the cellular problems are rather the problem of numerical mathematics then mechanics. The substantial progress in the analysis of the problem, as it appears in the practice, is related with the note concerning the magnitude of the initial stresses as compared with the elastic constants. There are naturally constraints on the initial stresses, i.e. σ_{ij}^* will not exceed the strength limit of the material. In turn, the strength limit for a real material is small by comparison with the elastic constants [4]. Then the initial stresses σ_{ij}^* are small compared with elastic constants c_{ijkl} and the coefficients A_{ijkl} introduced by formula (4) may be presented as

$$A_{ijkl}(\mathbf{x},\mathbf{y}) = c_{ijkl}(\mathbf{y}) + \mu b_{ijkl}(\mathbf{x},\mathbf{y}), \qquad (24)$$

where μ is a small parameter ($\mu \leqslant 0.01$) and the following notation is used:

$$B_{ijkl}(\mathbf{x},\mathbf{y}) = \delta_{ik}\sigma_{jl}^*(\mathbf{x},\mathbf{y}). \qquad (25)$$

This note, which is obvious for a mechanics, allows simplifying the homogenization procedure.

In order to solve the cellular problem (18) with coefficients (24) the classical method of small parameter can be used. This method is based on presentation of solution the cellular problem in the form

$$\mathbf{K}^{kl}(\mathbf{y}) = \mathbf{K}^{0kl}(\mathbf{y}) + \mu\mathbf{K}^{1kl}(\mathbf{y}) + \cdots$$

$$= \sum_{s=0}^{\infty}\mu^s\mathbf{K}^{skl}(\mathbf{y}). \qquad (26)$$

Substituting (26) into (18), we obtain

$$((c_{ijmn}(\mathbf{y}) + \mu b_{ijkl}(\mathbf{x},\mathbf{y}))\sum_{s=0}^{\infty}\mu^s K_{m,ny}^{skl}(\mathbf{y})$$

$$+ c_{ijkl}(\mathbf{y}) + \mu b_{ijkl}(\mathbf{x},\mathbf{y}))_{,jy} = 0. \qquad (27)$$

Equating the terms with identical power of μ in (27), we obtain an infinite sequence of problems, the first two of which have the following form

$$(c_{ijnm}(\mathbf{y})K_{m,ny}^{0kl} + c_{ijkl}(\mathbf{y}))_{,jy} = 0 \text{ in Y}, \qquad (28)$$

$$(c_{ijnm}(\mathbf{y})K_{m,ny}^{1kl} + b_{ijkl}(\mathbf{x},\mathbf{y})$$
$$+ b_{ijkl}(\mathbf{x},\mathbf{y})K_{m,ny}^{0kl}(\mathbf{y}))_{,jy} = 0 \text{ in Y} \qquad (29)$$

with the following conditions of periodicity:

$\mathbf{K}^{0kl}(\mathbf{y})$ and $\mathbf{K}^{1kl}(\mathbf{y})$ are periodic in $\mathbf{y}$

with periodicity cell Y.
Comprising (28) with (24), we have

$$\mathbf{K}^{0kl}(\mathbf{y}) = \mathbf{X}^{kl}(\mathbf{y}), \qquad (30)$$

where $\mathbf{X}^{kl}(\mathbf{y})$ is solution of the cellular problem for body having no initial stresses (the cellular problem (1.2.23) or, that is the same, the cellular problem (18) with $\sigma_{jl}^*(\mathbf{x},\mathbf{y}) = 0$).

Transform the formula (21) for computation the homogenized characteristics to a quadratic functional. For that we write the cellular problem (18) in the form

$$(A_{pqmn}(\mathbf{y})K_{m,ny}^{kl} + A_{pqkl}(\mathbf{y}))_{,jy} = 0 \text{ in Y}$$

Multiply this equations by $K_p^{ij}(\mathbf{y})$ and integrate by parts over the periodicity cell Y. As a result we obtain with allowance for periodicity of $\mathbf{K}^{ij}(\mathbf{y})$ and $\mathbf{K}^{kl}(\mathbf{y})$ the following equality:

$$\langle A_{pqnm}(\mathbf{y})K_{m,ny}^{kl}(\mathbf{y})K_{p,qy}^{ij}(\mathbf{y}) + A_{pqkl}(\mathbf{x},\mathbf{y})K_{p,qy}^{ij}(\mathbf{y})\rangle = 0. \qquad (31)$$

Subtracting (31) from (21), we obtain with regard for definition b_{ijkl} (25) and symmetry of the elastic constants c_{ijkl} the following formula for calculation of the homogenized constants:

$$C_{ijkl}(\sigma) = \langle -A_{pqmn}(\mathbf{y})K_{m,ny}^{ij}(\mathbf{y})K_{p,qy}^{kl}(\mathbf{y})$$
$$- A_{mnjj}(\mathbf{x},\mathbf{y})K_{m,ny}^{kl}(\mathbf{y})$$
$$+ A_{ijmn}(\mathbf{x},\mathbf{y})K_{m,ny}^{kl}(\mathbf{y}) + A_{ijkl}(\mathbf{x},\mathbf{y})\rangle$$
$$= \langle -A_{pqmn}(\mathbf{y})K_{m,ny}^{ij}(\mathbf{y})K_{p,qy}^{kl}(\mathbf{y})$$
$$- b_{mnjj}(\mathbf{x},\mathbf{y})K_{m,ny}^{kl}(\mathbf{y})$$
$$+ b_{ijmn}(\mathbf{x},\mathbf{y})K_{m,ny}^{kl}(\mathbf{y}) + A_{ijkl}(\mathbf{x},\mathbf{y})\rangle. \qquad (32)$$

Proposition 2. *Let the initial stresses $\sigma_{ij}^*(\mathbf{x},\mathbf{y})$ satisfy the following equation*

$$\sigma_{ij,jy}^* = 0 \text{ in Y}. \qquad (33)$$

Then, $\langle\sigma_{ij}(\mathbf{x},\mathbf{y})K_{i,jy}^{pq}(\mathbf{y})\rangle = 0$.

By virtue of proposition 2.2 and symmetry of the initial stresses σ_{ij}^* with respect to indices i and j and definition of $b_{ijkl}(\mathbf{x},\mathbf{y})$, we have

$$\langle b_{mnij}(\mathbf{x},\mathbf{y})K_{m,ny}^{kl}(\mathbf{y})\rangle = 0 \qquad (34)$$

and

$$\langle b_{ijmn}(\mathbf{x},\mathbf{y})K_{m,ny}^{kl}(\mathbf{y})\rangle = 0.$$

463

From (30) and (34) we obtain

$$C_{ijkl}(\sigma) = \langle -A_{pqmn}(\mathbf{y})K_{m,ny}^{ij}(\mathbf{y})K_{p,qy}^{kl}(\mathbf{y}) + A_{ijkl}(\mathbf{y})\rangle. \tag{35}$$

Substituting decomposition (26) into formula (35), we can write (35) as

$$C_{ijkl}(\sigma) = C_{ijkl}(0) + \mu \langle b_{ijkl}(\mathbf{x},\mathbf{y})\rangle + \mu C_{ijkl}^1(\sigma) + \cdots$$
$$= C_{ijkl}(0) + \mu \langle \sigma_{jl}^*(\mathbf{x},\mathbf{y})\rangle \delta_{ik} + \mu C_{ijkl}^1(\sigma) + \cdots, \tag{36}$$

where

$$C_{ijkl}^1(\sigma) = \langle -b_{pqmn}(\mathbf{x},\mathbf{y})K_{p,qy}^{0ij}(\mathbf{y})K_{m,ny}^{0kl}(\mathbf{y})$$
$$- c_{pqmn}(\mathbf{y})K_{p,qy}^{0ij}(\mathbf{y})K_{m,ny}^{1kl}(\mathbf{y})$$
$$- c_{pqmn}(\mathbf{y})K_{m,ny}^{1ij}(\mathbf{y})K_{p,qy}^{0kl}(\mathbf{y})\rangle. \tag{37}$$

Taking into account that $\langle \sigma_{jl}^*(\mathbf{x},\mathbf{y})\rangle = \sigma_{jl}$ (see (1.1.29)), we can write (36) as

$$C_{ijkl}(\sigma) = C_{ijkl}(0) + \mu\sigma_{jl}\delta_{ik} + \mu C_{ijkl}^1(\sigma) + \cdots, \tag{38}$$

where σ_{jl} are the initial stresses determined from the homogenized problem for body with no initial stresses (see for details [2, 5, 6]).

By resorting to equation (29) we can take the opportunity to rule out functions $\mathbf{K}^{1kl}(\mathbf{y})$ from (38). Multiplying (29) by $K_i^{0pq}(\mathbf{y})$ and integrating by parts over the cellular Y, we obtain with allowance for periodicity of $\mathbf{K}^{0ij}(\mathbf{y})$ and $\mathbf{K}^{1kl}(\mathbf{y})$:

$$\langle c_{ijnm}(\mathbf{y})K_{m,ny}^{1kl}(\mathbf{y})K_{i,jy}^{0pq}(\mathbf{y}) + b_{ijkl}(\mathbf{x},\mathbf{y})K_{i,jy}^{0pq}(\mathbf{y})$$
$$+ b_{ijkl}(\mathbf{x},\mathbf{y})K_{m,ny}^{0kl}(\mathbf{y})K_{i,jy}^{0pq}(\mathbf{y})\rangle = 0 \tag{39}$$

From (39) we have (after changing the indices ij $\leftrightarrow$ pq)

$$\langle c_{pqmn}(\mathbf{y})K_{m,ny}^{1kl}(\mathbf{y})K_{p,qy}^{0ij}(\mathbf{y})\rangle = -\langle b_{pqkl}(\mathbf{x},\mathbf{y})K_{p,qy}^{0ij}(\mathbf{y})$$
$$+ b_{pqmn}(\mathbf{x},\mathbf{y})K_{m,ny}^{0kl}(\mathbf{y})K_{p,qy}^{0ij}(\mathbf{y})\rangle. \tag{40}$$

In the similar way we obtain

$$\langle c_{pqnm}(\mathbf{y})K_{m,ny}^{0kl}(\mathbf{y})K_{p,qy}^{1ij}(\mathbf{y})\rangle = -\langle b_{pqij}(\mathbf{x},\mathbf{y})K_{p,qy}^{0kl}(\mathbf{y})$$
$$+ b_{pqmn}(\mathbf{x},\mathbf{y})K_{m,ny}^{0kl}(\mathbf{y})K_{p,qy}^{0ij}(\mathbf{y})\rangle. \tag{41}$$

Substituting (40) and (41) into (38), we obtain

$$C_{ijkl}^1(\sigma^*) = \langle b_{pqmn}(\mathbf{x},\mathbf{y})K_{m,ny}^{0kl}(\mathbf{y})K_{p,qy}^{0ij}(\mathbf{y})$$
$$+ b_{pqkl}(\mathbf{x},\mathbf{y})K_{p,qy}^{0ij}(\mathbf{y}) + b_{pqij}(\mathbf{x},\mathbf{y})K_{p,qy}^{0ij}(\mathbf{y})\rangle. \tag{42}$$

The following equalities take place

$$\langle b_{pqkl}(\mathbf{x},\mathbf{y})K_{p,qy}^{0ij}(\mathbf{y})\rangle = 0$$

and

$$\langle b_{pqij}(\mathbf{x},\mathbf{y})K_{p,qy}^{0ij}(\mathbf{y})\rangle = 0. \tag{43}$$

The equalities (43) can be derived in a manner similar to one used to derive equalities (34).

From (42) and (43) we have

$$C_{ijkl}^1(\sigma) = \langle b_{pqmn}(\mathbf{x},\mathbf{y})K_{m,ny}^{0kl}(\mathbf{y})K_{p,qy}^{0ij}(\mathbf{y})\rangle. \tag{44}$$

Substituting $b_{ijmn} = \delta_{jn}\sigma_{jn}^*$ in accordance with the definition (4), we obtain

$$C_{ijkl}^1(\sigma) = \langle \sigma_{qn}^*(\mathbf{x},\mathbf{y})K_{p,ny}^{0kl}(\mathbf{y})K_{p,qy}^{0ij}(\mathbf{y})\rangle. \tag{45}$$

In accordance with (30) $\mathbf{K}^{0kl}(\mathbf{y}) = \mathbf{X}^{kl}(\mathbf{y})$, where $\mathbf{X}^{kl}(\mathbf{y})$ is solution of the cellular problem for composite with no initial stresses (see [2, 3]). Then

$$C_{ijkl}^1(\sigma) = \langle \sigma_{qn}^*(\mathbf{x},\mathbf{y})X_{p,ny}^{kl}(\mathbf{y})X_{p,qy}^{ij}(\mathbf{y})\rangle. \tag{46}$$

Let the initial stresses σ_{qn}^* be determined from solution of the homogenized elasticity problem with no initial stresses. Then approximation of the stresses σ_{qn}^* is given by the formula

$$\sigma_{qn}^*(\mathbf{x},\mathbf{y}) = (c_{qnrs}(\mathbf{y}) + c_{qncd}(\mathbf{y})X_{c,dy}^{rs}(\mathbf{y}))u_{r,sx}^{(0)}, \tag{47}$$

where $\mathbf{u}^{(0)}$ is solution of the homogenized elasticity problem with no initial stresses. Substituting (46) into (45), we obtain expression for the corrector $C_{ijkl}^1(\sigma)$, which does not contain the local initial stresses in the explicit form:

$$C_{ijkl}^1(\sigma) = l_{ijklrs}(\mathbf{X})u_{r,sx}^{(0)}, \tag{48}$$

where

$$l_{ijklrs}(\mathbf{X}) = \langle (c_{qnrs}(\mathbf{y}) + c_{qncd}(\mathbf{y})X_{c,dy}^{rs}(\mathbf{y}))X_{p,ny}^{kl}(\mathbf{y})X_{p,qy}^{ij}(\mathbf{y})\rangle$$
$$= \langle c_{qnrs}(\mathbf{y})X_{p,ny}^{kl}(\mathbf{y})X_{p,qy}^{ij}(\mathbf{y})$$
$$+ c_{qncd}(\mathbf{y})X_{c,dy}^{rs}(\mathbf{y})X_{p,ny}^{kl}(\mathbf{y})X_{p,qy}^{ij}(\mathbf{y})\rangle. \tag{49}$$

As a result, we conclude that the homogenized characteristics of composite body with initial stresses have the form

$$C_{ijkl}(\sigma) = C_{ijkl}(0) + \mu\sigma_{jl}\delta_{ik} + \mu C_{ijkl}^1(\sigma) + \cdots. \tag{50}$$

The sum $C_{ijkl}(0) + \mu\sigma_{jl}\delta_{ik}$ in (50) exactly coincides with the formula for homogeneous solid with initial stresses (see Section 1.1) and the term $\mu C^1_{ijkl}(\sigma)$ is the corrector related with the inhomogeneity of body. As known (see [2, 3]) solution of the cellular problem for a homogeneous body with no initial stresses is $X^{ij}_{p,qy}(\mathbf{y}) = 0$. Thus, $C^1_{ijkl}(\sigma) = 0$ for a homogeneous body.

The formula (48) can be written in the term of the homogenized stresses (stresses determined from the homogenized elasticity problem). Write the formula relating the homogenized stresses and deformations (see [2, 3]) in the form

$$u^{(0)}_{r,sx} = \{C_{rskl}(0)\}^{-1}\sigma_{kl}, \tag{51}$$

where $C_{rskl}(\mathbf{0})$ tensor of the homogenized elastic constants for the body with no initial stresses and "-1" means the inversion of tensor. Substituting this formula into (51) into (48), we arrive at the formula establishing relationship between the corrector $C^1_{ijkl}(\sigma)$ and the initial homogenized stresses.

REFERENCES

1. Marcellini, P. (1975) Un teorema di passagio de limite per la sommma di funzioni convesse. *Bull. Unione Mat. Ital.* 4, 107–124.
2. Kalamkarov, A.L. and Kolpakov, A.G. (1997) *Analysis, Design and Optimization of Composite Structures*. John Wiley & Sons, Chichester, New York.
3. Bensoussan, A., Lions, J.-L. and Papanicolaou, G. (1978) *Asymptotic Analysis for Periodic Structures*. North-Holland, Amsterdam.
4. Timoshenko and Goodier. (1970) *Theory of Elasticity*. McGraw-Hill, New York.
5. Kolpakov, A.G. (1989) Stiffness characteristics of stressed inhomogeneous bodies. *Izvestiy Akad. Nauk SSSR* (translated in English as *Mechanics of Solids*) 3, 66–73.
6. Kolpakov, A.G. (1992) On the dependence of the velocity of elastic waves in composite media on initial stresses. *Computers & Structures* 44, 97–101.

Geotechnical Measurements and Modelling, Natau, Fecker & Pimentel (eds)
© 2003 Swets & Zeitlinger, Lisse, ISBN 90 5809 603 3

Testing and modelling the ductility of buried jetgrout structures

P. Kudella
Institute of Soil Mechanics and Rock Mechanics, University of Karlsruhe, Germany

P.M. Mayer
Ed. Züblin AG, Stuttgart

G. Möller
University of Applied, Sciences, Neubrandenburg, Germany

ABSTRACT: Material inhomogeniety and local stress concentrations in buried jetgrout structures are a matter of concern for overall safety and watertightness. Based on various laboratory tests of jetgrout core samples a more or less pronounced material ductility is demonstrated. Several constitutive models to describe the viscous properties are discussed, and the consequences of this more realistic material description are shown with a finite element calculation of an example excavation with jetgrout bottom slab.

1 FACING THE PROBLEMS

1.1 *Safety and serviceability demands*

Buried jetgrout structures are widely used for static or hydraulic purposes or a combination of both:

- bottom-sealing slabs with only hydraulic function,
- slabs in a medium or a high position combining sealing and static propping function (tied-back by piles or anchors against uplift, or activating dead-weights by vaulting),
- massive slabs, props or frames made of jetgrout columns in impermeable and soft soils with only static function.

With rising the level and emphasizing the static use, the risk of defects shifts from a serviceability problem to a question of ultimate limit state (Hoffmann et al. 2000, Kudella 2001). Based on today's experience with serious damages the conventional design describing the jetgrout structure as a homogenious weak concrete plate with constant strength and permeability cannot be regarded as satisfactory. Whenever such buried structures are loaded by retaining walls, the contact area between the two substructures formes a hinge due to wall rotation. Uplift piles introduce additional shear and spatial bending, while neither the geometry nor the strength of the structure is well-defined. To model these effects more accurately, the load-deformation characteristics of jetgrout material is questioned.

1.2 *Geometrical uncertainty*

Even for proper quality control such as automatic geodetic survey of drilling points and inclinations the mean deviation of a column position has to be expected as 0,5 to 1% of drilling depth (Stein 2000), but may also reach as much as 3%. The uncertainty in the radius of the columns produced reaches variation coefficients of 5–15% (van Riel et al. 2000). Design philosophies use either a fixed lateral overlapping of columns or a flexible spacing of wet-in-wet filling columns between prefabricated primary columns in dependence of their measured radii. The combined position and radius uncertainty causes

- leaks due to lateral shift or locally smaller diameters,
- shadowing of subsequent jetting due to locally larger diameters.

Both effects can be evaluated on a statistical basis, only the latter may be reduced by vertical overlapping (Kudella 2001, Lesnik 2002). Lasting uncertainty of the outer boundary and completeness of cross section is a problem not addressed in this paper.

1.3 *Material inhomogeneity*

But even without a defect, density and strength of the jetgrout vary. As the jet reaches its local limit width, the soil is not completely liquefied, but extensively sheared and penetrated by suspension. Inside the core

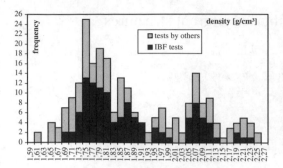

Figure 1. Typical density distribution of cored jetgrout samples.

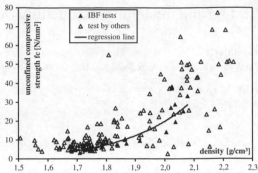

Figure 2. Unconfined compressive strength vs. jetgrout density.

column, there seems to be no clear correlation of density and radius.

One hundered and twenty one small samples were taken from 5 core borings at a large bottom sealing slab in homogenious dense sands in Northern Germany (Gudehus & Kudella 2000). They showed density variations of up to 20% within a column and similar between neighbouring ones. Typical variation coefficients were between 7 and 10%. Density distributions are asymmetric and difficult to describe by a normal distribution (Fig. 1). Setting of coarse-grained aggregates and floating of organic matter are responsible for systematic density variations with depth.

2 LABORATORY TESTS

2.1 Uniaxial compression

Twenty nine uniaxial compression tests with cored samples of 50 mm diameter and H/D = 2:1 have been made. All samples were loaded with 1%/h to the nominal strength of 5 Mpa, then unloaded and reloaded with 10%/h and loaded further until failure. The compression strength varied between 3 and 37 Mpa and is closely density correlated with a regression (Fig.2)

$$f_c = 0,028 \, \rho^{9,45} \qquad (1)$$

Axial failure strain is much influenced by preexisting microcracks and varies between 0,2 and 1,3% for all densities.

Some samples were cored in lateral direction, but no anisotropy was observed. Nor had the age of the jetgrout beyond 28 days any significant influence.

2.2 Triaxial compression

Following a similar procedure, 14 pairs of samples with identical density were subjected to triaxial tests with cell pressures of 0,9 and 1,8 Mpa. Again, deviatoric

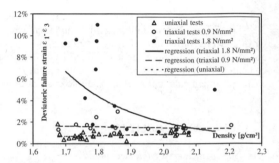

Figure 3. Characteristic failure strain vs. jetgrout density.

ultimate strengths were density correlatet following a potential regression of the form eq. (1).

Failure strain is significantly higher, however, reaching from 0,8% to 11% (Fig. 3). For weak jetgrout (low density), small lateral support is already sufficient to wake a high ductility. This behaviour is associated to shear fractures with an extended deviatoric strain plateau reminding on loose sands. For dense jetgrout with high strength under deviatoric load, tensile fracture still dominates, and the stress–strain relation has a clear peak like for dense granulates.

Furthermore, a triaxial test with velocity jumps demonstrates the viscous nature of the jetgrout material (Fig. 8).

2.3 Uniaxial tension

Six centric tension tests have been made stressing tension steel bars glued to the sample ends. Tensile strength vary between 0,2 and 1,4 Mpa, about 5% of compression strength. Although the number of tests is too small for statistical evidence, a similar density dependence as above is likely.

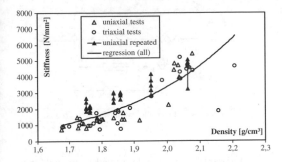

Figure 4.   Characteristic secant stiffness vs. jetgrout density.

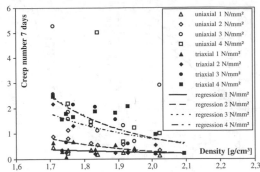

Figure 5.   Creep numbers for different stress levels vs. density.

2.4   Quasi-elastic stiffness

Stiffness data were available from all the uniaxial and triaxial compression tests above and from additional 11 uniaxial tests which were not loaded to failure but repeated several times in three weeks intervals. In general, the stress–strain-relation do not have well-defined elastic ranges. All tests observed a slow steady stiffness increase at the beginning of compression, which can be attributed to non-uniform contact pressures. No constant preload had been applied as usual for concrete testing because the samples were much weaker and some had failed in a stress or strain range where others still had contact problems. Prior to failure, the behaviour was again strongly non-linear, in some cases leaving no space for a linear portion in between. Four ways of stiffness data evaluation were used:

a) first loading between 1 and 5 Mpa
b) first loading 5 to 30% of failure load
c) first loading 0,15 to 0,25% strain
d) steepest secant with 0,3% strain difference.

All options have in common, that the measured stiffness is density correlated. But the parameters are quite different: Options a) and b) deliver values of 300–8600 Mpa, which is significantly less in comparison with 100 mm – samples (H/D = 1:1) tested by concrete institutions; c) seems completely useless and only d) with values of 800–5400 Mpa and a variation coefficient of 34% is referred to:

$$E_s = 30\,\rho^{6,77} \tag{2}$$

During some triaxial tests, the circumferential strain was measured, but with very limited accuracy. From those, an average poisson number decreasing from 0,35 to 0,10 with growing density can be estimated.

2.5   Creep and relaxation

The viscous nature of jetgrout was demonstrated in creep tests. Ten samples were tested in oedometer devices with controlled loads over 4–19 weeks. Loads were applied either in consecutive steps from 0 to 4 Mpa or after unloading and relaxation. It was attempted to separate a quasi-elastic spontaneous settlement from the creep portion, but this proved very difficult.

It is evident, however, that the delayed viscous strain in the first 7 days is 20–300% of the estimated spontaneous deformation. The viscosity expressed as "creep number" $\varepsilon_{vis}/\varepsilon_{el}$ (Fig. 5) depends

- strongly on density (more pronounced than quasi-elastic stiffness),
- significantly on the degree of strength mobilization, i.e. σ/f_c,
- not much on loading history and sample age,
- on temperature as expected.

Six further samples were tested in triaxial cells under 1 Mpa cell pressure and 1,8–4,8 Mpa vertical dead load. The results are qualitatively similar. Creep strain is higher, but that can be traced back to the fact, that the rubber sleeve around the samples could not avoid long-term diffusion. So the uniaxial samples dried out, while cell water was forced into the triaxial ones. Sample saturation and related suction obviously plays an important role beside density.

Relaxation steps after reloading of the uniaxial compression tests showed stress reductions between 1,5 and 9% in 30 min., again density dependent.

3.   CONSTITUTIVE MODELLING

3.1   Elasto-plastic flow

One of the aims of combined uniaxial and triaxial tests was the finding of Mohr-Coulomb shear parameters for a material description as simple as possible. Based on the assumption

$$|\sigma_1 - \sigma_3| = |\sigma_1 + \sigma_3| \sin \varphi(\rho) + 2 \cdot c(\rho) \cos \varphi(\rho) \tag{3}$$

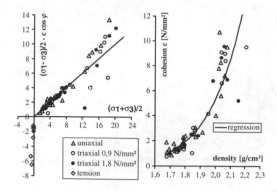

Figure 6. Fitting of shear parameters $\Psi = 30°$ and c vs. density.

linear regression analyses were carried out for different group densities. Almost the best fit (Fig. 6) results for the simple dependency

$$\varphi = 30° \neq f(\rho) \quad c = 0{,}0067\,\rho^{9{,}64}. \tag{4}$$

The cohesion measured in the individual tests has a variation coefficient of 28% against this regression.

The description with coulomb friction makes sense, as long as the jetgrout disintegrates to a granular material with "critical" density after sufficient shear. In this formulation dilatancy effects like peak friction as expected for the dense samples are hidden in the empirical density dependence of cohesion.

Possible tension stresses using this failure criterion overestimate the measured tensile resistance (Fig. 6). A separate tensile cutoff must be prescribed.

3.2 Concrete creep

Elastoplastic descriptions do not yet account for the pronounced viscosity. At first it seemed likely to adopt contemporary concrete creep models like the ones in DIN 1045 or EC 2:

$$\varepsilon_{cr} = \frac{\sigma_{cr}}{E_{el}} \cdot A \cdot \left(\frac{t - t_0}{B + t - t_0}\right)^C \tag{5}$$

containing constants A, B, C related to strength, age, humidity, structure thickness, water-cement ratio, and for high stresses also additional fit functions. They have not been followed here because:

– some of the required parameters are unknown difficult to define for jetgrout,
– descriptions have been developed for beams and plates and are not validated for 3D-problems,
– guiding idea was the deformation accumulation under constant dead weight, not the continuous

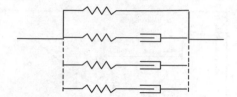

Figure 7. Visco-elastic model with Maxwell chains.

and variable loading under mixed stress and strain control.

3.3 Visco-elasticity

The empirical use of rheological chains separates the description from the underlying physical phenomens. This can be justified here, because an adequate description of viscous behaviour is required only for a certain time interval corresponding to the construction and service time of the jetgrout structure (days to months).

A series of one spring and one parallel Maxwell chain is sufficient to describe creep. The dashpot defines the creep rate at a certain time, the springs the spontaneous-elastic and the final deformation. If a material exhibits several inherent time scales, this can be accounted for by further Maxwell chains (Fig. 7). In our case, the observed stress relaxation over 30 min. and creep deformation over months could not be described with the same parameter set. One second chain seemed sufficient to meet all tests. The numerical description is straightforward:

$$\dot{\sigma}_{mi} = E_{mi}\dot{\varepsilon} - \frac{\sigma_{mi}}{T_{mi}} \quad \text{for Maxwell chain i}$$

and

$$\dot{\sigma} = \left(E_{el} + \sum E_{mi}\right)\dot{\varepsilon} - \sum \frac{\sigma_{mi}}{T_{mi}} \tag{6}$$

for the system (compression negative) with the relaxation time $T_{mi} = D_{mi}/E_{mi}$. A three-dimensional formulation of this rheological law exists already in many FE-codes. The required parameters are identified:

– E_{el} estimated from the final deformation in long-time creep,
– E_2, T_2 from fitting the relaxation curve,
– $E_1 = E - E_{el} - E_2$ from taking the spontaneous-elastic portion as the quasi-elastic deformation of Sec. 2.4 due to E,
– T_1 from fitting the creep curvature,
– $\nu_{el} = \nu_1 = \nu_2$ from Sec. 2.4.

Evaluation of the various tests with different densities lead again to a clear density dependence of the parameters E_1, E_2, while the correlation of T_1, T_2

remained rather vague. Characteristic values are given in Gudehus & Kudella 2000.

3.4 Visco-elasticity combined with MC-plasticity

A more realistic description is achieved if the visco-elastic model (Sec. 3.3) is combined with the Mohr-Coulomb plasticity (Sec. 3.1). Some FE codes like TOCHNOG allow for this possibility treating elastic, viscous and plastic properties as independent modules as far as possible.

Viscous deformations increase prior to plastic failure. Therefore, the Maxwell parameters (Sec. 3.3) are only applicable for stress levels beyond say 60% of ultimate stress. To account for higher stress levels, it was suggested to "soften" the Maxwell chain stiffnesses according to

$$E_m = E_{m0}\left(1 - F\left(\frac{|\sigma_1 - \sigma_3|}{|\sigma_1 + \sigma_3| \sin\varphi + 2 \cdot c \cdot \cos\varphi}\right)^n\right)$$

(7)

F and n are empirical constants.

3.5 Visco-plasticity

From the background of soil mechanics, there are alternative models which do better reflect the physical background of the phenomens and avoid purely empirical parameters: Oedometric creep and relaxation of clays have been modelled based on a rate process theory and using the concept of an "equivalent stress" σ_e (Niemunis & Krieg 1996). σ_e is an additional inner state variable describing the actual overconsolidation ratio $OCR = -\sigma_e/\sigma$. Stiffness parameters λ, κ are stress-proportional for the primary loading (compression negative)

$$\dot{\varepsilon} = \frac{\lambda}{1 + e_0} \cdot \left(\frac{-\dot{\sigma}}{\sigma}\right) = \frac{\lambda}{1 + e_0} \cdot \left(\frac{-\dot{\sigma}_e}{\sigma_e}\right)$$

(8)

and for unloading/reloading:

$$\dot{\varepsilon} = \frac{\kappa}{1 + e_0} \cdot \left(\frac{-\dot{\sigma}}{\sigma}\right) + \dot{\varepsilon}_{cr} \quad \text{with} \quad \dot{\varepsilon}_{cr} = -\dot{\varepsilon}_{ref} \cdot \left(\frac{-\sigma}{\sigma_e}\right)^{1/I_v}$$

(9)

I_v is a viscosity index.

A three-dimensional formulation has been developed by Niemunis (2002) combining this compression law with the tensorial framework of hypoplasticity and a Cam-clay definition for non-oedometric OCR. The concept has neither a purely non-viscous nor an elastic locus. Different from rheology, this kind of creep never comes to an end, but viscous effects are negligible for OCR > 2. Plastic flow occurs if the critical friction angle is reached independent of OCR. The cohesion is modelled by adding a fictituous isotropic stress $\sigma_c \approx c/\tan\varphi$.

3.6 Visco-hypoplasticity

A further viscoplastic description has been given recently by Gudehus (2003). It takes up the original hypoplastic stiffness dependency on relative density, models also strength increase due to constraint dilatancy, and creep comes to an end after limited time. With this model, we expect the behaviour of jetgrout material to be described still better. The ten required parameters are difficult to measure or have unusual values in compare with soft soils. Namely the physical meaning of variable limit void ratios gets lost with the solid jetgrout matrix. Assuming a degree of matrix saturation (>95%) and a specific weight of the soil-cement grain mixture (2,7–2,9 g/cm³) we used the known sample densities ρ to calculate "limit void ratios" and initial "relative density" of a given sample according to

$$e \approx \frac{\rho_s - \rho}{\rho - S_r}$$

(10)

4 MODEL VERIFICATION

A few representative laboratory tests have been back-calculated regarding them as element tests and using the different constitutive laws outlined in 3. Measurements and calculated results are sketched in Figure 8 for comparison. The reverse procedure – parameter variation until laboratory test curves are sufficiently met – holds for the parameter identification.

Visco-elastic parameters have been adopted following Section 3.4 according to sample density. The marked differences between test results and calculation are due to the scattering of real behaviour and due to nonlinear effects which cannot be expressed by the density-related characteristic values. Back calculations using a TOCHNOG finite-element mesh have also been made, which are reported elsewhere (Möller 2002).

For the visco-plastic law (Sec. 3.5), creep tests could be matched quite satisfactory, if the parameters λ, κ, φ_c, I_v, σ_c and the initial value for σ_e were chosen in appropriate combinations to fit the laboratory tests. Quasi-elastic regimes and failure states (test 58R) are not matched satisfactory by the cam clay model incorporated. There are not yet enough back-calculated tests to identify clear relations between these parameters and sample density.

471

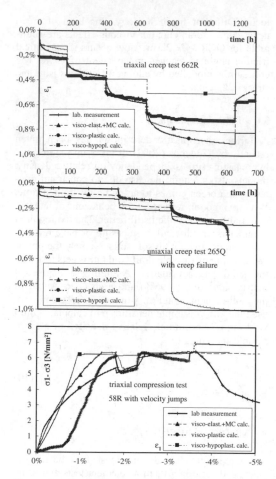

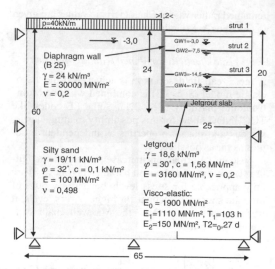

Figure 9. Excavation model with 3 strut levels and bottom slab.

Figure 8. Comparison of three lab tests with back calculations.

Also for the visco-hypoplastic description, the laboratory tests could be matched at least qualitatively (using $e_{d0} = 0,56$ resp. $e_{c0} = 1,6$). In compare to the Niemunis-type visco-plasticity (Sec. 3.5), quasi-elastic initial stiffness is modelled better and beginning creep failure (test 265Q) can not be predicted better by any other model. Viscous effects, however, are much underestimated. Obviously, this model still needs more consideration in regard to appropriate parameters and numerical stability.

5 NUMERICAL APPLICATION

The practical application of modelling the viscous nature of jetgrout is demonstrated by a deep excavation with a 2D finite-element model. The excavation in silty sand under groundwater has a depth of 17,3 m and a width of 50 m. A propped concrete diaphragm wall is constructed to 24 m depth, followed by a tied-back jet-grout bottom slab of 2 m thickness 20 m below ground level (Fig. 9). The FE mesh comprises 2091 quadrilinear continuum elements and uses the code TOCHNOG. The soil was modeled as a standard Mohr-Coulomb material. Two calculation runs are compared:

– elastoplastic jetgrout (Sec. 3.1); spontaneous excavation in four consecutive steps as time-independent calculation,
– visco-elasto-plastic jetgrout (Sec. 3.4); time-dependent calculation; the elastic properties of jet-grout were taken for a mean density of 1,86 g/cm³ such as to produce the same spontaneous strain as for the elastic case.

As the main difference to conventional analyses, the visco-elastic constitutive law requires to assume the construction sequence realistically in time-scale. In this case, 70 days have been modelled: 1 day was allowed for dewatering, 2 weeks for each of the four excavation phases and 3 days each for the installation of the three strut levels. The system behaviour results different, if the construction schedule is altered from this design assumption. But also the effects of a changed schedule can be predicted in the sense of the observational method.

During excavation, the peak stresses in viscous jet-grout are considerably relaxed due to creep (here abt. 25% lower, see Fig. 10) Earth pressures are reduced by 10%, while bottom heave is slightly higher and small load portions are transferred from the bottom slab to the struts (not shown here). The effect of local density variations inside the jetgrout still needs further discussion.

472

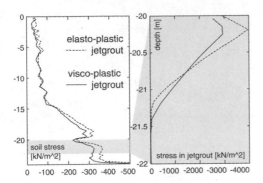

Figure 10. Horizontal stresses along the wall in the soil mass (left) and in the jetgrout 0,9 m from the wall (detail, right).

The stress peaks in Fig. 10 develop due to slab rotation in a short distance from the wall and do not indicate plastification. The strength of jetgrout in this example was also sufficient to avoid failure directly at the contact of wall and slab (with lateral stresses of 6500 resp. 5300 kN/m²). But for other cases with reduced slab thickness or higher water load, elasto-plastic jetgrout plastifies due to high contact pressures between wall and upper side of the slab, while the same system with creeping material can be shown to experience no failure.

REFERENCES

Gudehus, G., Kudella, P.·2000. Mechanische Eigenschaften von HDI-Material, hergestellt mit dem HDI-2/3-Verfahren in sandig- kiesigem Baugrund, unpubl.
Gudehus, G. 2003. A Visco-hypoplastic constitutive relation for soft soils. *Soils and Foundations,* in prep.
Hoffmann, H., Katzenbach, R., Quick, H., Weidle, A. 2000. Überlegungen zum Risk assessment beim Düsenstrahlverfahren auf der Basis aktueller Entwicklungen. *Beiträge zum 15. Veder Kolloquium, Gruppe Geotechnik Graz,* 7: 31 ff
Kudella, P. 2001. Wie sicher sind Düsenstrahl-Injektionssohlen? *Bautechnik* 12(78): 841 ff
Lesnik, M. 2002. Ermittlung der Reichweite beim Düsenstrahlverfahren unter Berücksichtigung der Herstellparameter und der Bodeneigenschaften mittels Rückflussanalyse. *Gruppe Geotechnik Graz,* in press
Möller, G. 2002. Bericht über die Prüfung eines Stoffgesetzes bezüglich seiner Anwendbarkeit auf DS-Material, unpubl.
Niemunis, A., Krieg, S. 1996. Viscous behaviour of soil under oedometric conditions. *Canadian Geotechnical Journal* 1 (33): 159 ff
Niemunis, A. 2002. Extended hypoplastic models for soils. *Schriftenreihe IGB Ruhr-Uni Bochum,* in prep.
van Riel, A.J.E., van Tol, A.F., Vrijling, J.K. 2000. Op weg naar en veraantword ontwerp van jetgroutlagen. *GeoTechniek* 4: 48 ff
Stein, J. 2000. Bohrabweichungen bei der Herstellung einer Düsenstrahlsohle. *Mitteilungen IGB TU Braunschweig* 62: 209 ff

Geotechnical Measurements and Modelling, Natau, Fecker & Pimentel (eds)
© 2003 Swets & Zeitlinger, Lisse, ISBN 90 5809 603 3

A numerical model for the site response analysis and liquefaction of soil during earthquakes

V.A. Osinov
Institute for Soil Mechanics and Rock Mechanics, University of Karlsruhe, Karlsruhe, Germany

ABSTRACT: The paper presents a numerical approach to the level ground response analysis and liquefaction of saturated soil during strong earthquakes. The approach is based on the solution of a one-dimensional dynamic problem for layered dry or saturated soil with the use of a hypoplastic constitutive equation for the description of the plastic behaviour of soil under cyclic loading. The dynamic problem is formulated within the so-called u–p approximation and is solved by a finite-difference technique. The paper contains a description of the boundary value problem and the numerical algorithm, and presents numerical examples of the dynamic liquefaction of saturated sand by periodic shear disturbances.

1 CONSTITUTIVE RELATIONS

The modelling of the dynamic deformation of soil during strong earthquakes is one of the topical problems in earthquake geotechnics (Ishihara 1996, Kramer 1996). This paper presents a numerical model for the level ground response analysis and liquefaction of soil based on the solution of the dynamic problem for a horizontal soil layer. The model is an extension of the approach developed by Osinov (2000) and Osinov & Gudehus (2003) for the numerical solution of the dynamic problems of hypoplasticity.

The model is applicable to both dry and saturated soils. In the latter case, the soil is assumed either to be fully saturated or to contain a small amount (few volume percent) of free gas so that the effective stress principle

$$\sigma = \sigma' - p_f \mathbf{I} \qquad (1)$$

can be used. Here σ is the total stress, σ' is the effective stress (compressive stresses are taken to be negative), p_f is the fluid pressure (positive for compression), and $\mathbf{I}$ is the unit tensor.

The evolution of the effective stress is described by a hypoplastic constitutive relation (Niemunis & Herle 1997). This relation describes plastic deformations of a solid skeleton under monotonic and cyclic loading. It incorporates the critical state concept of soil mechanics and the dependence of the stiffness on the current stresses, the density and the history of deformation. The rate of the effective stress $\dot{\sigma}'$ is determined by the

rate of strain $\dot{\varepsilon}$, the current stress σ', the void ratio e and the so-called intergranular-strain tensor δ which takes into account the influence of the recent deformation history on the current response of the material. The constitutive relation is written as two tensor-valued functions

$$\dot{\sigma}' = \mathbf{H}(\sigma', \dot{\varepsilon}, \delta, e), \qquad (2)$$

$$\dot{\delta} = \mathbf{F}(\dot{\varepsilon}, \delta), \qquad (3)$$

supplemented with the evolution equation for the void ratio (with incompressible grains):

$$\dot{e} = (1 + e) \operatorname{tr} \dot{\varepsilon}. \qquad (4)$$

The parameters of the constitutive equation (2), (3) are independent of the state of the material, so that the behaviour of a given material is modelled in a wide range of stresses and densities with the same set of parameters. A detailed description of the hypoplastic relation is given by Niemunis & Herle (1997) (see also Osinov 2000 and Osinov & Gudehus 2003).

Figures 1–4 show examples of the cyclic deformation of dry and saturated sand calculated with the constitutive parameters taken from Osinov & Gudehus (2003). Figure 1 shows the strain-controlled simple shearing of dry sand with a constant normal stress $\sigma'_{11} = -100\,\text{kPa}$, initial stresses $\sigma'_{22} = \sigma'_{33} = -50\,\text{kPa}$, $\sigma'_{12} = 0$ and an initial void ratio of 0.8. The shear stress σ'_{12} and the volumetric deformation ε are shown as functions of the shear strain $\gamma = 2\varepsilon_{12}$.

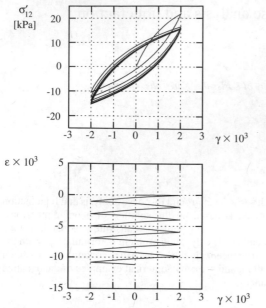

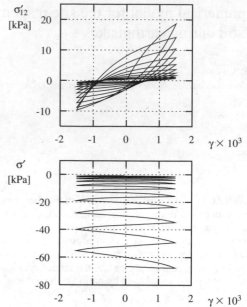

Figure 1. Calculated cyclic simple shearing of dry sand.

Figure 3. Calculated strain-controlled cyclic shearing of saturated sand.

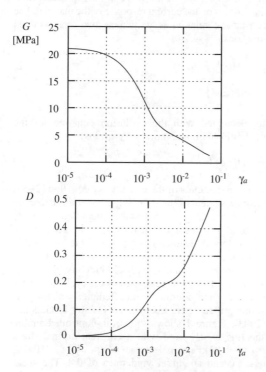

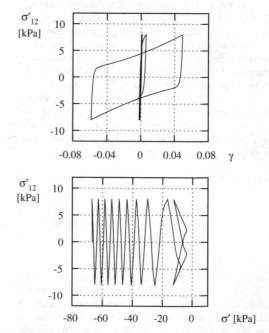

Figure 2. Calculated secant shear modulus and damping ratio.

Figure 4. Calculated stress-controlled cyclic shearing of saturated sand.

476

The response of a soil to the multi-cycle shear loading is usually quantified through the dependence of the secant shear modulus $G = \tau_a/\gamma_a$ and the damping ratio $D = A/2\pi\tau_a\gamma_a$ on the strain amplitude γ_a (Ishihara 1996, Kramer 1996). Here τ_a is the amplitude of the shear stress σ'_{12}, and A is the area of the hysteresis loop in the (γ, σ'_{12})-plane. Figure 2 shows the secant shear modulus and the damping ratio as functions of the strain amplitude calculated for the cyclic simple shearing with a constant normal stress $\sigma'_{11} = -100\,\text{kPa}$ and the same initial conditions as in Figure 1. The secant shear modulus and the damping ratio are calculated for the third cycle.

The cyclic shearing of a dry granular material is known to result in its gradual compaction, which is well described by the present constitutive equation (Fig. 1). If a granular material is saturated with a fluid, the latter restrains any changes in the volume that would occur in a dry material due to the rearrangements of the grains during cyclic shear. In this case, instead of compaction, repeated shearing without drainage is accompanied by the reduction of the effective pressure. If the shear amplitude is large enough, the shearing may result in the vanishing of the effective pressure and, as a consequence, in the loss of the shear stiffness. The manner in which the effective pressure is reduced depends on the type of loading. In the strain-controlled case, the effective pressure approaches zero in a gradual manner. In the stress-controlled case, after a number of cycles, the effective pressure abruptly falls to a low value and then oscillates with an amplitude determined by the amplitude of the applied shear stress. The strain amplitude increases to several percent ('cyclic mobility', Seed et al. 1976).

Both these cases are correctly modelled by the hypoplastic equation. Figure 3 shows the strain-controlled cyclic shearing of a fully saturated material calculated with the same initial state as in Figure 1. Shown are the shear stress σ'_{12} and the mean effective stress $\sigma' = (\sigma'_{11} + \sigma'_{22} + \sigma'_{33})/3$ versus the shear strain γ. After several cycles of loading, the mean effective stress is reduced to zero, thus resulting in the vanishing of the shear stiffness, and the material turns into the state of 'initial liquefaction' (Seed et al. 1976). The transition to cyclic mobility during stress-controlled shearing is shown in Figure 4.

2 BOUNDARY VALUE PROBLEM

The dynamic problem for a saturated soil is formulated within the framework of the so-called u–p approximation (Zienkiewicz et al. 1980, 1999). In this approximation, the solid and fluid phases have different velocities but the same acceleration. Under this assumption, one can eliminate the fluid velocity from the system of equations and thus reduce the number of unknown functions. This approximation is widely used in soil dynamics and is shown to be adequate for the earthquake modelling. The system of equations consists of the equation of motion

$$\frac{\partial \sigma'_{ji}}{\partial x_j} - \frac{\partial p_f}{\partial x_i} + \varrho g_i = \varrho \dot{v}_i, \qquad (5)$$

the constitutive equations for the effective stress (2), for the intergranular strain (3), for the void ratio (4), and the constitutive equation for the fluid phase

$$-\frac{n}{K_f}\dot{p}_f = \frac{\partial v_i}{\partial x_i} + \frac{1}{g}\frac{\partial}{\partial x_i}\left[k\left(-\frac{1}{\varrho_f}\frac{\partial p_f}{\partial x_i} + g_i - \dot{v}_i\right)\right].$$
$$(6)$$

In these equations, v_i are components of the velocity of the solid phase, ϱ is the density of the soil, ϱ_f is the density of the fluid phase, K [m/s] is the permeability of the soil, K_f is the compression modulus of the fluid, n is the porosity, and g_i are components of the acceleration due to gravity whose absolute value is g.

Replacing $\dot{v}_i$ with the use of (5), equation (6) can be written in the form

$$-\frac{n}{K_f}\dot{p}_f = \frac{\partial v_i}{\partial x_i} + \frac{1}{g}\frac{\partial}{\partial x_i}\left[\frac{k}{\varrho}\left(-\frac{\partial \sigma'_{ji}}{\partial x_j} + \left(1 - \frac{\varrho}{\varrho_f}\right)\frac{\partial p_f}{\partial x_i}\right)\right].$$
$$(7)$$

In the case of a dry body, the system of equations consists of the equation of motion (5) without the fluid pressure term, and the same constitutive equations except for (7).

The boundary value problem is formulated for a horizontal soil layer as shown in Figure 5. All unknown functions depend on x_1 and time. The initial vertical stress and the pore pressure are distributed in accordance with the densities ϱ, ϱ_f and the gravity force. The initial horizontal stresses are connected with the vertical stress through the earth pressure coefficient. The upper boundary $x_1 = 0$ is free of traction. At the lower boundary $x_1 = -h$, the horizontal (v_2) and the vertical (v_1) velocity components are prescribed as functions of time, which imitates the influence of an excitation wave coming from below. In the case of saturated soil, the lower boundary is assumed to be impermeable, and the pore pressure at the water table is assumed to remain zero.

3 NUMERICAL ALGORITHM

The system of the dynamic equations is solved by a finite-difference algorithm which is derived from a

477

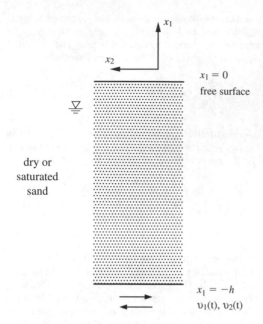

Figure 5.   Boundary value problem.

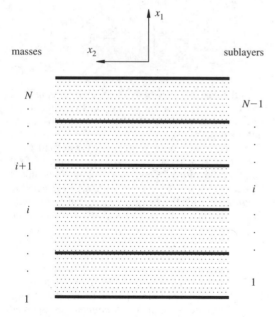

Figure 6.   The discrete model of a layer.

discrete model of the layer. The whole layer is divided into $N-1$ sublayers as shown in Figure 6. The mass of the soil is concentrated at the boundaries $x_{1(i)}$ of the sublayers (lumped masses). A mass $m_{(i)}$ per unit area is equal to $\varrho_{(i-1)}\,(x_{1(i)} - x_{1(i-1)})$, where $\varrho_{(i-1)}$ is the

soil density in the sublayer below, and $m_{(1)} = 0$. The stresses and the pore pressure inside a sublayer are assumed to be spatially uniform. The velocity is continuous at the boundaries of the sublayers and is assumed to vary linearly inside. This gives a uniform deformation of each sublayer and is consistent with the uniformity of the stresses.

Let $\sigma'_{11(i)}$, $\sigma'_{12(i)}$, $p_{f(i)}$ stand for the stress components and the pore pressure in the i-th sublayer, and let $v_{1(i)}$, $v_{2(i)}$ stand for the velocity components of the i-th mass. The motion of the i-th mass for $i = 2, \ldots, N-1$ is governed by the equations

$$m_{(i)}\dot{v}_{1(i)} = \sigma'_{11(i)} - \sigma'_{11(i-1)} + p_{f(i-1)} - p_{f(i)} - m_{(i)}g,$$
(8)

$$m_{(i)}\dot{v}_{2(i)} = \sigma'_{12(i)} - \sigma'_{12(i-1)}.$$
(9)

The motion of the mass $m_{(1)}$ is determined immediately from the boundary condition for velocity at the lower boundary of the layer. The upper boundary is free of traction, which gives the equations of motion for the N-th mass:

$$m_{(N)}\dot{v}_{1(N)} = -\sigma'_{11(N-1)} + p_{f(N-1)} - m_{(N)}g,$$
(10)

$$m_{(N)}\dot{v}_{2(N)} = -\sigma'_{12(N-1)}.$$
(11)

The components of the strain rate tensor $\dot{\varepsilon}_{(i)}$ in the i-th sublayer are

$$\dot{\varepsilon}_{11(i)} = \frac{v_{1(i+1)} - v_{1(i)}}{x_{1(i+1)} - x_{1(i)}}, \quad \dot{\varepsilon}_{12(i)} = \frac{v_{2(i+1)} - v_{2(i)}}{2(x_{1(i+1)} - x_{1(i)})}.$$
(12)

The time derivatives of the components $\sigma'_{11(i)}$, $\sigma'_{12(i)}$, $\sigma'_{22(i)}$, $\sigma'_{33(i)}$, $\delta_{11(i)}$, $\delta_{12(i)}$ in the i-th sublayer are expressed through the corresponding values in the same sublayer according to the constitutive equations (2), (3). The same is true for the void ratio, equation (4).

In the case of a saturated soil, in addition to the above equations, evolution equations for the pore pressure in each sublayer are required. They can be derived as follows.

According to Darcy's law, the seepage velocity w in the x_1-direction is given as

$$w = -\frac{k}{\varrho_f g}\frac{\partial \tilde{p}_f}{\partial x_1},$$
(13)

where $\tilde{p}_f$ is the excess pore pressure defined as the difference between the actual fluid pressure and an equilibrium pressure which would cause no seepage. For a fluid flowing with an acceleration $\dot{v}$ in a gravity field whose acceleration is $-g$, the gradient of the equilibrium pressure is equal to $-\varrho_f(g + \dot{v})$. Since in the present formulation the acceleration of the fluid is assumed to be equal to that of the solid, we can use the equation of motion (5) to express this acceleration through the vertical stress and the pore pressure. The law (13) is then written as

$$w = \frac{k}{\varrho g}\left[-\frac{\partial \sigma'_{11}}{\partial x_1} + \left(1 - \frac{\varrho}{\varrho_f}\right)\frac{\partial p_f}{\partial x_1}\right]. \qquad (14)$$

Turning back to the discrete model, we assign each mass $m_{(i)}$ below the water table a seepage velocity

$$w_{(i)} = \frac{k_{(i-1)}}{g m_{(i)}}\left[\sigma'_{11(i-1)} - \sigma'_{11(i)} \right.$$
$$\left. + \left(1 - \frac{\varrho_{(i-1)}}{\varrho_f}\right)(p_{f(i)} - p_{f(i-1)})\right]. \qquad (15)$$

The seepage velocity $w_{(1)}$ is put equal to zero according to the boundary condition (impermeable boundary). It is easily seen that (15) is a discrete version of (14).

Assuming the grains to be incompressible, the rate of change in the fluid pressure in the i-th sublayer is given as

$$\dot{p}_{f(i)} = -\frac{K_f(w_{(i+1)} - w_{(i)} + v_{1(i+1)} - v_{1(i)})}{n_{(i)}(x_{1(i+1)} - x_{1(i)})}, \qquad (16)$$

where $n_{(i)}$ is the porosity of the i-th layer. The term $w_{(i+1)} - w_{(i)}$ in (16) is responsible for the change in the volume of the pore fluid in the sublayer caused by the inflow and outflow of the fluid, while the term $v_{1(i+1)} - v_{1(i)}$ gives the change in the pore space caused by the deformation of the skeleton. Formula (16) is used for all sublayers below the water table except for the upper one, for which we put $\dot{p}_f = 0$ according to the boundary condition (constant pore pressure).

Close inspection of equations (8)–(12), (15), (16) reveals that they constitute a finite-different approximation of the original differential equations. This spatial discretization reduces the original system of partial differential equations to a system of ordinary differential equations of the form $\dot{z} = f(z)$, where $z = (z_1, z_2, \ldots)$ is a finite set of unknown functions of time. The time integration of the system is performed implicitly as

$$z^{j+1} = z^j + \left[\alpha f(z^j) + (1 - \alpha)f(z^{j+1})\right]\Delta t, \qquad (17)$$

where z^j stands for the value of z at time $t = t_j$, $\Delta t = t_{j+1} - t_j$ is the time increment, and α is the

parameter of the scheme, $0 \leqslant \alpha \leqslant 1$. Equations (17) are solved for z^{j+1} by successive approximations.

4 NUMERICAL EXAMPLES

In this section we consider numerical solutions for a layer of saturated sand subjected to a periodic shear disturbance at the base.

In problems with saturated soil we are primarily interested in the evolution of the effective stress and in liquefaction of the soil. Liquefaction is understood here as the state of vanishing effective pressure caused by cyclic shearing ('initial liquefaction' in the terminology of Seed et al. 1976). It should be mentioned that the reduction of the effective pressure to zero during cyclic shearing can be achieved in both loose and dense saturated sand. When occurring in the field, this does not necessarily lead to an observable failure or flow of the soil: the shear strength may be restored in monotonic shearing if the sand is dense enough. However, the loss of the shear stiffness in the liquefied zones strongly influences the response of the soil mass to dynamic excitation.

There are many factors which influence the solution to the problem under study. Among them are the boundary conditions (amplitude, frequency content), the constitutive behaviour of the solid skeleton and the thickness of the layer. In the case of saturated soil there are two additional factors which influence the dynamic response, namely, the permeability of the soil and the compressibility of the pore fluid. Note that the compressibility may differ significantly from that of pure water in the presence of a small amount of free gas. Besides, the final distribution of the effective pressure depends on the duration of the disturbance. In the examples below we will change the permeability and the compressibility of the fluid, with all other parameters being fixed.

Consider a 25 m thick layer of saturated soil with a sinusoidal boundary condition at the base prescribed for the horizontal velocity component starting at $t = 0$. The vertical velocity component at the base is equal to zero. The amplitude of the boundary velocity is 5 cm/s with a frequency of 4 Hz. The soil has an initial void ratio of 0.8 and an earth pressure coefficient of 0.5. The behaviour of the soil is described by equations (2), (3) with the constitutive parameters taken from Osinov & Gudehus (2003).

In the first example, the permeability is equal to 10^{-6} m/s which corresponds to fine sand. The compressibility of the fluid is taken to be 20 MPa; this approximately corresponds to a degree of saturation of 0.99. Figure 7 shows the distribution of the mean effective stress in the layer at two different times. After few seconds of the excitation we observe the formation of thin separate zones in which the effective pressure is

479

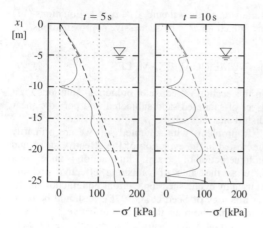

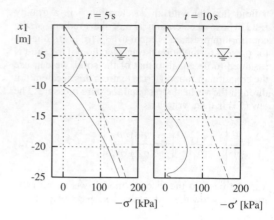

Figure 7. Mean effective pressure in the layer at two times. Dashed line: the initial distribution. $K_f = 20\,\text{MPa}$, $k = 10^{-6}\,\text{m/s}$.

Figure 9. The same as in Figure 7 for $K_f = 20\,\text{MPa}$, $k = 10^{-3}\,\text{m/s}$.

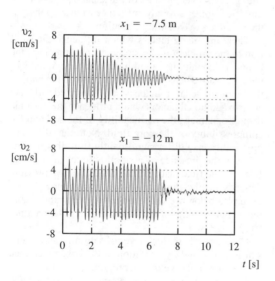

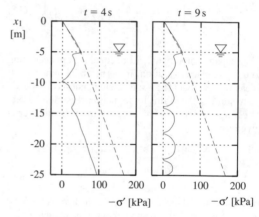

Figure 10. The same as in Figure 7 for $K_f = 2\,\text{GPa}$, $k = 10^{-6}\,\text{m/s}$.

Figure 8. Horizontal velocity versus time at two different depths in the case shown in Figure 7.

reduced to zero and the soil becomes liquefied. This localization of liquefaction can be explained as a manifestation of the instability of inhomogeneous effective stresses during cyclic shearing under undrained conditions (Osinov 2002). The instability occurs because of the fact that a higher shear amplitude produces a higher rate of the reduction of the effective stress and the stiffness, resulting in an increase in the amplitude and thus in positive feedback.

The vanishing shear stiffness and the high hysteretic damping in the liquefaction zones do not allow shear waves to propagate through these zones. Each newly emerging liquefaction zone isolates the above situated soil from further excitation. This is illustrated in Figure 8. In spite of the continuing boundary disturbance, the oscillations of the soil at a certain depth decay when a liquefaction zone emerges below that depth, and the duration of the motion is thus shorter than the actual duration of the excitation. The dynamic motion of the whole layer terminates when the soil is liquefied in the immediate vicinity of the base.

Calculations with the present model show that, if the coefficient of permeability is below $10^{-5}\,\text{m/s}$, the solution is practically the same as that under locally undrained conditions. An increase in the permeability up to $10^{-3}\,\text{m/s}$ (coarse sand) leads to a different

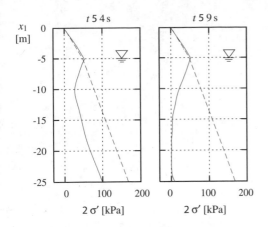

Figure 11. The same as in Figure 7 for $K_f = 2\,$GPa, $k = 10^{-3}\,$m/s.

liquefaction pattern for large times (Fig. 9). Although at the beginning the liquefaction is similar (a narrow zone at a depth of 10 m), eventually the soil becomes liquefied over a finite thickness in the middle of the layer.

In the case of full saturation the compression modulus of the fluid is equal to that of pure water (2 GPa). This increases the rate of the reduction of the effective pressure. The solution with low permeability is shown in Figure 10. Although the liquefaction is localized in separate zones, the average value of the effective pressure in the layer is much lower than in Figure 7. High permeability prevents the localization of liquefaction and results in the situation shown in Figure 11: the lower half of the layer becomes uniformly liquefied.

ACKNOWLEDGEMENT

The study was carried out within the framework of SFB 298 supported by the Deutsche Forschungsgemeinschaft.

REFERENCES

Ishihara, K. 1996. *Soil Behaviour in Earthquake Geotechnics*. Clarendon Press, Oxford.

Kramer, S. L. 1996. *Geotechnical Earthquake Engineering*. Prentice Hall, Upper Saddle River.

Niemunis, A., Herle, I. 1997. Hypoplastic model for cohesionless soils with elastic strain range. *Mech. Cohesive-frictional Mater.* 2(4), 279–299.

Osinov, V. A. 2000. Wave-induced liquefaction of a saturated sand layer. *Continuum Mech. Thermodyn.* 12(5), 325–339.

Osinov, V. A. 2002. Cyclic shear of saturated soil: the evolution of stress inhomogeneity. *Continuum Mech. Thermodyn.* 14(2), 191–205.

Osinov, V. A., Gudehus, G. 2003. Dynamics of hypoplastic materials: theory and numerical implementation. In: *Dynamic Response of Granular and Porous Materials under Large and Catastrophic Deformations*, K. Hutter and N. Kirchner (eds.), Springer, Berlin, 265–284.

Seed, H. B., Martin, P. P., Lysmer, J. 1976. Pore-water pressure changes during soil liquefaction. *J. Geot. Eng. Div., Proc. ASCE.* 102, GT4, 323–346.

Zienkiewicz, O. C., Chang, C. T., Bettess, P. 1980. Drained, undrained, consolidating and dynamic behaviour assumptions in soils. *Géotechnique* 30(4), 385–395.

Zienkiewicz, O. C., Chan, A. H. C., Pastor, M., Schrefler, B. A., Shiomi, T. 1999. *Computational Geomechanics with Special Reference to Earthquake Engineering*. John Wiley, Chichester.

A database for the performance of numerical modelling of retaining wall

Y. Riou
Ecole Centrale de Nantes, Nantes, France

Ph. Mestat
Laboratoire Central des Ponts et Chaussées, Paris, France

P. Kudella
Universität Karlsruhe, Karlsruhe, Germany

P. von Wolffersdorff
Baugrund Dresden, Dresden, Germany

ABSTRACT: Professional competency in geotechnical engineering is based both on the knowledge of fundamental principles which govern the behaviour of the works and the relevant experience on projects. This in-situ experience can be supplemented by information contained in the scientific literature and reports relating to comparisons of numerical results and in-situ or laboratory measurements. The conclusions drawn from these studies are indispensable to the understanding of the behaviour of geotechnical works. The Laboratoire Central des Ponts et Chaussées (LCPC) in partnership with the Ecole Centrale de Nantes is working out two databases gathering all the comparisons measurements – numerical results with reference to any type of geotechnical works. In this paper, an analysis of the information contained in the databases regarding retaining walls is proposed. It intends to provide a preliminary analysis of the modelling error dispersion.

1 INTRODUCTION

Over the last twenty years, geotechnical engineers concerned with the modelling of works are aware of the importance of benchmarks for the validation of numerical codes and simulation procedures, and for the justification of design assumptions and input parameters. A benchmark provides predictions before construction (i.e. class A predictions; Lambe 1973).

A test analysis has been generally conducted by the authors of each benchmark. This analysis consists of comparison of numerical results and measurements related to the modelling assumptions. However, this analysis is fairly tricky due to the large number of parameters with respect to the number of benchmark participants. For an evaluation and a hierarchy of error or for some modelling guidelines, a synthesis of the whole of previous benchmarks and a data collection process for new benchmarks are needed. In this prospect, the Laboratoire Central des Ponts et Chaussées (LCPC-Paris) and the Ecole Centrale de Nantes (ECNantes) decided the constitution of a computerized database relating to the geotechnical benchmarks.

In addition, a bibliographic database, called MOMIS (acronym for "Modélisation des Ouvrages et Mesures In Situ"), dealing meanly with class C predictions (i.e. posteriori predictions; Lambe 1973) and containing a total of 416 case studies is maintained by LCPC. Among these cases studies, 66 deal with sheet-piled retaining structures and 102 with diaphragm walls.

2 DATABASE FOR CLASS A PREDICTIONS: BENCHMARKS RELATING TO RETAINING WALLS

Due to the large amount of information and the need for further investigation, especially correlation between numerical results, measurements, modelling assumptions and parameters, a specific database for class A predictions was conceived in a computerized form. This database gathers not only information mentioned previously but some comments relating to the geotechnical work itself, its construction, the benchmark characteristics. All the information relating to the structure of this database is provided in the references (Mestat et al. 2001, Mestat et al. 2002).

The part of this database, dealing with retaining works, is introduced in this paper. The three benchmarks included in this base are mentioned in Table 1. It will be noticed that these three cases proposed various modelling conditions likely to provide useful indications on the different error types. Various modelling conditions mean various requirements for calculation.

A benchmark can be "academic": all the modelling assumptions are imposed by the benchmark author (dimension, mesh, limit conditions, interface conditions, constitutive relationship, rheological parameters, phases of constructions). The comparison concerns solely the numerical results. Only the modelling implementation and the code (numerical problem and wrong instruction) are tested. A benchmark can be "industrial": the information is typical of project conditions. No recommendation for modelling are given. The results dispersion of numerical predictions refers as well to modelling error, or alternative modelling (unsuitable mechanical model, parameters, interaction conditions, loading conditions, initial stress state, etc.). The

numerical values can be placed in relation with measurements.

All types of benchmark can be imagined between these two configurations, especially confrontations between numerical predictions and physical modelling (centrifuge test) in order to validate one aspect of modelling (constitutive relationship). The benchmarks included in this database are distinguished by three classes: academic, semi academic and industrial benchmarks.

For information purposes, Table 2 shows the numbers of measurements and computed values provided by the 74 modellings captured in the database for class A predictions.

The database is currently in progress. So, a complete synthesis of the whole of this information is not available. The main concern is now the retrieval and the capture of information relating to all types of geotechnical works subjected to a benchmark (embankment, foundation (deep, surface), underground works, retaining wall). However, a preliminary thought about modelling error, in complement of existing conclusions provided by some benchmark authors, is proposed below.

Table 1. Benchmarks dealing with retaining walls and included in the database.

Retaining wall type	Test type	Author	Participants number	Modelling requirements		
				Domain	Constitutive model	Mechanical parameters
Anchored	Semi-industrial Semi-academic	Graz University	18	Proposed	Free	Proposed

Aim:
* to assess the range of numerical solutions under conditions typically found in practice
* to provide a test for validation of numerical models and solution procedure
* to test interface element

Comments:
Benchmark related to a project in Berlin
Parameters taken from the literature and from some classical geotechnical tests

| Strutted | Academic | Graz University | 12 | Imposed | Imposed | Imposed |

Aim:
* to check codes and its use

| Strutted | Industrial | Karlsruhe University | 44 | Free | Free | Free |

Aim:
* to test retaining wall modelling in professional practice conditions

Comments:
19 FEM predictions, 25 traditional predictions (mainly Subgrade Reaction Method)
Parameter identification based on various geotechnical tests (unusual condition in practice)

Table 2. Number of captured values in the data base relating to the retaining walls.

	Lateral displacement	Settlement	Bending moment	Earth pressure	Strut force
Number of measurements	31	25	20	37	4
Number of calculated values	850	680	350	650	60

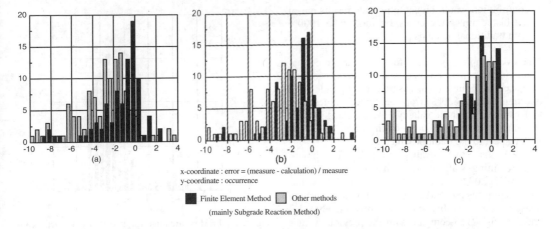

x-coordinate : error = (measure - calculation) / measure
y-coordinate : occurrence

■ Finite Element Method □ Other methods

(mainly Subgrade Reaction Method)

Figure 1. Prediction error histogram – lateral displacement benchmark – retaining wall of Karlsruhe, (a) 5 m excavation, (b) 5 m excavation + surface loading, (c) 5 m excavation + surface loading + reduction of strut length (ultimate limit state).

3 MODELLING ERROR: A PRELIMINARY SYNTHESIS

One of the first concerns for a geotechnical engineer is the relevance and place of a Finite Element Method calculation beside traditional methods (mainly Subgrade Reaction Method) in professional practice conditions. An element of reply is given by the benchmark of Karlsruhe (Hochstetten) organised by Dutch Center for Civil Engineering and the Karlsruhe University (Von Wolffersdorff 1994a, b). This first study relates to the lateral displacements, representative information of the general wall behaviour. Figure 1 shows this lateral displacement for all the points up to 6 m depth according to a step of 1 m.

From these three histograms, we can make the following observations[1].

- For the calculation of a strutted wall, in serviceability state, FEM provides results more in agreement with traditional approaches: the largest occurrence (19/90 results) is found for an error range of [0% to −50%] (conservative calculation) compared with 14/124 results in the error range of [−150% to −100%] related to other methods (fig. 1a). The mean value with FEM is −1.7 compared with −3.1 provided by other methods. The calculated values discrepancy is less with FEM (standard deviation

of 3.2 compared with 4.1 for traditional methods[2]). This result confirms the conservative character of the Subgrade Reaction Method commonly used in the retaining wall design and based on very large practice and experience feedback. FEM proposes a more realistic calculation but involves a potential risk of minimisation of lateral displacements.

- For a more complex calculation, including a surface loading, this tendency is confirmed (fig. 1b). The maximum of occurrence is located for an error range of [0 to −50%] with FEM and [−300% to −250%] with traditional methods. The mean value of error is −1.8 with the FEM and −3.8 with traditional methods. The standard deviations are respectively 3.3 and 7.5. However, 20% of the FEM results are not conservative compared with 10% for other methods. A similar study was carried out on an embankment test, benchmark of Haarajoki (FNRA 1997, Aalto et al. 1998). This distinction between numerical methods and traditional methods is not so marked. So, that confirms a more complex configuration makes more efficient the FEM approach.

- For a ultimate limit state, caused by a reduction of strut length, and if the error more than 1000% is neglected, the error distribution is similar for the two approaches (fig. 1c). This fact leads us to think that FEM and the constitutive law used for soils do not have yet the capacity to correctly reproduce the failure mechanism of such a geotechnical work. It would seem that FEM modelling, in this particular stage of construction, is not obvious in regard of traditional methods relying on observations. The only

---

[1] All the computed values with an error exceeding 10 (1000%) are ignored in this study. It was considered that these calculations involve a big error beyond the traditional framework of modelling error (error in implementation procedure of modelling, or non significant error on small displacement).

[2] These values are indicatives. The error distribution error is not a Gaussian distribution but similar in the two cases.

contribution would be not to provide results out of the histogram: some errors are more than 1000% with traditional methods.

This over-simple mode used for the error representation, without consideration of main points (maximum displacement point, and its position) can be called into question. So, the 600% errors mentioned in these figures can be of doubtful validity. However, this mode is revealing of the contribution of the FEM modelling with respect to the traditional methods.

A similar study dealing with the benchmark of a retaining wall in Berlin has been performed (Schweiger 2002). The error mean value and the standard deviation, using FEM calculation, for the excavation phase in serviceability state, was respectively -1.2 and 2.3 (fig. 2) to compare with the previous values -1.7 and 3.2 (fig. 1a). This lateral displacements analysis confirms, or rather doesn't invalidate, the observations made previously on the benchmark of Karlsruhe. The error mean value is appreciably weaker in this case (-1.2 against -1.7) but this latter is not guaranteed because based on an estimated (and not measured) rigid-body translation of 10 mm. A translation value of 7 mm would have led to a error mean value similar to that of the benchmark of Karlsruhe. So the main result is, in this analysis, the error dispersion. The error standard deviation is slightly smaller on the benchmark of Berlin (30%: 2.3 compared to 3.2). This histogram presents two distribution families: one whose error is lower than 150% and the other one whose error mean value is 300%. According to the benchmark author, H.F. Schweiger, these two distributions are to be connected with two different options for the identification of rheological parameters: the first based on the parameters provided in the literature, the second one based on parameters deduced from some geotechnical tests but whose validity was to be called into question. According to the benchmark author, the results out of the diagram, would be related to a big error in the setting up of the modelling procedure and to the use of a constitutive model unsuited to the excavation problem.

In any case, the results interpretation of the benchmark of Karlsruhe requires more time for confirmation of the benchmark conclusions of Berlin. This work will constitute the next stage of this study.

In its conclusions of the benchmark of Berlin, H.F. Schweiger warned participants against gross modelling errors, wrong setting up of modelling, and non-respect of calculation instructions (for example use of contact elements when not required). These two last error sources (wrong setting up and non respect of specification) are illustrated on Figure 3 with numerical results provided by the academic benchmark organised by the University of Graz (Schweiger 1998). Further analysis is needed in order to quantify each error con-

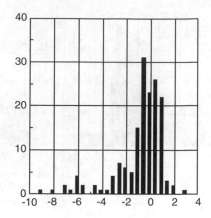

Figure 2. FEM prediction error, lateral displacement semi-academic benchmark – Berlin Wall Graz University.

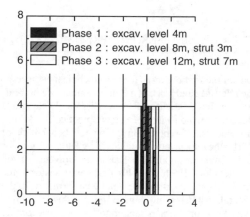

Figure 3. FEM prediction error, lateral displacement academic benchmark Graz University.

tribution. But, it was shown that various soil/structure interaction conditions can represent 30% of the whole discrepancy in some points, even more in the first stages of excavation.

There is now a need to examine more closely the three benchmarks in relation with constitutive models, input parameters, modelling procedures, etc. The database presented here is expected to effectively help in this work.

4 DATABASE FOR CLASS C PREDICTIONS: MOMIS

The MOMIS database can be used to highlight modelling principles (in order to provide a guide for good finite element practice to users), to provide informations about current practice and to estimate the deviations

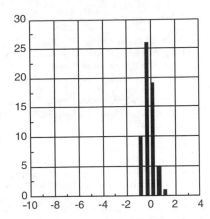

Figure 4. Prediction error histogram, maximal lateral displacement at the end of construction retaining wall – class C predictions.

between results given by numerical models and measured values (calculation after measurement, Lambe 1973). The primary objective is to preserve the record of these models and their comparisons.

Comparisons between measured and computed values were made at the end of construction. The parameters are the maximum horizontal displacement of the wall; the maximum surface settlement behind the wall; the maximum heave of the bottom of the excavation; forces in the struts of ground anchors and the bending moments in the wall.

Regarding the maximum horizontal displacement of wall at the end of construction (fig. 4), the agreement is fairly good: most class C predictions have led to relatively satisfactory results. The maximum of occurrence is located for an error range of [0 to −50%] with FEM. One occurrence concerns one case study: for class C predictions, we have considered only the maximum horizontal displacement. 41% of the case studies stored in MOMIS database have a relative error which is less than 25% and 61% provide a relative error less than 50%. The foregoing analysis makes no allowance for the specific geotechnical conditions and technical construction.

It may be useful to recall that the measured values are themselves subject to various errors inherent to the monitoring process and that it is therefore pointless to look for a perfect agreement between computations and measures.

This histogram, compared with the previous one (fig. 1a) in relation with class A predictions, gives some information about the possibility of improvement in terms of retaining wall modelling.

Many studies can be made with the MOMIS database and following remarks deduced on other variables. For the estimation of the maximum settlement

behind the wall, results seem not so good: computed settlements are smaller than measured values. 13.5% of references give a relative error less than 25%. From a general standpoint, the models are not yet able to describe the surface settlements and often underestimate them. This may be explained by the fact that simple elastoplastic models (e.g. Mohr-Coulomb or Drucker-Prager without hardening) overestimate the plastic volume increase, which leads to underestimating the settlement. Another potential source of inaccuracy is the fact that the elastic part of the constitutive law is generally linear, which leads to inappropriate estimation of the strains in the lower part of the mesh.

The previous analyses have considered relative errors with respect to the variables in an independent manner. An "effective model" however must be able to simultaneously predict all of the key aspects in the response of a geotechnical structure. Model error estimation must take into account the entire array of measured variables (various displacements and pressures).

For sheet-piling problems, we may define a "cumulative error" which is equal to the sum of the absolute values of relative errors on surface settlements and horizontal displacements. For a computation-measurement comparison at the end of construction of sheet-piles, 4% of the models analysed provide a cumulative error of less than 25% and just 16% show a cumulative error of below 50%. In contrast, 48% of the predictions reveal a cumulative error of above 100%. These errors are generated by the unsatisfactory numerical modelling of soil movements behind the wall.

Other aspects concerning sheet-piled retaining structures and diaphragm walls may soon be open to analysis, such as the simulation of bending moments or heave at the bottom of the excavation. Other prospective applications for MOMIS include the introduction of computation-measurement comparisons for foundation support structures and reinforced soils.

5 CONCLUSION

The aim of this paper was to report two database related to predictions dealing with mechanical behaviour of geotechnical works. A preliminary analysis of retaining wall was presented. A series of standard figures displaying error distributions, provided some quantitative elements of appreciation about error in practical and academic situations. The authors of this database are aware that much remains to be done in order to draw recommendation and guidelines for the modelling of geotechnical works:

• bibliographic study: the data and results mentioned in the paper are incomplete. For an efficient database, all the benchmark information is required. This information has to be collected from benchmarks authors;

- data capture work: if the number of class A benchmarks is limited, the content capture is tedious and time-consuming, due to the large amount of information for a relevant analysis of all the benchmarks. In addition, the information structure is not always adapted to computerised data analysis. Sometimes, this information is only available in graphical form;
- synthesis work: the main interest of this base lies in the synthesis which will emerge from the whole of the benchmarks. For now, indications on errors and their hierarchy are only fragmentary. Much work still remains to do, consisting in relating prediction error with modelling assumptions (constitutive model, rheological parameters, etc). This work will require a global analysis of the set of benchmarks for consolidation of recommendations or new ones.

However, we hope, at medium-term, to be able to deal with several benchmarks together, supplementing the current analysis related to a particular benchmark. With the participation of all researchers concerned about numerical methods, the geotechnical community will have a tool:

- overviewing current practices in modelling (mesh, boundary conditions, construction stages, type of analysis, constitutive models, rheological parameters, etc.);
- highlighting modelling biases;
- reporting the problems encountered by the benchmarks authors and participants.

Interesting collective initiatives were taken with regard to the code and modelling validations. There is a need now for a common analysis and synthesis.

REFERENCES

Aalto A., Rekonen R., Lojander M. (1998) The calculation on Haarajoki test embankment with the finite element program Plaxis 6.31. *Application of Numerical Methods to Geotechnical Problems*, NUMGE98, Cividini (ed.), Springer Wien, NewYork, pp. 37–46.

Finnish National Road Administration (1997) Competition to calculate settlements at Haarajoki. Competition program: http://www.tieh.fi/pailas/

Lambe T.W. (1973) Prediction in soil engineering. *Géotechnique*, 23, 2, pp. 149–202.

Mestat Ph., Riou Y. (2001) *Modélisation numérique en géotechnique et mesures sur ouvrages en vraie grandeur*, 1st A. Caquot International Conference, Modelling and simulation in Civil Engineering: from practice to theory, Paris, 3–5 octobre 2001, cédérom.

Mestat Ph., Riou Y. (2002) Database for class A and C predictions – Comparison between FEM results and measurements in excavation problem. *5th Europ. Conf. on Numerical Methods in geotechnical Engineering – NUMGE 2002*, Paris, Presses de l'ENPC/LCPC, pp. 703–710.

Schweiger H.F. (1998) *Results from two geotechnical benchmark problems*, Aplications of numerical methods to geotechnical problems, *Proceeding of the 4th Europ. Conf. on Numerical in Geotechnical Engineering* – Numge 98. Springer Wien, New York, Udine, pp. 645–654.

Schweiger H.F. (2002) *Benchmarking in geotechnics*, Institute for Soils Mechanics and Foundation Enginneering, Graz University of Technology, Austria, Computational geotechnics group, mars, CGG_IR006_2002.

Von Wolffersdorff P.A. (1994a) *The Results of the Sheetpile Wall Field Test in Hochstetten*, Université de Karlsruhe, Civieltechnish Centrum, Avril 1994a.

Von Wolffersdorff P.A. (1994b) *The Results and predictions*, Workshop Sheet Pile Test Karlsruhe, Delft University, Octobre 1994b.

Geotechnical Measurements and Modelling, Natau, Fecker & Pimentel (eds)
© 2003 Swets & Zeitlinger, Lisse, ISBN 90 5809 603 3

Stability of deep wellbores in mobile saliferous layers

S. Rübel, E. Kunz & O. Natau
Lehrstuhl für Felsmechanik Universität Karlsruhe, Germany

ABSTRACT: The operation of a gas storage facility at a harvested natural gas field is handled by a number of wellbores for shut-in and production. After running the storage facility for several years deformations of the casings have been detected in some wells. The suspicious cross-sections have been located in the Upper Permian saliferous layers. The deformations ranged from a slight ovality to the collapse of the casings accompanied by the effect of a reduced flow-rate and minor performance in production. To prevent the slightly ovalized casings from total collapse and to save the cost of expensive sidetracks numerical models have been developed to back-analyze the deformations and to find optimized solutions that guarantee the economical long-term stability of the wells. The back-analysis of slightly deformed cross-sections can be successfully performed with the material properties determined at our rock-testing laboratory. The calculations have shown that the deformations are time-dependent. It is possible that they increase up to collapse depending on the casing's resistance. With a variety of numerical simulations it is possible to evaluate the risk of collapse for existing casings as well as for projected wellbores. In the presented example it has been possible to determine the critical cross-sections and to recommend a solution for a long-term-stable well.

1 MEASUREMENTS

Caliper logs have been made 8,5 years after the wellbore was deepened. In some sections the logging-data showed that the shape of the casing became oval.

The deformed sections were located below the Bunter sandstone in the Upper Permian saliferous layer that covers the porous rock of the gas reservoir.

In a depth of 2100 m to 2120 m the casing was deformed with a radial shortening and lengthening of about 2 mm (see Figure 1).

In the depth of 2260 m to 2290 m the deformation of the cross section has also become to an oval shape. A radial shortening and lengthening of about 3.7 mm has been measured.

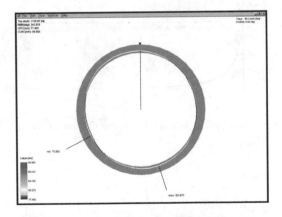

Figure 1. Example wellbore, 2100 m to 2120 m, about 2 mm radial shortening and lengthening.

2 LABORATORY TESTING

In numerical models at constant temperature the NORTON's-creep-law can be implemented:

$$\dot{\varepsilon} = C \cdot \left(\frac{\sigma_{eff}}{\sigma_0} \right)^n$$

Specimen from the saliferous layer were tested in uniaxial creep tests at the "Lehrstuhl für Felsmechanik". So the creep-parameter C in NORTON's law was determined for the relevant range of temperature with about $C = 5.0 \cdot 10^{-9} d^{-1}$ at an exponent of $n = 5$.

This parameter describes an average creep behavior of the saliferous rock.

3 BACK-ANALYSIS OF A DEFORMATION

The finite-element modelling implements the composite system of viscoelastic saliferous rock, the elastoplastic cementation and the elastoplastic casing. The back-analysis of a deformed casing is practiced at the following typical sections.

In a depth of 2100 m to 2120 m the time-dependent deformation for a local anisotropic state of stress distribution with its ratio λ and the creep-parameter C is calculated as shown in Figure 2. Within a period of about 10 years maximum deformation comes up to the same magnitude as the measured in-situ deformations. The results of the finite-element-analysis show that there exists no significant increase of the deformations after this time.

To perform the back-analysis for the measured in-situ deformations of the cross-section in the depth of 2260 m to 2290 m the boundary conditions are depth-corrected. The calculated deformations are shown in Figure 3. About 8 years after completing the wellbore the deformation's progress accelerates and results in

the collapse of the casing after about 12 years. With the results of the simulation this section of the casing can not be predicted as long-term-stable.

As a result of the calculations at the specified boundary conditions the following statements can be derived for the example wellbore:

– The sections passing the saliferous layer in the depth of 2100 m to 2120 m are prospectively long-term-stable.
– The sections passing the saliferous layer in the depth of 2260 m to 2290 will prospectively collapse in the near future.

The existing measurement-program has been extended to calibrate the results of the back-analysis on a large data set.

4 OPTIMIZING THE WELLBORE IN THE DEPTH OF 2260 M TO 2290 M

To optimize the wellbore in the critical section a variety of possibilities was simulated by using the finite-element method. The numerical simulation basing on the parameters from the back-analysis of the logged in-situ deformations are applied to predict the long-term-stability of a recompletion.

The following 3 alternatives to recomplete a wellbore are examples of the various possibilities that were calculated and analyzed.

4.1 Recompletion with a 4 1/2″-liner applied into the outer-liner without cementation

The results of the calculation are shown in Figure 4. By plotting radial lengthening and shortening versus time it is distinct that the deformation of the 7″-casing keeps increasing after the 4 1/2″-liners was applied (3100 days after deepening the well) until the time of contact (t = 4560 days). After this point of time the rate of deformation decreases slightly as the deformations keep increasing continuously.

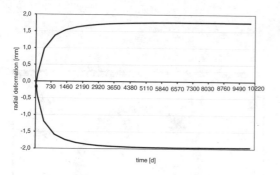

Figure 2.   Radial deformation vs. time (2100 m–2120 m).

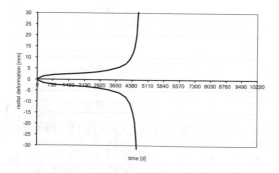

Figure 3.   Radial deformation vs. time (2260 m–2290 m).

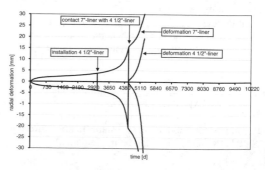

Figure 4.   Recompletion with a 4 1/2″-liner applied into the outer-liner without cementation.

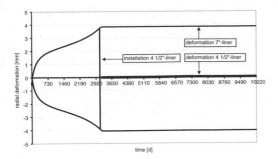

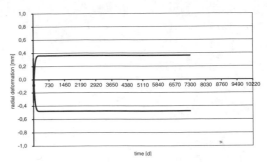

Figure 5. Recompletion with a 4 1/2″-liner applied into the outer-liner with cementation.

Figure 6. Realized reconstruction with a 4 1/2″-liner as a new single-casing.

The prospective deformations will result in a collapse of the pipe-in-pipe casing.

4.2 *Recompletion with a 4 1/2″-liner applied into the outer-liner with cementation*

The results of the calculation are shown in Figure 5. By plotting radial lenghthening and shortening versus time it is distinct that the deformation of the 7″-liner increases only insignificantly after the 4 1/2″-liners was applied (3100 days after deepening the well). The rate of deformation reduces significantly and becomes decreasing. The prospective deformations will result in steady-state deformation of the pipe-in-pipe casing.

Due to the analysis this system of pipes and cementation can be regarded as long-term-stable.

4.3 *Realized recompletion with a 4 1/2″-liner as a new single-casing*

This calculation postulates that the existing 7″-liner was removed completely by a milling-tool.

Figure 6 shows the radial deformation of the cross-section's axes versus time starting with the point of time when the reconstruction has begun. As it can be clearly seen the deformation of the casing which results from the loading by the surrounding saliferous rock stagnates to a constant magnitude after the wellbore was recompleted.

Due to the analysis the realized alternative of recompletion can be regarded as long-term-stable.

5 SUCCESS AND CONCLUSIONS

The success of the realized alternative has been verified by multiple caliper-logs after the recompletion of the wellbore.

These measurements will be proceeded in a period of one year to control and document the casing's geometry continuously.

By modelling the interaction between casing, cementation and surrounding saliferous rock it is possible to simulate the detected deformations as well as to predict the long-term-stability of existing and projected wellbores.

REFERENCES

Lehrstuhl für Felsmechanik 1998a. Feldesentwicklung Mittelplate/Dieksand – Bericht über Felsmechanische Gesteinsproben aus den Bohrungen Mittelplate A8 und Dieksand 1 und 1a. Unveröffentlichter Projektbericht.

Lehrstuhl für Felsmechanik 1998b. Feldesentwicklung Mittelplate/Dieksand – Literaturstudie zum Zechsteinsalz. Unveröffentlichter Projektbericht.

Lehrstuhl für Felsmechanik 1998c. Programm zusätzlicher Arbeiten zur Gewährleistung der Sicherheit des Forschungsbergwerkes Asse im Rahmen der "Fortsetzung der Verfüllmaßnahme Südflanke" (Teil 9 des Nachtrags Nr. 2 zur HU/AFU-Bau – Bericht zum Arbeitspaket 6, Aufgabe 6.4.1 "Parameterabsicherung für das Deckgebirge". Unveröffentlichter Projektbericht.

Hampel, A., Hunsche, U., Plischke, I. & Schulze, O. 1996. Thermomechanisches Verhalten von Salzgesteinen – Abschlußbericht zum Forschungsvorhaben mit dem Förderkennzeichen 02 E 8542 0. Hannover: Bundesanstalt für Geowissenschaften und Rohstoffe. Archivnummer 114805.

Geotechnical Measurements and Modelling, Natau, Fecker & Pimentel (eds)
© 2003 Swets & Zeitlinger, Lisse, ISBN 90 5809 603 3

Effect of cyclic deformations on the evolution of localizations in granular bodies

J. Tejchman
Gdansk University of Technology, Gdansk, Poland

ABSTRACT: In the paper, shear localization in granular materials is investigated during cyclic deformation. Two rate boundary value problems are analyzed with a finite element method and a polar hypoplastic constitutive law: cyclic shearing of a narrow layer of sand between two very rough boundaries under conditions of free dilatancy and cyclic biaxial compression–extension of a sand specimen under a constant lateral pressure. Attention is laid on the influence of the cyclic loading amplitude on the thickness of a shear zone inside of sand.

1 INTRODUCTION

The mechanism of the creation and evolution of spontaneous and induced localisations of deformation in granular bodies under monotonous shearing in the form of shear zones is well recognized on the basis of model tests and numerical calculations (Vardoulakis 1980, Yoshida et al. 1994, Desrues et al. 1996, Tejchman et al. 1999, Tejchman & Gudehus 2001, Tejchman 2002). The results show that the thickness of shear zones depends on many different factors as: pressure level, void ratio, direction of deformation, specimen size, mean grain diameter, grain roughness, grain hardness and boundary conditions surrounding the granulate (roughness and deformability). However, shear localisation under cyclic loading was rarely investigated both theoretically and experimentally. Since shear localisation is a precursor of the loss of stability of soils, the effect of cyclic loading on the strength of soils is important during earthquake, liquefaction of slopes and dynamic impacts from machines.

In this paper, shear localisation in the form of an induced shear zone is investigated during cyclic shearing of a narrow layer of sand between two very rough boundaries under conditions of free dilatancy. In addition, shear localisation in the form of a spontaneous shear zone is studied during a cyclic biaxial compression–extension test of a sand specimen under a constant lateral pressure. The analyses are performed with a finite element method and a polar hypoplastic constitutive law. The polar hypoplastic law is able to describe the essential properties of granular bodies during shear localisation. Due to the presence of a characteristic length in the form of a mean grain diameter, the law can describe the formation of shear zones

with a certain thickness and spacing. The FE-results converge to a finite size of a shear zone via mesh refinement, and initial and boundary value problems become mathematically well-posed at the onset of localisation when using constitutive laws with softening (Mühlhaus 1989, de Borst et al. 1992).

The FE-calculations are carried out with initially dense cohesionless sand subject to a large cyclic deformation amplitude.

2 CONSTITUTIVE LAW

The polar hypoplastic constitutive law (Tejchman et al. 1999, Tejchman & Gudehus 2001, Tejchman 2002) can reproduce essential features of granular bodies during shear localisation in dependence on the void ratio, pressure level, deformation direction, mean grain diameter and grain roughness. It was formulated by an extension of a hypoplastic law proposed by Gudehus (1996) and Bauer (1996) within a polar continuum (Schäfer 1962, Mühlhaus 1989). It is characterised by simplicity and a very wide range of applications. The material constants can be found by means of standard element tests and simple index tests. They can be estimated from granulometric properties encompassing grain size distribution curve, shape, angularity and hardness of grains (Herle & Gudehus 1999). The capability of this law has already been demonstrated in solving boundary value problems involving localisation such as biaxial test, shearing of a narrow granular layer, silo filling and silo flow, furnace flow, footings and sand anchors. A close agreement between calculations and experiments was achieved.

3 CYCLIC SHEARING

The FE-calculations of cyclic shearing with free dilatancy were performed for a sand strip of $h = 20$ mm height. To simulate shearing of an infinite layer, the calculations were performed with only one element column with a width of $b = 10$ cm, consisting of 20 quadrilateral horizontal elements. Thus, the height of the elements was $5 \times d_{50}$ for the mean grain diameter considered in the analysis ($d_{50} = 0.5$ mm). As the initial stress state in the granular strip, a K_o-state without polar quantities ($\sigma_{22} = -10$ kPa, $\sigma_{11} = \sigma_{33} = -3$ kPa, $\sigma_{12} = \sigma_{21} = m_1 = m_2 = 0$) was assumed ($\sigma_{11}$ – horizontal normal stress, σ_{22} – vertical normal stress, σ_{33} – lateral normal stress perpendicular to the plane of deformation, σ_{12} – horizontal shear stress, σ_{21} – vertical shear stress, m_1 – horizontal couple stress, m_2 – vertical couple stress). Gravity was neglected.

Both bottom and top of the sand layer were very rough. The boundary conditions were along the bottom: $u_1 = 0$, $u_2 = 0$ and $\omega^c = 0$, and along the top: $u_1 = n\Delta u$, $\omega^c = 0$, and $\sigma_{22} = p$ (n denotes the number of time steps, Δu is the constant displacement increment in one step and p is the vertical uniform pressure prescribed to the top of the strip). Thus, full shearing of sand along both boundaries was assumed.

The quasi-static shear deformation was initiated through constant horizontal displacement increments Δu prescribed at the nodes along the top of the strip. Initially, the sand layer was subject to compression by pressure p and then subject to shearing in one direction up to $u_1^t/h = 100\%$ to obtain a residual (stationary) state (u_1^t – total horizontal displacement of the top boundary, h – layer height). Afterwards, the direction of shearing was repeatedly changed applying a large ($u_1^t/h = \pm200\%$) shear amplitude. The six full shear cycles were carried out.

The calculations were performed with an initial void ratio equal to $e_0 = 0.60$. The vertical pressures was $p = 500$ kPa and the mean grain diameter was $d_{50} = 0.5$ mm ($h/d_{50} = 40$).

Figures 1–5 show results for an initially dense sand strip ($h = 20$ mm, $e_o = 0.60$, $d_{50} = 0.5$ mm, $p = 500$ kPa) subject to cyclic shearing with a large magnitude of the shear amplitude. Figure 1 presents the evolution of normalized stress components σ_{ij}/h_s (h_s – granular hardness) at the mid-point of the strip against the normalised horizontal displacement of the top u_1^t/h (σ_{33} – lateral normal stress perpendicular to the plane of deformation). In Figure 2, the evolution of the normalized vertical couple stress $m_2/(h_sd_{50})$ at the layer boundary and the evolution of the mobilized wall friction angle $\phi_w = \arctan(\sigma_{12}/\sigma_{22})$ versus u_1^t/h is shown. The mobilised wall friction angle φ_w is related to the entire granular layer (stresses σ_{12} and σ_{22} are constant along both the height and length of the layer). Figure 3 demonstrates the evolution of the

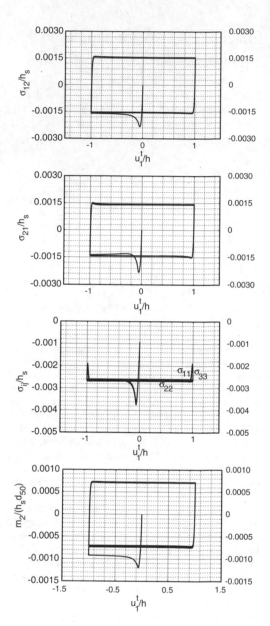

Figure 1. Cyclic shearing with dense sand (high shear amplitude): evolution of normalised stresses σ_{ij}/h_s at the mid-point and normalised wall couple stress $m_2/(h_sd_{50})$ versus u_1^t/h.

void ratio e in 20 elements along the layer height from the bottom ($x_2/d_{50} = 1$) up to mid-height ($x_2/d_{50} = 19$) versus u_1^t/h. Figures 4 and 5 present the distribution of normalised stresses σ_{ij}/h_s, couple stress $m_2/(h_sd_{50})$, Cosserat rotation ω^c and void ratio e along the normalised height x_2/d_{50} after the initial shearing and after six shear cycles.

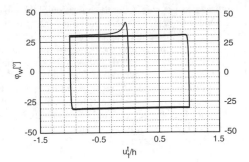

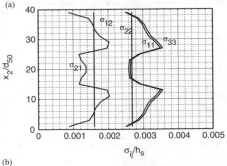

Figure 2. Cyclic shearing with dense sand (high shear amplitude): evolution of wall friction angle φ_w versus u_1^t/h.

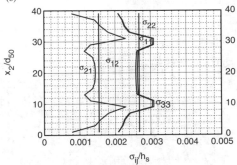

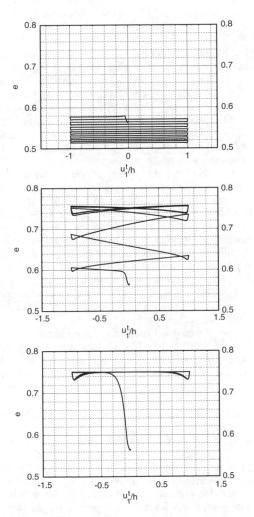

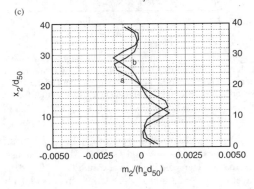

Figure 4. Shearing with dense sand (high shear amplitude): (a) distribution of normalised stresses σ_{ij}/h_s after the initial shearing, (b) distribution of normalised stresses σ_{ij}/h_s after the 6th full shear cycle, (c) normalised wall couple stress $m_2/(h_s d_{50})$ across the normalised height x_2/d_{50} (a – after the initial shearing, b – after the 6th full shear cycle).

All state variables (stress, couple stress and void ratio) tend to asymptotic values. The shear stresses insignificantly decrease during cyclic shearing. The shear stress σ_{12} is slightly larger than σ_{21} in the middle of the layer. During each reversal shearing, the normal stresses σ_{11} and σ_{33} diminish by 25%. The behaviour of the couple stress m_2 is similar as for the shear stresses. The maximum wall friction angle is 42° (obtained during initial shearing). The residual wall friction angle is 30° and almost independent of the

Figure 3. Cyclic shearing with dense sand (high shear amplitude): evolution of void ratio e across the layer height versus u_1^t/h at x_2/d_{50}: 7, 13 and 19.

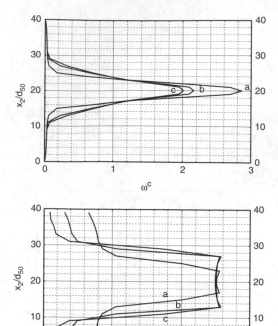

Figure 5. Shearing with dense sand (high shear amplitude): distribution of Cosserat rotation ω^c and void ratio e across the normalised height x_2/d_{50} (a – after the initial shearing, b – after the 3rd full shear cycle, c – after the 6th full shear cycle).

number of shear cycles. The void ratio close to the boundaries ($x_2/d_{50} \leqslant 9$ and $x_2/d_{50} \geqslant 31$) continuously decreases. The void ratio in the middle of the shear layer ($x_2/d_{50} \geqslant 13$ and $x_2/d_{50} \leqslant 27$) increases globally during each cycle and tends towards the pressure-dependent critical value (e = e_c = 0.75). During each reversal shearing, contractancy occurs in the entire shear layer. After this additional compaction, the void ratio in the middle of the shear zone increases and reaches again the critical value.

In the middle of the layer, the deformation is localised in a shear zone, characterised by the appearance of Cosserat rotations and a strong increase of the void ratio. At the boundaries of the shear zone, a strong jump of the curvature, stresses and couple stress takes place.

The thickness of the shear zone, as visible from the Cosserat rotation and the stress jump at the shear zone edges, is about $14 \times d_{50}$ after the initial shearing and $18 \times d_{50}$ after six full shear cycles. Thus, during the cyclic shearing, the thickness of the shear zone grows. The increase of the thickness is pronounced within the

first three shear cycles. After the transition of the top of the granular layer by its initial position ($u_1^t/h = 0$), the distribution of horizontal displacements is not uniform, i.e. a zig-zag occurs across the height of the localized zone. With continuing shearing, the displacement field becomes s-shaped similar to that after initial shearing. The distribution of stresses σ_{11}, σ_{33} and σ_{21} across the shear zone is strongly non-linear. In the middle of the shear zone, the stresses σ_{11} and σ_{33} show their minimum and the stress σ_{21} its maximum. The stress ratios σ_{11}/σ_{22} and σ_{11}/σ_{33} become equal to 1.

4 CYCLIC COMPRESSION–EXTENSION

The FE-calculations of a biaxial compression–extension test were performed with a sand specimen which was h = 100 mm high and b = 20 mm wide. In total, 320 quadrilateral elements (2.5 mm $\times$ 2.5 mm), i.e. 1440 triangular elements were used. As the initial stress state, the K_0-state with the stresses $\sigma_{22} = \sigma_c + \gamma_d x_2$ and $\sigma_{11} = \sigma_c + K_0\sigma_{22}$ was assumed in the sand specimen, where σ_c denotes the confining pressure, x_2 is the vertical coordinate measured from the top of the specimen, γ_d denotes the initial density and $K_0 = 0.40$ is the pressure coefficient at rest. The quasi-static deformation in sand was initiated through a constant vertical displacement increment prescribed to the nodes along the upper edge of the specimen. The boundary conditions of the sand specimen were: along the vertical sides traction and moment free while the top and the bottom smooth, i.e. $u_2 = n\Delta u$, $\sigma_{12} = 0$, $m_2 = 0$ (top) and $u_2 = 0$, $\sigma_{12} = 0$, $m_2 = 0$ (bottom), n denotes the number of the time step. To preserve the stability of the specimen against the sliding along the top boundary, the node in the middle of the bottom was kept fixed. To numerically obtain a shear zone inside the specimen, a weak element with a large initial void ratio, $e_0 = 0.9$, was inserted in the middle of the left side of the specimen.

Initially, the sand specimen was subject to vertical compression up to $u_2/h = 10\%$ to obtain a residual state (u_2 – total vertical displacement of the top, h – specimen height). Afterwards, the direction of vertical deformation was repeatedly changed applying a large ($u_2/h = \pm20\%$) cyclic amplitude. The five full biaxial compression–extension cycles were carried out.

The numerical results with dense specimen ($d_{50} = 0.50$ mm, $e_o = 0.60$), confining pressure, $\sigma_c = 200$ kPa and large deformation amplitude are presented in Figures 6–10. Figure 6 shows the evolution of the normalised vertical force on the top $P/(\sigma_c bl)$ versus the normalised vertical displacement of the top u_2/h (l = 1 m – specimen length). In Figure 7, the deformed meshes during deformation are presented. The evolution of the void ratio inside a shear zone is demonstrated in Figure 8. The distribution of void ratio and Cosserat

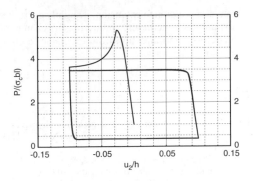

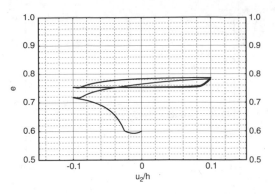

Figure 6. Normalised load–displacement curve during a cyclic compression–extension test.

Figure 8. Evolution of void ratio in the middle of the shear zone during a cyclic biaxial compression–extension test.

(a)

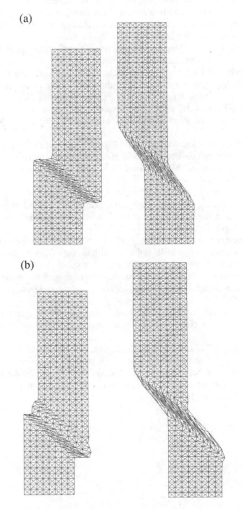

(b)

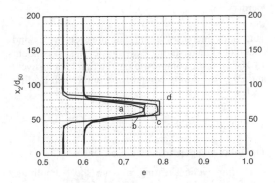

Figure 9. Distribution of void ratio along the normalized specimen height during compression and extension cycles: a – first cycle ($u_2/h = -0.1$), b – first cycle ($u_2/h = 0.1$), c – fifth cycle ($u_2/h = 0.1$), d – fifth cycle ($u_2/h = -0.1$).

rotation along the height in the middle of the specimen at the initial position is shown in Figures 9 and 10.

The vertical force on the top tends to an asymptotic value during compression and extension. It insignificantly decreases during each loading cyclic shearing. The maximum global internal friction angle calculated from the principle stresses $\sigma_1 = P/(bl)$ and $\sigma_2 = \sigma_c$ using the Mohr's criterion ($\phi = \arcsin(\sigma_1 - \sigma_2)/(\sigma_1 + \sigma_2)$) is during compression $\phi_{max} = 43°$ (initial shearing). The residual wall friction angle is $\phi_{cr} = 33.7°$ (compression) and $\phi_{cr} = 29°$ (extension). During the first cycle, the void ratio in the shear zone increases during compression and extension. During the next cycles, it decreases during compression (due to larger mean pressure) reaching the pressure-dependent critical value ($e = e_c = 0.75$), and increases during extension (due to smaller mean pressure) reaching $e = e_c = 0.78$.

The thickness of a shear zone expanding from the imperfection towards the bottom and going across the

Figure 7. Deformed meshes during a biaxial compression–extension test: (a) during first loading cycle, (b) during 4th loading cycle.

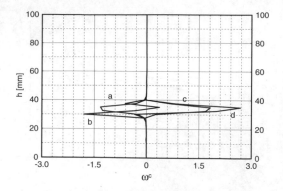

Figure 10. Distribution of the Cosserat rotation ω^c along the specimen height h during a biaxial cyclic compression–extension test (a – compression, first cycle; b – compression, fourth cycle; c – extension, first cycle; d – extension, fourth cycle).

sand body increases during compression cycles and does not change during extension cycles. It is about $13 \times d_{50}$ during compression in the first loading cycle and $20 \times d_{50}$ during compression in the fifth loading cycle.

5 CONCLUSIONS

The following conclusions can be drawn:

– During cyclic shearing, a large shear amplitude influences the thickness of an induced shear zone. The growth of the number of shear cycles increases the thickness of an induced shear zone in dense sand.
– During a biaxial compression–extension test, a large deformation amplitude causes an increase of the thickness of a spontaneous shear zone inside of the specimen.
– During cyclic shearing, the void ratio inside of the shear zone continuously increases.
– Each change of the shear direction causes contractancy in the shear zone.
– During cyclic biaxial compression–extension, the void ratio in the shear zone globally increases. During the change of the deformation direction

from compression to extension, it slightly decreases first and then increases, and during the change of the deformation direction from extension to compression, it decreases.

REFERENCES

Bauer, E. 1996. Calibration of a comprehensive hypoplastic model for granular materials. *Soils Found.* 36(1): 13–26.
de Borst, R., Mühlhaus, H.B., Pamin, J. & Sluys, L. 1992. Computational modelling of localization of deformation. In D.R.J. Owen., H. Onate & E. Hinton (ed.), *Proc. of the 3rd Int. Conf. Comp. Plasticity*, Swansea: Pineridge Press, 483–508.
Desrues, J., Chambon, R., Mokni, M. & Mazerolle, F. 1996. Void ratio evolution inside shear bands in triaxial sand specimens studied by computed tomography. *Géotechnique* 46(3): 529–546.
Gudehus, G. 1996. A comprehensive constitutive equation for granular materials. *Soils Found.* 36(1): 1–12.
Herle, I. & Gudehus, G. 1999. Determination of parameters of a hypoplastic constitutive model from properties of grain assemblies. *Mech. Cohes. Frict. Mat.* 4(5): 461–486.
Mühlhaus, H.B. 1989. Application of Cosserat theory in numerical solutions of limit load problems. *Ing. Arch.* 59: 124–137.
Schäfer, H. 1962. Versuch einer Elastizitätstheorie des zweidimensionalen ebenen Cosserat-Kontinuums, *Miszellaneen der Angewandten Mechanik*, Festschrift W. Tolmien, Berlin, Akademie-Verlag.
Tejchman, J., Herle, I. & Wehr, J. 1999. FE-studies on the influence of initial void ratio, pressure level and mean grain diameter on shear localization. *Int. J. Num. Anal. Meth. Geomech.* 23(15): 2045–2074.
Tejchman, J. & Gudehus, G. 2001. Shearing of a narrow granular strip with polar quantities. *Int. J. Num. Anal. Meth. Geomech.* 25: 1–28.
Tejchman, J. 2002. Patterns of shear zones in granular materials within a polar hypoplastic continuum. *Acta Mechanica* 155(1–2): 71–95.
Vardoulakis, I. 1980. Shear band inclination and shear modulus in biaxial tests. *Int. J. Num. Anal. Meth. Geomech.* 4: 103–119.
Yoshida, T., Tatsuoka, F. & Siddiquee, M. 1994. Shear banding in sands observed in plane strain compression. In R. Chambon, J. Desrues & I. Vardoulakis (ed.), *Localisation and Bifurcation Theory for Soils and Rocks*, Rotterdam: Balkema, 165–181.

Author index